Photon Correlation and Light Beating Spectroscopy

NATO ADVANCED STUDY INSTITUTES SERIES

A series of edited volumes comprising multifaceted studies of contemporary scientific issues by some of the best scientific minds in the world, assembled in cooperation with NATO Scientific Affairs Division.

Series B: Physics

Volume 1 – Superconducting Machines and Devices
edited by S. Foner and B. B. Schwartz

Volume 2 – Elementary Excitations in Solids, Molecules, and Atoms
(Parts A and B)
edited by J. Devreese

Volume 3 – Photon Correlation and Light Beating Spectroscopy
edited by H. Z. Cummins and E. R. Pike

The series is published by an international board of publishers in conjunction with NATO Scientific Affairs Division

A	**Life Sciences**	Plenum Publishing Corporation
B	**Physics**	New York and London
C	**Mathematical and Physical Sciences**	D. Reidel Publishing Company Dordrecht and Boston
D	**Behavioral and Social Sciences**	Sijthoff International Publishing Company Leiden
E	**Applied Sciences**	Noordhoff International Publishing Leiden

Photon Correlation and Light Beating Spectroscopy

Edited by

H. Z. Cummins

Department of Physics
New York University
New York, New York

and

E. R. Pike

Royal Radar Establishment
Great Malvern, Worcestershire
England

SPRINGER-SCIENCE+BUSINESS MEDIA, LLC

Library of Congress Cataloging in Publication Data

NATO Advanced Study Institute, Capri, 1973
Photon correlation and light beating spectroscopy.

(The NATO Advanced Study Institutes series (Series B: Physics, v.3))
Includes bibliographical references.
1. Light–Scattering–Congresses. 2. Photon correlation–Congresses. 3. Light beating spectroscopy–Congresses. I. Cummins, Herman Z., 1933- ed. II. Pike, Edward Roy, 1929- ed. III. Title. IV. Series.
QC427.N37 1973 539.7 74-938
ISBN 978-1-4615-8908-2 ISBN 978-1-4615-8906-8 (eBook)
DOI 10.1007/978-1-4615-8906-8

Lectures presented at the NATO Advanced Study Institute, Capri, Italy,
July 16-27, 1973

Originally published by Plenum US in 1974.
Softcover reprint of the hardcover 1st edition 1974

PREFACE

This volume contains the invited lectures and seminars and abstracts of the contributed seminars presented at the NATO Advanced Study Institute on Photon Correlation and Light Beating Spectroscopy held at the Centro Caprense Di Vita E Di Studi Ignazio Cerio , Capri, Italy, July 16-27, 1973.

The Institute was organized to provide a comprehensive presentation of this new and rapidly developing field for those interested in applying these techniques to problems in many areas including Physics, Biology, Engineering and Chemistry. The lectures were divided into three principal categories: the first Basic Theory (Photon Statistics and Correlation, Scattering Theory), secondly Instrumentation (Correlation Techniques, Light Beating), and the third Areas of Application (Gas and Liquid Dynamics, Critical Phenomena, Biology). The seminars provided detailed presentations of applications to a number of specific problems.

Although the selection of topics was inevitably limited, it was the hope of the organizing committee that the lectures would provide a broad coverage appropriate for the needs of the interdisciplinary audience represented by the participants, and that this volume would serve for some years to come as a useful introduction for those entering the field.

The members of the Organizing Committee were:

E.R. Pike, RRE, Malvern U.K.	Co-directors
H.Z. Cummins, New York University	
M. Bertolotti, Universita di Roma -	Local Organizer
J.M. Vaughan, RRE, Malvern, U.K.	Secretary
H. Swinney, New York University	Treasurer
P. Lallemand, Ecole Normale Superieure, Paris	
H. Haken, Universitat Stuttgart, Germany.	

We wish to express our appreciation to the NATO Scientific Affairs Division whose generous support made this Institute possible, and to the following organizations which provided Fellowships and travel grants to a number of students:

The National Science Foundation
The Science Research Council
The French Ministry of Foreign Affairs
The Rank Organisation
Messrs Precision Devices and Systems
The Saicor-Honeywell Company and
Messrs Ealing Beck

We wish to thank Professor Mario Bertolotti whose efforts as local organizer contributed enormously to the success of the Institute, and also to thank Dr Michael Vaughan and Professor Harry Swinney for their extensive efforts as Secretary and Treasurer. Finally, we wish to express our particular gratitude to Donna Laetitia Cerio Holt, President of the Centro Caprense I.C. for making the facilities of the Centro available to this Institute.

HZC (New York)
ERP (Malvern)

October, 1973

CONTENTS

LECTURES

INTRODUCTORY LECTURE

E R Pike

Royal Radar Establishment, Malvern, Worcs, England

1 SUBJECT OF THE STUDY INSTITUTE

The main purpose of the lectures to be presented in this school is to collect together and teach in a systematic way for the first time the basic ideas and practical techniques of photon correlation methods in light scattering studies and spectroscopy. These ideas and methods have sprung up and have been pursued in a number of centres in the last three or four years and offer great potential for application in many branches of physical and biological science. They follow a number of years of development of what we shall call 'light-beating spectroscopy', in which detected optical signals are analysed by analogue methods in the frequency domain, and to some extend have galvanised further activity in light scattering by the offer of greatly increased speed and accuracy over these earlier methods. The techniques and background of light-beating spectroscopy will be discussed in such detail as is necessary to put the main topic in context by Professor Cummins, who also gives the main course on biological applications.

It is, of course, the brightness and coherence of the laser as an optical source on which the widespread application of these new spectroscopic methods depends, although we should note that the laser is not necessary, in principle, for photon correlation experiments. [The coherence requirements in many of the experiments to be discussed would be met by any source for which the value of $\Delta \underset{\sim}{K} \cdot \Delta \underset{\sim}{r}$, the scalar product of the range of scattering vectors $\underset{\sim}{K}$ (see the first lecture on scattering theory for a definition) and the spatial extent of the scattering volume, is less than 2π].

The impact of these new methods, as we shall see, has been widespread; in effect, they open up new fields in which optical methods can be applied to the study of almost any moving or temporally varying object. In scattering experiments the photon correlations are independent of phase instability of the source and fluctuation frequencies can be measured down to the order of one Hertz. Historically, photon correlation methods were preceded by much detailed and accurate work on single-interval photon counting statistics. It is of direct concern to review these fundamentals and this will be done in Professor Bertolotti's lectures. The statistical properties of the laser itself will be described in Professor Haken's seminar lecture.

Other topics of overlapping interest with our main subject are the comparison of classical interferometric spectroscopy with photon correlation in the region of optical resolution of around one part in 10^8, where both are applicable, and the use of classical light scattering photometry for the study of macromolecules. These two topics will be covered in seminar lectures by Drs Vaughan and Eisenberg respectively.

Again not directly concerned with the methods of photon correlation will be my own course of lectures on the theory of light scattering itself. We should, of course, need no excuse to interest ourselves in a basic theoretical understanding of the spectra to be studied by these new methods and, as the reader will find, there is much new and challenging material for the theoretically minded researcher in this area to develop and clarify using modern many-body theory. Some of the lectures of Professor Lallemand will also cover general scattering theory and the alternative,and very popular, Langevin equation approach to similar problems which he will give is interesting to study and compare in juxtaposition with the field-theory methods.

Regrettable omissions due to lack of time are any discussion of neutron scattering and its relation to light scattering and any mention of the problems of describing a localised particle of zero rest mass and unit spin. The too frequent picture of the photon as a 'fuzzy ball' is not properly condemned and anywhere in this book where photons are invoked one should, to be accurate ,connote the associated photoelectrons wherever possible.

2 SPECTRA AND SCALES OF LENGTH TIME AND FREQUENCY

In order to introduce the concept of correlation in spectroscopy let us first think of the definition of the spectrum of a light source such as we would see, for instance, by exposing a photographic plate in a prism spectrometer. The electric field falling on the instrument has some instantaneous value $\mathcal{E}(t)$ (we neglect spatial variations for the present and assume a single polarised scalar component). This field is decomposed by the instrument into its frequency components whose mean squared envelope is registered as the spectral power density at each frequency. Mathematically therefore, we have for the spectrum, $S(\omega)$,

$$S(\omega) = \left< \lim_{T\to\infty} \left| \frac{1}{T} \int_{-T/2}^{T/2} \mathcal{E}^{+}(t) e^{-i\omega t}\, dt \right|^2 \right> \tag{1}$$

This can be expanded

$$S(\omega) = \lim_{T\to\infty} \frac{1}{T} \int_{-T/2}^{T/2} \int_{-T/2}^{T/2} < \mathcal{E}^{+}(t) \mathcal{E}^{-}(t') > e^{-i\omega(t-t')} dt\, dt', \tag{2}$$

which for a stationary field reduces to

$$S(\omega\ ) = \int_{-\infty}^{\infty} < \mathcal{E}^{+}(\tau) \mathcal{E}^{-}(0) > e^{-i\omega\tau}\, d\tau \quad , \tag{3}$$

that is, the Fourier transform of the field autocorrelation function defined by

$$g^{(1)}(\tau) = < \mathcal{E}^{+}(\tau) \mathcal{E}^{-}(0) > \tag{4}$$

We see, therefore, that the essential content of the spectrum is the relationship between values of the field at separated points in time. If the spectrum is broad, then its Fourier transform, $g^{(1)}(\tau)$, will be narrow, which implies that after short times, τ, the field $\mathcal{E}(t)$ has 'forgotten' its previous value $\mathcal{E}(t-\tau)$ and the correlation function has equal positive and negative contributions which average to zero. Conversely, a narrow spectrum implies a long 'correlation time'.

To quantify the scales of time and frequency under discussion let us consider a spectrum consisting of two sharp lines at frequencies f and $f + \Delta f$, where $\Delta f / f = 10^{-n}$. (This will not give a stationary time behaviour without some spreading but will serve for illustration). A classical interferometric or dispersive instrument will require an optical path differnce within the apparatus of the order of one wavelength, λ, in order to resolve the lines (Fig 1). It will, therefore, have a physical dimension, L, of $10^n\ \lambda$. The instantaneous values of the field due to such a pair of lines will have a fluctuating intensity of the form shown also in Fig 1, with the beat frequency, T, of value $1/\Delta f$ or $10^n/f$.

To investigate the spectrum, therefore, we have the option of measuring L or T. Taking an optical wavelength of $\frac{1}{2}$ μm, the respective values of L and T for various degrees of resolution $1{:}10^n$ are given in the following table

Resolution	L	T
$1{:}10^5$	5 cm	200 ps
$1{:}10^6$	50 cm	2 ns
$1{:}10^7$	5 m	20 ns
$1{:}10^8$	50 m	0.2 μs

TABLE 1 Scales of length and time for instruments of various optical resolving powers.

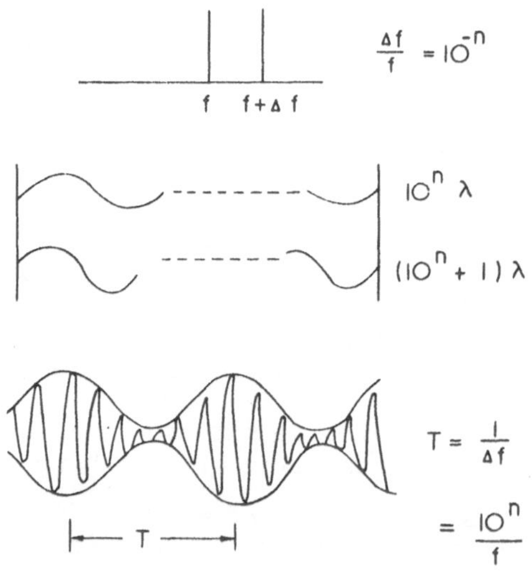

FIG 1 Scales of length, time and frequency for two-mode spectrum.

It is easy to see why temporal correlation plays no part in optical spectroscopy at resolution levels below, say, $1{:}10^6$. The corresponding fluctuation times are too short for convenient measurement in the laboratory, while the size of interferometric instruments is not inconveniently large. On the other hand, however, it is quite plain that as the size of classical instruments becomes unmanageably large for ease of operation and alignment, even allowing for multiple reflection paths as in the Fabry Perot interferometer, the corresponding time scales become more and more easy for laboratory measurements. Thus, at resolutions greater than say $1{:}10^7$ the situation is reversed and classical interferometric measurements can play little part, leaving the field to correlation spectroscopy in one or other of its forms to be discussed in this school.

It is a curious fact of nature that classical light sources, that is to say pre-laser light sources, rarely had linewidths demanding more resolution than could be achieved using classical interferometric instruments and thus the development of spectroscopy by temporal correlation awaited the era of the laser. The first experiments of this type, however, we may recall did in fact just predate the laser (Hanbury-Brown and Twiss, 1956; Forrester, et al 1955).

3 PHOTONS

A glance at the beat frequency spectrum depicted in Fig 1 will serve to remind us that there is nothing very new in looking at the spectral content of signals in the time domain. The arguments of the previous section are equally applicable to microwaves or radio waves or indeed even sound waves. Correlation methods are commonly employed in these fields to extract information from signals and we might imagine that the application of such methods in optics would involve only a simple transposition of well-known electromagnetic theory to the shorter wavelengths. This is not the case due to the fact that between the microwave and optical regions of the spectrum we cross the value of $h\nu$ equal to kT. In the optical region the detector is effectively cold and operates by annihilation only of photons. The result is that instead of measuring a continuous field $\mathcal{E}(t)$, as can be done in the lower frequency region†, the optical detector gives random impulses, corresponding to discrete absorption of quanta, at a mean rate proportional to the value $|\mathcal{E}^{+}(t)|^2$. The fluctuations of the field intensity discussed above, therefore, are manifest by increases or decreases in the rate of photodetections. The values and temporal correlations of numbers of

† Detection in this region employs 'off-diagonal' photon density operators.

photodetections recorded in short fixed times are thus the basic data of our experiments. This, of course, has always been the case in all optical experiments but in most circumstances the photon numbers involved have been sufficiently large for them to be viewed as providing continuous signals. The development, concurrently with the laser, of fast modern phototubes and digital electronics has allowed us at the present time to approximate ideal optical recording. The ways in which such trains of digital photon impulses are processed to give spectral and other types of information about radiating or scattering sources will be taken up in detail in the following courses and seminars.

REFERENCES

A T Forrester, R A Gudmunsen and P O Johnson, Phys Rev, **99**, 1691, 1955.

R Hanbury-Brown and R Q Twiss, Nature, **177**, 27; **178**, 1046, 1047 (1956).

THE THEORY OF LIGHT SCATTERING

E R Pike

Royal Radar Establishment, Malvern, Worcs., England

1 INTRODUCTION

The study of material media by the scattering of light has a long history, dating at least back to Tyndall's experiments of 1869. The advent of the laser, however, has given a great stimulus in recent years to this type of study, particularly with regard to fine spectral features at low energy in the scattered light. Such features were first predicted by Brillouin in 1914 to arise from thermally excited first-sound waves and first observed by Gross in 1930. Spontaneous fluctuations of entropy density give, similarly, low-energy scattering and a theory was given by Landau and Placzek in 1934. Raman scattering from optic phonons was also discovered around this time but the energy shifts are higher and we will not be concerned with such scattering in these lectures.

It is difficult to distinguish in a general theoretical treatment between scattering from solids, liquids or gases since they merge one into the other in various regions of temperature and pressure. For instance, first-sound fluctuations occur in all three cases, only embryonically in a rare gas in the longitudinal branch but through to the incipient appearance of transverse branches in high viscosity liquids and to a full panoply of sound-wave excitations in an arbitrary solid. We shall, therefore, carry the theory as far as possible without relating it to any particular medium. The properties of the medium required to describe the scattered spectrum will be elucidated in general terms and these can be investigated in specific cases at a later stage. We do assume that the medium is a dielectric and that the light frequencies are much higher than any mechanical reasonances in the medium. We can then neglect direct interactions of the light field with ionic charges and consider only interaction via electrons which can be treated by a linear electronic dielectric susceptibility, relating the local polarisation, $\underset{\sim}{P}(\underset{\sim}{r},t)$, in the medium to the incident electric field, $\underset{\sim}{E}_i(\underset{\sim}{r},t)$. We shall be concerned only with small changes of this susceptibility from an overall mean value since no scattering arises from this latter part. The contribution of the varying part to the local polarisation is

$$4\pi \ \Delta \underset{\sim}{P}(\underset{\sim}{r},t) = \Delta \underset{\sim}{\epsilon}(\underset{\sim}{r},t)\, \underset{\sim}{E}_i(\underset{\sim}{r},t), \tag{1}$$

where $\Delta\underset{\sim}{\epsilon}$ is the varying part of the local dielectric-constant tensor. We consider the scattering in the wave-zone where for media in which the uniform part of $\underset{\sim}{\epsilon}$ is isotropic the field is proportional to $d^2(\Delta P)/dt^2$. This will be the case discussed here. To calculate the total scattered field we imagine the medium to be divided up into small equal volumes, $\delta^3 r$ and sum the contributions from each. The incident field is assumed to be undisturbed by the small scattering (Rayleigh-Gans approximation). We shall not discuss the situation outside this limit where, except in the case of dilute solutions discussed by Professor Cummins in this volume, multiple scattering effects need to be considered.

1.1 The Scattered Field

Let us consider one such element at position $\underset{\sim}{r}$ with respect to an arbitrarily chosen origin, Fig 1, in a medium with mean refractive index n.

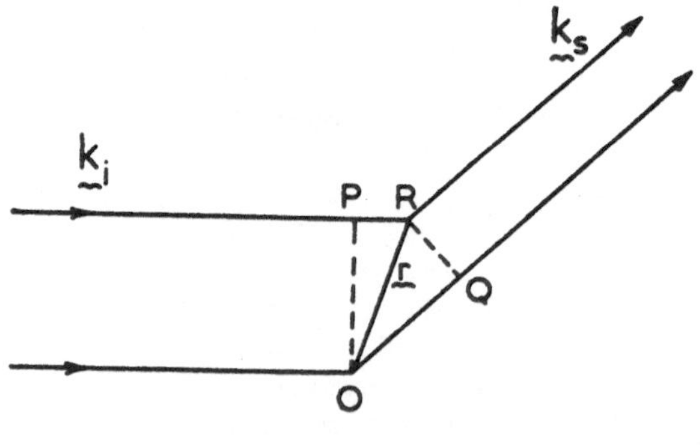

FIG 1 Scattering geometry

The incident and scattered fields have directions given by the wave vectors $\underset{\sim}{k}_i$ and $\underset{\sim}{k}_s$; the wavelength shift in the scattering is very small so that

$$\left|\underset{\sim}{k}_i\right| \simeq \left|\underset{\sim}{k}_s\right| = 2\pi n/\lambda ,$$

where λ is the wavelength in free space. The phase difference, $\Delta\phi$, in the wave zone between the contributions to the scattering from the points R and O is $2\pi n/\lambda$ times the path difference (OQ - PR).

$$\Delta\phi = \underset{\sim}{k}_s\cdot\underset{\sim}{r} - \underset{\sim}{k}_i\cdot\underset{\sim}{r} = (\underset{\sim}{k}_s-\underset{\sim}{k}_i)\cdot\underset{\sim}{r} = \underset{\sim}{K}\cdot\underset{\sim}{r}, \quad (2)$$

where we have defined the scattering wave-vector $\underset{\sim}{K}$

$$\underset{\sim}{K} = (\underset{\sim}{k}_s-\underset{\sim}{k}_i). \quad (3)$$

If we denote the scattering angle $k_i\widehat{\ }k_s$ by θ, reference to figure 2 shows that $\underset{\sim}{K}$ lies in the direction of the bisector of the incident and scattered beams and has magnitude $2\,|k|\sin(\theta/2)$.

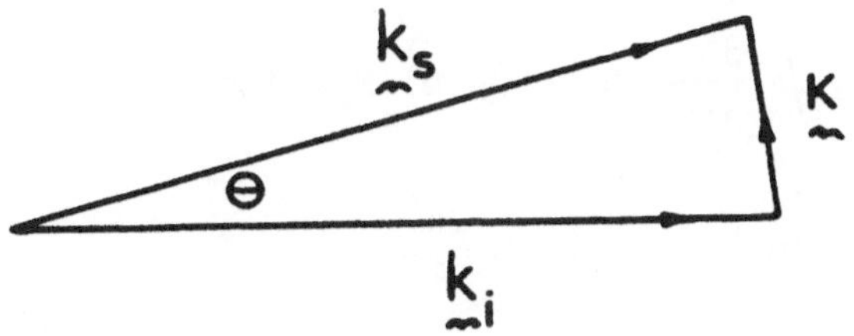

FIG 2 Geometrical construction for the scattering wave vector $\underset{\sim}{K}$

If we define a wavelength Λ such that

$$|K| = 2\pi/\Lambda \tag{4}$$

we find the Bragg relation

$$\lambda = 2n\,\Lambda \sin(\theta/2). \tag{5}$$

The contribution to the scattering in the wave-zone from the element at R will thus be

$$\delta \underset{\sim}{E}(\underset{\sim}{K},t) = C\,\frac{d^2}{dt^2}[\Delta\,\underset{\sim}{\epsilon}\,(\underset{\sim}{r},t)\;\underset{\sim}{E}_i\,(\underset{\sim}{O},t)]\;e^{i\Delta\phi}$$

$$= C\,\frac{d^2}{dt^2}[\Delta\,\underset{\sim}{\epsilon}\,(\underset{\sim}{r},t)\;\underset{\sim}{E}_i^o]\;e^{i\underset{\sim}{K}\cdot\underset{\sim}{r}}\,, \tag{6}$$

where we have written $\underset{\sim}{E}_i^o$ for the quantity

$$\underset{\sim}{E}_i(\underset{\sim}{O},t) = \underset{\sim}{E}_i\,e^{i\omega_o t + \phi_o} \tag{7}$$

and where C is a constant equal to $\delta^3 r/4\pi Lc^2$, L being the distance to the far field point. The total scattered field is therefore given by

$$\underset{\sim}{E}(\underset{\sim}{K}t) = C\int \frac{d^2}{dt^2}[\Delta\,\underset{\sim}{\epsilon}\,(\underset{\sim}{r},t)\;\underset{\sim}{E}_i^o]\;e^{i\underset{\sim}{K}\cdot\underset{\sim}{r}}\,d^3r. \tag{8}$$

The component in the cartesian direction α is

$$E_\alpha\,(\underset{\sim}{K},t) = C\int \frac{d^2}{dt^2}\left[\sum_\gamma \Delta\,\epsilon_{\alpha\gamma}(\underset{\sim}{r},t)\;E_{i\gamma}^o\right] e^{i\underset{\sim}{K}\cdot\underset{\sim}{r}}d^3r \tag{9}$$

and for a polarised incident wave $E_{i\beta}$ this becomes

$$E_{\alpha\beta}(\underset{\sim}{K},t) = C\int \frac{d^2}{dt^2}[\Delta\,\epsilon_{\alpha\beta}(\underset{\sim}{r},t)\;E_{i\beta}^o\,]\;e^{i\underset{\sim}{K}\cdot\underset{\sim}{r}}d^3r$$

$$= C\,\frac{d^2}{dt^2}[\Delta\,\epsilon_{\alpha\beta}\,(\underset{\sim}{K}t)\;E_{i\beta}^o\,]. \tag{10}$$

We have assumed the volume of the medium sufficiently large for boundary effects to be unimportant.

1.2 The Spectrum of the Scattered Field

The spectrum of this scattered field is by definition

$$S_{\alpha\beta}(\underset{\sim}{K}, \omega_{opt}) = \lim_{T\to\infty} \left\langle \frac{1}{T} \left| \int_{-T/2}^{+T/2} E_{\alpha\beta}(\underset{\sim}{K},t)e^{-i\omega_{opt}t}\,dt \right|^2 \right\rangle$$

$$= \left\langle \left| E_{\alpha\beta}(\underset{\sim}{K}, \omega_{opt}) \right|^2 \right\rangle , \qquad (11)$$

where the angle brackets denote an average over a stationary ensemble and ω_{opt} is the optical frequency. By equations (7) and (10) this gives, using simple properties of the Fourier transform,

$$S_{\alpha\beta}(\underset{\sim}{K}, \omega_{opt}) \propto \omega_{opt}^4 \left\langle \left| \Delta\epsilon_{\alpha\beta}(\underset{\sim}{K}, \omega) \right|^2 \right\rangle , \qquad (12)$$

where the spectral shift is

$$\omega = \omega_{opt} - \omega_o . \qquad (13)$$

There may be a number of distinct processes in the medium which affect the value of these tensor components; we have already mentioned, for instance, sound waves and entropy fluctuations. Let us suppose that a set of such identifiable disturbances from uniformity $A(\underset{\sim}{r},t)$, $B(\underset{\sim}{r},t)$.. exist. Then

$$\Delta\epsilon_{\alpha\beta} = \frac{\partial\epsilon_{\alpha\beta}}{\partial A} A + \frac{\partial\epsilon_{\alpha\beta}}{\partial B} B + \ldots\ldots \qquad (14)$$

and

$$S_{\alpha\beta}(\underset{\sim}{K},\omega) \propto \left[\left(\frac{\partial\epsilon_{\alpha\beta}}{\partial A} \right)^2 \left\langle \left| A(\underset{\sim}{K},\omega) \right|^2 \right\rangle + \left(\frac{\partial\epsilon_{\alpha\beta}}{\partial B} \right) \left\langle \left| B(\underset{\sim}{K},\omega) \right|^2 \right\rangle + \right.$$

$$\left. + \left(\frac{\partial\epsilon_{\alpha\beta}}{\partial A} \right)\left(\frac{\partial\epsilon_{\alpha\beta}}{\partial B} \right) \left\langle A(\underset{\sim}{K}, \omega)\, B^*(\underset{\sim}{K},\omega) \right\rangle + \ldots \right] \omega_{opt}^4 . \qquad (15)$$

We see that the calculation of light-scattering spectra, leaving aside the coupling constants represented by the partial derivatives, is now reduced to the evaluation of mean-square and cross fluctuations of components of the space-time Fourier transforms of the medium disturbances which have the general form

$$S_{AB^*}(\underset{\sim}{K},\omega) = \left\langle A(\underset{\sim}{K},\omega)\, B^*(\underset{\sim}{K},\omega) \right\rangle .$$

1.3 Correlation Function

The Fourier transforms of the above general functions are also useful

$$F_{AB*}(\underset{\sim}{K},t) = \int_{-\infty}^{+\infty} < A(\underset{\sim}{K},\omega)\ B^*\ (\underset{\sim}{K},\ \omega) > e^{-i\omega t}\ dt$$

$$= \lim_{T\to\infty} < \frac{1}{T} \int_{-T/2}^{+T/2} A(\underset{\sim}{K},\ t-t')\ B^*\ (-t')dt' >$$

$$= \quad < A(\underset{\sim}{K},t)\ B^*\ (\underset{\sim}{K},0) > , \qquad (16)$$

where we have used the theorem that the Fourier transform of a product is the fold of the separate transforms and the fact that the fluctuations are stationary and hence depend only on the time difference of their arguments. $F_{AB*}(\underset{\sim}{K},t)$ is a cross-correlation function of a general type and the above relation (16) is a generalization of the Wiener-Khintchine theorem which states that the autocorrelation function of a random stationary variable is the Fourier cosine transform of its power spectrum.

Before proceeding further to a full discussion of the evaluation of such functions let us consider some simple forms which they might take. We will only deal with a single mode of disturbance at this stage and so will speculate on possible simple forms for $< |A(\underset{\sim}{K},\ \omega)|^2>$ or $< A(\underset{\sim}{K}t)A^*(\underset{\sim}{K},0)>$. At a particular time a spontaneous fluctuation $A(\underset{\sim}{K})$ might take any time dependence but we invoke the so-called 'Onsager hypothesis' which states that on average such behaviour will follow the law of macroscopic transport. Two simple macroscopic types of behaviour are exponential decay given by

$$\frac{dA}{dt} = -\lambda A \qquad (17)$$

$$A(t) = A(0)e^{-\lambda|t|} \qquad (18)$$

and free propagation obeying

$$\frac{d^2A}{dt^2} = -\ \omega_o^2 A \qquad (19)$$

$$A(t) = A(0)e^{\pm i\omega_o t} \quad . \qquad (20)$$

In these cases we have

$$F_{AA*}(t) = < AA* > e^{-\lambda|t|} \tag{21}$$

$$S_{AA}*(\omega) = \frac{\lambda < AA* >}{\omega^2 + \lambda^2} \tag{22}$$

and

$$F_{AA*}(t) = < AA* > e^{\pm i\omega_o t} \tag{23}$$

$$S_{AA*}(\omega) = < AA*> \delta\,(\omega \pm \omega_o), \tag{24}$$

respectively. The averages are now at equal times.

Lorentzian spectra of the first type above are met in the cases of pure molecular rotational relaxation, scattering from macromolecular Brownian motion and thermal relaxation, while the pair of line spectra of the second type would be approximately realised in scattering from a long-lived sound wave, say in crystalline quartz or from entropy waves in liquid ^{4}He at very low temperature and high pressure. It is significant to consider this last example at low pressure, where although the entropy waves are strong they do not affect appreciably the value of the dielectric constant. The low value of $\partial\epsilon/\partial A$ in this case make the spectrum difficult to observe.

It is clear from these simple considerations that the evaluation of correlation functions and spectra in the general case will require a knowledge of the transport properties of the medium. Indeed it is a fact that the behaviour of these functions <u>define</u> the transport properties of the medium in the theory of irreversible thermodynamics, where the basis of the theory is to <u>postulate</u> that they have the simple forms of the above two examples (De Groot and Mazur 1962). The Navier-Stokes equations, for instance, are derived from assuming such forms for the pressure correlations. It would be, therefore, somewhat devious to calculate light scattering spectra by starting from these equations as has been done in the past (Mountain 1966) although the final answers should be the same. This distinction becomes more important as the complexity of the medium increases.

2 THERMODYNAMIC FLUCTUATIONS

2.1 The Green's Function

A more fundamental approach to the evaluation of correlation functions leads us into the heart of the subject of modern statistical mechanics and many-body theory. The correlation function and functions derived from it are primary objects of concern in modern mechanics both classical and quantum, having taken over the central role from that of the wave field which is only a useful concept in very simple situations. We shall try to steer a way in these lectures through the mass of relevant modern literature and will present a concise formal and general theory for the expression of correlation functions, and thus light scattering spectra, in terms of phenomenological parameters. The approach is most closely linked to those of Zubarev (1960) whose notation for the correlation function is used above, and of Tyablikov and Bonch Bruevich (1962), but overlaps the same ground as covered by many modern authors. We quote only Zwanzig (1961), Kadanoff and Martin (1963), Luttinger (1964), Felderhof and Oppenheim (1965) Hohenberg and Martin (1965) Mori (1965) Pike (1965) Kwok and Martin (1966) Kubo (1966a,b), Kawasaki (1970),(1973), Selwyn and Oppenheim (1971), Wehner and Klein (1972), Pike and Swain (1972), Weinberg et al (1973) and Martin et al (1973).

The theory will be a complete mechanics of statistically stationary systems and, as such, represents only a formal and exact restructuring of the basic equations of Newton or Schrodinger at zero temperature, or of Liouville or von Neumann at finite temperature. The time dependences are determined by a time independent Hamiltonian of the system which remains unspecified throughout the formal development and is only finally taken into account by comparing the resulting theory with phenomenological observations.

We shall deal with the quantum theory at finite temperature for maximum generality but the theory can easily be transposed into the classical limit if required. See § 5 of Professor Haken's lecture.

2.1.1 <u>Basic Functions.</u> The basic objects of study are then defined as <u>general correlation</u> functions

$$F_{AB}(t) = \mathrm{Tr}\,(\hat{\rho}\, A(t' + t)B(t')) = \langle A(t)\ B(0)\rangle, \tag{25}$$

where $\hat{\rho}$ is the thermal equilibrium density operator and $X(t)$ is a Heisenberg operator $e^{iHt}\,Xe^{-iHt}$, H being the system Hamiltonian. Spectral functions are defined by the Fourier transform (real ω)

$$S(\omega) = \frac{1}{2\pi}\int_{-\infty}^{+\infty} F(t)\ e^{i\omega t}dt \tag{26}$$

and Green's functions by the Hilbert transform (complex E)

$$G(E) = \frac{1}{2\pi} \int_{-\infty}^{+\infty} \frac{S(\omega)}{E-\omega} \, d\omega \tag{27}$$

The operators A and B can be taken in the variables $\underset{\sim}{r}$ or $\underset{\sim}{K}$ related by a spatial Fourier transform but unless specified otherwise we shall assume that they are written in the $\underset{\sim}{K}$ variable. A and B can be any operators or their Hermitian conjugates without restriction†.

The suffices AB are to be understood where not used.

The above three functions are related step by step in the reverse direction first by the Plemelj formula for the inversion of a Hilbert transform

$$S(\omega) = i \lim_{\eta \to 0} \left[G(\omega + i\eta) - G(\omega - i\eta) \right] \tag{28}$$

and then by the inverse Fourier transform

$$F(t) = \int_{-\infty}^{+\infty} S(\omega) \, e^{-i\omega t} \, dt. \tag{29}$$

Direct transforms from G to F and back can be derived. Under very general conditions on F(t) the only singularities of G(E) are poles on the real axis and in this case we may go directly from G(E) to F(t) by the simply derived formula

$$F(t) = 2\pi \ \Sigma \ \text{residues of } G(\omega) \, e^{-i\omega t}; \tag{30}$$

this follows from substituting (28) into (29), changing the variables $\omega + i\eta$ in the two terms to the real axis and applying the Cauchy residue theorem.

These three functions and their interrelationships described above are basic to the theory. The first two functions have obvious experimental significance. The relevance of the third function the Green's function, is not yet obvious but, in fact,

---

† Readers familiar with many body theories should note that we have not made use of the artifice of using the anti-commutator in the definitions to obtain the spectral population factors. Our approach is the simpler one of Pike and Swain (1972) which is applicable more generally to non-equilibrium problems. It should be noticed also that in the present optical situations, where the annihilation parts only are measured, the usual fluctuation-dissipation formula is not applicable (See Butcher and Ogg 1965)

its limiting values on the real axis are also measurable functions. We shall see later (eq (106) below) that certain transport parameters which we shall require to describe light scattering are themselves Green's functions.

2.1.2 Energy Representation. The detailed structure of these functions is best studied by going into an energy representation. The energy eigenstates of the system will be denoted by the kets $|n\rangle$ with eigenvalues E_n. In this representation the canonical density operator is

$$\hat{\rho} = Z^{-1} \sum_s e^{-\beta E_s} |s\rangle\langle s|, \tag{31}$$

where

$$Z = \mathrm{Tr}\,(e^{-\beta H}) \tag{32}$$

is the partition function and $\beta = 1/k_B T$. The correlation function takes the form

$$\begin{aligned} F_{AB}(t) &= \mathrm{Tr}\,(Z^{-1} \sum_s e^{-\beta E_s} |s\rangle\langle s| A(t)B(0)) \\ &= Z^{-1} \sum_{nm} \langle n| e^{-\beta E_s} |s\rangle\langle s| e^{iHt} A e^{-iHt} |m\rangle\langle m|B|n\rangle \\ &= Z^{-1} \sum_{mn} e^{-\beta E_n} A_{nm} B_{mn} e^{-iE_{mn}t}, \end{aligned} \tag{33}$$

where we have made the abbreviation

$$E_{mn} = E_m - E_n. \tag{34}$$

By applying (26) we find

$$S(\omega) = Z^{-1} \sum_{mn} e^{-\beta E_n} A_{nm} B_{mn}\, \delta(\omega - E_{mn}), \tag{35}$$

where we have used

$$\int_{-\infty}^{+\infty} e^{ixt}\, dx = 2\pi\, \delta(t).$$

The Hilbert transform (27) gives then, simply,

$$G(E) = \frac{Z^{-1}}{2\pi} \sum_{mn} e^{-\beta E_n} A_{nm} B_{mn} \frac{1}{E - E_{mn}} . \tag{36}$$

The poles of this function can be seen to lie at energy-level differences of the system. One thus sees at this point the role of the Hilbert transform. The reverse formulae can be seen to hold remembering that

$$\lim_{\eta \to 0} \left[\frac{1}{x - i\eta} - \frac{1}{x + i\eta} \right] = 2\pi i \delta(x). \tag{37}$$

The one-step forward formulae for the Green's function from the correlation function requires the identity

$$i \int_{\infty}^{+\infty} e^{ixt}\, \theta(t) dt = \frac{1}{x + i\epsilon}; \tag{38}$$

$\theta(t)$ is the function equal to zero for $t < 0$ and equal to unity for $t > 0$.

2.1.3 <u>Relationships with Ordinary Green's Operator</u>. Before we complete our study of the structure of these functions we need to derive an alternative form for $G(E)$ in terms of the ordinary Green's (or resolvent) operator, $\mathcal{G}(E)$, of the Hamiltonian which is defined by

$$(H-E)\, \mathcal{G}(E) = 1. \tag{39}$$

In the energy representation

$$\mathcal{G}(E) = - \sum_{m} \frac{| m \rangle \langle m |}{E-E_m} , \tag{40}$$

which can be shown by using

$$H = - \sum_{n} E_n | n \rangle \langle n | \tag{41}$$

and the definition (39).

We first write out G(E) as

$$G(E) = \frac{Z^{-1}}{2\pi} \sum_n e^{-\beta E_n} < n| A \left\{ \sum \frac{|m><m|}{E-E_{mn}} \right\} B | n > \quad (42)$$

then, using (40) we see that

$$G(E) = - \frac{Z^{-1}}{2\pi} \sum_n e^{-\beta E_n} < n| A \mathcal{G}(E+E_n) B | n > . \quad (43)$$

This form will be useful to find equations for the Green's function.

The problem now is to find the dependence of F on its argument t, the time difference variable, or equivalently to find the dependence of S on ω, ie to find the spectrum, or again equivalently, to find the E dependence of G. The basic dynamical statement controlling these dependencies is found from the Heisenberg equation of motion for A(t) (setting $\hbar = 1$),

$$i \frac{dA}{dt} = AH - HA \quad , \quad (44)$$

which gives us the equation for F

$$i \frac{d}{dt} F_{AB}(t) = < [A(t)H] \; B \; > , \quad (45)$$

the square brackets denoting the commutator.

We shall return to this equation later but use the argument that the form of the equation in terms of the Green's function is simpler in two ways to look first at the transformed equation. The first simplification is that differential equations turn into algebraic ones in the transform domain and the second is that boundary conditions are also automatically built in.

We could go through the transforms of the above equation to find the equation for the Green'sfunction but a simpler way is given below.

2.2.1 Equations for the Green's Function. It can be seen that the following equation is an identity by virtue of equation (39)

$$-\frac{Z^{-1}}{2\pi}\sum e^{-\beta E_n} < n\left|A\left\{\left[H-(E+E_n)\right]\mathcal{G}(E+E_n) - 1\right\} B\right|n > = 0. \tag{46}$$

Multiplying out

$$-\frac{Z^{-1}}{2\pi}\sum_n e^{-\beta E_n}\left\{< n\left|(AH-AE_n)\mathcal{G}E(E+E_n)B\right|n> - E < n\left|A\mathcal{G}(E+E_n)B\right|n > - < n\left|AB\right|n >\right\} = 0 \tag{47}$$

But

$$< n\left|AE_n\,\mathcal{G}(E+E_n)B\right|n >$$

$$= < n\left|E_nA\,\mathcal{G}(E+E_n)B\right|n >$$

$$= < n\left|HA\,\mathcal{G}(E+E_n)B\right|n > \tag{48}$$

Thus, using the definition (43) we obtain

$$EG_{AB}(E) = \frac{1}{2\pi} < AB> - G_{[HA]B}(E), \tag{49}$$

where we have now an equal-time average in the first term of the right-hand side. By writing a second identity

$$-\frac{Z^{-1}}{2\pi}\sum_n e^{-\beta E_n} < n\left|A\left\{\mathcal{G}(E+E_n)\left[H-(E+E_n)\right] - 1\right\}B\right|n > = 0, \tag{50}$$

which is true since $\mathcal{G}$ and H commute, we obtain similarly

$$EG_{AB}(E) = \frac{1}{2\pi} < AB > + G_{A[HB]}(E). \tag{51}$$

Replacing [HA] for A as the arbitrary first operator in this latter equation gives

$$EG_{[HA]B}(E) = \frac{1}{2\pi} < [HA]B> + G_{[HA][HB]}(E). \tag{52}$$

Combining this with (49) we have

$$EG_{AB}(E) = \frac{1}{2\pi} \langle AB\rangle + \frac{1}{2\pi E} \langle [HA]B\rangle - \frac{1}{E} G_{[HA][HB]}(E). \tag{53}$$

By a similar process using the equations in the other order we also can show that

$$EG_{AB}(E) = \frac{1}{2\pi} \langle AB \rangle + \frac{1}{2\pi E} \langle A[HB] \rangle - \frac{1}{E} G_{[HA][HB]}(E). \tag{54}$$

It follows, as can easily be shown directly, that

$$\langle A[HB] \rangle = - \langle [HA]B \rangle . \tag{55}$$

We will arbitrarily use the second form of equation.

2.2.2 <u>Notational Simplifications</u>. To discuss this equation for the Green's function we first make some simplifications in notation. We denote

$$\frac{1}{2\pi} \langle AB \rangle = N_{AB} \tag{56}$$

$$\frac{1}{2\pi} \langle A[HB] \rangle = \Omega_{AB} \tag{57}$$

$$G_{[HA][HB]}(E) = i\Gamma_{AB}(E) \tag{58}$$

and

$$\Omega_{AB}(E) - i\Gamma_{AB}(E) = M_{AB}(E). \tag{59}$$

Equation (54) thus reads

$$EG = N + \frac{1}{E} \Omega - \frac{i}{E} \Gamma = N + \frac{1}{E} M, \tag{60}$$

where we have dropped the suffices and arguments for convenience and consider this to be a matrix equation with rows and columns corresponding to the set of interacting modes of motion A,B,C ... under consideration. We shall henceforth also consider these as a column vector A. Further condensation of the formalism is obtained by multiplying (60) from the right by the matrix N^{-1}

$$E(GN^{-1}) = 1 + \frac{1}{E}(\Omega N^{-1}) - \frac{i}{E}(\Gamma N^{-1}). \tag{61}$$

We introduce the 'normalised' matrices

$$g = GN^{-1} \tag{62}$$

$$\vartheta = \Omega N^{-1} \tag{63}$$

$$\gamma = \Gamma N^{-1} \tag{64}$$

$$\mu = \vartheta - i\gamma = MN^{-1} \tag{65}$$

and arrive, therefore, at the compact result

$$E^2 g = E + \vartheta - i\gamma = E + \mu. \tag{66}$$

2.2.3 The Self-Energy Function. To elucidate the pole structure of g we perform the following manipulations. First add and subtract $Eg\mu$ to the left hand side of (66)

$$E^2 g + Eg\mu - Eg\mu = E + \mu \tag{67}$$

ie
$$Eg\,(E + \mu) - Eg\mu = E + \mu\,. \tag{68}$$

Multiply from the right by $(E + \mu)^{-1}$ to get

$$Eg - Eg\mu\,(E + \mu)^{-1} = 1 \tag{69}$$

or

$$g[E - E\mu\,(E + \mu)^{-1}] = 1. \tag{70}$$

Multiply again from the right by the inverse of the term in the square bracket

$$g = [E - E\mu\,(E + \mu)^{-1}]^{-1}. \tag{71}$$

We now introduce the notation

$$E\mu\,(E + \mu)^{-1} = \ell\,, \tag{72}$$

which we call the 'self-energy' matrix (normalised) in analogy with practice in the quantum theory of fields. In our application we shall see that the function plays a key role in the theory of transport.

Equation (72) is the basic equation of the theory. It was given in a slightly less general form by Landau and Lifshitz (Eq 124 loc cit) whose work in this area has been remarkably underestimated outside the USSR.

The equation may be inverted to give

$$\mu = E\ell\,(E - \ell)^{-1}. \tag{73}$$

Using the definition of ℓ, eq (72), in equation (71), we have for the normalised Green's function

$$g = (E - \ell)^{-1}. \tag{74}$$

Equations (72) and (73) were also given in one-dimensional form in a study of Brownian motion by Kubo in 1966 (his equations (62) and (63)).

Use of (72) and (74) gives a form for iterative determination of the self-energy

$$\ell = \mu - \ell g \ell\ . \tag{75}$$

We also note that a matrix Dyson equation

$$g = g_o - g_o\,(\ell - \ell_o)g \tag{76}$$

is easily proved for the normalised Green's function where the self energy has been split in such a way that

$$g_o = (E - \ell_o)^{-1}\ . \tag{77}$$

2.2.4 The Self-Energy in Transport Theory. Our theory of light scattering has now reached the point where the evaluation of correlation functions and spectra devolve upon the determination or calculation of the self-energy matrix ℓ. The significance both experimentally and theoretically of the self-energy matrix is best seen by tracing its effect back to eq (45) for the time dependence of the correlation function. We start from eq (74) and multiply from the left by $E - \ell$

$$(E - \ell)g = 1. \tag{78}$$

Now multiply from the right by N and use eq (62)

$$EG - N = \ell\,G. \tag{79}$$

The Fourier transform of this equation along a line approaching the real axis from above is

$$i \frac{d}{dt} \tilde{G}(t) - \delta(t) < AB > = \int_{-\infty}^{+\infty} \tilde{\ell}(s) \tilde{G}(t-s)ds, \qquad (80)$$

where we have used the Faltung theorm to transform the product on the right-hand side and where the Fourier transforms are denoted by a tilde. We have stated above, however, that

$$\tilde{G}(t) = - i\theta (t) F(t). \qquad (81)$$

Performing the differentiation and remembering that

$$\theta(t) = \int_{-\infty}^{t} \delta(t)dt, \qquad (82)$$

we find that

$$\theta(t) \frac{dF}{dt} = -i \int_{-\infty}^{t} \tilde{\ell}(s)F(t-s)ds \qquad (83)$$

or
$$i \frac{dF}{dt} = \int_{-\infty}^{t} \tilde{\ell}(s)F(t-s)ds, \quad t > 0. \qquad (84)$$

From this equation for the correlation function, which implies, referring back to (45) that

$$< [A(t)H] B > = \sum_{C} \int_{-\infty}^{t} \tilde{\ell}_{AC}(s) < C(t-s)B > ds, \ t > 0, (85)$$

we may find the equation of motion for macroscopic values of a variable A(t) by invoking the Onsager hypothesis in the form

$$< \frac{dA(t)}{dt} B(0) > = \frac{d}{dt} F_{AB}(t). \qquad (86)$$

This gives then from (85) for macroscopic disturbances

$$< \left[i \frac{dA}{dt} - \sum_{C} \int_{-\infty}^{t} \tilde{\ell}_{AC}(s) C(t-s)ds \right] B(0) > = 0, \quad (87)$$

which since true for arbitrary B imples that

$$i\,\frac{d\underset{\sim}{A}}{dt} = \int_{-\infty}^{t} \tilde{\ell}(s)\, \underset{\sim}{A}(t-s)ds, \tag{88}$$

where we have reverted to the matrix notation. If we restrict the set $\underset{\sim}{A}$ then any term C(t) for which $< C(t)A(0)> = 0$ for all A can be added to the right-hand side. A change of variable t − s to t' leads to

$$i\,\frac{d\underset{\sim}{A}}{dt} = \int_{0}^{\infty} \tilde{\ell}(t-t')\, \underset{\sim}{A}(t')dt'.$$

From the form of this equation we see that by physical considerations $\tilde{\ell}(t)$ must be zero for −ve values of t. This is also apparent from eq (83). The integral then becomes

$$i\,\frac{d\underset{\sim}{A}}{dt} = \int_{0}^{t} \tilde{\ell}(t-t')\, \underset{\sim}{A}(t')dt'. \tag{89}$$

Calling the left hand side of this equation a generalized current $\underset{\sim}{J}$ we have

$$\underset{\sim}{J}(t) = \int_{-\infty}^{t} \tilde{\ell}(s)\, NN^{-1}\, \underset{\sim}{A}(t-s)ds$$

or

$$\underset{\sim}{J}(t) = \int_{-\infty}^{t} \tilde{L}(s)\, \underset{\sim}{X}\,(t-s)ds, \tag{90}$$

where we have defined the transport matrix L and generalized forces $\underset{\sim}{X}$ by the relations

$$\ell = LN^{-1} \tag{91}$$

and

$$\underset{\sim}{X} = N^{-1}\underset{\sim}{A}. \tag{92}$$

Equation (90) is a non-local generalisation of the fundamental matrix equation of linear transport theory in irreversible thermodynamics (de Groot and Mazur 1962) where L is a matrix of phenomenological transport coefficients.

The usual local equations arise if

$$\tilde{L}(s) = L^o \delta(s), \tag{93}$$

where L^o is a matrix of constant coefficients. In this case, substituting (93) in (90) we have thus

$$\underset{\sim}{J}(t) = L^o \underset{\sim}{X}(t) \tag{94}$$

and we have thus equated the self-energy matrix (unnormalised) to the set of Onsager transport coefficients. In the more general case, for sufficiently large values of t compared with the range of $\underset{\sim}{\ell}(t)$, we can write equation (88) as

$$i \frac{d\underset{\sim}{A}}{dt} = \int_{-\infty}^{+\infty} \tilde{\ell}(s)\underset{\sim}{A}(t-s)ds,$$

which gives on Fourier transforming

$$\omega\underset{\sim}{A}(\omega) = \ell(\omega)\underset{\sim}{A}(\omega) \tag{95}$$

for frequencies low compared with the inverse relaxation times. This equation can be used to identify the self-energy function from phenomenological transport equations. The rate of entropy production in the system, or the dissipation, in the local linear case is (de Groot and Mazur 1962)

$$\dot{\sigma} = \underset{\sim}{J}\cdot\underset{\sim}{X} = (L^o\underset{\sim}{X})\cdot\underset{\sim}{X} \tag{96}$$

and we have thus in this limit, related the fluctuations F, S, in which we are interested, to the dissipation, $\dot{\sigma}$, via the self-energy matrix of the formal theory. There is hardly a unique way of writing a fluctuation-dissipation 'theorem' (Kubo 1966) the entire theoretical structure going backwards from (96) through (91), (74), (62-65), (27) and (26) relates fluctuation to dissipation.

We have imposed, as yet, no particular energy or wave-vector dependence on the elements of the self-energy matrix, indeed the theory allows for the Hamiltonian to be given in any basis set, not necessarily even orthonormal. This generality, among other differences, makes the present approach considerably simpler than that, for instance, of Mori (1965) or Kawasaki (1970). Equations (72) and (74) are a new exact starting point for the statistical mechanics of arbitrary systems.

3 MODE COUPLING

3.1 Diagonal Green's Function

We wish now to consider the effect of the presence of other modes on the Green's function $G_{AA^\dagger}$ of the mode A. This we do by going back to equation (60); multiplying by E,

$$E^2 G = EN + M \tag{97}$$

giving for the $AA^\dagger$ element

$$E^2 G_{AA^\dagger} = EN_{AA^\dagger} + M_{AA^\dagger} \tag{98}$$

or

$$G_{AA^\dagger} = \frac{(EN_{AA^\dagger} + M_{AA^\dagger})N_{AA^\dagger}}{E^2 N_{AA^\dagger} + EM_{AA^\dagger} - EM_{AA^\dagger}}$$

$$= \frac{N_{AA^\dagger}}{E - \dfrac{EM_{AA^\dagger}}{EN_{AA} + M_{AA}}} = \frac{N_{AA^\dagger}}{E - \ell_A}\,, \tag{99}$$

where we have defined a 'mode self-energy'

$$\ell_A = \frac{EM_{AA^\dagger}/N_{AA^\dagger}}{E + M_{AA^\dagger}/N_{AA^\dagger}}\,. \tag{100}$$

It should be noted that $M_{AA^\dagger}/N_{AA^\dagger}$ is not $(MN^{-1})_{AA^\dagger}$ and that ℓ_A is not $\ell_{AA^\dagger}$ except in the single-mode case. The influence of other modes is thus felt only through their interaction energy with A as expressed in the Hamiltonian, which affects the commutator [HA] appearing in M_{AA}. Equation (88) can therefore be reduced to a single dimension or 'master' equation

$$i\frac{dA}{dt} = \int_{-\infty}^{t} \tilde{\ell}_A(s)\, A(t-s) \tag{101}$$

in which the effect of all other modes is included in the behaviour of ℓ_A. Cross terms are also given by a formula similar to (99)

$$G_{AB} = \frac{N_{AB}}{E - \dfrac{EM_{AB}/N_{AB}}{E + M_{AB}/N_{AB}}} \tag{102}$$

The way in which the theory is applied to complicated systems is to attempt to separate out from M_{AA} terms which describe the effects of particularly important couplings $< B[HA]>$ in any given region of energy or wave vector; we shall investigate this later.

3.2. Hydrodynamic Modes and Kubo Formula

In many cases a long-lived excitation (quasi-particle), A,exists which only couples significantly to a small energy range of the same excitation. In such cases we obtain an energy independent ℓ_A with a small complex part and we speak of a 'single-mode' theory. If this excitation describes a locally conserved variable then its lifetime will increase with wavelength and we have a 'hydrodynamic' mode. For such modes at energies sufficiently large compared with the inverse lifetime an expansion of ℓ_A can be made in powers of the wave vector giving

$$\ell_A^{(1)} = \omega^{(1)} \tag{103}$$

$$\ell_A^{(2)} = \omega^{(2)} - \frac{[\omega^{(1)}]^2}{E} + i\gamma^{(2)}, \tag{104}$$

where

$$\ell_A = \ell_A^{(1)} K + \ell_A^{(2)} K^2 + \ldots. \tag{105}$$

The well-known Kubo formula for transport coefficients comes from this expansion where $\omega = 0$ and is

$$\ell = i\gamma^{(2)} + O(K^3). \tag{106}$$

The possibility of ℓ having an imaginary part requires discussion since we have stated that G cannot have poles off the real axis. The energy dependence giving rise to the real poles however, can be described by analytic continuation of the function into a second sheet of the complex plane (Fig 3). A densely spaced set of poles with a Lorentzian distribution of residues, for instance, is the continuation of a single pole in the second sheet.

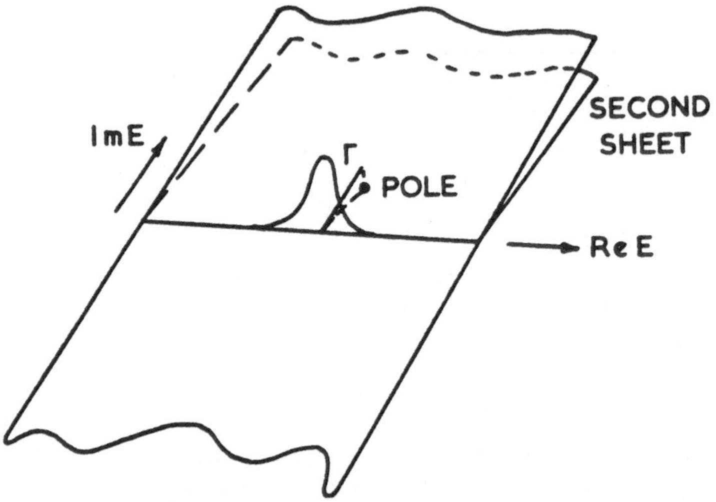

FIG 3 Position of a single simple pole on the second sheet of the complex energy plane giving rise to a Lorentzian distribution of residues on the real axis.

A single freely propagating excitation will be described by a purely imaginary Γ and hence purely real self-energy. It will have the following structure

$$F(t) \sim e^{i\omega_o t} \tag{107}$$

$$S(\omega) \sim \delta\,(\omega - \omega_o) \tag{108}$$

$$G(E) \sim \frac{1}{2\pi\,(E-\omega_o)} \tag{109}$$

$$\ell = \omega_o \tag{110}$$

$$\vartheta = \omega_o \tag{111}$$

$$\gamma = \frac{i\omega_o^2}{E-\omega_o} \tag{112}$$

$$\mu = \frac{E\omega_o}{E-\omega_o} . \tag{113}$$

For coupled hydrodynamic or other modes in regions where $M_{AA^\dagger}/N_{AA^\dagger}$ is small compared with E we have

$$\ell_A \sim M_{AA^\dagger}/N_{AA^\dagger} . \tag{114}$$

3.3 The Mode Self-Energy

The effect of coupling to a mode B can be investigated by expanding the Hamiltonian as follows

$$H = c_{AA^\dagger} AA^\dagger + c_{BB^\dagger} BB^\dagger + c_{AB^\dagger} AB^\dagger + c^*_{AB^\dagger} BA^\dagger + \dots . \tag{115}$$

Then

$$[HA] = c_{AA^\dagger} \mathcal{N}_{AA^\dagger} A - c^*_{AB^\dagger} \mathcal{N}_{AA^\dagger} B + c_{BB^\dagger}[BB^\dagger A] + c_{AB^\dagger}[B^\dagger A]A - c^*_{AB^\dagger}[AB]A^\dagger + \dots \tag{116}$$

where

$$\mathcal{N}_{AA^\dagger} = [AA^\dagger] . \tag{117}$$

Multiplying from the left firstly by A and secondly by B and taking the thermal trace

$$< [HA] A^\dagger > = -c_{AA^\dagger} \mathcal{N}_{AA^\dagger} < AA^\dagger > + \text{higher order terms} \tag{118}$$

$$< [HA] B^\dagger > = - c^*_{AB^\dagger} \mathcal{N}_{AA^\dagger} < BB^\dagger > + \text{higher order terms.} \tag{119}$$

We have assumed the value of $< AB^\dagger >$ to be small.

Thus, using (55) and (57),

$$[HA] = - \frac{\Omega_{AA^\dagger}}{N_{AA^\dagger}} A - \frac{\Omega_{AB^\dagger}}{N_{BB^\dagger}} B + \dots \tag{120}$$

Similarly we find

$$[HA^{\dagger}] = \frac{\Omega_{AA^{\dagger}}}{N_{AA^{\dagger}}} A^{\dagger} + \frac{\Omega_{BA^{\dagger}}}{N_{BB^{\dagger}}} B + \ldots \tag{121}$$

In the approximation (114) we then have for the self-energy, using (58), (63), (64) and (65)

$$\ell_A = \frac{\langle A[HA^{\dagger}]\rangle}{\langle AA^{\dagger}\rangle} - \frac{G_{[HA][HA^{\dagger}]}}{\langle AA^{\dagger}\rangle} \tag{122}$$

$$\simeq \vartheta_{AA^{\dagger}} - i\gamma_{AA^{\dagger}} + \left|\vartheta_{BA^{\dagger}}\right|^2 g_{BB^{\dagger}}, \tag{123}$$

where we have used

$$\vartheta^*_{AB^{\dagger}} = -\vartheta_{BA^{\dagger}} . \tag{124}$$

The extra term due to the coupling of the mode B is the Kawasaki-Ferrell result used to explain light scattering in critical systems.

3.4 Kawasaki Formula

We will just take the extra trouble to write this term in a form which may be compared directly with Kawasaki's work. Let us denote it by ℓ_A^K, then

$$\ell_A^K = \frac{\left|\Omega_{BA^{\dagger}}\right|^2}{N_{AA}\ N_{BB}} g_{BB^{\dagger}}$$

$$= \frac{N_{BB^{\dagger}}}{N_{AA^{\dagger}}} \left|\frac{\Omega_{BA^{\dagger}}}{N_{BB^{\dagger}}}\right|^2 g_{BB^{\dagger}} . \tag{125}$$

We take **A** to be the mode a_j and B to be the combination $a_\ell\ a_m$ then

$$\ell_A^K \sim \frac{N_{\ell\ell}\ N_{mm}}{N_{jj}} \left| \frac{\langle \dot{a}_j\ a_\ell\ a_m \rangle}{N_{\ell\ell}\ N_{mm}} \right|^2 \int_0^\infty g_\ell(t)\ g_m(t) e^{i\omega t} dt, \tag{126}$$

where we have factorized both the normalisation functions and the Green's function. A sum of these contributions over ℓ and m gives the form used by Kawasaki (1971). This formula will be further pursued by Professor Swinney in this School.

With the result (123) we have come as far as we shall go with the development of our formal theory of scattering. We shall spend the rest of the course in the discussion of some practical examples.

4 APPLICATIONS

4.1 Brillouin Scattering

We consider as a first example of the detailed application of the theory the case of Brillouin scattering from longitudinal momentum fluctuations in an isotropic fluid.

We take as our single-mode variable the longitudinal momentum $A = \rho \underset{\sim}{v}_\ell$. The phenomenological equation is that of Newton

$$\Gamma_\ell + \eta_\ell \nabla \cdot \underset{\sim}{v} = \rho , \tag{127}$$

where Γ_ℓ is the longitudinal momentum current and η_ℓ the longitudinal viscosity $(\frac{4}{3}\eta + \zeta)$. This is a satisfactory description at energies large compared with the damping which latter comes from mode interactions responsible for thermal expansion. Using the law of conservation of momentum

$$\frac{\partial}{\partial t} (\rho \underset{\sim}{v}_\ell) = - \nabla \Gamma_\ell , \tag{128}$$

taking the gradient and differentiating with respect to time we obtain

$$- \frac{d^2}{dt^2} (\rho \underset{\sim}{v}) + \eta_\ell \frac{d}{dt} \nabla^2 \underset{\sim}{v} = \nabla \frac{dp}{dt} = \nabla \frac{dp}{d\rho} \nabla . (\rho \underset{\sim}{v}) = c^2 k^2 \rho \underset{\sim}{v} \tag{129}$$

or

$$\omega^2 \rho \underset{\sim}{v} - \frac{\eta_\ell k^2}{\rho} i \omega \rho \underset{\sim}{v} = \omega_1^2 \rho \underset{\sim}{v} . \tag{130}$$

This is of the form of equation (95) with

$$\ell_A = \frac{\omega_1^2}{E - 2i\Gamma_1} , \tag{131}$$

where we have used the notation for the first-sound damping

$$2\Gamma_1 = \frac{\eta_\ell k^2}{\rho} \tag{132}$$

and ω_1/k is the adiabatic first-sound velocity.

We have thus identified the self-energy function and can construct the normalised Green's function according to (74)

$$g_{AA^\dagger}(E) = \frac{E - 2i\Gamma_1}{E^2 - \omega_1^2 - 2i\,\Gamma_1 E} . \tag{133}$$

The denominator may be factorised to give

$$g_{AA^\dagger}(E) = \frac{E - 2i\Gamma_1}{(E - i\Gamma_1 + \sqrt{\omega_1^2 - \Gamma_1^2})\,(E - i\Gamma_1 - \sqrt{\omega_1^2 - \Gamma_1^2}\,)} . \tag{134}$$

To find the residue at the pole

$$E = i\Gamma_1 - \sqrt{\omega_1^2 - \Gamma_1^2} \tag{135}$$

we rearrange (134) to give

$$g = \frac{1}{E - i\Gamma_1 - \sqrt{\omega_1^2 - \Gamma_1^2}} + \frac{-i\Gamma_1 - \sqrt{\omega_1^2 - \Gamma_1^2}}{(E - i\Gamma_1 + \sqrt{\omega_1^2 - \Gamma_1^2})(E - i\Gamma_1 - \sqrt{\omega_1^2 - \Gamma_1^2})} .$$

The residue is therefore

$$R_- = \frac{-i\Gamma_1 - \sqrt{\omega_1^2 - \Gamma_1^2}}{-2\sqrt{\omega_1^2 - \Gamma_1^2}} = \frac{1}{2}\left(1 - \frac{i\Gamma_1}{\sqrt{\omega_1^2 - \Gamma_1^2}}\right). \tag{136}$$

Similarly the second pole gives a residue

$$R_+ = \frac{1}{2}\left(1 + \frac{i\Gamma_1}{\sqrt{\omega_1^2 - \Gamma_1^2}}\right). \tag{137}$$

Use of equation (30) now gives

$$F_{AA^\dagger}(t) \sim \left(1 + \frac{i\Gamma_1}{\sqrt{\omega_1^2-\Gamma_1^2}}\right) e^{-i\left(\sqrt{\omega_1^2-\Gamma_1^2}\right)t - \Gamma_1 t}$$

$$+ \left(1 - \frac{i\Gamma_1}{\sqrt{\omega_1^2-\Gamma_1^2}}\right) e^{+i\left(\sqrt{\omega_1^2-\Gamma_1^2}\right)t - \Gamma_1 t} \qquad (138)$$

The spectrum follows from equation (26) and consists of a positive-frequency part

$$S_{AA^\dagger}(\omega) \sim \frac{\Gamma_1}{\left(\omega - \sqrt{\omega_1^2-\Gamma_1^2}\right)^2 + \Gamma_1^2} \left(1 - \frac{\omega - \sqrt{\omega_1^2-\Gamma_1^2}}{\sqrt{\omega^2 - \Gamma_1^2}}\right) \qquad (139)$$

and its mirror image. This has the form of the difference of a Lorentzian and the same Lorentzian weighted by a linear factor. An experimental result of Brillouin scattering from benzene is shown in Fig 4; this was obtained at RRE by Pike and Vaughan (unpublished) using a plane, piezoelectrically scanned Fabry Perot interferometer. The sharp central component is the Rayleigh line to be discussed in the next section.

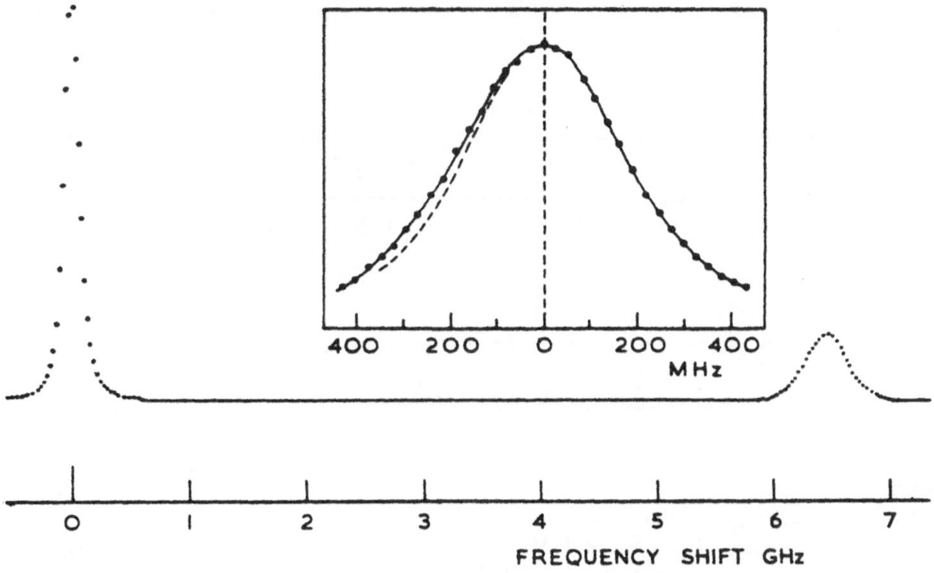

FIG 4 Light-Scattering Spectrum of Benzene at High Resolution Showing Asymmetric Brillouin Line Profile

4.2 Rayleigh Scattering

Our second example is Rayleigh scattering which arises when we consider the longitudinal momentum fluctuations (phonons) of the previous example at low energy. In this region the interaction responsible for the thermal expansion of the medium not only gives rise to the single-mode damping but couples the low energy phonon, $A(\underset{\sim}{k})$ to pairs of phonons, $A(\underset{\sim}{q})\,A(\underset{\sim}{k}-\underset{\sim}{q})$, with the same total wave vector. This phonon-density or 'second-sound' mode is treated by the mode-coupling theory above as an interacting mode $B(\underset{\sim}{q})$.

The self energy now is augmented by a contribution of the form (125) so that the Green's function is, in the same approximation as the Kawasaki formula,

$$G_{AA^\dagger} = \frac{N_{AA^\dagger}}{E - \dfrac{\omega_T^2}{E} - \dfrac{\left|\langle A_q A_{k-q}[HA_k]\rangle\right|^2 G_{BB^\dagger}}{N_{AA^\dagger} N_{BB^\dagger}^2}} \,. \qquad (140)$$

The first-sound damping is now explicitly related to the interactions and $\omega_T k^2$ is the isothermal first sound velocity. The second-sound Green's function G_{BB} can be treated phenomenologically by assuming it has the form (133). If we ignore the overlap N_{AB} we obtain

$$G_{AA^\dagger} = \frac{EN_{AA^\dagger}}{E^2-\omega_T^2 - \dfrac{E(E-2i\Gamma_2)\left|\vartheta_{AB^\dagger}\right|^2}{E^2 - \omega_2^2 - 2i\,\Gamma_2 E}} \qquad (141)$$

$$= \frac{EN_{AA}}{E^2 - \omega_1^2 - \dfrac{\omega_2^2\,\omega_T^2 \quad R_{LP}}{E^2 - \omega_2^2 - 2i\Gamma_2 E}} \,, \qquad (142)$$

where

$$\omega_1^2 = \omega_T^2 + \left|\vartheta_{AB^\dagger}\right|^2 \qquad (143)$$

This Green's function has four poles, the two original Brillouin components but now at the adiabatic frequencies and two new 'second-sound' contributions. These latter merge together to form an over-damped single 'Rayleigh' component at the high values of Γ_2 which occur in normal liquids. In superfluids however they are visible as two separate components. The result (142) is similar to that obtained using two-fluid models, (Hohenberg and Martin 1965) and generalized Boltzmann equations (Kwok and Martin 1966, Wehner and Klein 1972). The intensity ratios, which depart from Landau Placzek theory, are given by the latter authors and follow from (142). Experimental spectra of light scattered from liquid helium are shown in the seminar by Dr Vaughan. The central Rayleigh component, when sufficiently decoupled from the first-sound can be treated phenomenologically as a single-mode theory of the second operator B in its own right as shown below.

4.3 Simplified Theory of Rayleigh Scattering

In this treatment the variable is ρs, the entropy density, with conservation law

$$\frac{\partial}{\partial t}(\rho s) = - \nabla . J_s . \tag{145}$$

Fourier's law for the phenomenological transport equation is

$$J_q = - \lambda \nabla T \tag{146}$$

$$\text{but} \quad J_s = \frac{1}{T} J_q \quad \text{and} \quad \nabla T = \frac{T}{c_p} \nabla s , \tag{147}$$

$$\text{thus} \quad J_s = - \frac{\lambda}{c_p} \nabla s \tag{148}$$

$$\text{and} \quad \frac{\partial}{\partial t}(\rho s) = - kJ_s = \frac{\lambda k^2}{\rho c_p} (\rho s). \tag{149}$$

Comparing this with (95) we see that

$$\ell = \frac{i\lambda k^2}{\rho c_p} + O(\Delta\rho) \tag{150}$$

The correlation function follows by inverting the Green's function and inserting N from equilibrium thermodynamics

$$F(t) = \langle \hat{\rho s}(t) \ \hat{\rho s}(0) \rangle = \rho^2 k c_p e^{-\frac{\lambda k^2}{\rho c_p} t} . \qquad (151)$$

The above general theory can also be applied to the problem of thermally relaxing molecular fluids (Weinberg, Kapral amd Desai, 1973) when frequency and wavenumber dependent generalised transport coefficients are required.

It is interesting to note that the theory is sufficiently general to describe light scattering itself as a coupled-mode problem. For this purpose the incoming radiation mode, A, is coupled to an outgoing scattered mode in the direction of B in the usual way and the coupling Ω_{AB} is just the dielectric-constant fluctuation ϵ (Kω).

4.4 Conclusions

In conclusion we should note that although the theory presented was exact in its general form and its application to single-mode cases, a number of approximations were involved, starting at equation (114), to derive the explicit mode-coupling formula of equation (123). These approximations were also used again in equation (142). In both cases we made contact with existing literature and can thus be precise about the nature of approximations previously made in these cases. The validity of these approximations is not at all clear at the present time in spite of the striking success of the Kawasaki formula in critical scattering and the apparently useful new approach to the problem of Rayleigh scattering. Improvements should result, for example, by replacing our approximation (114) by the exact (100). For further discussion of these difficulties we refer the reader, for instance, to the recent work of Martin et al (1973).

REFERENCES

Brillouin L, Computes Rendues, 158, 1331, 1914.
Butcher P N and Ogg N R, Proc Phys Soc, 86, 699, 1965.
Felderhof B V and Oppenheim I, Physica, 31, 1441, 1965.
De Groot S R and Mazur P, Irreversible Thermodynamics (North Holland 1962).
Gross E, Z Phys, 63, 685, 1930, Nature, 126, 201, 400, 603, 1930.
Hohenburg P C and Martin P, Ann Phys (NY), 34, 291, 1965.
Kadanoff L P and Martin P C, Ann Phys, 24, 419, 1963.
Kawasaki K, Ann Phys, 61, 1, 1970.
Kawasaki K, Prog Th Phys, 45, 1691, 1971.
Kawasaki K, J Phys A, 6, 1289, 1973.
Kubo R, Rep Prog Phys, 29, 255, 1966a.
Kubo R, Many Body Theory Part 1 (Syokabo Tokyo and Benjamin New York) 1966b.

Kwok P C and Martin P, Phys Rev, 142, 495, 1966.
Landau L D and Lifshitz E M, Statistical Physics (Pergamon Press, 1958) Ch 12.
Landau L D and Placzek G, Z Phys Sowjetunion, 5 172, 1934.
Luttinger J M, Phys Rev A, 8, 423, 1973.
Martin P C, Siggia E D, and Rose H A, Phys Rev A, 8, 423, 1973.
Mori H, Prog Th Phys, 33, 423, 1965.
Mountain R D, J Res Nat Bur Stand, 70A, 207, 1966.
Pike E R and Swain S, J Phys A, 4, 555, 1972.
Pike E R, Physica, 31, 461, 1965.
Selwyn P A and Oppenheim I, Physica, 54, 161, 1971.
Tyablikov S and Bonch Bruevich V L, Adv in Physics, 11, 317, 1962.
Tyndall J, Proc Roy Soc, 17, 223, 1969.
Wehner R K and Klein R, Physica, 62, 161, 1972.
Weinberg M, Kapral R, and Desai R C, Phys Rev A, 7, 1413, 1973.
Zubarev D N, Soviet Physics, Uspekhi, 3, 320, 1960.
Zwanzig R, Phys Rev, 124, 983, 1961.

PHOTON STATISTICS

M.Bertolotti

Istituto di Fisica-Facolta' di Ingegneria-

Universita' di Roma - Roma - Italy

1. Introduction

In the present lectures we will consider the relationship between the statistics of the photoelectric counting and that of the light falling on the detector. More specifically we will study the probability distribution p(n,T) of emission of n photoelectrons and some of the statistical constants associated with it (e.g. the mean, the variance, the factorial moments, etc) as a function of the detecting time T, the coherence time, the coherence area, and a few other parameters. Examples pertaining to light from thermal and laser sources will be discussed and a few cases of non-gaussian statistics will be considered in connection with some problems relevant in scattering theory.

2. The photon detection process

One of the most practical methods of studying fluctuations of light beams including the determination of their spectral and statistical properties is by means of photoelectric detectors.

The simplest counting apparatus of this kind can be schematized by a single atom free to undergo photoelectric effect, after which the ejected electron is observed.

The detection process is an intrinsically quantum mechanically one since it is related to the individual nature of the single photon. A quantitative relationship between intensity and the probability of photoemission (and, hence, of counts) may be deduced from the conventional time dependent perturbation theory of quantum mechanics. The calculation was performed by MANDEL et al. [1964] . The method was semiclassical, since the radiation field was considered to be a classical stochastic process and its interaction with the electrons of the detector was treated quantum mechanically. They considered first a single realization of the electromagnetic field interacting with one atom detector. Then assuming that in practice, a photoelectric detector could be considered as a group of independent atoms, interacting with the radiation field, they were able to show that the probability of photoemission of an electron in a time interval $t, t+\Delta t$ is proportional to the classical measure of the light intensity defined in terms of the complex analytic signal

$$p(t)\,\Delta t = \alpha\, I(t)\,\Delta t \quad . \tag{1}$$

Here

$$I(t) = V^*(t)\,V(t), \tag{2}$$

$V(t)$ being the analytic signal (GABOR [1946] , MANDEL and WOLF [1965]) associated with the real vector potential $A(t)$ of the electromagnetic field, and α the quantum efficiency of the detector which depends on the geometry and various other parameters of the detector.

It was also assumed in this derivation that

the light falling on the detector is a quasi-monochromatic plane wave and that the time interval Δt is much smaller than the coherence time of the light but much longer than the period of the light.

We are here interested to the relationship between the statistics of the photoelectron counting and that of the light falling on it. More specifically, we will study the probability $p(n,t,T)$ of counting n photons in the time interval t to $t+T$. If, as is usually the case, the radiation field is stationary and ergodic the dependence on t can be dropped being the counting probability only a function of the counting interval 0–T, i.e. $p(n,T)$.

We will not consider the exact derivation of the final Mandel's formula but rather give an intuitive justification of it. To this purpose let us suppose firstly that there are no random fluctuations in the intensity. We then assume on plausible physical grounds that counts in distinct time intervals are statistically independent. Each electron ejection is therefore an independent process which has a small and constant probability of happening [given by Eq. (1)]. The resultant distribution is therefore a Poisson distribution

$$p(n,T) = \frac{\mu^n}{n!} e^{-\mu}, \tag{3}$$

where

$$\mu = \alpha \int_0^T I(t')\, dt' = \alpha\, I\, T\,. \tag{4}$$

In actual cases the wave fields $V(t)$ and, thereby, the intensity $I(t)$ are random or stochastic variables. Then Eq. (3) refers to the counting distribution appropriate to a single realization of the intensity ensemble. To account for the stochastic property of $I(t)$ we must now take average of the Poisson counting distribution $p(n,T)$ over the relevant distribution of the intensity. Let us put

$$U = \int_0^T I(t')\,dt' , \qquad (5)$$

which, in virtue of the random nature of $I(t')$ is itself a random variable with some distribution $p(U)$. The modified counting distribution taking account of the variable U is given by Mandel's formula

$$p(n,T) = \left< \frac{(\alpha U)^n}{n!} e^{-\alpha U} \right>_U$$

$$= \int_0^\infty \frac{(\alpha U)^n}{n!} e^{-\alpha U} p(U)\,dU \qquad (6)$$

where $< \ldots . >$ is the ensemble average over the U distribution, which is no longer a Poisson distribution. This formula was first derived by Mandel [1958] [1959] [1963] by classical arguments, and then rederived by Mandel et al. [1964] using the semiclassical approach and first order perturbation theory. As a conclusion we observe that the fluctuations in the photoelectric emission may be regarded as due to two causes :

1) intrinsic fluctuations in the detection process. This is due to the random ejection of the photoelectrons, even when there are no fluctuations in the intensity of the light beam falling on the detector and it results in a Poisson distribution of photoelectric counts.

2) fluctuations in the intensity of light falling on the detector.

A complete quantum mechanical derivation of the photo electron counting formula was first derived by GLAUBER (see GLAUBER [1966]) and KELLEY and KLEINER [1964] . In the quantum mechanical treatment (GLAUBER [1966]) the probability of photoemission of an electron in a time interval 0–T from an assembly of atoms in a photocounter is written as

$$p(1,T) = \int_c d\vec{r}\,' \int_c d\vec{r}\,'' \int_0^T dt' \int_0^T dt''\, S(t''-t')\, G^{(1)}(\vec{r}\,',t',\vec{r}\,''\,t'') . \qquad (7)$$

In Eq.(7) the first integration is made in space over the photon-counter surface ; $G^{(1)}(\vec{r}',t',\vec{r}'',t'')$ is the first order quantum mechanical correlation function of the electric field defined as

$$G^{(1)}(\vec{r}'t',\vec{r}''t'') = \left\{ \langle i| E^{(-)}(\vec{r}'t')E^{(+)}(\vec{r}''t'') |i\rangle \right\}_{\text{av. over } i} \quad (8)$$

firstly introduced by GLAUBER [1963], where $E^{(-)}$ and $E^{(+)}$ are respectively the operators corresponding to the negative and positive frequency part of the electric field operator E, taken in the state $|i\rangle$ of the field before the measurement.

The quantity S(t) in Eq. (7) summarizes the response of the detecting atomic system. If one considers a broad-band detector, corresponding to a sensitivity independent from frequency in the relevant frequency range, S(t) behaves as a δ-function.

Starting from Eq.(7) by a rather complicated calculation one finally obtains a formula analogous to Eq. (6) in which $p(U) \Rightarrow p_Q(U)$ must now be re-interpreted as a generalized function which is not a positive definite function. However, for radiation fields produced from most of the available sources one may interpret it as a probability function. In general, one must regard it as a generalized function. We thus find that the basic formula of photoelectron counting is essentially unaffected by field quantization.

We now observe that in treating with photocounting distributions it is often convenient to consider the so called factorial moments defined as

$$\left\langle \frac{n!}{(n-m)!} \right\rangle = \langle n(n-1)\ldots(n-m+1)\rangle = \sum_{n=0}^{\infty} n(n-1)\ldots(n-m+1)p(n,T). \quad (9)$$

One easily finds that the mth factorial moment of n is simply proportional to the mth moment of U

$$< n\ (n-1)\ \dots\ (n-m+1) > = \alpha^m < U^m > . \qquad (10)$$

So that

$$< n > = \alpha < U > ; \qquad (11)$$

$$< n\ (n-1)> = < n^2> - < n > = \alpha^2 < U^2 >$$

$$\therefore \quad < n^2> = \alpha < U > + \alpha^2 < U^2> . \qquad (12)$$

From Eqs. (12) and (11) one can write

$$< (\Delta n)^2> = < n^2> - < n>^2 = < n> + \alpha^2 < (\Delta U)^2> , \qquad (13)$$

where

$$< (\Delta U)^2> = < U^2> - < U >^2 . \qquad (14)$$

Formula (13) shows that the variance of the fluctuations in photoelectrons may be regarded as consisting of two parts :

1) The fluctuations in the number of classical particles obeying Poisson distribution (term $< n>$)

2) The fluctuations in the classical wave field (the wave interference term $\alpha^2 < (\Delta U)^2>$).

We already observed that in the quantum mechanical derivation of Eq. (6) $p_Q(U)$ is, in general, not a positive definite function. Thus the result that $< (\Delta U)^2> \geq 0$ which holds for all classical probability distributions $p(U)$, is not necessarily true in general. One may therefore expect to find cases where the variance of the fluctuations in the number of photoelectric emissions becomes smaller than that expected from classical particle statistics.

For example, when the radiation field has a well defined number of photons $<(\Delta n)^2> = 0$, and from Eq. (13) we see that $<(\Delta U)^2>$ is then negative.

For radiation fields from a well stabilized laser, the intensity is essentially constant and $<(\Delta U)^2> = 0$. The variance $<(\Delta n)^2>$ is than seen to be equal to the same as that for a system of classical particles.

For fields obtained from thermal sources, p(U) is always positive so that for such fields $<(\Delta U)^2> \geqslant 0$, and hence the variance of the fluctuations in the number of photoelectrons is always greater than $<n>$.

3. Examples of photon-statistics

The basic quantity which enters in the formula for the photoelectron counting distribution is the probability density of the integrated light intensity U. Here we will investigate the form of this probability density for the two most typical cases of interest thermal and laser sources, and derive the corresponding p (n,T).

a) Polarized thermal light

Let us assume that the light beam is completely polarized. In this case we can describe the wave field by a scalar random process V(t) in the form of an analytic signal. If the light originates from a thermal source, the random function V(t) may then be represented as a stationary complex Gaussian process (MANDEL and WOLF [1965]). The istantaneous intensity I (t) is given by Eq.(2). Since V(t) is distributed according to a Gaussian distribution, the probability density of I is an exponential function

$$p(I) = \frac{1}{\langle I \rangle} \exp \left\{ - I / \langle I \rangle \right\} , \tag{15}$$

where

$$\langle I \rangle = \lim_{T\to\infty} \frac{1}{T} \int_0^T I\,(t')\,dt' ,$$

is the mean intensity.
We are interested in the statistical properties of the integrated light intensity U given by Eq. (5). The mean value of U and its variance can be readily obtained by Eqs. (2) and (5). They are

$$\langle U \rangle = \langle I \rangle \, T , \tag{16}$$

and

$$\langle (\Delta U)^2 \rangle = \int_0^T\!\!\int_0^T \left| \Gamma(t-t') \right|^2 \, dt\, dt', \tag{17}$$

where

$$\Gamma(\tau) = \langle V^* (t) \, V \, (t+\tau) \rangle$$

is the time correlation function of the field, and use has been made of the Gaussian properties of the field.

It is very difficult to derive an exact expression for the probability density of U if T is arbitrary. In fact no simple expression for P(U) is known for any case of direct physical interest.However it is not difficult to derive asymptotic expressions for p(U) when the parameter T is either very small or very large.

i) T very small

When T is very small compared to the coherence time

T_c, the intensity I(t) may be considered to be constant in the time interval of duration T and U is then approximately equal to IT. Hence from Eq. (15) we may write

$$p(U) = \frac{1}{< U >} e^{-U/ < U >}, \tag{18}$$

where < U > is given by Eq. (16). In this case

$$< U^m > = \int_0^T U^m p(U) = m ! < U >^m. \tag{19}$$

Substitution of the distribution (18) in Mandel's formula (6) gives the resultant photocounting distribution

$$p (n) = \frac{< n >^n}{(1+ < n >)^{1+n}} , \tag{20}$$

where $< n > = \alpha < I > T$. This is the well known Bose-Einstein distribution for n identical particles in one quantum state. This can be understood as follows : during the time of measurement the photons contained in a volume of length T_c are counted. Because $T << T_c \simeq \frac{1}{\Delta\nu}$ where $\Delta\nu$ is the bandwidth of the radiation, as a consequence of the uncertainty principle only one cell of phase space is spanned.

Therefore all the counted photons belong to the same phase space cell and fill it like Bose-Einstein particles. The variance of the distribution is given by Eqs. (13) and (19) as

$$< (\Delta n)^2 > = < n > \left[1 + < n > \right] , \tag{21}$$

and the factorial moments are

$$< \frac{n!}{(n-m)!} > = \alpha^m < U >^m = \alpha^m m! < U >^m = m! < n >^m. \tag{22}$$

ii) T very large

When T is very large compared to T_c, we may divide this interval into a large number of subintervals each of which is greater than or of the order of T_c. The contributions to U from each of these subintervals are random variables and may be considered statistically independent. From the central limit theorem, one may then conclude that U is approximately normally distributed (RICE [1945]).

$$p(U) = \left\{2\pi <(\Delta U)^2>\right\}^{-1/2} \exp\left\{-\frac{1}{2}(U-<U>)^2/<(\Delta U)^2>\right\}.$$

where $<(\Delta U)^2>$ is given by Eq. (17).
In the limit when T becomes very large, all the fluctuations in the intensity may be expected to be smoothed out on integration. U may then be regarded as a constant corresponding to a δ-function distribution

$$p(U) = \delta(U - < U >) .$$

The resulting photoelectron distribution is

$$p(n,T) = \frac{<n>^n}{n!} e^{-<n>} \qquad (23)$$

which is the poisson distribution, with

$$<(\Delta n)^2> = <n> \qquad (24)$$

and

$$<\frac{n!}{(n-m)!}> = <n>^m \qquad (25)$$

iii) T arbitrary

An approximate expression for p(U) when T is arbitrary which agrees very well with the exact distribution in the two limits when either T is very small or when it is very large is given (RICE [1945] ,

MANDEL [1959], TROUP [1965] , BEDARD et al [1967]) as

$$p(U) = \frac{a^N}{2^N} \frac{U^{N-I}}{(N-I)!} e^{-1/2\, a\, U}$$

where the constants a and N may be determined from the requirement that the mean and the variance of U should agree with the values given by Eqs. (16) and (17)

$$N = T^2 \Big/ \int_0^T \int_0^T | \gamma(t-t') |^2 \, dt \, dt', \qquad (26)$$

$$a = 2N \Big/ \langle U \rangle , \qquad (27)$$

$$\gamma(\tau) = \Gamma(\tau) \,/\, \Gamma(0).$$

This distribution agrees very well also with the exact distribution in the two extreme limits when either $T \ll T_c$ or $T \gg T_c$.

The corresponding expression for p(n,T) is

$$p(n,T) = \frac{\Gamma(n+N)}{n!\,\Gamma(N)} \frac{1}{\left(1+\frac{\langle n \rangle}{N}\right)^N} \frac{1}{\left(1+\frac{N}{\langle n \rangle}\right)^n} \qquad (27\text{ bis})$$

where Γ is the γ-function.

This formula is encountered in statistical mechanics in connection with the fluctuation of n bosons in N cells of phase space.

One has

$$\langle (\Delta n)^2 \rangle = \langle n \rangle \left(1+ \frac{\langle n \rangle}{N} \right) \qquad (28)$$

The number $\langle n \rangle /N$ represents the degeneracy

parameter.

A remarque can be done on Eqs. (21) (24) and (28) (WEBB [1972]). From Planck formula the mean energy of monochromatic radiation of frequency ν in the interval $d\nu$ contained in a volume V is

$$\bar{E}_\nu = V \frac{8\pi h\nu^3}{c^3} \left(\frac{1}{e^{h\nu/kT} - 1} \right) d\nu \tag{29}$$

This formula contains, as it is evident in the Einstein's derivation using A and B coefficients, the contributions from spontaneous and stimulated transitions. Application of the Einstein-Fowler equation for statistical fluctuations

$$< (\Delta E_\nu)^2 > = k\, T^2 \left(\frac{\partial \bar{E}_\nu}{\partial T} \right) \tag{30}$$

leads to

$$< (\Delta E_\nu)^2 > = h\nu\, \bar{E}_\nu + \frac{c^3}{8\pi\nu^2}\, V\, \bar{E}_\nu^2$$

This expression can be written in terms of the number of quanta $< n >$ by setting $\bar{E}_\nu = < n > h\nu$ thus obtaining Eq. (28) where $N = 8\pi h\nu^2 V/ c^3$ is the number of normal modes of the radiation in the volume V per unit frequency interval, or the number of phase space cells occupied by the radiation. For one phase cell $N = 1$ thus giving Eq. (21).

The presence of the term $< n >^2$ in Eq. (28) or (21) has a precise meaning which can be seen in the following way. If the energy $\bar{E}_\nu$ is rederived following Einstein's treatment but dropping the presence of stimulated transitions, one obtains instead of Eq.(29),

$$\bar{E}_\nu = V \frac{8\pi h\nu^3}{c^3} \exp\left(- \frac{h\nu}{kT} \right) d\nu ,$$

which inserted in Eq. (30) gives

$$< (\Delta E_\nu)^2 > = \bar{E}_\nu \; h\nu$$

or Eq. (24). In this case the fluctuations of radiation derived with the stimulated emission term omitted, give only the term < n > which is characteristic of classical particles alone.

It is therefore apparent that the process of stimulated emission is responsible for photons obeying Bose-Einstein statistics. In other words the number of photons already occupying a phase cell influences the probability of an additional photon entering that cell thus preventing photons from behaving as independent entities.

b) Laser light

An ideal laser emits light of a well stabilized intensity. This implies that, even if fluctuations in the phase are possible, the amplitude of the wave field remains constant. The probability density of the instantaneous intensity may now be approximated by a delta function

$$p\,(I) = \delta(I - < I >)\;, \tag{31}$$

from which it results

$$p\,(n\,,\,T\,) = \frac{< n >^n}{n\,!} \quad e^{-<n>}. \tag{32}$$

However, the probability density p(I) given above is not completely consistent with experimental observations. A probability density of the laser output intensity has been derived by Risken in his treatment of the laser as a Van der Pohl oscillator (HAKEN [1970])

$$p(I) = \frac{2}{\sqrt{\pi}\, I_o} \frac{1}{1+\operatorname{erf} w} \exp\left\{-\left(\frac{I}{I_o} - w\right)^2\right\} \qquad (33)$$

where

$$I_o = \frac{< I >}{w + \frac{1}{\sqrt{\pi}} \frac{e^{-w^2}}{I+\operatorname{erf} w}} \qquad (34)$$

and the parameter w varies from large negative values to large positive values as the laser is brought from well below threshold to well above threshold.

Finally from Eq. (13) we see that the variance of the photoelectric counts is in general different from, and is usually greater than that expected from a Poisson distribution. This is a reflection of the fact that photons do not arrive at random, but they possess a certain bunching property characteristics of Boson particles (Mandel [1963]) .
Bunching effects may more clearly be understood in terms of the conditional probability $p_c(t/\tau)\, d\tau$ that a photoelectric count be registered in a time interval $d\tau$ at $t+\tau$, given that one count has been registered at time t. Assuming that the light is stationary, it may readily be shown (Mandel [1963] , Mandel and Wolf [1965]) that this conditional probability density is given by

$$p_c\,(t/\tau) = \alpha < I(t)\, I\,(t+\tau) > / < I(t) > . \qquad (35)$$

In the case of an ideal laser stabilized in intensity is

$$p_c\,(t/\tau) = \alpha < I > , \qquad (36)$$

that is a constant independent of τ .
For polarized thermal light, using the moment theorem for Gaussian distribution, we find that

$$p_c\ (t/\tau) = \alpha < I > \left[1 + \ |\gamma(\tau)|^2\right]. \qquad (37)$$

In fig.1 the counting rates illustrating the phenomena

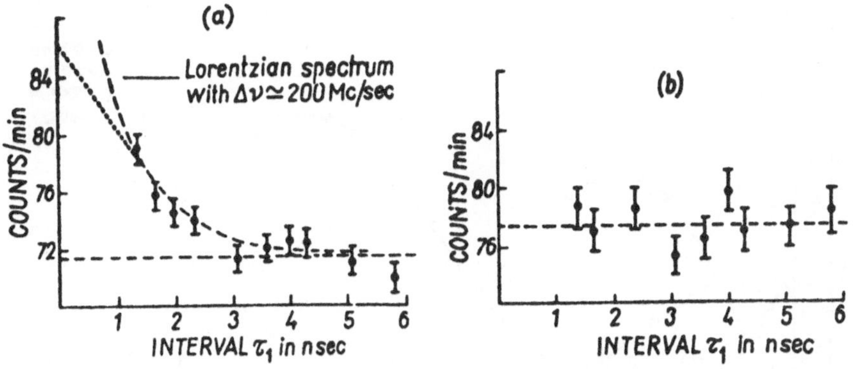

Fig. 1

of photon bunching with gaussian light are shown from MORGAN and MANDEL [1967] in a) with light from a ^{198}Hg source, in b) with light from a tungsten lamp. The ordinates represents essentially a quantity which is proportional to the integral $\int_{\tau_1}^{\tau_2} p_c(\tau)\, d\tau$ where τ_2 is a constant. In part b) the wide frequency spectrum of the conventional tungsten lamp leads to intensity correlations in an immeasurably short time interval. A laser light had given an orizontal straight line at every time.

A comparison of approximate formula (27 bis) for various values of T/T_c has been done , e.g. by Pearl and Troup [1969] by using light scattered by a rotating ground glass disc.

Typical curves are shown in Fig.2. It is seen from these curves that when $q = \frac{T}{T_c}$ is very small the curve representing p(n,T) is well fitted by the hypergeometric series of Eq. (20), which changes towards Poisson distribution for very large values of the parameter q .

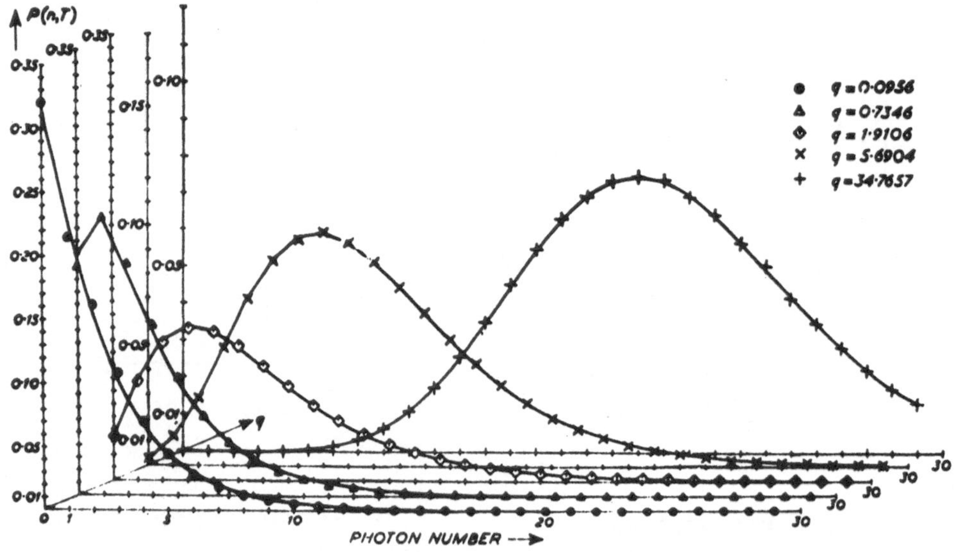

Fig.2

A similar result is obtained if $\frac{T}{T_c} << 1$ but the area of the photon-counter is increased so that the number of the seen coherence areas increases. In this case also, Eq. (28) applies because the number of phase space cell which is N = 1 for a single coherence area (or less) is equal to the number of coherence areas seen at the receiver.

The effects of spatial coherence on intensity fluctuation distribution of Gaussian light have been considered in detail by JAKEMAN, OLIVER and PIKE [1970].

4. Dead time effects

In the derivation of Mandel's formula it has been assumed that the photoelectric emissions in a photodetector are statistically independent of each other. However, in practice, this condition is not strictly satisfied. Most of the detectors require some characteristic time, called the recovery or dead time τ, after each registration of a count, during which the detector does not respond to any external field, i.e. during which no further photoemission may be registered. Consequently the number of events registered during the counting interval T will be smaller than the actual number of events and the measured photocounting distributions must be corrected. These problems have been studied by Johnson et al. [1966] and Bedard [1967a].

BEDARD [1967a] has found an exact expression for p(n,T) taking into account the dead time effects. By retaining only the first-order correction we have

$$p(n,T,\tau) = \left\langle \frac{(\alpha U)^n}{n!} e^{-\alpha U} \left[1+n\,(\alpha U-n+1)\,\frac{\tau}{T}\right]\right\rangle \qquad (38)$$

Figure 3 (JOHNSON et al. [1966]) shows an example of the effect of dead time in an experiment in which a light with Poisson statistics was studied. The quantity

$$F(n) = \frac{(n+1)p(n+1,T)}{p(n,T)} \simeq \langle n \rangle \left[1+ \langle n \rangle \frac{\tau}{T} - \frac{2n\tau}{T}\right]$$

is here plotted as a function of n. The presence

of dead-time gives a negative slope $-\ 2n\,\frac{\tau}{T}$ to F(n).

The experimental points obtained with light from a tungsten lamp with $T = 10^{-6}$ sec are on the full

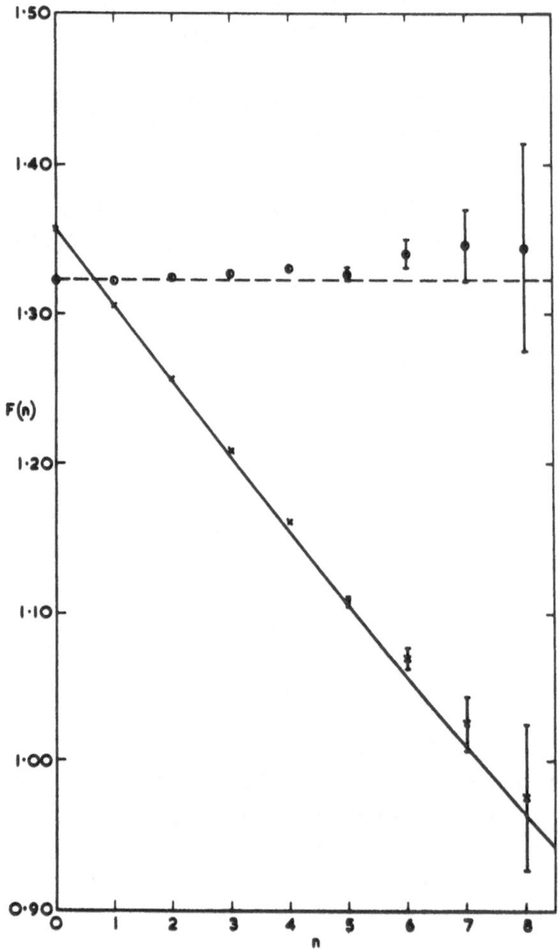

Fig.3

inclined line. Corrected data with a nominal photomultiplier dead-time $\tau = 2.10^{-8}$ sec are on the dashed line which represents the true Poisson distribution.

5. Inversion Problem

We will treat here briefly the problem of determining the probability density $p(U)$ from a given $p(n,T)$ (inversion problem).

We will first show that there exists a relation between the cumulants k_l of the photoncount distribution and the correlation functions of the intensity distribution. (MANDEL [1959]).

The cumulants k_l of a random variable are defined by

$$\sum_{l=1}^{\infty} \frac{s^l}{l!} k_l = \ln \langle e^{s\xi} \rangle_\xi = k(s) \tag{39}$$

where $\langle \dots \rangle_\xi$ is the average over the distribution of ξ . The expression of Eq. (39) is the generating function of the cumulants k_l . In our case the random variable is n and the distribution function of n is given by $p(n,T)$. Then

$$\langle e^{sn} \rangle_n \equiv \sum_{n=0}^{\infty} e^{sn} p(n,T), \tag{40}$$

and the cumulants are

$$\sum_{l=1}^{\infty} \frac{s^l}{l!} k_l = \ln \left\{ \sum_{n=0}^{\infty} e^{sn} p(n,T) \right\} \equiv k(s). \tag{41}$$

We further consider the random variable $U = \int_0^T I(t)dt$. The cumulants of the distribution of U are denoted by K_l , thus the generating function of these cumulants is

$$\sum_{\ell=1}^{\infty} \frac{s^{\ell}}{\ell!} K_1 = \ln < e^{sU} >_U \equiv K(s) \cdot \tag{42}$$

Inserting Eq. (6) into Eq. (41) yields

$$k(s) = \ln \left\{ < \sum_{n=0}^{\infty} \frac{(\alpha e^{s} U)^{n}}{n!} \quad e^{-\alpha U} >_U \right\} . \tag{43}$$

The evaluation of the sum lead to

$$k(s) = \ln \left\{ < \exp \left[\alpha \ (e^{s}-1) \ U \right] >_U \right\} . \tag{44}$$

On the other hand, when we replace s in Eq. (42) by $\alpha(e^{s}-1)$ a comparison with Eq. (44) yields

$$k(s) = K\left[\alpha \ (e^{s}- 1)\right] . \tag{45}$$

This equation establishes the relation between the cumulants , k , of the discrete distribution p(n,T) and those of the continuous intensity distribution.

Equating the coefficients of the same powers of s allows us to determine the cumulants K explicitly .

$$\begin{aligned} k_1 &= \alpha K_1 \\ k_2 &= \alpha K_1 + \alpha^2 K_2 \\ k_3 &= \alpha K_1 + 3\alpha^2 K_2 + \alpha^3 K_3 \\ k_4 &= \alpha K_1 + 7\alpha^2 K_2 + 6\alpha^3 K_3 + \alpha^4 K_4 \qquad \text{etc.} \end{aligned} \tag{46}$$

It remains to show how the K's can be determined. For this purpose we raise Eq. (42) to the exponential and expand e^{sU} into a power series of U . By taking into account Eq. (5) this leaves us with

$$\sum_{m=0}^{\infty} \frac{1}{m!} s^m \int_0^T dt_m \int_0^T dt_{m-1} \cdots \int_0^T dt_1 < I(t_m) \cdots I(t_1) > =$$

$$= \exp \left\{ \sum_{l=1}^{\infty} \frac{s^l}{l!} K_l \right\}. \tag{47}$$

After expanding the right hand side into a power series and equating the coefficients of equal powers of s, we find

$$K_1 = \int_0^T < I(t_1) > dt_1$$

$$K_2 = \int_0^T \int_0^T < \left[I(t_2) - < I(t_2) > \right] \left[I(t_1) - < I(t_1) > \right] > dt_1 dt_2$$

$$K_3 = \int_0^T \int_0^T \int_0^T < \left[I(t_3) - < I(t_3) > \right] \left[I(t_2) - < I(t_2) > \right] \times$$

$$\left[I(t_1) - < I(t_1) > \right] > dt_1 dt_2 dt_3 . \tag{48}$$

The higher K's are somewhat more complicated.

Let us now face more in general the problem of obtaining the complete probability density $p(U)$. To this purpose we set (WOLF and METHA [1964])

$$F(x) = \int_0^\infty e^{i\,x\,U} p(U)\, e^{-\alpha U} dU . \tag{49}$$

Then by means of the Fourier inversion formula

$$p(U) = \frac{1}{2\pi} \quad e^{\alpha\, U} \int_{-\infty}^{+\infty} F(x) \quad e^{-\,ixU}\, dx. \tag{50}$$

Now from Eq. (49) and (6), we have formally

$$F(x) = \sum_{n=0}^{\infty} \left(\frac{i\,x}{\alpha} \right)^n p(n,T) . \tag{51}$$

Thus the required probability density p(U) may be obtained from the knowledge of p(n,T) by first evaluating F(x) from Eq. (51) and then by employing Eq. (50).

If the counting time can be adjusted so that it is much shorter than the coherence time of the light, then

$$U \simeq I\, T ,$$

and Eq. (50) yields the probability density of the instantaneous intensity. Some care must be taken in this case because, when the counting interval is small, the dead time effects of the counter become appreciable and one must take these corrections into account.

The method just outlined is only useful when a chosed analytic expression for p(n,T) is known. In practice, however, we can determine p(n,T) experimentally, only for a few small values of n. Even if p(n) is small for larger values of n, we cannot neglect p(n), since

F(x) in that case becomes a polynomial and the Fourier inversion then gives a singular function for P(U).

In such case one must use some approximation techniques (BEDARD [1967 b]) .

If one knows only the first moments of n, Eq.(10) gives the corresponding moments of U . From these moments, one can determine some general features and approximate expression for p(U) (BEDARD [1967 b]) .

If we only search for whether a field is fully coherent or not, it is simply necessary to measure first and second order correlation function.

Glauber and Titulaer [1965] have indeed demonstrated that for all " classical " fields (fields capable of being generated by a macroscopic current density, including therefore a changing polarization in an atomic ensemble) a first-order coherent field (i.e. one giving Young's fringes with full contrast) which is coherent to second order (i.e. giving no Hanbury and Twiss [1956] correlation)is coherent to all orders.

In some other cases a knowledge of first and second-order correlation functions is of great help. If we expect that the field has a particular statistics (e.g. a gaussian one) it can be sufficient to verify that the second order moment satisfies the appropriate relations.

6. Gaussian-Gaussian scattering

We wish here to make some remarks on photon-counting experiments which are relevant in the theory of scattering.

It is usually said that the statistical properties of a medium change the statistical properties of the scattered light and this fact is often formalized by saying that a laser source will acquire the statistical properties of the medium. If the medium is gaussian the scattered light will be gaussian.

One may ask what happens if a gaussian light is made to impinge on a gaussian medium. At a first view one would be tempted to say that the scattered light will remain gaussian since a gaussian distribution corresponds to maximum entropy.

This answer would however be wrong because due to the scattering process the incoming light is divided into many modes, a process which by itself increases entropy. The entropy of scattered light in one single mode can therefore be less then the one of the initial single mode.

It can easily be shown that actually the statistics of the diffused field in this case is not gaussian. We remember that the intensity of quasi-monochromatic light scattered by a fluctuating medium, whose typical frequencies are assumed to be much smaller than the mean frequency ω_0 of the incident light, can be written at point R as (see, for example, Mandel [1969])

$$I(t) = \beta I_0 \left|\Delta\varepsilon(k,t-\frac{R}{c})\right|^2 , \qquad (52)$$

where I_0 is the intensity of the incident beam and β a constant associated with the geometry of the experiment ; $\Delta\varepsilon\ (k,t-\frac{R}{c})$ is the space-Fourier transform of the susceptibility fluctuations of the medium, which is supposed to be homogeneous and isotropic, evaluated at the wave number $\vec{k} = \vec{k}_0 - k_0\vec{R}/R$. Thus U results as the product of two independent statistical variables I_0 and $J_1 = \left|\Delta\varepsilon\ (k,t-\frac{R}{c})\right|^2$, the first relative to the incident field and the second to the scattering medium. The expression of the probability distribution $p(n,T)$ reads then, for time intervals T shorter than the coherence time

$$p(n,T) = \frac{1}{n!}\int_0^\infty \gamma^n T^n I_o^n J_1^n \exp(-\gamma I_o J_1 T) P_o(I_o) P_1(J_1) dI_o\, dJ_1 , \tag{53}$$

where $\gamma = \alpha\beta$.

The intensity distribution of the incident radiation can be taken of the form [see Eqs. (31) and (15)]

$$P_o(I_o) = \delta(I_o - \langle I_o \rangle) , \tag{31}$$

or

$$P_o(I_o) = \frac{1}{\langle I \rangle} \exp\left\{ - \frac{I}{\langle I \rangle} \right\} , \tag{15}$$

according to whether the source is a laser one or a Gaussian one. Furthermore, if the susceptibility fluctuations are produced by random motions of particles, the probability distribution of $|\Delta\varepsilon(k,t)|^2$ is given by the expression (Mandel [1969])

$$P_1(J_1) = \frac{1}{\langle J_1 \rangle} \exp\left\{ - \frac{J_1}{\langle J_1 \rangle} \right\} . \tag{54}$$

In the case of a laser beam impinging on a Gaussian medium one obviously obtains through equations (53)(31) and (54) the usual Bose-Einstein distribution for p(n,T) :

$$p(n,T) = \frac{< n >^n}{(1+ < n >)^{1+n}} \tag{20}$$

where $< n > = \gamma < I_o > < J_1 > T$ represents the average number of counts recorded in the time interval T.

In the case of a Gaussian beam scattered by a Gaussian medium, equations (53),(15) and (54) furnish

$$p(n,T) = \frac{n!}{< n >^{1/2}} \exp \left(\frac{1}{2< n >} \right) W_{-(n+1/2),o} \left(\frac{1}{< n >} \right) \tag{55}$$

with $< n > = \gamma < I_o > < J_1 > T$, the $W_{k,n}$ being the Whittaker functions (Whittaker and Watson [1963]).

The expression of the m-th factorial moment associated with the distribution given in equation (55) can also be worked out, thus getting

$$< \frac{n!}{(n-m)!} > = (m!)^2 < n >^m, \tag{56}$$

which can be compared with the corresponding expression for the Bose-Einstein distribution :

$$< \frac{n!}{(n-m)!} > = m! \quad < n >^m \tag{22}$$

This example can have an application in the case of multiple scattering of a laser light. In this case the statistics of the scattered light will depend on whether the scattering process is a single or multiple one (KELLY [1973]).

7. An example of non-Gaussian statistics. The rotating disk.

Another example of non-gaussian statistics is obtained when considering a rotating ground glass disk.

A rotating ground glass screen containing a random distribution of scattering centers, has been for the first time considered by Martienssen and Spiller [1964] in order to obtain radiation fields with extremely long coherence time and gaussian statistics.

The Gaussian nature of such radiation has been experimentally tested by means of photoncounting experiments and proved on theoretical grounds (ARECCHI [1965] , JOHNSON et al. [1967] , ROUSSEAU [1971]) . On the other hand, from a very heuristic point of view, if the incident light is assumed to illuminate many surface inhomogeneities, the central-limit theorem ensures that the probability-density function of the scattered electric- field amplitude is a gaussian with zero mean.

The expression of the electric field scattered by the disk can be written (in the wave zone and in the single scattering approximation)

$$E\ (R,t) = A(R)\ e^{-\omega_o t} \int_V e^{i\,\vec{k}.\vec{r}}\ \Delta\varepsilon(\vec{r},t)\ d\vec{r}$$

here $A(R)$ is a geometrical factor, ω_o is the frequency of the plane polarized monochromatic field of wave number $\vec{k}_o$, $\vec{k} = \vec{k}_o - \frac{\omega_o}{c}\frac{\vec{R}}{R}$ and the integral ranges

over the scattering volume V (s.f.e. CROSIGNANI et al. [1971]) .

We can choose a geometry (see Fig. 4) in which the incident field propagates along the z direction orthogonal to the plane of a disk of thickness d, and the sections of V parallel to the xy plane are assumed to be rectangles H of sides l_2, l_1, $o \leq x \leq l_2$, $L \leq y \leq L+l_1$.

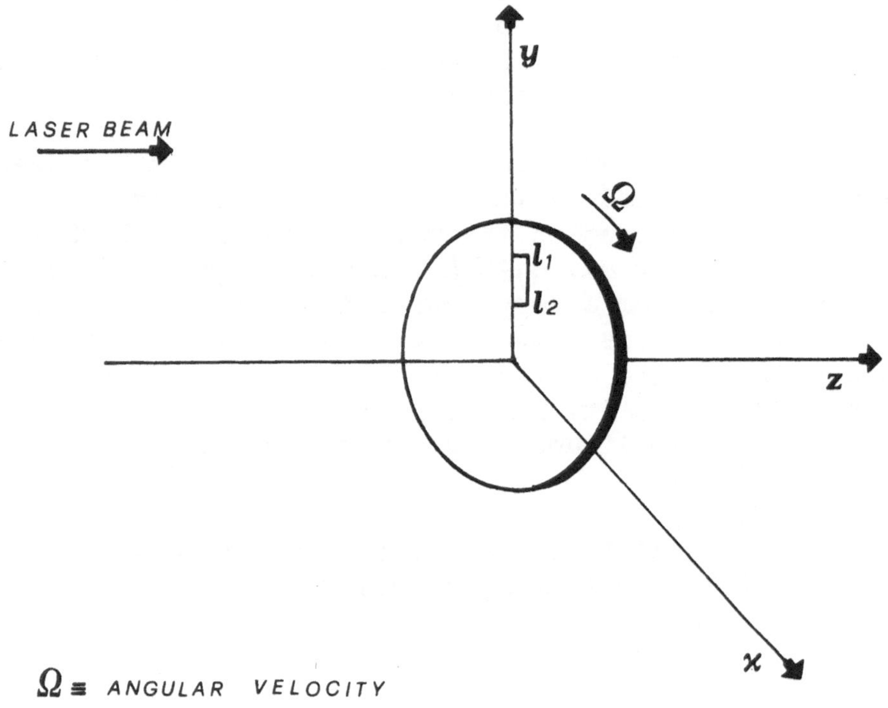

Fig.4

Therefore the correlation function of the scattered electric field reads

$$< E(t)\, E^*(t+\tau)> = |A(R)|^2 \int_H dxdy \int_H dx'dy' \int_0^d dz \int_0^d dz'\, e^{ik(r-r')}$$

$$< \Delta\varepsilon(x,y,z,t)\; \Delta\varepsilon(x',y',z',t+\tau) >$$

The explicit form of the correlation function of the dielectric fluctuations
$G(r,r',\tau) = < \Delta\varepsilon(x,y,z,t)\ \Delta\varepsilon\ (x',y',z',t+\tau) >$ can now be explicitly evaluated and therefore the spectrum of scattered light calculated (CROSIGNANI et al. [1971] , ESTES et al. [1971]).

If however the surface irregularities of the disc are of the order of the size of the illuminated region, the central-limit theorem can no more be invoked and the statistics is no longer gaussian. The quantitative reason for this will be given more ahead in this course when treating scattered light by moving particles and by very few particles (s.Crosignani).

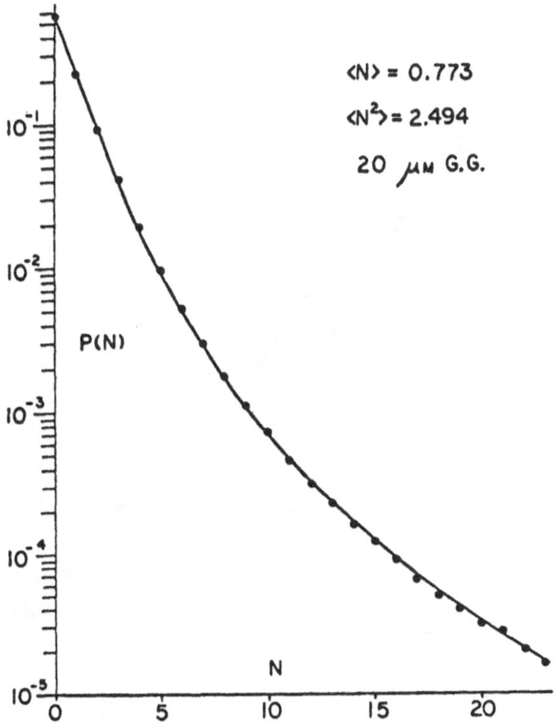

Fig.5

Bluemel et al. [1972] by using a coarseground glass with inhomogeneities of the size of 20μm, have measured the resultant p(n,T). In fig.5 their results are shown. The solid curve through the point represents the computer calculated distribution by assuming a log-normal distribution for I of the form

$$p(I) = \frac{1}{2I}\,\frac{1}{(2\pi\sigma^2)^{1/2}}\,\exp\left\{-\,\frac{\ln\left(\frac{I}{I_o}\right)^2}{4\,\sigma^2}\right\}.$$

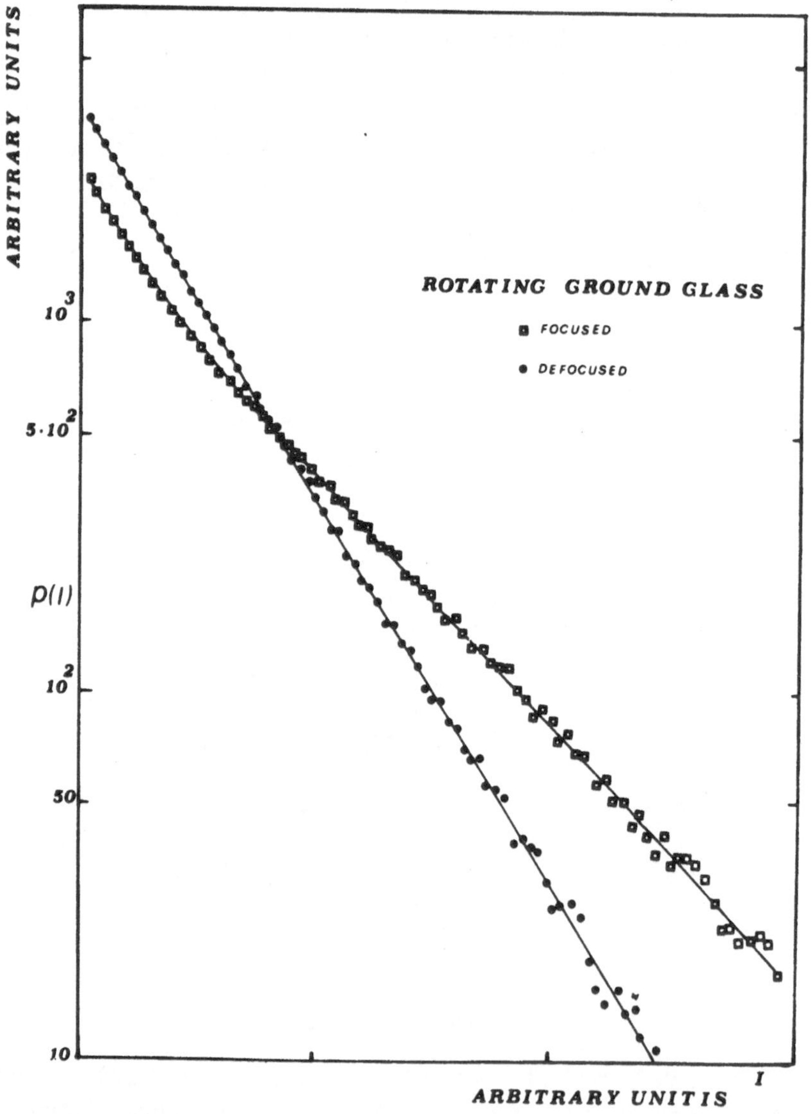

Fig.6

In a more recent experiment we have measured directly the p(I) distribution for a disk enligthened with a variable spot size. When the spot size was decreased down to a dimension of the same order of the inhomogeneities, the statistical distribution of I was no more an exponential (s.fig.6).

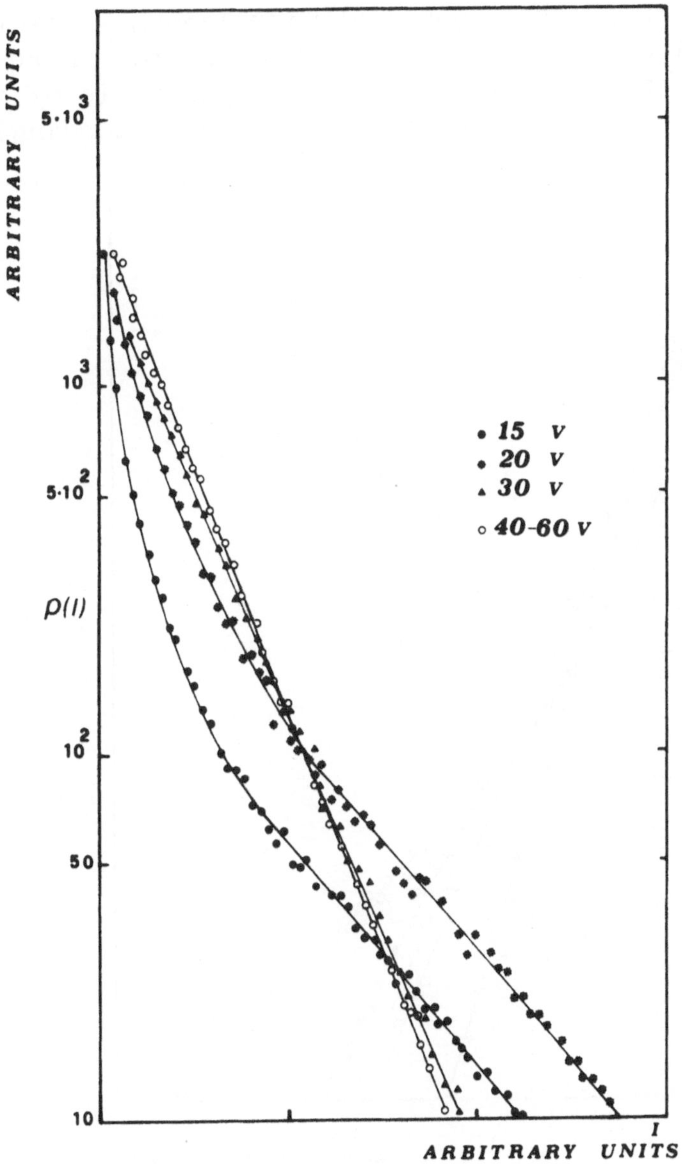

Fig.7

A similar experiment was also performed with a nematic liquid crystal. In this case the inhomogeneity dimensions can be changed by varying the value of a d.c. electric field applied to it. Fig.7 shows the change in the light intensity statistics when the laser light was focussed (spot size ≃30 μm) and the electric field decreased.

In this case the correlation length of the index of refraction fluctuations in the liquid crystal increases and when the applied field is low enough, it is of the same order of dimension of the spot size. Accordingly

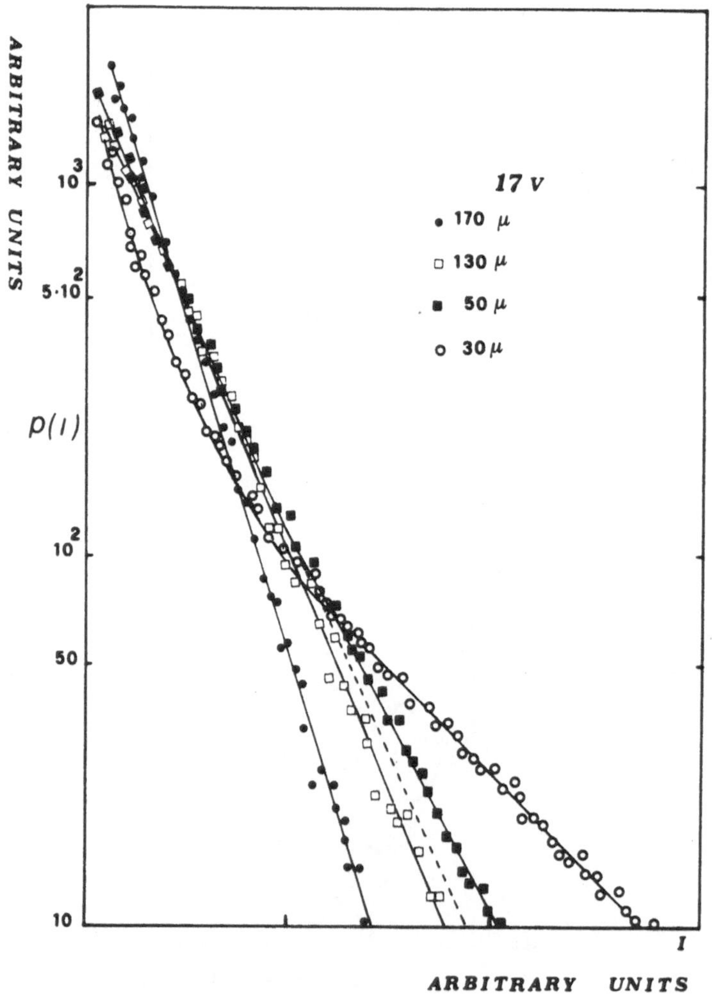

Fig.8

the light intensity distribution is no more exponential (in fig.7 this occurs at an applied voltage of about 20 Volts). In fig.8 the applied voltage is taken fixed at about 17 Volts and the spot size l_s changed. Again a change in the intensity distribution occurs when $l_s \simeq 30$ μm.

R E F E R E N C E S

Arecchi, F.T. 1965, Phys.Rev.Letters 15, 912

Bedard, G. 1967 a, Proc.Phys.Soc. 90, 131

Bedard, G. 1967 b, Phys.Rev. 161, 1304

Bedard, G. Chang, J.C. and Mandel, L. 1967, Phys.Rev. 160, 1496

Bluemel, V. Narducci, L.M.Tuft, R.A. 1972, J.Opt.Soc. America 62, 1309

Crosignani, B. Daino B., Di Porto, P. 1971, J.Appl.Phys. 42, 339

Estes, L.E.Narducci, L.M. Tuft, R.A. 1971, J.Opt.Soc. America 61, 1301

Gabor, D. 1946, J.Inst.Elec.Engrs. 93, 429

Glauber, R.J. 1963, Phys.Rev. 131, 2766

Glauber, R.J. 1966, in Physics of the Quantum Electronics eds. P.L.Kelly, B.Lax and P.E. Tannenwald (Mc Graw-Hill Boox Co. Inc.New York) pag.788

Glauber, R.J.Titulaer, O. 1965, Phys. Rev. 140, B 676

Haken, H. 1970, Handb.der Physik, S.Flugge ed. vol.25/2c, Springer-Verlag, Berlin

Hanbury Brown, R. and R.Q.Twiss 1956, Nature 178, 1046 s. also Proc.Roy.Soc.1958 A248, 199, 222

Jakeman, E. C.J.Oliver, E.R.Pike 1970, J.Phys. A (Gen. Phys.) L45

Johnson, F.A. Jones, R.McLean, T.P. and Pike, E.R. 1966, Phys.Rev. Letters 16,589

Johnson, F.A. Jones, R.McLean T.P., Pike, E.R. 1967, Opt. Acta 14, 35

Kelley, P.L. and W.H.Kleiner, 1964, Phys.Rev. 136 A 316

Kelley H.C. 1973, J.Phys.A (Gen.Phys.) 6, 353

Mandel, L. 1958, Proc.Phys.Soc.(London) 72, 1037

Mandel, L. 1959, Proc.Phys.Soc.(London) 74, 233

Mandel, L. 1963, Progress in Optics, vol. 2 (North-Holland Publishing Co.)
Mandel, L. 1969,Phys.Rev. 181, 75
Mandel, L. Sudarshan, E.C.G., Wolf, E. 1964,Proc.Phys. Soc. (London) 84, 435
Mandel, L. and Wolf, E. 1965,Rev.Mod.Phys. 37, 231
Martienssen, W. and Spiller, E. 1964, Am.J.Phys. 32,919
Morgan, B.L. and Mandel, L. 1967 , Phys.Rev.Letters 16, 1012
Pearl, P.R. and G.J.Troup 1969 , Opto Electronics 1,151
Rice, S.O. 1945,Bell System Techn. J. 23, 282
Rousseau, M. 1971 ; J.Opt.Soc.America 61, 1307
Troup, G.J. 1965, Proc.Phys. Soc.86, 39
Webb, J.H. 1972,Am. J.of Phys. 40, 850
Whittaker, E.T., and Watson, G.N. 1963 , A Course of Modern Analysis (London : Cambridge University) pp. 339-40
Wolf, E. and C.L.Metha 1964, Phys.Rev.Letters 13, 705

PHOTON CORRELATION

E Jakeman

Royal Radar Establishment

Malvern, Worcestershire, England

1. INTRODUCTION

In this paper the development of post-detection signal processing techniques for the analysis of optical frequency electromagnetic radiation will be reviewed. Analogue post-detection signal processing, commonly employed at microwave frequencies, was first demonstrated for visible radiation in the mid-fifties(1). It was not until the invention of the laser, however, that such methods were able to make an impact in the field of optical spectroscopy(2). The digital techniques to be described here are of a somewhat more recent origin. They have evolved from photon-counting experiments which were devised, after the advent of the laser, for the investigation of the statistical properties of various light sources(3). Spurred on by a desire to extend the frequency range of optical spectroscopy below the MHz limit-recently achieved with a Fabry-Perot interferometer(4), an efficient, real time, digital autocorrelator capable of rapid spectroscopic measurement over the entire range $1\text{-}10^8$ Hz has been developed(5). The principles which lie behind the design and performance of this instrument form the main subject of this presentation.

The first three sections of the paper deal with the design and performance of a digital correlation system. Section 2 is an introduction to the necessary statistical theory and notation to be used and the third section deals with correlator design.

The fourth section is devoted to an analysis of the correlator performance when it is used in spectral linewidth measurements. Much of the analysis in these sections is confined to the gaussian (thermal) light case. Although optical fields having this type of statistical property are commonly met with in practice, non-gaussian situations are also of considerable interest and in the first part of section 5 modification of the processing techniques to cope with this case is considered. The section concludes with a discussion of experimental and theoretical results on a scattering system which can usefully be studied in a non-gaussian regime[6].

2 PROPERTIES OF THE SIGNAL BEFORE DETECTION

This section is intended to act as an introduction to the mathematical formalism and notation to be used in the rest of the presentation. The first subsection is a brief introduction to the statistical description of any quantity which fluctuates randomly with time[7]. The second subsection concentrates more specifically on the properties of an optical field before detection, particularly those of the quantity usually called the intensity which is measured by a photo-electric detector.

2.1 Signal Statistics

Consider any time dependent process characterised by a real variable $V(t)$ (fig 1). Although the functional dependence of V on time may be random, the result of a measurement of V can be expressed statistically in terms of the single interval probability density function $P(V(t))$ which defines the likelihood of obtaining a value V at time t. Similarly the double interval distribution $P(V(t_1),V(t_2))$ defines the joint probability of obtaining values V_1 and V_2 by making measurements at times t_1 and t_2 respectively.

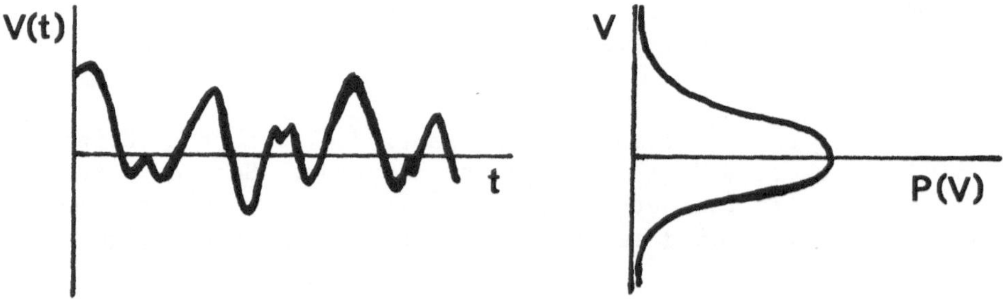

FIG 1. A FLUCTUATING SIGNAL AND ITS DISTRIBUTION

A knowledge of the entire set of multiple joint distributions $P(\{V(t_i)\})$ provides a complete statistical description of the quantity V(t). Equivalent information is contained in the moments

$$\langle V^n(t)\rangle = \int_{-\infty}^{+\infty} V^n\, P(V)dV \tag{2.1}$$

and the correlation functions

$$\langle \prod_i V^{n_i}(t_i)\rangle = \int_{-\infty}^{+\infty} P(\{V_i\}) \prod_i V_i^{n_i}\, dV_i \tag{2.2}$$

When the probability distributions do not change with time so that

$$P(\{V(t_i)\}) = P(\{V(t_i + \tau)\}) \tag{2.3}$$

the process is said to be stationary. This implies, for example, that

$$\langle V(t)V(t + \tau)\rangle = \langle V(0)V(\tau)\rangle \tag{2.4}$$

When the ensemble averages (2.1) and (2.2) are equal to the corresponding time averages eg

$$\langle f(v)\rangle = \int_{-\infty}^{+\infty} f(V)P(V)dV = \lim_{T\to\infty} \frac{1}{T}\int_{-T/2}^{T/2} f(V(t))dt \tag{2.5}$$

then the system is said to be ergodic. This relation may also be written

$$\langle f(V)\rangle = \int_{-\infty}^{+\infty} fP(f)df \tag{2.6}$$

using the simple variable transformation

$$P(f) = P(V)\,\frac{dV}{df} \tag{2.7}$$

Thus for a sine wave with phase ϕ uniformly distributed from zero to 2π (fig 2)

$$V = A\sin(\omega t + \tfrac{1}{2}\phi),\quad P(\phi) = \frac{1}{2\pi}\quad 0 < \phi < 2\pi$$

we have

$$P(V) = \frac{1}{\pi\sqrt{A^2 - V^2}} \tag{2.8}$$

If $f = V^2$ and P(V) is a gaussian distribution (fig 3)

$$P(V) = \frac{1}{\sigma\sqrt{2\pi}}\exp(-V^2/2\sigma^2) \tag{2.9}$$

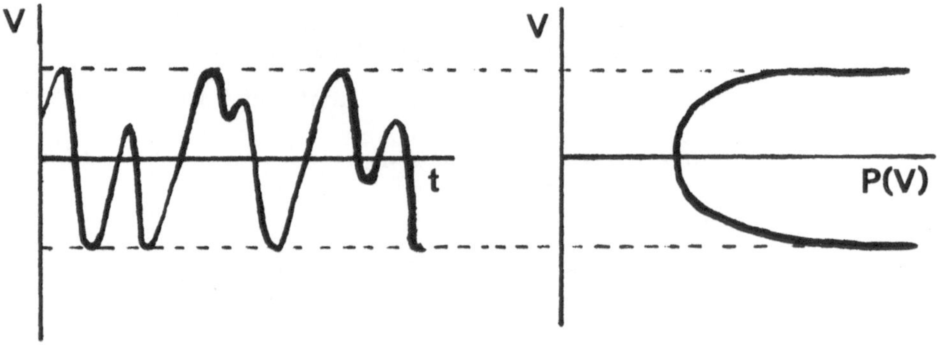

FIG 2 A RANDOMLY PHASED SINE WAVE AND ITS DISTRIBUTION

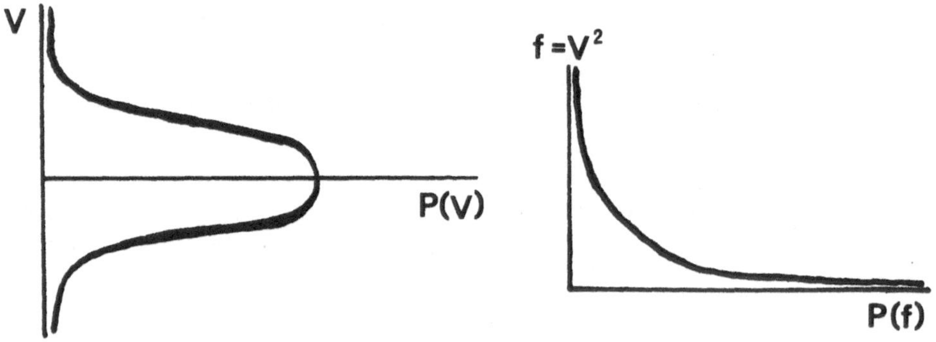

FIG 3 DISTRIBUTION BEFORE AND AFTER SQUARE-LAW DETECTION OF GAUSSIAN NOISE

then

$$P(f) = \frac{1}{2\sigma\sqrt{2\pi f}} \exp(-f/2\sigma^2) \tag{2.10}$$

The output of a square law device detecting gaussian noise would have the distribution (2.10) which is of considerable importance in the radar field. Equation (2.7) can be generalised to effect a transformation between probability distributions of many variables

$$P(\{V_i\}) = |J|\; P(\{f_i\}) \tag{2.11}$$

where J is the jacobian for the transformation.

Consider now the autocorrelation function (stationary, ergodic, system).

$$G(\tau) = \langle V(0)V(\tau)\rangle = \lim_{T\to\infty} \frac{1}{T} \int_{-T/2}^{T/2} V(t)V(t+\tau)dt \tag{2.12}$$

Evidently $G(\tau) = G(-\tau)$. For an entirely random process all memory vanishes for large τ so that

$$\lim_{\tau\to\infty} \langle V(0)V(\tau)\rangle = \langle V\rangle^2 . \tag{2.13}$$

As τ approaches zero, on the other hand, the autocorrelation function reduces to the second moment of the single interval probability distribution (equation 1.1)

$$\lim_{\tau\to o} \langle V(0)V(\tau)\rangle = \langle V^2\rangle \tag{2.14}$$

It is often convenient to define the normalised function

$$g(\tau) = \langle V(0)V(\tau)\rangle / \langle V^2\rangle \tag{2.15}$$

Since

$$\langle (V(0) \pm V(\tau))^2 \rangle = 2\langle V^2\rangle \pm 2\langle V(0)V(\tau)\rangle \tag{2.16}$$

we have

$$\langle V^2\rangle \geqslant |\langle V(0)V(\tau)\rangle| \geqslant 0 \tag{2.17}$$

and $$-1 < g(\tau) < 1 \tag{2.18}$$

From the above properties we can sketch the form for a typical autocorrelation function, assuming for simplicity $\langle V\rangle = 0$ (fig 4).

An arbitrary function V(t) may be expanded in a Fourier series in the interval [-T/2,T/2] provided that

$$\int_{-T/2}^{T/2} V(t)dt < \infty \tag{2.19}$$

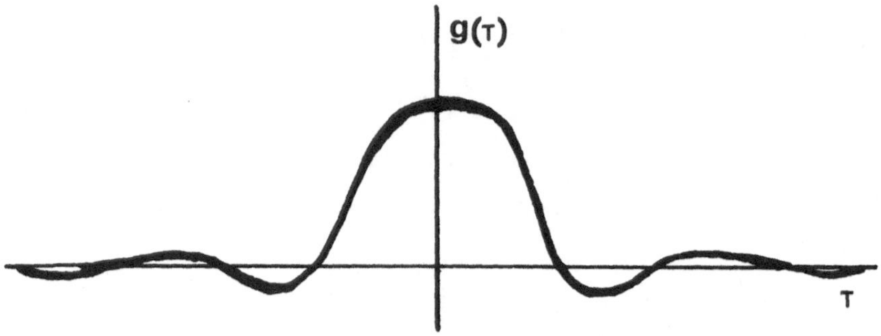

FIG 4 A TYPICAL AUTOCORRELATION FUNCTION

We may therefore write

$$V(t) = \sum_{n=-\infty}^{+\infty} v_n e^{in\omega t} \tag{2.20}$$

$$\text{where } \omega = \frac{2\pi}{T} \tag{2.21}$$

$$\text{and} \quad v_n = \frac{1}{T}\int_{-T/2}^{T/2} V(t)e^{-in\omega t}dt \tag{2.22}$$

If V(t) is a random function of time, then Fourier analysis of samples of the signal of duration T taken at different times will lead to an ensemble of values for the coefficients v_n and these, too, will be random variables. In the limit of large times T they are, moreover, uncorrelated for

$$\lim_{T\to\infty} T\langle v_n v_m^*\rangle = \lim_{T\to\infty} \frac{1}{T}\int_{-T/2}^{T/2} dt \int_{-T/2}^{T/2} dt' \; G(t-t')e^{i(m-n)\omega t}e^{im\omega(t'-t)}$$

$$= \delta_{nm}\int_{-\infty}^{+\infty} G(\tau)\; e^{-in\omega\tau}d\tau \tag{2.23}$$

When n = m we obtain the <u>Wiener-Khintchine theorem</u>

$$s(\omega) = \lim_{T\to\infty} \left\langle \frac{1}{T} \left| \int_{-T/2}^{T/2} V(t)e^{-i\omega t}dt \right|^2 \right\rangle = \int_{-\infty}^{+\infty} G(\tau)e^{-i\omega\tau}d\tau \quad (2.24)$$

which relates the power spectral density $s(\omega)$ of the signal to the autocorrelation function $G(\tau)$. Thus the power spectrum and autocorrelation function are a Fourier transform pair. For example, if $V(t)$ has a lorentzian power spectrum of half width at half height Γ, then the autocorrelation function is a negative exponential of decay, or correlation time $\tau_c = \Gamma^{-1}$ (fig 5).

The characteristic function

$$C(s) = \langle \exp isV \rangle = \int_{-\infty}^{+\infty} \exp(isV)P(V)dV \quad (2.25)$$

which is the exponential Fourier transform of the single interval probability distribution is often used to facilitate mathematical calculations. However, when $0 < V < \infty$ it is convenient to work with the moment generating function

$$Q(s) = \langle \exp(-sV) \rangle = \int_{0}^{\infty} \exp(-sV)P(V)dV \quad (2.26)$$

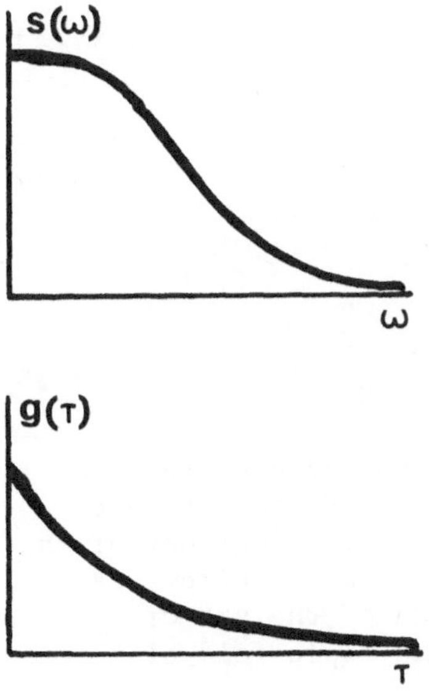

$$s(\omega) = \frac{\Gamma}{\omega^2 + \Gamma^2} \quad (2.27)$$

$$g(\tau) = \exp(|\tau|/\tau_c) \quad (2.28)$$

FIG 5 AUTOCORRELATION FUNCTION AND SPECTRUM IN THE LORENTZIAN CASE which is the Laplace transform of P(V). By differentiating (2.26) with respect to s through the integral sign it may be shown that

$$\left(-\frac{d}{ds}\right)^n Q(s)\Big|_{s=0} = \langle V^n\rangle \tag{2.29}$$

Generating functions corresponding to multiple joint probability distributions can also be defined by analogy with (2.26).

$$Q(\{s_i\}) = \langle \exp(-\sum_i s_i V_i) \rangle = \int_0^\infty P(\{V_i\}) \prod_i \exp(-s_i V_i) dV_i \tag{2.30}$$

Differentiation with respect to the s_i leads to the correlation functions:

$$\prod_i \left(\frac{-d}{ds_i}\right)^{n_i} Q(\{s_i\})\Big|_{s_i=0} = \langle \prod_i V_i^{n_i}\rangle \tag{2.31}$$

Another useful property of generating functions may be demonstrated as follows.

Let $$V = \sum_i U_i \tag{2.32}$$

where the U_i are independent contributions with $U_i > 0$. Then

$$Q_V(s) = \int_0^\infty P(\{U_i\}) \prod_i \exp(-sU_i)\, dU_i \tag{2.33}$$

but

$$P(\{U_i\}) = \prod_i P_i(U_i) \tag{2.34}$$

since the U_i are independent, so that

$$Q_V(s) = \prod_i \int_0^\infty P_i(U_i) \exp(-sU_i) dU_i = \prod_i Q_{U_i}(s) \tag{2.35}$$

Thus the generating function corresponding to the distribution of a sum of independent contributions is equal to the product of generating functions corresponding to the distributions of the individual components. Since Q(s) and P(V) are a Laplace transform pair this implies that the distribution of the sum of independent contributions is equal to the multiple fold of the distributions of the individual components.

2.2 Optical Fields

We now concentrate more specifically on the properties of optical frequency radiation. A single polarised component of the electromagnetic field can be represented at the point $\underset{\sim}{r}$ and time t by the scalar quantity $\mathcal{E}(\underset{\sim}{r},t)$. For the moment we shall assume that the detector area will be vanishingly small. Thus we need only consider the behaviour of the field at a single space point and can drop the $\underset{\sim}{r}$-coordinate from our notation. Spatial integration effects will be treated in section 4. The radiation from an ideal, amplitude-stabilised, single-mode laser may then be expressed in the form

$$\mathcal{E}_c(t) = \mathcal{E}_o \cos(\omega_o t + \phi) \tag{2.36}$$

Amplitude fluctuations are of small magnitude so that $\mathcal{E}_o$ can be regarded as constant. ϕ is randomly fluctuating with time and gives rise to a spread of frequencies perhaps as large as GHz about the optical frequency ω_o. As we shall see later, however, this random phase variable is eliminated by the detection process and for present purposes $\mathcal{E}_c(t)$ may be thought of as a pure sine wave. Thus in a typical laser light scattering experiment the incident field has a single delta-function spectrum at the optical frequency and this is broadened by the modulating effect of the scattering process (fig 6). The spectral width of the scattered radiation will be of order τ_c^{-1} where the correlation or coherence time τ_c is a measure of the modulation period.

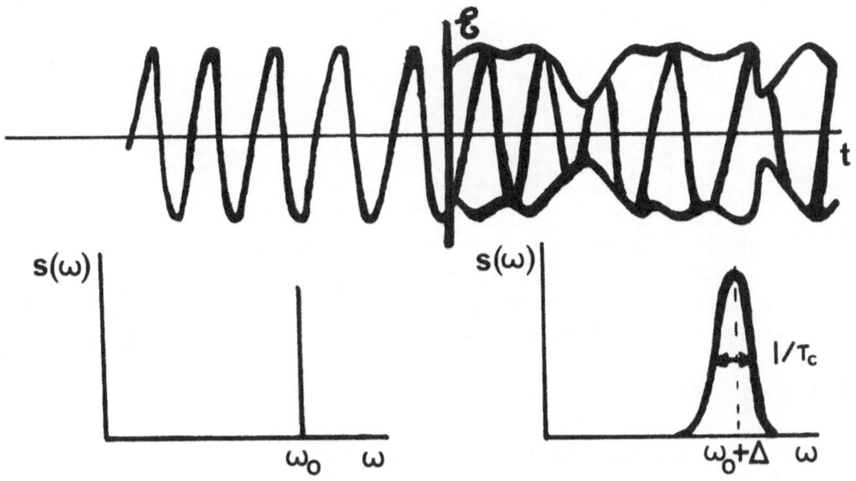

FIG 6 THE SCATTERING OF LASER LIGHT

Any overall motion of the scatterer will, of course, lead to a doppler shift Δ from the incident frequency ω_o. Both Δ and the spectral broadening or correlation time reflect properties of the scattering process and are therefore of interest. However, since a knowledge of the entire set of correlation functions (2.1) and (2.2) is necessary to completely specify the scattered light it is evident that the power spectrum or autocorrelation function usually represents only a small fraction of the information carried by the signal. It is, nevertheless, the quantity one would naturally choose to measure first for two reasons. Firstly, apart from the single interval statistics (2.1) it is the simplest characteristics of the field that one can predict theoretically or measure experimentally. Secondly, in the case of gaussian fields, commonly met with in practice and to be discussed later, all the higher order spectral properties can be expressed in terms of the autocorrelation function.

Our desire at this point, then, is to measure the power spectrum or autocorrelation function of an optical field by post-detection signal processing. The first problem is to decide exactly what an optical detector measures. A quantum mechanical approach[(8)] leads to the conclusion that a detector which operates by photon annihilation (eg a photomultiplier or photo-diode) responds to a quantity which may be expressed in classical terms as

$$I(t) = \mathcal{E}^+(t)\,\mathcal{E}^-(t) \tag{2.37}$$

where the positive and negative frequency parts of the field are defined by the Fourier decomposition

$$\begin{aligned}\mathcal{E}(t) &= \sum_{\omega>o} a^*_\omega\, e^{-i\omega t} + \sum_{\omega\geqslant o} a_\omega e^{i\omega t} \\ &= \mathcal{E}^+(t) + \mathcal{E}^-(t)\end{aligned} \tag{2.38}$$

with $$a_\omega = \frac{1}{T}\int_{-T/2}^{T/2} \mathcal{E}(t)\, e^{-i\omega t} = a^*_{-\omega} \tag{2.39}$$

The definitions (2.38) and (2.39) are entirely analogous to (2.20)-(2.22). Evidently I(t) is constant for the coherent field (2.37). Moreover, taking advantage of the narrow-band nature of the spectrum of scattered laser radiation we can write

$$\mathcal{E}_s^+(t) = \mathcal{E}_o \exp\left[-i(\omega_o t + \phi)\right] f(t) \tag{2.40}$$

where f(t) is the modulation produced by the scattering process. Thus neither the phase nor the frequency of the incident field will enter the expression (2.37). I(t) is in fact the <u>square of</u>

the envelope of the field, although it is usually referred to as the intensity. A simple classical argument justifying (2.37) as the quantity measured is that the response time of the detector is so long ($> 10^{-9}$s) that it integrates over many cycles of the optical frequency.

For an arbitrary randomly varying field, we may write

$$\mathcal{E}^{+} = \tfrac{1}{2}(\mathcal{E} + i\mathcal{E}') \tag{2.41}$$

so that $$I = \tfrac{1}{4}\,(\mathcal{E}^2 + \mathcal{E}'^2) \tag{2.42}$$

where $$\mathcal{E}' = i\,(\mathcal{E}^{-} - \mathcal{E}^{+}) \tag{2.43}$$

As we take the limit $T \to \infty$ in (2.39) the orthogonality condition (2.23) is satisfied and we also have

$$\langle \mathcal{E}\mathcal{E}' \rangle = 0 \tag{2.44}$$

The intensity is therefore the sum of the squares of two independent variables. If, in addition the distributions of $\mathcal{E}$ and $\mathcal{E}'$ are identical

$$P(\mathcal{E}) \equiv P(\mathcal{E}') \tag{2.45}$$

application of the result (2.35) implies that

$$P(I) = \int_0^I P(I - f)\,P(f)\,df \tag{2.46}$$

where $f = \mathcal{E}^2$. For the coherent field (2.36)

$$P(I) = \delta(I - \mathcal{E}_0^2) \tag{2.47}$$

If $\mathcal{E}$ is gaussian distributed, on the other hand:

$$P(\mathcal{E}) = \exp\,(-\mathcal{E}^2/2\sigma^2)/\,\sigma\sqrt{2\pi} \tag{2.48}$$

(2.46) may be evaluated using (2.10) to give

$$P(I) = \exp\,(-I/2\sigma^2)/2\sigma^2 \tag{2,49}$$

It is interesting to compare this result applicable to a square law envelope detector with the expression (2.10) for a straight square law detector (shown by the broken curve in fig 7).

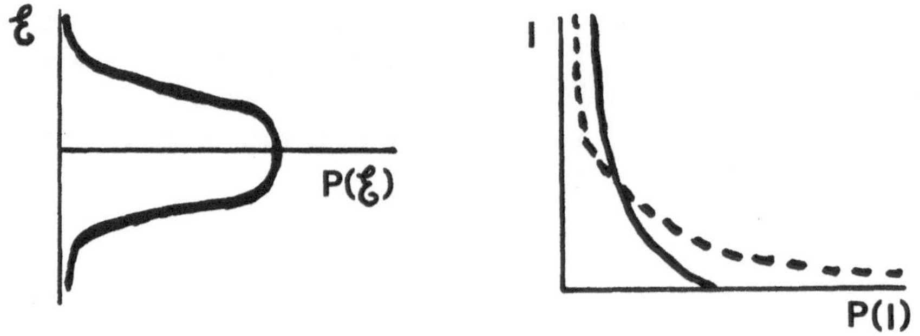

FIG 7 DISTRIBUTIONS BEFORE AND AFTER SQUARE LAW ENVELOPE DETECTION OF GAUSSIAN LIGHT (STRAIGHT SQUARE LAW DETECTION SHOWN DOTTED)

The normalised moments of the distribution P(I)

$$n^{(r)} = \langle I^r \rangle / \langle I \rangle^r \tag{2.50}$$

may easily be evaluated for the two cases (2.47) and (2.49).

From the distribution (2.47) we obtain the results

$$Q(s) = \exp(-s\mathcal{E}_o^2) \tag{2.51}$$

and
$$n^{(r)} = 1 \tag{2.52}$$

whilst for the distribution (2.49) it is not difficult to show that

$$Q(s) = (1 + 2s\sigma^2)^{-1} \tag{2.53}$$

$$n^{(r)} = r! \tag{2.54}$$

Returning now to the problem of spectral measurement, we define the first order or field correlation function as follows

$$G^{(1)}(\tau) = \langle \mathcal{E}^+(0)\mathcal{E}^-(\tau)\rangle = \sum_{\omega \geqslant o} s(\omega)\, e^{i\omega t} \tag{2.55}$$

by virtue of the orthogonality condition (2.23) which holds for large T. We shall often use the normalised form

$$g^{(1)}(\tau) = \langle \mathcal{E}^{+}(0)\mathcal{E}^{-}(\tau) \rangle \; / \; \langle I \rangle \tag{2.56}$$

where I is defined by (2.37). It is clearly not possible to make a direct measurement of $G^{(1)}(\tau)$ after detection owing to the square-law nature of the detection process. In fact the lowest order correlation function that we can determine from the post detection signal is the intensity autocorrelation function

$$G^{(2)}(\tau) = \langle I(0)I(\tau) \rangle \tag{2.57}$$

or, in normalised form

$$g^{(2)}(\tau) = \langle I(0)I(\tau) \rangle \; / \; \langle I \rangle^2 \tag{2.58}$$

The relationship between the second order correlation function (2.58) and the first order correlation function (2.56) will be dealt with in the next section.

3 DIGITAL CORRELATION

In the last section we established $g^{(1)}(\tau)$ or its Fourier transform $s(\omega)$ as the simplest spectral information carried by an optical field. We also found, however, that the nature of the detection process prevented direct measurement of this quantity and that the simplest correlation function that we could hope to determine experimentally was the intensity correlation function $g^{(2)}(\tau)$. In the first part of this section we shall investigate the relationship between these two quantities and show that a measurement of $g^{(2)}(\tau)$ can, in certain circumstances, be used to establish properties of $g^{(1)}(\tau)$. The effect of the shot noise added by the detector will also be considered in this sub-section. A brief discussion of the techniques available for post-detection signal processing will be presented in the second subsection and the final part of the section concentrates on the design of a fast digital autocorrelator system.

3.1 The Intensity Correlation Function

Consider the Fourier expansion in section 2.2 for the scalar field $\mathcal{E}(t)$

$$\mathcal{E}(t) = \sum_{\omega>0}^{\infty} a^*_\omega \, e^{-i\omega t} + \sum_{\omega\geq 0}^{\infty} a_\omega \, e^{i\omega t} \tag{3.1}$$

According to sections (2.1) and (2.2), for a randomly fluctuating function $\mathcal{E}(t)$ the requirement of stationarity implies that the a_ω are uncorrelated. If the somewhat stronger assumption of statistical independence can be made, however, a number of important results can be deduced. In this case (3.1) is the sum of a large number of statistically independent random contributions and the central limit theorem[9] states that such a sum will be gaussian distributed as the number of contributions increases without limit.

$$P(\mathcal{E}) = \exp\left(-\mathcal{E}^2/2<\mathcal{E}^2>\right)/\sqrt{2\pi<\mathcal{E}^2>} \qquad (3.2)$$

The linear combination

$$M = \sum_{i=1}^{k} s_i \mathcal{E}(t_i) \qquad (3.3)$$

is also gaussian by the same argument

$$P(M) = \exp\left(-M^2/2<M^2>\right)/\sqrt{2\pi<M^2>} \qquad (3.4)$$

so that the characteristic function corresponding to the joint distribution of the field at different times is given by

$$C(\{s_i\}) = <\exp iM> = \exp\left(-\tfrac{1}{2}<M^2>\right) \qquad (3.5)$$

Fourier transformation of (3.5) leads to the joint-gaussian distribution

$$P(\{\mathcal{E}(t_i)\}) = (2\pi)^{-k/2}\,|G|^{-\frac{1}{2}}\exp\left[-\tfrac{1}{2}\mathcal{E}G^{-1}\tilde{\mathcal{E}}\right] \qquad (3.6)$$

where the vector $\mathcal{E}$ is defined by

$$\mathcal{E} = (\mathcal{E}(t_1), \mathcal{E}(t_2) \ldots. \mathcal{E}(t_k)) \quad , \qquad (3.7)$$

$\tilde{\mathcal{E}}$ is its transpose, and G is a matrix whose elements are the correlation functions

$$G_{ij} = <\mathcal{E}(t_i)\mathcal{E}(t_j)> \qquad (3.8)$$

The quantity $\mathcal{E}'$ defined in section (2.2) is also distributed according to (3.6) so that distributions of the intensity

$$I(t) = \mathcal{E}^2(t) + \mathcal{E}'^2(t) \qquad (3.9)$$

can be calculated from the above formulae. In particular the joint distribution of intensities is given by[10]

$$P(I,I') = \frac{1}{\langle I\rangle^2(1-|g^{(1)}(\tau)|^2)} \exp\left[-\left\{\frac{(I+I')}{\langle I\rangle(1-|g^{(1)}(\tau)|^2)}\right\}\right] \times$$

$$\times\, I_o\left(\frac{2|g^{(1)}(\tau)|\sqrt{II'}}{\langle I\rangle\,(1-|g^{(1)}(\tau)|^2)}\right) \tag{3.10}$$

corresponding to the generating function[11]

$$Q(s,s') = [(1 + s\langle I\rangle)(1 + s'\langle I\rangle) - ss'\langle I\rangle^2|g^{(1)}(\tau)|^2]^{-1} \tag{3.11}$$

In equation (3.10) I_o is a modified Bessel function of order zero. The intensity autocorrelation function can easily be evaluated from (3.11):

$$\langle I(t)I(t+\tau)\rangle = \frac{d}{ds}\frac{d}{ds'}Q(ss')\Big|_{s=s'=0} = \langle I\rangle^2(1+|g^{(1)}(\tau)|^2) \tag{3.12}$$

or more concisely

$$g^{(2)}(\tau) = 1 + |g^{(1)}(\tau)|^2 \tag{3.13}$$

This is the Siegert relation[12] and may be derived directly from (3.1) using the statistical independence of the a_ω's. The formula (3.13) is only one manifestation of the property implicit in equation (3.6) that the higher order statistical properties of the gaussian field (3.2) are functions of the first order correlation function only. This factorisation property is a direct consequence of the statistical independence of the a_ω's. It is not necessary to assume for example that these variables are gaussian distributed (although if they are uncorrelated and gaussian distributed then they are also statistically independent). Indeed, as Rice[13] has pointed out if the a_ω are independent random variables which can take only two values each with a probability of one half then $\mathcal{E}(t)$ defined by (3.1) would be gaussian.

The importance of the Siegert relation in the present context lies in the fact that it enables first order spectral properties to be deduced from the second order or intensity correlation function $g^{(2)}(\tau)$ which, as we have already seen, is the simplest spectral property that we can hope to measure owing to the nature of the detection process. Since any field which is composed of a large number of statistically independent contributions will possess this factorisation property, (3.13) will obviously be quite widely applicable. There are situations, however, (for example when the number of scattering centres is small - see section 5) when the field will not be gaussian and (3.13) will not hold. In contrast to the gaussian situation, when the Siegert relation merely expresses the fact that the field is made up from many statistically independent contributions, the relation between $g^{(1)}(\tau)$ and $g^{(2)}(\tau)$ in the non-gaussian case may contain extra useful information about the scattering process. Although formulae connecting these two quantities may be established from theoretical considerations, it will normally be desirable to make an independent measurement of $g^{(1)}(\tau)$ if at all possible.

This may be accomplished by using a heterodyne technique: a method in which the scattered field is mixed coherently on the photocathode of the detector with light from a local oscillator source (usually the same laser as is used for the scattering). The requirement of coherent mixing is that the wavefronts of the scattered light and reference beam must be parallel over the detector surface. This condition may lead to experimental difficulties but if it is not satisfied spatial averaging of information contained in the interference terms between the two fields will occur. As we shall see, these terms are proportional to the desired optical spectrum. Assuming, then, that the scattered and local-oscillator fields appear to originate from the same space point, we can write

$$\mathcal{E}^+(t) = \left[f(t)\mathcal{E}_1 e^{-i(\omega_1 t+\phi_1(t))} + \mathcal{E}_2\, e^{-i(\omega_2 t+\phi_2(t))} \right] \tag{3.14}$$

where the subscripts 1 and 2 refer to the scattering and reference laser sources, respectively, and f(t) is the modulation produced by a (stationary) scattering process (section 2.2). The first and second order correlation functions characterizing the scattered light are

$$g_s^{(1)}(\tau) = \langle f(0)f(\tau)\rangle \tag{3.15}$$

and $$g_s^{(2)}(\tau) = \langle f^2(0)f^2(\tau)\rangle \tag{3.16}$$

assuming that $\langle f^2\rangle$ is normalised to unity for convenience. The intensity of the heterodyned signal from (3.14) is

$$I(t) = \mathcal{E}_2^2 + \mathcal{E}_1^2 f^2(t) + 2\mathcal{E}_1\mathcal{E}_2\, f(t)\cos(t\Delta\omega + \Delta\phi(t)) \tag{3.17}$$

where $\Delta\omega = \omega_1 - \omega_2$ and $\Delta\phi = \phi_1 - \phi_2$. (3.18) and the intensity correlation function is given by

$$<I(0)I(\tau)> = (\mathcal{E}_1^2+\mathcal{E}_2^2)^2 + \mathcal{E}_1^4(g_s^{(2)}(\tau)-1) + 2\mathcal{E}_1^2\mathcal{E}_2^2 g^{(1)}(\tau) \times$$

$$\times < \cos\,(\tau\Delta\omega+\Delta\phi(0)-\Delta\phi(\tau))> \tag{3.19}$$

In order to obtain (3.19) we have taken advantage of the random nature of the phases ϕ_1 and ϕ_2. The last term in (3.19) also vanishes unless the same source is used for the scattering and reference beams. In this case $\Delta\phi$ is simply related to the difference in path taken by the scattered and reference beams. Provided that thsi difference is less than the distance over which the phase remains correlated then $\Delta\phi$ will not change with time and (3.19) reduces to

$$g^{(2)}(\tau) = 1 + \left(\frac{<I_1>}{<I>}\right)^2 (g_s^{(2)}(\tau) - 1) + 2\left(\frac{<I_1>}{<I>}\right) g_s^{(1)}(\tau)\cos\,\tau\Delta\omega \tag{3.20}$$

A non-zero frequency difference $\Delta\omega$ can arise due to overall motion of the scatterer leading to a doppler shift in ω_1 or to deliberately introducing a shift in the frequency of the reference beam. If this frequency shift is large enough the power spectrum of the intensity can be resolved into two parts corresponding to the $g_s^{(2)}$ and $g_s^{(1)}$ terms in (3.20) (Fig 8). One component, the <u>intensity fluctuation spectrum</u>, is centred at zero frequency and corresponds to the self beat (second) term whilst the other component, the <u>doppler spectrum</u>, is centred at $|\Delta\omega|$ and is just the power spectrum of the field. If these two features overlap the doppler term can be made to dominate by simply turning up the reference beam power. This procedure also improves the signal to noise characteristics of the technique in certain circumstances (for example if the detector is an important source of noise). Formula (3.20) holds whatever the statistical properties of the scattered light, so that the first order correlation function of the scattering process can always be measured in principle by heterodyning.

Thus it is evident that, in principle, one can always extract information about $g^{(1)}(\tau)$ from a measurement of $g^{(2)}(\tau)$ either using the Siegert relation (3.13) for gaussian light or by heterodyning when the statistics of the field are non-gaussian.

In practice, of course, the signal is digital in nature because of the shot noise photo-electric detection process. We cannot therefore construct $g^{(2)}(\tau)$ directly but only correlate

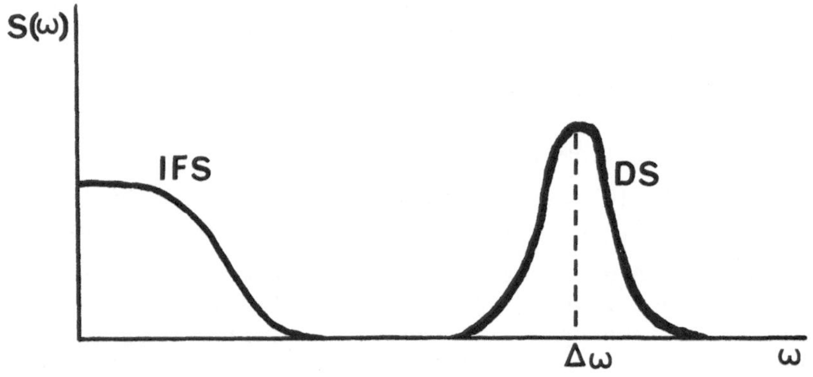

FIG 8 HETERODYNE SPECTRUM

the train of photo-electrons emitted in response to the intensity I(t) falling on the detector. A detailed discussion of the statistical properties of such a signal is presented elsewhere,(14) here we shall state just a few of the more important results.

The emission of photoelectrons is a poisson process - the distribution of counts registered in time T due to a constant intensity I falling on the detector being

$$p(n;T) = \frac{(\alpha\langle IT\rangle)^n}{n!} e^{-\alpha\langle IT\rangle} \tag{3.21}$$

where α is the efficiency of the detection process and $\langle IT\rangle$ is the integrated intensity. It is convenient henceforth to absorb T into the definition of α. If the intensity fluctuates with time then the photon counting distribution will be the compound poisson distribution(15)

$$p(n;T) = \int_0^\infty \frac{(\alpha I)^n}{n!} e^{-\alpha I} P(I)\, dI \tag{3.22}$$

It follows from the Mandel formula (3.22) that the normalised factorial moments of the photon-counting distribution are equal to the normalised moments of P(I):

$$\sum_{n=r-1}^{\infty} n(n-1)\ \ldots\ (n-r+1)\ p(n;T)/\ \langle n\rangle^{r} = \int_{0}^{\infty} (\alpha I)^{r} P(I)\,dI\,/\langle\alpha I\rangle^{r}$$

$$= n^{(r)}(T) \qquad (3.23)$$

Moreover, the definition

$$Q(s;T) = \langle e^{-sI}\rangle = \int_{0}^{\infty} e^{-sI} P(I)\,dI \qquad (3.24)$$

of the generating function corresponding to P(I) not only implies the formula

$$\left(-\frac{\alpha}{\langle n\rangle}\ \frac{d}{ds}\right)^{r} Q(s;T)\Bigg|_{s=0} = n^{(r)}(T) \qquad (3.25)$$

for the factorial moments of p(n;T) but also the relation

$$\frac{1}{n!}\left(-\alpha\ \frac{d}{ds}\right)^{n} Q(s;T)\Bigg|_{s=\alpha} = p(n;T) \qquad (3.26)$$

for the distribution itself so that definition (3.24) may be rewritten in the form

$$Q(s;T) = \sum_{n=0}^{\infty} (1 - s/\alpha)^{n}\ p(n;T) = \langle(1 - s/\alpha)^{n}\rangle \qquad (3.27)$$

Using the results of section (2.2) equation (3.26) leads to the distributions (Fig 9)

$$p(n) = \frac{\langle n\rangle^{n}}{n!}\ e^{-\langle n\rangle} \qquad \text{coherent light} \qquad (3.28)$$

$$p(n) = \frac{\langle n\rangle^{n}}{(1 + \langle n\rangle)^{n+1}} \qquad \text{gaussian light } (T \ll \tau_{c}) \qquad (3.29)$$

The geometric distribution (3.29) is a monotonic decreasing function of n but (3.28) is peaked near $n = \langle n\rangle$ for large $\langle n\rangle$.

Formula (3.22) can be generalised to give the joint distribution of counts in two samples separated by a delay τ

$$p(n(t),n(t+\tau);T) = p(n,m;T) = \int_{0}^{\infty} \frac{e^{-\alpha(I+I')}}{n!m!}\ (\alpha I)^{n}(\alpha I')^{m} P(I,I')\,dI\,dI' \qquad (3.30)$$

An important result which follows immediately is

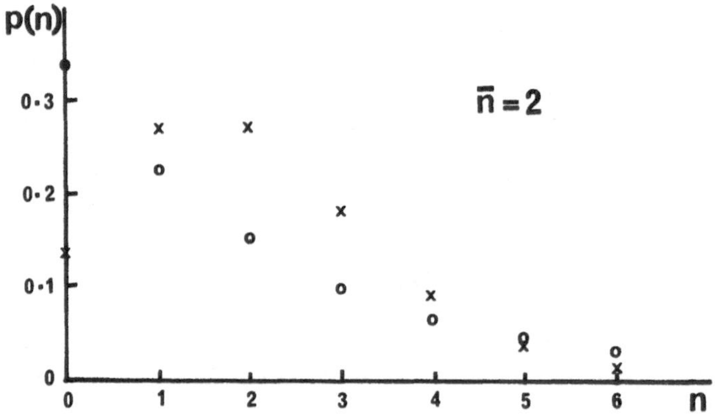

FIG 9 PHOTON COUNTING DISTRIBUTIONS: X COHERENT, O GAUSSIAN LIGHT

$$\frac{\langle n(t;T)n(t+\tau;T)\rangle}{\langle n\rangle^2} = \frac{\langle I(t;T)I(t+\tau;T)\rangle}{\langle I\rangle^2} = g^{(2)}(\tau;T) \quad \tau \neq 0$$

$$= n^{(2)}(T) + \frac{1}{\langle n\rangle} \quad \tau = 0 \qquad (3.31)$$

The left hand side of this relation is the correlation function of the number of counts arriving in a sample time T at time t with the number arriving in the same sample time at time t+τ. Thus the normalised autocorrelation function of the train of photo-electrons is equal to the normalised autocorrelation function of the intensity of light falling on the detector.

We have therefore established that a measurement of the autocorrelation function of the digital post detection signal can be used to obtain spectral information through relations (3.31) and either (3.20) or (3.13), and can now proceed to the practical implementation of such a form of processing.

3.2 Processing Techniques

These may be classified according as they operate in the time or frequency domains and according to whether they are single or parallel channel methods.

DOMAIN / TYPE	TIME	FREQUENCY
Single Channel	Delayed Coincidence	Wave Analyser
Parallel Channels	Autocorrelator	Bank of Filters

Consider first the single channel instruments. By single channel we mean that essentially only one piece of information is extracted from each sample of the signal. Take, for example, the wave analyser or scanning electrical filter. It takes a sample of the signal, filters out one frequency component, takes another sample of the signal, filters out another frequency component, and so on over a range of frequencies. The process is then cycled so as to take averages and construct the intensity spectrum

$$S(\omega) = \lim_{T\to\infty} \left\langle \frac{1}{T} \left| \int_{-T/2}^{T/2} I(t)\, e^{i\omega t}\, dt \right|^2 \right\rangle \tag{3.32}$$

The equivalent instrument in the time domain is the delayed coincidence counter. This takes a sample of the signal and multiplies the intensity at one time by that at some later time. It then takes another sample of the signal and carries out the procedure for a different delay time and so on for a whole range of delay times. By cycling the procedure the intensity autocorrelation function is constructed. Both these techniques makes inefficient use of the signal since only one frequency or one time delay component is taken from each sample. This may be an important consideration if the optical field of interest is rather weak or has a long correlation time so that long experiments are needed to achieve a reasonable degree of statistical accuracy (section 4). It is perhaps worth mentioning that because of the inherently digital nature of the signal, the delayed coincidence technique may have an advantage over the analogue spectrum analyser, particularly at low light intensities.

Passing on to the parallel channel methods: the operation of the bank of filters is fairly obvious. A range of frequency components is now filtered from each sample of the signal so that the experiment time is reduced considerably. The basic disadvantage of this technique is its inflexibility. Retuning a a bank of filters to a new frequency range is not an easy task. Moreover the analogue nature of the store may also restrict the use of a particular instrument to a particular kind of job. Let us therefore consider the last alternative. The autocorrelator is the analogue of the bank of filters in the time domain, and may be constructed in principle using a shift register and digital store (fig 10).

The signal is passed continuously through the shift register. Every shift time a new bit of signal is fed into the first channel whilst the contents of the remaining channels are shuffled down a channel, the contents of the last one being discarded. During the shift time the contents of the first channel is multiplied by the contents of each of the other channels and stored. Continuous cycling of the process builds up the autocorrelation function in the store. Clearly every time delayed product is constructed from each sample of the signal equal to the total length of the shift register. The amount of information not used is determined by the resolution of the shift register (ie the shift time). The multiplication and storage can be carried out by an analogue method, of course, but the autocorrelator comes into its own when the original digital form of the signal is retained throughout. The frequency range covered by the instrument is governed by the total sample length, normally chosen to be a few correlation times. This can easily be changed by altering the shift clock and hence the shift time. Unfortunately, if we assume modestly that we want a resolution of 1/10 of the coherence time of the signal then perhaps a hundred pairs of numbers must be multiplied together

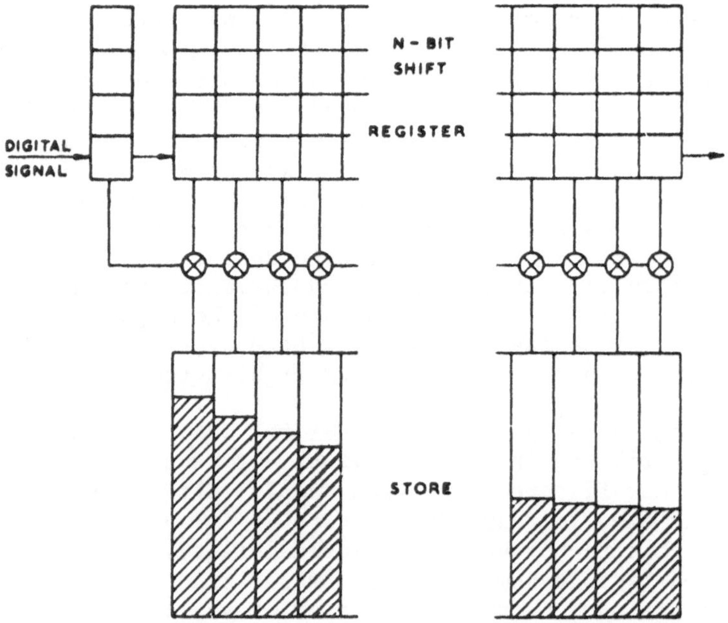

FIG 10 USE OF A SHIFT REGISTER FOR AUTOCORRELATION

digitally in times which may be as small as 10^{-7}s. This would normally require expensive computing facilities and therefore raises a serious practical objection to digital autocorrelation. The method by which this problem can be overcome will now be discussed in some detail.

3.3 The Clipping Correlator

In order to overcome the practical difficulties of fast real time digital autocorrelation we make use of a technique which was originally developed to effect the "jamming" of radar and communication systems.[16] This is the procedure of <u>clipping</u> or one-bit quantization in which the analogue signal V(t) is replaced by a two level scheme $V_c(t)$ according to some preset amplitude criterion. In an extreme form of clipping known as <u>hard limiting</u>, for example, the signal is set equal to 1 if it lies above the mean value and -1 if it lies below. An interesting relation exists between the autocorrelation functions of the original and unclipped signals when the signal V(t) is gaussian distributed about the mean. As we have seen in section (3.1) the probability distribution P(V(t),V(t')) is given in this case by a joint-gaussian distribition (equation 3.6)

$$P(V,V') = \frac{1}{2\pi\sqrt{1-g^2(\tau)}} \exp\{-(V^2+V'^2-2VV'g(\tau))/2(1-g^2(\tau))\} \qquad (3.33)$$

where

$$g(\tau) = \langle V(t)V(t+\tau)\rangle \qquad (3.34)$$

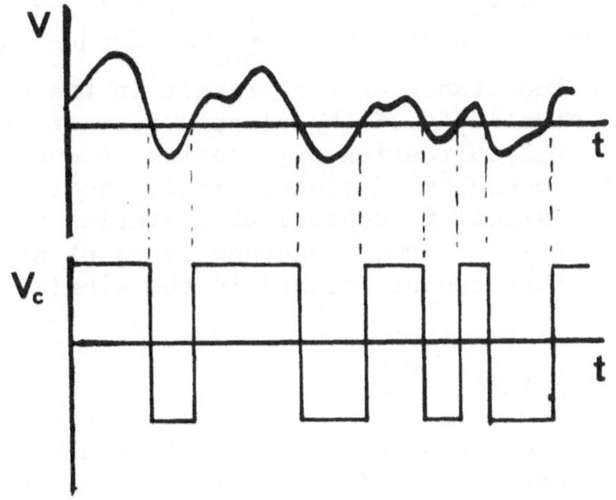

FIG 11 HARD LIMITING

and we have taken

$$\langle V \rangle = 0 \qquad \langle V^2 \rangle = 1 \tag{3.35}$$

for convenience. If we now define the hard limited signal by

$$\begin{aligned} V_c(t) &= +1 \quad \text{for } V(t) > 0 \\ &= -1 \quad \text{for } V(t) < 0 \end{aligned} \tag{3.36}$$

then the double clipped (ie both channels clipped) autocorrelation function

$$g_{dc}(\tau) = \langle V_c(t) V_c(t+\tau) \rangle \tag{3.37}$$

is given by

$$\begin{aligned} g_{dc}(\tau) &= \int_{-\infty}^{+\infty}\int_{-\infty}^{+\infty} V_c V_c' P(V,V')\, dV dV' \\ &= 4\int_0^{\infty}\int_0^{\infty} P(V,V')\, dV dV' - 1 \end{aligned} \tag{3.38}$$

using the normalisation of the probability distribution and (3.36). Substituting into (3.38) from (3.33) leads after integration to the Van Vleck theorem[17]

$$g_{dc}(\tau) = \frac{2}{\pi} \arcsin g(\tau) \tag{3.39}$$

relating the correlation functions of the original and double clipped signals. Thus a measurement of $g_{dc}(\tau)$ can be used to determine $g(\tau)$. The importance of this result in the present context lies in the simplicity of the clipped form of V(t) defined by (3.36). Multiplication of a series of ones and minus ones is easily accomplished electronically so that autocorrelation of such a signal is considerably simpler than autocorrelation of the original form. Another types of clipped correlation function that can be defined is the single clipped version of (3.37)

$$g_{sc}(\tau) = \langle V_c(t) V(t+\tau) \rangle \tag{3.40}$$

in which only one channel is hard limited. Although rarely mentioned in the literature this function is related by an even simpler formula to the unclipped correlation function. If V_c is defined by (3.36) then $g_{sc}(\tau)$ vanishes but if we take

$$V_c(t) = +1 \quad \text{for } V > 0 \qquad (3.41)$$
$$= 0 \quad \text{for } V < 0$$

then

$$g_{sc}(\tau) = \int_0^{\infty} dV' \int_{-\infty}^{+\infty} V\, P(V,V')dV \qquad (3.42)$$

and again using (3.33)

$$g_{sc}(\tau) = \frac{1}{\sqrt{2\pi}}\; g(\tau) \qquad (3.43)$$

The existence of formulae such as (3.39) and (3.43) is at first sight somewhat surprising but is entirely due to the a priori assumption of gaussian statistics. Since all the higher order statistical properties of the light can be expressed in terms of the lowest order correlation function for this type of signal it should obviously be possible to express the correlation function of the hard limited form of the signal, which is related to the distribution of zero crossings, to $g(\tau)$. The fact that higher order correlation functions are involved in this relationship means that some loss of statistical accuracy may be expected, but this turns out to be a relatively small effect.

The Van Vleck theorem cannot be taken over directly in the form (3.39) for use at optical frequencies because (a) the square of the envelope of the signal is detected and (b) the signal is digital and corrupted by shot noise. However, it is possible to devise an analogous technique for the gaussian light case. This is based on the clipped count rate defined by (fig 12)

$$n_k(t;T) = 1 \qquad n(t;T) > k$$
$$= 0 \qquad n(t;T) \leqslant k \qquad (3.44)$$

The mean value of this quantity is

$$< n_k > = < n_k^r > = \sum_{n=k+1}^{\infty} p(n;T)$$

Substituting for $p(n;T)$ from equation (3.19) leads to

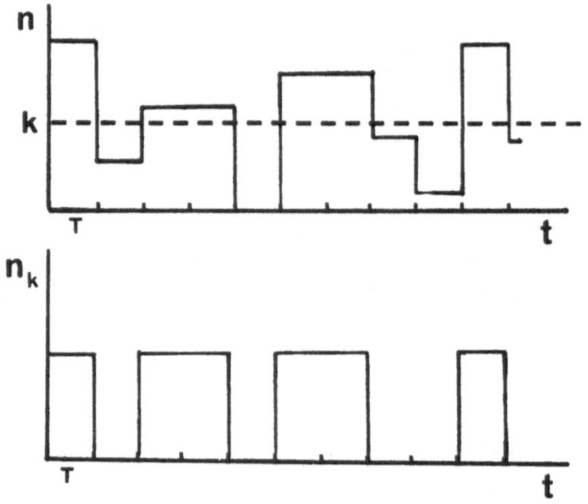

FIG 12 CLIPPING OF PHOTON COUNTING DISTRIBUTIONS

$$n_k = \left(\frac{\bar{n}}{1 + \bar{n}} \right)^{k+1} \tag{3.45}$$

This is the mean clipped count rate for gaussian light in the short sample time limit $T \ll \tau_c$ (we have adopted the notation $\bar{n} \equiv \langle n \rangle$). In the coherent case the sum cannot be evaluated analytically although the properties of the resulting exponential sum are well known and tabulated. As in the analogue case, we can define two types of normalised clipped correlation function (assuming throughout that $\tau \ll \tau_c$)

$$g_k^{(2)}(\tau) = \langle n_k(t) n_k(t+\tau) \rangle / \langle n_k \rangle \langle n \rangle \tag{3.46}$$

and $$g_{kk'}^{(2)}(\tau) = \langle n_k(t) n_{k'}(t+\tau) \rangle / \langle n_k \rangle \langle n_{k'} \rangle \tag{3.47}$$

The single clipped function (3.46) can be evaluated and expressed in closed form for gaussian light using the expression (3.11) for the generating function of the joint distribution of intensities and the relation

$$g_k^{(2)}(\tau) = \frac{1}{\bar{n}}\left(\frac{1+\bar{n}}{\bar{n}}\right)^{k+1} \sum_{n=k+1}^{\infty}\sum_{m=1}^{\infty} mp(n,m)$$

$$= \frac{1}{\bar{n}}\left(\frac{1+\bar{n}}{\bar{n}}\right)^{k+1} \sum_{n=k+1}^{\infty} \frac{1}{n!}\left(-\alpha\frac{d}{ds}\right)^{n}\left(-\alpha\frac{d}{ds'}\right)Q(ss')\Bigg|_{\substack{s'=0\\ s=\alpha}} \tag{3.48}$$

This leads to the formula[18]

$$g_k^{(2)}(\tau) = 1 + \frac{1+k}{1+\bar{n}}\,|g^{(1)}(\tau)|^2 \tag{3.49}$$

In general no such reduction is possible for the double clipped case (3.47), which may be written in terms of a triple sum as follows

$$g_{kk'}^{(2)}(\tau) = p^{k+1}+p^{k'+1} - 1 + \frac{(1-p)^2}{1-p^2|g^{(1)}(\tau)|^2}\sum_{n=0}^{k}\sum_{m=0}^{k'}\sum_{r=0}^{m}\binom{n+r}{r}\binom{m}{r}x^n y^m z^r \tag{3.50}$$

where

$$x = \frac{p(1-p\,|g^{(1)}(\tau)|^2)}{(1-p^2\,|g^{(1)}(\tau)|^2)}\,,\quad y = \frac{p(1-|g^{(1)}(\tau)|^2)}{(1-p\,|g^{(1)}(\tau)|^2)} \tag{3.51}$$

and $$z = \frac{(1-p)^2|g^{(1)}(\tau)|^2}{(1-|g^{(1)}(\tau)|^2)(1-p^2|g^{(1)}(\tau)|^2)}$$

and $$p = \frac{\bar{n}}{1+\bar{n}} \tag{3.52}$$

In the special case k' = 0 ie clipping at zero in one of the channels (3.50) reduces to

$$g_{ko}^{(2)}(\tau) = \frac{1}{p} - \frac{1-p}{p}\left(\frac{1 - p|g^{(1)}(\tau)|^2}{1 - p^2|g^{(1)}(\tau)|^2}\right)^{k+1} \tag{3.53}$$

and when k is also set equal to zero we obtain[18]

$$g_{oo}^{(2)}(\tau) = \frac{1 + |g^{(1)}(\tau)|^2(1-2p)}{1 - p^2|g^{(1)}(\tau)|^2} = \frac{1 + \frac{1-\bar{n}}{1+\bar{n}}|g^{(1)}(\tau)|^2}{1 - \left(\frac{\bar{n}}{1+\bar{n}}\right)^2 |g^{(1)}(\tau)|^2} \tag{3.54}$$

A comparison of (3.49) with the Siegert relation (3.13) shows that single clipping has not distorted the time dependence of the intensity autocorrelation function . In fact if we set $k \sim \bar{n}$ the two relations are virtually identical. Any form of double clipping, on the other hand, does produce distortion. Since k can be chosen at will it appears that the "information" term of (3.49) is amplified by clipping at $k > \bar{n}$ giving an improved signal to noise ratio. However it is intuitively obvious that as k is increased beyond the mean count rate the clipped count rate will decrease and the statistical accuracy will be impaired. As we shall see in the next section the error in the measurement of some parameter characterising $g^{(1)}(\tau)$ depends both on the size of the $|g^{(1)}(\tau)|^2$ term in (3.49) and on the statistical fluctuations in measurements of $g_k^{(2)}(\tau)$ which increase as k is increased beyond $\bar{n}$.

Thus we see that by correlating the clipped gaussian signal defined by (3.44) we can, at least in the single clipping case, obtain undistorted spectral information. In the next section we shall discuss the performance of an autocorrelation system based on (3.49) or (3.50) but it should be mentioned at this point that the effect of clipping on a Heterodyne signal can also be worked out if the scattered component is gaussian,[19] and that clipping formulae have recently been derived for a number of other model optical fields.[20] In the heterodyne case we can make use of the generating function

$$Q_D(s,s') = \exp(-\langle W\rangle Q(s,s')[s+s'+2ss'\langle E\rangle\{g^{(1)}(0)-|g^{(1)}(\tau)|\cos\omega\tau\}])$$
$$\times\, Q(s,s') \tag{3.55}$$

where Q(s,s') is the generating function corresponding to the joint distribution of the scattered component E and W is the intensity of the reference beam. This enables the single clipped correlation function to be written in the form(21)

$$g_k^{(2)}(\tau) = 1 + \frac{2\bar{n}_s\bar{n}_c}{\bar{n}\bar{n}_k} |g^{(1)}(\tau)| \cos \omega\tau \frac{\partial \bar{n}_k}{\partial \bar{n}_c} +$$

$$+ \bar{n}_s^{-2} \frac{|g^{(1)}(\tau)|^2}{\bar{n}\bar{n}_k} \frac{\partial}{\partial \bar{n}_c} \bar{n}_c \left(\frac{\partial \bar{n}_k}{\partial \bar{n}_c} \right) \qquad (3.56)$$

where(22)

$$\bar{n}_k = 1 - \exp \left(\frac{-\bar{n}_c}{1+\bar{n}_s} \right) \sum_{n=0}^{k} \frac{\bar{n}_s^{-n}}{(1+\bar{n}_s)^{n+1}} L_n \left(\frac{-\bar{n}_c}{\bar{n}_s(1+\bar{n}_s)} \right) \qquad (3.57)$$

($\bar{n}_s = \alpha\langle E \rangle$, $\bar{n}_c = \alpha \langle W \rangle$). The ratio of the IFS to the DS terms in (3.56) is not simply governed by the ratio $\bar{n}_s/(\bar{n}_c+\bar{n}_s)$ as in the unclipped case. Indeed only when $\bar{n}_s \ll 1$ can the doppler term be made to dominate (3.56) by ensuring that $\bar{n}_c \gg \bar{n}_s$. Otherwise a rather complicated condition relating k, $\bar{n}_c$ and $\bar{n}_s$ must be satisfied to achieve this end. In the shot noise limit $\bar{n}_s \ll 1$ (3.56) is approximately given by

$$g_k^{(2)}(\tau) \simeq 1 + 2\bar{n}_s n_c^k \exp(-\bar{n}_c) |g^{(1)}(\tau)|\cos\omega\tau|/\gamma(1+k,\bar{n}_c) \qquad (3.58)$$

where $\gamma(\alpha,\beta)$ is the incomplete gamma function.

Finally, let us consider briefly how an instrument can be designed to operate on the basis of (3.49) Fig 13. The shift register is used as previously described but now simple "and" gates control the input to the store. The clipped signal is fed into the shift register as a set of "0"s and "1"s every shift time and the value of the unclipped signal is simultaneously fed to each of the gates. These are controlled by the contents of the shift register, allowing storage only in channels where there is a "one".

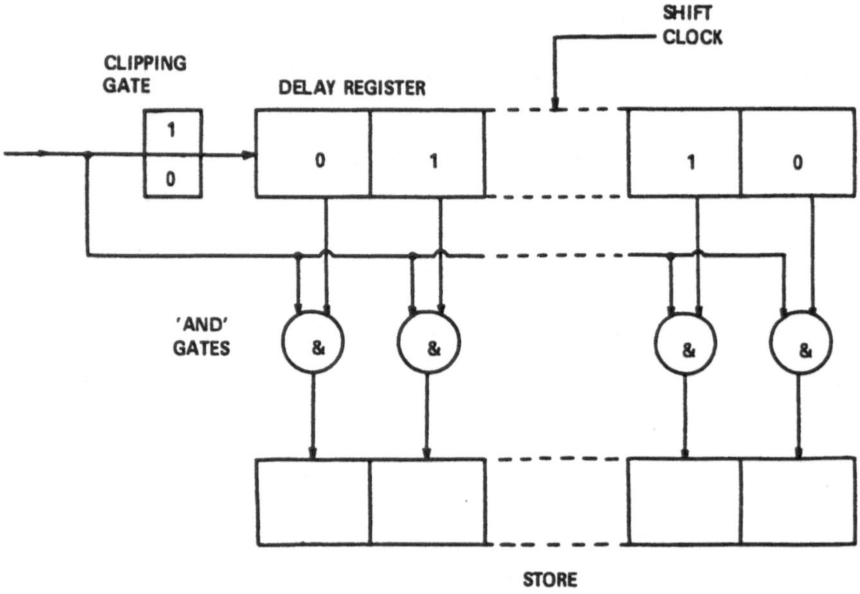

FIG 13 A SINGLE CLIPPING CORRELATOR

4 SPECTRAL LINEWIDTH MEASUREMENTS

The performance of the clipping correlator is controlled by two types of factor. Firstly those of instrumental origin such as spatial and temporal integration and dead time effects, and secondly those of statistical origin which arise from the finite duration of any real experiment. In this section we shall examine the effect of both of these types of factor on the performance of an autocorrelation system used to measure a spectral linewidth. Only the gaussian light case will be considered and any detailed analysis will be confined to a Lorentzian spectral profile in order to simplify the mathematics. Nevertheless, several general conclusions which should not depend strongly on the detailed structure of the spectrum can be drawn regarding the optimum mode of operation of the correlator.

In the first subsection the effect of temporal and spatial integration on the form of true and clipped correlation functions will be considered. A brief discussion of dead-time effects will be given in subsection 2 whilst in the third part of the section a logical method of approach to the problem of statistical errors will be developed. A fairly detailed account of such calculations will be given for the simple case of the moments of photon counting distributions, and the method by which these techniques are generalised to deal with autocorrelation functions will be discussed. In the final subsection fitting procedures will be described and the results of some linewidth error calculations for the heterodyne case will be presented and their implications briefly discussed.

4.1 Temporal and Spatial Integration

We have already mentioned in section (2.2) the integration effected by the detection process. Temporal integration will also occur in the post detection signal processing and may indeed be desirable. In effect, the quantity measured in such circumstances will be the integrated intensity

$$E(t;T) = \frac{1}{T}\int_{t-T/2}^{t+T/2} I(t')dt' \tag{4.1}$$

$$\text{with } \langle E \rangle = \langle I \rangle \tag{4.2}$$

The moments and correlation functions of E will be defined in the usual way, and the same notation used as for the instantaneous intensity apart from inclusion of the integration or sample time T. In the limit $T \to 0$ the distributions of integrated and instantaneous intensity coincide, whereas in the limit $T \to \infty$ integration will average out fluctuations in the signal and the integrated intensity will become constant. All fields thus begin to look coherent if the sample time is large enough. Since the characteristic fluctuation time of the signal is the coherence time τ_c we may write, in somewhat more quantitative terms (Fig 14).

$$P(E) \to P(I) \qquad T \ll \tau_c \tag{4.3}$$

$$P(E) \to \delta(I - \langle I \rangle) \qquad T \gg \tau_c \tag{4.4}$$

Thus the measurement of changes in statistical properties with sample time can be used to estimate τ_c. Since any source can be made to appear coherent by temporal integration it should in principle be possible to use white light for scattering experiments rather than laser radiation. This is indeed possible in certain circumstances (see for example section 5) when the scattering process is sufficiently strong for the high power per unit bandwidth of the laser to be unnecessary, and when the scatterer introduces path differences which are short compared to the distance $c\tau_c$ over which the incident light propagates coherently.

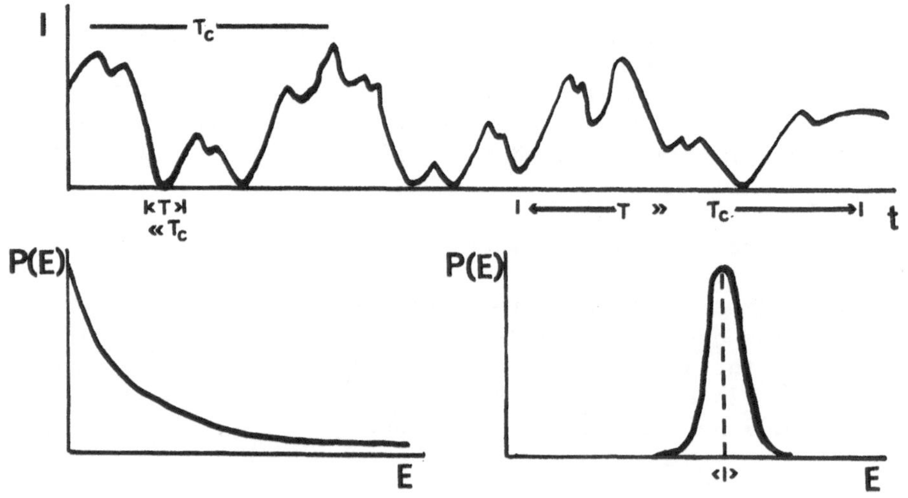

FIG 14 THE EFFECT OF A FINITE INTEGRATION TIME ON THE DISTRIBUTION OF INTENSITY

The autocorrelation function of the integrated intensity E(t;T) may be written

$$g^{(2)}(\tau;T) = \frac{1}{T^2}\int_{t-T/2}^{t+T/2} dt' \int_{t-T/2+\tau}^{t+T/2+\tau} dt''\, g^{(2)}(t'-t'') \tag{4.5}$$

For gaussian light this takes the form

$$g^{(2)}(\tau;T) = 1 + \frac{1}{T^2}\int_{t-T/2}^{t+T/2} dt' \int_{t-T/2+\tau}^{t+T/2+\tau} dt'' \left| g^{(1)}(t'-t'') \right|^2 \tag{4.6}$$

When the spectrum is Lorentzian

$$\left| g_L^{(1)}(\tau) \right|^2 = \exp(-2\Gamma\tau) \tag{4.7}$$

(4.6)may be evaluated to give

$$g^{(2)}(\tau;T) = 1 + \frac{\sinh^2\gamma}{\gamma^2} \exp(-2\Gamma\tau) \qquad \tau > T \tag{4.8}$$

where $\gamma = \Gamma T$. When this is compared with the Siegert relation (Section 3.1 equation 3.13) we observe that the dependence on τ is not distorted by the process of integration. Note however that when the spectrum is gaussian

$$|g^{(1)}(\tau)|^2 = \exp(-\alpha^2\tau^2) \tag{4.9}$$

we obtain

$$g^{(2)}(\tau;T) = 1 + \frac{\exp[-\alpha^2(\tau^2+T^2)]}{\alpha^2T^2} \sinh 2\alpha^2\tau T + \\ + \frac{1}{\alpha T^2}[(\tau+T)\,\mathrm{erf}\,\alpha(\tau+T) + (\tau-T)\mathrm{erf}\,\alpha(\tau-T) - 2\tau\,\mathrm{erf}\,\alpha\tau] \tag{4.10}$$

so that distortion certainly does take place and (4.8) must be regarded as rather a special case.

The effect of temporal integration on the clipped correlation functions cannot be evaluated so easily. However, in the single-clipped case we can obtain a useful result using the generating function corresponding to the joint intensity distribution of gaussian-lorentzian light[19]

$$Q(s,s') = Q(s)\,Q(s')\Big/\left[1 - \frac{Q(s)Q(s')}{P(s)P(s')}\,|g_L^{(1)}(\tau)|^2\right] \tag{4.11}$$

$$\text{where } Q(s) = e^{\gamma}\left[\cosh y + \frac{1}{2}\left(\frac{\gamma}{y} + \frac{y}{\gamma}\right)\sinh y\right]^{-1}$$

$$P(s) = \left\{\frac{1}{2}\left(\frac{\gamma}{y} - \frac{y}{\gamma}\right)\sinh y\right\}^{-1} \tag{4.12}$$

$$y^2 = \gamma^2 + 2\gamma s < I >$$

and $\gamma = \Gamma T$

It is not difficult to show that $P(s) \to \infty$ and $dP/ds \to P^2$ as $s \to 0$. Now the single clipped correlation function may be expressed in the form

$$g_k^{(2)}(\tau) = \sum_{n=k+1}^{\infty} \frac{1}{n!}\left(\frac{-\partial}{\partial s'}\right)^n\left(\frac{-\partial}{\partial s}\right) Q(ss')\Bigg|_{\substack{s=0\\ s'=1}} \Bigg/ < n_k > \bar{n} \tag{4.13}$$

(taking the quantum efficiency to be unity for convenience so that $\bar{n} = \langle I \rangle$). But, using the above mentioned properties of P(s) it is not difficult to show that

$$- \left. \frac{dQ(s,s')}{ds} \right|_{s=0} = \bar{n}\, Q(s') \left[1 + \frac{\sinh\gamma}{\gamma P(s')} \left| G_L^{(1)}(\tau) \right|^2 \right] \tag{4.14}$$

Thus from 4.13)

$$g_k^{(2)}(\tau;T) = 1 + \left[\frac{\sinh\gamma}{\gamma \langle n_k \rangle} \sum_{n=k+1}^{\infty} \left(- \frac{d}{ds'} \right)^n \left(Q(s')/P(s') \right) \right] \left| g_L^{(1)}(\tau) \right|^2 \tag{4.15}$$

and the single clipped correlation function is not distorted by temporal integration effects if the spectrum is lorentzian. Again, a property characteristic of this particular spectral shape.

It is perhaps worth mentioning that the effect of temporal integration on heterodyning can be evaluated in the unclipped case using formula (4.5). The IFS term (section 3.1) is modified according to (4.8) for a lorentzian spectrum but care must be taken with the doppler term if there is an overall frequency shift $\Delta\omega$ between incident and scattered radiation. If $\Delta\omega=0$ this component will simply be multiplied by a factor $4\sinh^2\frac{1}{2}\gamma/\gamma^2$ by analogy with (4.8), however.

We have so far concentrated exclusively on the temporal characteristic of the scalar field $\mathcal{E}(\underset{\sim}{r},t)$. An analogous treatment of the spatial structure of the field may also be given. Just as in the time domain, the dependence of $\mathcal{E}(\underset{\sim}{r},t)$ on $\underset{\sim}{r}$ may be deterministic or random in some sense. We therefore introduce spatial correlation or coherence functions such as

$$\Gamma^{(1)}(\underset{\sim}{r},\underset{\sim}{r}') = \langle \mathcal{E}^+(\underset{\sim}{r})\, \mathcal{E}^-(\underset{\sim}{r}') \rangle \tag{4.16}$$

$$\Gamma^{(2)}(\underset{\sim}{r},\underset{\sim}{r}') = \langle I(\underset{\sim}{r}) I(\underset{\sim}{r}') \rangle \tag{4.17}$$

omitting the time variable for convenience. For example, if $\underset{\sim}{r}$ and $\underset{\sim}{r}'$ are points on the plane wave front emitted by an ideal coherent source of intensity I_o then

$$\Gamma^{(1)}(\underset{\sim}{r},\underset{\sim}{r}') = I_o = \text{constant} \tag{4.18}$$

On the other hand a simple model which can be used for the light emitted by a number of identical independently re-radiating scattering centres is

$$\Gamma^{(1)}(\underset{\sim}{r},\underset{\sim}{r}') = \langle I_o \rangle \; \delta(\underset{\sim}{r} - \underset{\sim}{r}') \qquad (4.19)$$

where I_o is the intensity from each scatterer. Spatial correlation functions change during the course of propagation - their behaviour being governed, of course, by Maxwell's equations. It is not appropriate in this presentation to give a detailed account of propagation theory[23] and we shall confine ourselves at this point to a discussion of the solution of a particularly relevant problem. This is the coherence of points in a plane at a distance Z from a source characterised by (4.19) (Fig 15). It may be shown in this case, for a circular source of radius S, that the normalised spatial correlation function is given by[24]

$$\gamma^{(1)}(\underset{\sim}{r},\underset{\sim}{r}') = \frac{\Gamma^{(1)}(\underset{\sim}{r},\underset{\sim}{r}')}{\langle I \rangle} = \frac{2J_1\left(\frac{k_o S}{Z}\,|r-r'|\right)}{\frac{k_o s}{Z}\,|\underset{\sim}{r}-\underset{\sim}{r}'|}\exp[\,ik_o(\mathbf{r}^2-\mathbf{r}'^2)/2Z] \qquad (4.20)$$

where k_o is the wave vector of the light. This quantity is often referred to as the complex degree of coherence. Evidently, then, the initially incoherent light becomes partially coherent during propagation. This well known effect is exploited in the Michaelson stellar interferometer[25]. Young's interference experiment is a useful demonstration of (4.20)(fig 16). An

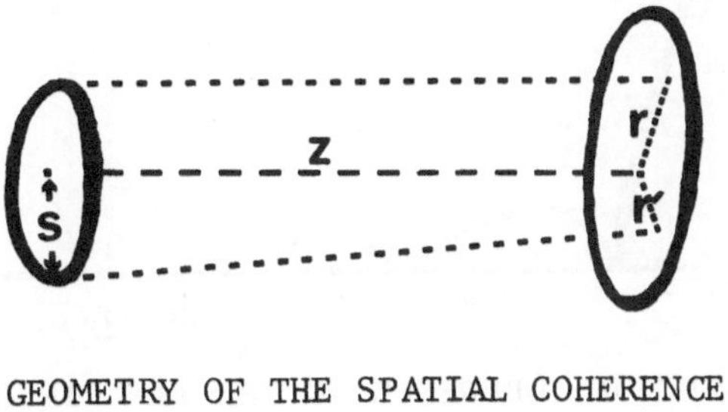

FIG 15 GEOMETRY OF THE SPATIAL COHERENCE PROBLEM

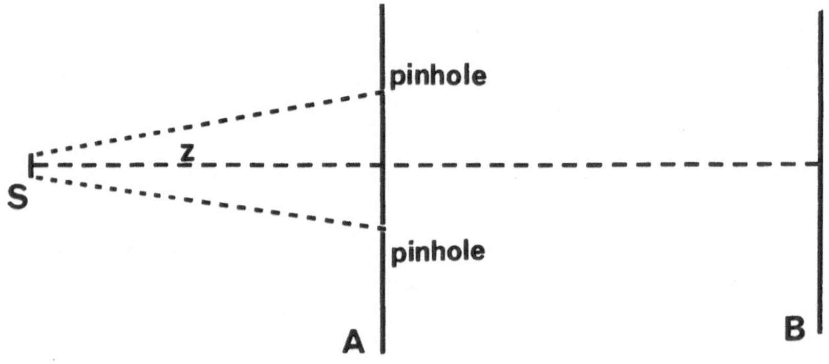

FIG 16 YOUNG'S INTERFERENCE EXPERIMENT

interference pattern is only visible at the plane B if the plane A containing the pinholes (which act as secondary sources) is placed at a distance Z from the extended incoherent source such that $|r-r'| < Z/k_o S$ ie such that the pinholes lie within an area $A = Z^2/k_o^2 A_s$. This ensures through (4.5) that light emerging from the two pinholes possess the reasonable degree of

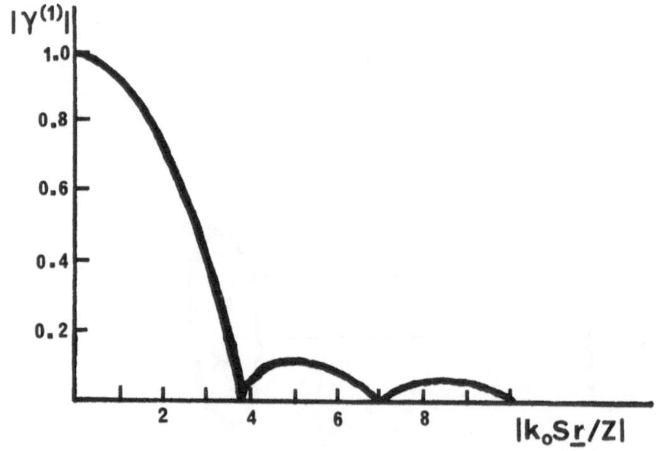

FIG 17 THE DEGREE OF COHERENCE IN THE FAR FIELD OF AN INCOHERENT SOURCE

coherence necessary to obtain interference. A plot of $|\gamma^{(1)}|$ is shown in fig 17. The first zero occurs when

$$|\underset{\sim}{r}-\underset{\sim}{r}'| = \frac{1.22Z\lambda_o}{2S} \tag{4.21}$$

This may be used to define an area

$$A_c = \pi\left(\frac{1.22Z\lambda_o}{2S}\right)^2 \tag{4.22}$$

over which the light can be regarded as coherent. The importance of spatial coherence in the present context lies in its implications for real experiments involving a detector of finite area. It is clear that if this area is larger than A_c then spatial averaging of the signal will occur.

In order to take into account both temporal and spatial integration we generalise the definition (4.5) to include the effect of a finite detector area as follows

$$E(\underset{\sim}{r},t;A,T) = \frac{1}{AT}\int_{t-T/2}^{t+T/2} dt' \int_A d^2r' I(\underset{\sim}{r}',t') \tag{4.23}$$

For gaussian light, this leads to

$$g^{(2)}(\tau;A,T) = 1 + \frac{1}{A^2T^2}\int_{-T/2}^{T/2} dt \int_{\tau-T/2}^{\tau+T/2} dt' \int_A d^2r \int_A d^2r' |g^{(1)}(\underset{\sim}{r}-\underset{\sim}{r}',t-t')|^2 \tag{4.24}$$

For the spatially incoherent source model (4.19) which will apply to many scattering systems we can assume that the light will be cross-spectrally pure[24] so that the time and space coordinates factorise

$$g^{(1)}(\underset{\sim}{r}-\underset{\sim}{r}', t-t') = \gamma^{(1)}(\underset{\sim}{r}-\underset{\sim}{r}')\, g^{(1)}(t-t') \tag{4.25}$$

and

$$g^{(2)}(\tau;A,T) = 1 + [g^{(2)}(\tau;0,T) - 1]\, f(A) \tag{4.26}$$

where $$f(A) = \frac{1}{A^2} \int_A \int_A d^2r d^2r' \; |\gamma^{(1)}(\underset{\sim}{r} - \underset{\sim}{r}')|^2 \tag{4.27}$$

We have already seen that $g^{(2)}(\tau;0,T)$ has the same τ-dependence as $g^{(2)}(\tau;0,0)$ for gaussian-lorentzian light so that spatial integration also causes no distortion in this case. $f(A)$ may be evaluated for a cylindrically symmetric system using the formula (4.20). This leads to [26]

$$f(A) = \sum_{s=0}^{\infty} \left[\frac{(2s+2)!}{\{(s+1)!\}^2 \; (s+2)!} \right]^2 (-1)^s \; (\tfrac{1}{2}\kappa R)^{2s} \tag{4.28}$$

where $\kappa = k_o S/Z$ and R is the radius of the detector surface, assumed circular. In practice both S and R would be defined by aperture stops.

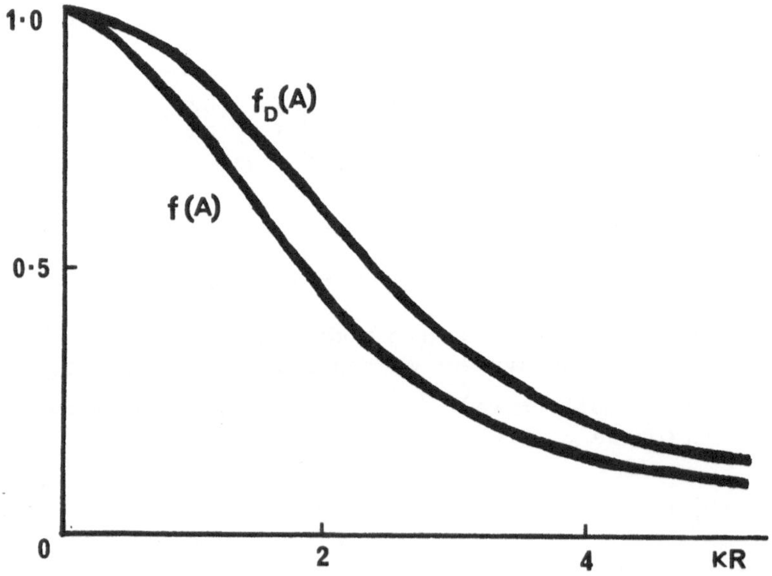

FIG 18 THE EFFECT OF FINITE DETECTOR AREA ON THE INTENSITY FLUCTUATION SPECTRUM (f(A)) AND DOPPLER SPECTRUM (f_D(A)) FOR GAUSSIAN LIGHT.

Thus as the coherence condition $\kappa R \ll 1$ is violated f(A) decreases from unity and the second term in (4.26) is averaged out. For large κR, f(A) is inversely proportional to the detector area. The doppler term of the heterodyne spectrum (see section 3.1) is similarly reduced by a factor[27]

$$f_D(A) = \frac{4}{\pi R^4} \int_o^R \int_o^R r_1 r_2 dr_1 dr_2 \int_o^{2\pi} d\phi \, \frac{J_1\left(\kappa\sqrt{r_1^2+r_2^2-2r_1r_2\cos\phi}\right)}{\kappa\left(r_1^2+r_2^2-2r_1r_2\cos\phi\right)^{\frac{1}{2}}} \tag{4.29}$$

when there is spatial averaging at the detector. This integral can be expressed in the form

$$f_D(A) = \frac{4}{(\kappa R)^2}\left[1 - J_o^2(\kappa R) - J_1^2(\kappa R)\right] \tag{4.30}$$

which falls off somewhat less rapidly than f(A) (fig 18)

For cross-spectrally pure gaussian lorentzian light in fact both temporal and spatial integration effects cause no distortion of the spectral shape of the full intensity autocorrelation function owing to the factorisation (4.26). The generalised Siegert relation may in this case be written

$$g^{(2)}(\tau;A,T) = 1 + f(A)\,\frac{\sinh^2\gamma}{\gamma^2}\exp(-2\Gamma\tau) \tag{4.31}$$

where f(A) is given by (4.28).

It may also be shown[28][29] for gaussian-lorentzian light that the single clipped autocorrelation function remains undistorted in the presence of spatial and temporal integration, although the dependence of the constant of proportionality involved (analogous to the f(A) $\sinh^2\gamma/\gamma^2$ in (4.31) is a complicated function of clipping level, T and A and must be evaluated numerically. As we have already remarked, the double clipped autocorrelation function shows distortion even in the absence of time and space averaging and is generally not a useful quantity unless $\bar{n} \ll 1$ and k = 0 when it reduces to (4.31).

4.2 Dead Time Effects

These arise due to the fact that, after each 'photon' registration by the detection system there exists a time interval of duration τ_D during which no further photo-emission can be registered. Only at the upper frequency limit of the correlator is this likely to prove a significant effect. We shall not give a detailed account of the theory of dead time effects here as there is a good deal of literature on the subject which is of interest to workers in the fields of nuclear and particle physics as well as in photon counting[30][31]. Here we shall present a few of the more important results and in particular those which are relevant to the process of digital autocorrelation.

It has been shown[32] that the poisson emission probability of photo-electrons due to an intensity I falling on the photo-cathode of a detector is modified by the presence of dead-time effects as follows

$$\left(\frac{\alpha I}{n!}\right)^n \exp(-\alpha I) \rightarrow \left(\frac{\alpha I}{n!}\right)^n \exp(-\alpha I) \left\{1+n(\alpha I-n+1)\frac{\tau_D}{T}+0\left(\frac{\tau_D}{T}\right)^2\right\} \tag{4.32}$$

where we have assumed that $\tau_D \ll T \ll \tau_c$ and $A \ll A_c$. Equation (4.32) and the Mandel formula (section 3.1) imply that

$$n^{(r)} = n_o^{(r)} - r\left(\frac{\tau_D}{T}\right)\left[(r-1)n_o^{(r)} + n_o^{(r+1)}\right] \tag{4.33}$$

$$\text{where } n_o^{(r)} = \langle n_o(n_o-1)\ \dots\ (n_o-r+1)\rangle = \langle I^r\rangle / \langle I\rangle^r \tag{4.34}$$

is the factorial moment of the true photon counting distribution and $\bar{n}_o$ is the true mean count rate. Thus, we obtain for coherent light

$$\bar{n} = \bar{n}_o - \bar{n}_o^2\left(\frac{\tau_D}{T}\right) \qquad \tau_D \ll T \tag{4.35}$$

whilst in the gaussian case

$$\bar{n} = \bar{n}_o - 2\bar{n}_o^2\left(\frac{\tau_D}{T}\right) \qquad \tau_D \ll T \ll \tau_c \tag{4.36}$$

As expected the measured count rate is reduced by the presence of a finite dead time in the apparatus particularly at high means. The single interval statistics for gaussian light can be evaluated from (4.32) and the Mandel formula using the usual exponential form for P(I). This gives

$$p(n;T) = \frac{\bar{n}_o^{\,n}}{(1+\bar{n}_o)^{n+1}}\left\{1 - n\left[(n-1) - (n+1)\frac{\bar{n}_o}{1+\bar{n}_o}\right]\left(\frac{\tau_D}{T}\right)\right\} \tag{4.37}$$

Equation (4.37) together with the definition

$$\bar{n}_k = \sum_{n=k+1}^{\infty} p(n;T) \tag{4.38}$$

leads to the clipped count rate

$$\bar{n}_k = \left(\frac{n_o}{1+\bar{n}_o}\right)^{k+1}\left\{1 - \frac{k(k+1)}{1+\bar{n}}\left(\frac{\tau_D}{T}\right)\right\} \tag{4.39}$$

The effect of a dead-time is therefore enhanced by clipping at high levels. Relations (4.32) can be generalised to cope with double interval statistics as follows:

$$p(n,m;T) = \int_o^{\infty} dI \int_o^{\infty} dI' \,\frac{(\alpha I)^n}{n!}\,\frac{(\alpha I')^m}{m!}\exp\left\{-\alpha(I+I')\right\} P(I,I') \times$$

$$\times\left(1 + \frac{\tau_D}{T}\left\{n(\alpha I - n + 1) + m(\alpha I' - m + 1\right\}\right) \tag{4.40}$$

Inspection of (4.40) indicates that the measured full autocorrelation function is related to third order correlation functions in the presence of dead time effects. In order to evaluate the single clipped correlation function it is convenient to work in terms of generating functions. Thus from (4.40)

$$\bar{n}\bar{n}_k\, g_k^{(1)}(\tau) = \bar{n} - \sum_{n=o}^{\infty}\sum_{m=o}^{\infty} np(n,m;T)$$

$$= \bar{n} - \sum_{n=o}^{k}\left[-\frac{\partial}{\partial s}\left(\frac{-\partial}{\partial s'}\right)^m \frac{Q(ss')}{m!} + \frac{\tau_D}{T}\left\{\frac{\partial^2}{\partial s^2}\left(-\frac{\partial}{\partial s'}\right)^m \frac{Q(s,s')}{m!}\right.\right.$$

$$\left.\left.- \frac{m}{m!}\frac{\partial}{\partial s}\left(\frac{-\partial}{\partial s'}\right)^{m+1} Q(s,s') + \frac{m(m-1)}{m!}\frac{\partial}{\partial s}\left(\frac{-\partial}{\partial s'}\right)^m Q(ss')\right\}\right]_{\substack{s=o\\ s'=\alpha}} \tag{4.41}$$

In the gaussian light case we can use the results of section (3.1) and equation (4.39) to obtain finally[33]

$$g_k^{(2)}(\tau) = 1 + \left(\frac{1+k}{1+\bar{n}_o}\right)\left|g^{(1)}(\tau)\right|^2 - \left(\frac{\tau_D}{T}\right) \frac{\bar{n}_o(1+k)}{(1+\bar{n}_o)^2} \left|g^{(1)}(\tau)\right|^2$$

$$\left\{2 + 2\bar{n}_o - k + (k-2\bar{n}_o)\left|g^{(1)}(\tau)\right|^2\right\} \qquad (4.42)$$

It is clear from the form of (4.42) that distortion of $g_k^{(2)}(\tau)$ will occur in general if dead-time effects are important, due to the presence of the fourth power of $g^{(1)}(\tau)$. However, for small $\bar{n}_o$ the effect may be small. The problem does not arise in the case of double clipping at zero since the correction term in (4.40) vanishes if n=m=0. In fact it is fairly obvious that $g_{oo}^{(2)}(\tau)$ will not be distorted if $\tau_D < T$ (dead times do not overlap samples) since only one count in a sample is necessary to produce a clipped signal of unity so that the presence of a dead time following that count will not affect the clipped count rate.

In practice dead times are not a serious limitation of correlator operation although if high accuracy is required in measurements of linewidth of the order of MHz, for example, corrections may be necessary.

4.3 Statistical Accuracy

Consider a simple photon counting experiment in which the mean count rate is to be measured. In such an experiment the number of counts $n(t_i;T)$ in each of a large number N of samples of duration T seperated by intervals T_p (Fig 19) will be registered and an estimate of the mean calculated as follows[34]

$$\hat{N}^{(1)} = \frac{1}{N}\sum_{i=1}^{N} n(t_i;T) \qquad (4.43)$$

Taking the average of both sides

$$<\hat{N}^{(1)}> = \frac{1}{N}\sum_{i=1}^{N}\bar{n} = \bar{n} \qquad (4.44)$$

so that $\hat{N}^{(1)}$ is an unbiassed estimate of the mean count rate per sample time. $\hat{N}^{(1)}$ can in fact be thought of as one member of an ensemble of similar measurements each of which will give a slightly different estimate of the mean due to the statistical nature of the signal. The spread of such a set of measurements

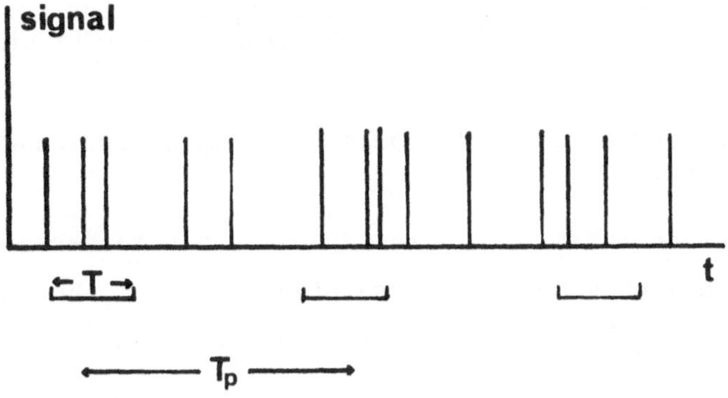

FIG 19 SAMPLING SCHEME IN A TYPICAL PHOTON COUNTING EXPERIMENT

can be used to characterise the error to be expected in a single measurement due to statistical fluctuations. Thus we can write

$$\mathrm{Var}\ \hat{N}^{(1)} = \frac{1}{N^2}\sum_{ij=1}^{N}\left[\langle n(t_i)n(t_j)\rangle - \bar{n}^2\right] \tag{4.45}$$

The diagonal elements may be summed and this leads to[35]

$$\mathrm{Var}\ \hat{N}^{(1)} = \frac{\mathrm{Var}\ n}{N} + \frac{2\bar{n}^2}{N}\sum_{k}\left(1-\frac{k}{N}\right)\left(g^{(2)}(kT_p;\tau)-1\right) \tag{4.46}$$

For gaussian-lorentzian light the sum can be evaluated without approximation to give finally ($\gamma = \Gamma\tau$)

$$\mathrm{Var}\ \hat{N}^{(1)} = \frac{1}{N}\left\{ \bar{n}+\bar{n}^2\left(\frac{1}{\gamma}-\frac{1}{2\gamma^2}+\frac{e^{-2\gamma}}{2\gamma^2}\right) + \frac{2\bar{n}^2 e^{-2\Gamma T_p}}{1-e^{-2\Gamma T_p}} + \frac{2\bar{n}^2\, e^{-2\Gamma T_p}(1-e^{-2N\Gamma T_p})}{(1-e^{-2\Gamma T_p})^2 N}\right\} \tag{4.47}$$

The last term is usually negligible for typical values of N. Several interesting conclusions can be drawn from this simple formula.

(1) The error decreases as the number of samples increases as might be expected.

(2) The error decreases as T_p increases. This is because the information in each sample is correlated if $T_p < \tau_c$ but becomes independent for $T_p \gg \tau_c$ when the sum on the right hand side of (4.46) vanishes.

(3) (4.47) is dominated by the leading term when $\bar{n}$ is sufficiently small. This expresses the importance of shot noise when the mean number of counts per sample becomes small. This is the only contribution to Var $\hat{N}^{(1)}$ when the light is coherent.

(4) In the limit $\Gamma T_p \to 0$ for fixed $\bar{n}$ the right hand side of (4.47) is inversely proportional to the number of correlation times per experiment time.

The approach described above has been used to evaluate statistical errors in measurements of higher moments of the single interval photon counting distribution[34] and is easily extended to deal with correlation functions. Thus the error in the estimator

$$\hat{G}^{(2)}(\tau) = \frac{1}{N}\sum_{r=1}^{N} n(t_r+\tau)n(t_r) \tag{4.48}$$

of the un-normalised photon-correlation function may be expressed in the form[36]

$$\text{Var}\,\hat{G}^{(2)}(\tau) = \frac{1}{N}\,\text{Var}(n(\tau)n(0)) + \frac{2}{N}\sum_{k=1}^{N-1}\left(1-\frac{k}{N}\right)\text{x}$$

$$\text{x}\left\{R(kT_p) - (G^{(2)}(\tau))^2\right\} \tag{4.49}$$

$$\text{where } R(kT_p) = \langle n(kT_p+\tau)n(kT_p)n(\tau)n(0)\rangle \tag{4.50}$$

Here T_p is again the time taken between consecutive samples of the signal and n(t) is the number of counts recorded during the integration time T at time t. For the correlator system described in the last section the sampling period is equal to the resolution time so that $T = T_p$ (Fig 20).

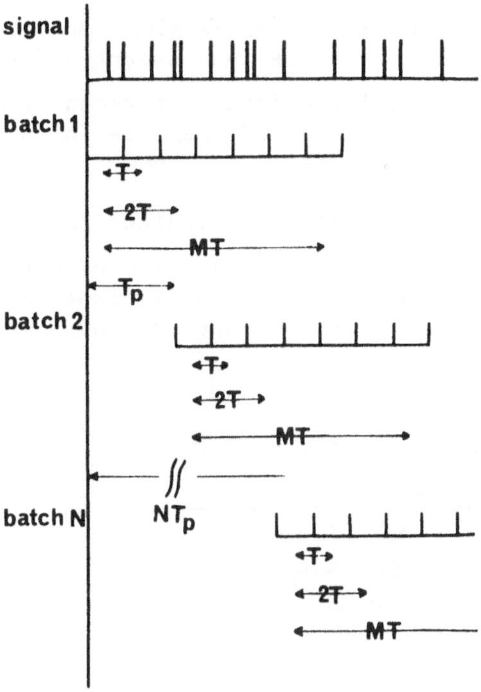

FIG 20 SAMPLING SCHEME USED IN AUTOCORRELATION

The right hand side of (4.49) can be evaluated exactly for gaussian-lorentzian light using results and techniques that we have already described. The answer is a fairly complicated formula and need not be given here. Perhaps it is worth mentioning, however, that a very simple expression is obtained in the shot noise limit when $\bar{n} \ll 1$. In this case (4.49) will be dominated by terms containing low powers of $\bar{n}$. These arise from the first term on the right hand side which may be expanded as follows

$$\begin{aligned} \mathrm{Var}\ (n(\tau)n(0)) &= \langle n^2(\tau)n^2(0)\rangle - \langle n(\tau)n(0)\rangle^2 \\ &= \langle n(\tau)[n(\tau)-1]\ n(0)-1]\rangle + 2\langle n(\tau)n(0)[n(0)-1]\rangle \\ &+ \langle n(0)n(\tau)\rangle - \langle n(0)n(\tau)\rangle^2 \\ &= G^{(2)}(\tau) + 0(\bar{n}^3) \end{aligned} \qquad (4.51)$$

Thus

$$\mathrm{Var}\ \hat{G}^{(2)}(\tau) = \frac{G^{(2)}(\tau)}{N} \qquad (4.52)$$

for low counting rates. Equation (4.49) also simplifies when $T_p \gg \tau_c$. In this case samples are independent and the sum can be neglected. However, for a fixed experiment time the number of samples will be reduced and there is in fact always something to be gained by reducing T_p.

In practice it has often proved convenient to measure the normalised correlation function $g^{(2)}(\tau)$. The effect of slow variations in the mean count rate due to fluctuations in the mains supply etc can be minimised by such a ploy. The estimator for this quantity is

$$\hat{g}^{(2)}(\tau) = \frac{N^{-1} \sum_{r=1}^{N} n(t_r+\tau)n(t_r)}{\left\{N^{-1} \sum_{r=1}^{N} n(t_r)\right\}^2} \tag{4.53}$$

However

$$< \hat{g}^{(2)}(\tau) > \neq g^{(2)}(\tau) \tag{4.54}$$

and $\hat{g}^{(2)}(\tau)$ is therefore a biassed estimate of the normalised correlation function. The importance of the biassing can be calculated using the identity

$$N^{-1} \sum_{r=1}^{N} n(t_r) = \bar{n} + \frac{1}{N}\sum_{r=1}^{N} (n(t_r) - \bar{n}) \tag{4.55}$$

If the sum on the right hand side is regarded as a small deviation from the mean the denominator of (4.53) can be expanded in powers of this quantity. Retaining only the first two terms of such an expansion, we obtain after averaging

$$< \hat{g}^{(2)}(\tau) > = 1 - \frac{2}{N}\left(1 + \frac{2}{\bar{n}} + \frac{2}{\Gamma T_p}\right) + \left| g^{(2)}(\tau) \right|^2 \times$$

$$\times \left[1 - \frac{2}{N}\left(3 + \frac{2}{\bar{n}} + \frac{2}{\Gamma T_p} + \frac{2\tau}{T_p}\right)\right] \quad . \tag{4.56}$$

In any real experiment $N \gg 1$, $NT_p/\tau \gg 1$ so that (4.56) may be reduced to

$$< \hat{g}^{(2)}(\tau)> = \left[1 - \frac{4}{N}\left(\frac{1}{\bar{n}} + \frac{1}{\gamma}\right)\right]\left(1 + \left|g^{(1)}(\tau)\right|^2\right) \tag{4.57}$$

and no distortion is incurred by biassing. Moreover, the deviation of (4.57) from the ideal value $g^{(2)}(\tau)$ is of the order of the square of the standard deviation of $\hat{g}^{(2)}(\tau)$ and may therefore normally be neglected.

The statistical error in measurements of $g^{(2)}(\tau)$ may be evaluated to first order in N^{-1} using (4.55) and an analogous relation for the numerator of (4.53):

$$N^{-1}\sum_{r=1}^{N} n(r_r+\tau)n(t_r) = G^{(2)}(\tau) + N^{-1}\sum_{r=1}^{N}\left[n(t_r+\tau)n(t_r) - < n(0)n(\tau)>\right] \tag{4.58}$$

Expanding (4.53) to second order in the deviations from the mean values of $\hat{n}$ and $\hat{G}^{(2)}$ eventually leads to the formula[36]

$$\mathrm{Var}\left(\frac{\hat{G}^{(2)}(\tau)}{\hat{n}^2}\right) \simeq \frac{1}{\bar{n}^4}\,\mathrm{Var}\,\hat{G}^{(2)}(\tau) + 4\,\frac{[g^{(2)}(\tau)]}{\bar{n}^2}\,\mathrm{Var}\,\hat{n} - \frac{4g^{(2)}(\tau)}{\bar{n}^3}\, \times$$

$$\times \left[< \hat{n}\hat{G}^{(2)}(\tau) > - \bar{n}^3\, g^{(2)}(\tau)\right] \quad . \tag{4.59}$$

This expression is valid provided that N is so large that

$$\left|\frac{\hat{n} - \bar{n}}{\bar{n}}\right| \ll 1 \tag{4.60}$$

Since the bias terms in (4.59) are proportional to $\frac{1}{\bar{n}}$ or independent of the mean (4.59) and (4.52) lead, in the shot noise limit, to

$$\mathrm{Var}\,\hat{g}^{(2)}(\tau) = \frac{g^{(2)}(\tau)}{\bar{n}^2 N} \qquad n \ll 1 \tag{4.61}$$

In the useful case $T = T_p$ and $\gamma \ll 1$ we obtain the more general result

$$\text{Var } \hat{g}^{(2)}(\tau) \simeq \frac{1}{N\gamma}\left[\frac{1}{2}\left\{1+8e^{-x}-e^{-2x}(5+2x)\right\} + \frac{2}{r}(1+e^{-x})^2 + \frac{1+e^{-x}}{r^2\gamma}\right] \quad (4.62)$$

where $x = 2\Gamma\tau$ and $\bar{n} = r\gamma$. The quantity $N\gamma$ is proportional to the total experiment time and may be regarded as fixed. The number of counts per coherence time r will also be fixed by external considerations such as laser power, scattering efficiency etc. However γ may be chosen at will. Thus the shot noise limit (4.61) may be reached by reducing the sample time until $r^2\gamma \ll 1$.

Expression (4.59) for the variance of the normalised photon-correlation function can in fact be evaluated without further approximation. Analysis of the clipped case can also be carried out analytically using the same approach when clipping is carried out at zero, but the calculations become somewhat laborious. At higher clip levels computer simulation has proved a more appropriate method of analysis. As we have seen in section 2.2 the positive frequency part of the optical field can be expressed as the sum of two components

$$\mathcal{E}^{+}(t) = \mathcal{E}(t) + i\mathcal{E}'(t) \quad (4.63)$$

In the case of gaussian-lorentzian light the values of $\mathcal{E}$ and $\mathcal{E}'$ at times t and t-τ are related by first order Markoff processes given by

$$\mathcal{E}(t) = g\,\mathcal{E}(t-T) + \alpha_1 \quad (4.64)$$

$$\mathcal{E}'(t) = g\,\mathcal{E}'(t-T) + \alpha_2 \quad (4.65)$$

where α_1 and α_2 are independent random numbers with gaussian probability distributions of equal variance about zero mean. From (4.64) for example

$$\langle\mathcal{E}(t)\,\mathcal{E}(t-T)\rangle = g\,\langle\mathcal{E}^2\rangle \quad (4.66)$$

and

$$\frac{\langle\mathcal{E}(t)\,\mathcal{E}(t-nT)\rangle}{\langle\mathcal{E}^2\rangle} = g^n = \left(\frac{\langle\mathcal{E}(t)\,\mathcal{E}(t-T)\rangle}{\langle\mathcal{E}^2\rangle}\right)^n \quad (4.67)$$

so that if T is the delay time between successive samples

$$g = e^{-\Gamma T} \quad (4.68)$$

Using independent series of gaussian random numbers for α_1 and α_2 an intensity distribution can be simulated by setting

$$I(t) = \mathcal{E}^2(t) + \mathcal{E}'^2(t) \tag{4.69}$$

The intensity at each sample may then be used as the mean value of a poisson random number generator whose output n(t) is an integer equal to the photocounts occurring in a sample at time t. Sets of the corresponding photon-correlation functions can thus be constructed whose values will show the desired statistical fluctuations. These can then be analysed by the methods normally used in experimental spectral linewidth measurements now to be described.

4.4 Linewidth Errors

Turning now to the experimental determination of the linewidth Γ of a simple lorentzian spectral feature we first consider how this information should be extracted from measurements of $g^{(2)}(\tau)$. Various types of curve fitting procedures may be used but here we shall investigate the two parameter model

$$g^{(2)}(\tau) = 1 + C \exp(-2\Gamma\tau) \tag{4.70}$$

since we know that this form of the Siegert relation is preserved in the presence of spatial and temporal integration, single clipping, biassing and small amounts of heterodyning (which is unavoidable in some experiments). We shall assume that C is an arbitrary constant independent of linewidth. Suppose then that in our experiments we have made measurements of the quantity $g^{(2)}(\tau_i)$ for M values of the delay time. A least squares fit of the measured points to the curve (4.70) is accomplished by minimising

$$s = \sum_{i=1}^{M} \left\{ \hat{g}^{(2)}(\tau_i) - 1 - C \exp(-2\hat{\Gamma}\tau_i) \right\}^2 \chi_i \tag{4.71}$$

with respect to variations in C and the linewidth estimator $\hat{\Gamma}$. χ_i is the weighting which will subsequently be chosen to make the variation of $\hat{\Gamma}$ a minimum.

Performing variations in C and $\hat{\Gamma}$, and eliminating C from the equations we find that Γ must be chosen to satisfy

$$\frac{\sum_i \chi_i \tau_i e^{-2\hat{\Gamma}\tau_i} \left(\hat{g}^{(2)}(\tau) - 1 \right)}{\sum_i \chi_i \left(\hat{g}^{(2)}(\tau_i) - 1 \right) e^{-2\hat{\Gamma}\tau_i}} = \frac{\sum_i \chi_i \tau_i e^{-4\hat{\Gamma}\tau_i}}{\sum \chi_i e^{-4\hat{\Gamma}\tau_i}} \tag{4.72}$$

Expanding this relation about the mean values Γ and $g^{(2)}(\tau)$ to first order in the difference $\delta\hat{\Gamma} = \hat{\Gamma}-\hat{\Gamma}$ and $\delta\hat{g} = \hat{g} - \hat{g}$ leads to

$$4\langle\delta\hat{\Gamma}^2\rangle = 4 \text{ Var } \hat{\Gamma}$$

$$= \frac{\sum_{ij} \chi_i\chi_j \exp -2\Gamma(\tau_i+\tau_j) \ (\tau_i-\bar{\tau})(\tau_j-\bar{\tau})\langle\delta\hat{g}^{(2)}(\tau_i)\delta\hat{g}^{(2)}(\tau_j)\rangle}{\sum \chi_i\tau_i \exp(-2\Gamma\tau_i)(g^{(2)}(\tau) - 1)(\tau_i - \bar{\tau})^2} \tag{4.73}$$

where $\bar{\tau}$ is the right hand side of (4.72) and we have assumed that

$$\text{Var } \hat{g}^{(2)}(\tau) = \langle\left[\delta\hat{g}^{(2)} \ (\tau_i)\right]^2\rangle \ll (g^{(2)}(\tau_i) -1)^2 \tag{4.74}$$

which can always be satisfied for sufficiently large N. At this point the calculations can be simplified by assuming the errors in $\hat{g}^{(2)}(\tau_i)$ to be uncorrelated[36]

$$\langle\delta\hat{g}^{(2)}(\tau_i) \ \delta\hat{g}^{(2)}(\tau_j)\rangle = \left\langle\left[\delta \ \hat{g}^{(2)}(\tau_i)\right]^2\right\rangle \delta_{ij} \tag{4.75}$$

(4.75) will certainly hold in the shot noise limit $\bar{n} \ll 1$ which we have mentioned earlier in the section. For fixed r (counts per coherence time) it will also hold for sufficiently large values of γ since signal statistics are the dominant noise process if $\bar{n} = r\gamma \gg 1$ and these too are uncorrelated from sample to sample if $T_p = T \gg \tau_c$. Outside these regions one might expect the approximation (4.75) to lead to a lower bound on the errors in Γ. This conjective has been verified[37] by numerical computation using the 'exact' result (4.73). Substituting from (4.75) into (4.73) and minimising the resulting expression with respect to χ_i leads to

$$\chi_i = [\text{Var } \hat{g}^{(2)}(\tau_i)]^{-1} \left\{ \sum_i e^{-\Gamma\tau_i}/\text{Var } \hat{g}^{(2)}(\tau_i)\right\}^{-1} \tag{4.76}$$

and a minimum value of Var Γ given by

$$\frac{\text{Var}\hat{\Gamma}}{\Gamma^2} = \left\{\left(b - \frac{c^2}{a}\right) N\gamma\right\}^{-1} \tag{4.77}$$

$$\text{where} \quad b = \frac{4\Gamma^2}{N\gamma} \sum_{i=1}^{M} \tau_i^2 \, (g^{(2)}(\tau_i) - 1)^2 / \text{Var}\, \hat{g}^{(2)}(\tau_i)$$

$$c = \frac{2\Gamma}{N\gamma} \sum_{i=1}^{M} \tau_i \, (g^{(2)}(\tau_i) - 1)^2 / \text{Var}\, \hat{g}^{(2)}(\tau_i) \tag{4.78}$$

$$a = \frac{1}{N\gamma} \sum_{i=1}^{M} (g^{(2)}(\tau_i) - 1)^2 / \text{Var}\, \hat{g}^{(2)}(\tau_i)$$

The dependence of χ_i on τ_i is not strong over the range of values where (4.74) is satisfied (see equation (4.61) and 4.62)) and for most purposes a flat weighting can be taken. Relation (4.77) has been evaluated for various values of M,γ and r using both the analytic and computer simulation approaches outlined earlier in this section. In addition assessment of the effects of biassing and clipping have been made in the small sample time limit. The implications of these calculations for correlator operation are fully discussed elsewhere together with supporting experimental data(38). It is perhaps worth mentioning here that the effect of spatial coherence on the accuracy of spectral measurements is rather difficult to calculate except in the two limits $\bar{n} \ll 1$ and $\bar{n}$, $A/A_c \gg 1$. In the first case we have already shown that

$$\text{Var}\, \hat{g}^{(2)}(\tau_i) = g^{(2)}(\tau_i) / \bar{n}^{2} N \tag{4.79}$$

and since the effects of clipping can be neglected for $\bar{n} \ll 1$ the relative variance used in equations (4.78) takes the form ($\gamma \ll 1$)

$$\frac{\text{Var}\, \hat{g}^{(2)}(\tau)}{[g^{(2)}(\tau_i) - 1]^2} = \frac{1 + f(A)e^{-2\Gamma\tau_i}}{f^2(A)\bar{n}^2 \, N e^{-4\Gamma\tau_i}} \tag{4.80}$$

and (4.77) may be expressed in the form

$$\left.\sqrt{\frac{\text{Var}\,\Gamma}{\Gamma^2}}\right|_A = \frac{1}{f(A)} \; \frac{4.606}{r\sqrt{N\gamma}} \qquad \bar{n} \ll 1 \tag{4.81}$$

Since in the limit $\bar{n} \gg 1$, $A \gg A_c$ little information is lost by taking samples far apart ie $T_p \gg \tau_c$ only the first term on the right hand side of (4.49) need be considered. A similar reduction can be made of the bias terms in (4.59), and for large areas it may be shown that(39)

$$\sqrt{\frac{\text{Var }\Gamma}{\Gamma^2}} \simeq \left(\frac{A}{N\gamma}\right)^{\frac{1}{2}} \quad \bar{n} \gg 1,\ A \gg A_c \tag{4.82}$$

Finally, let us consider the effect of statistical fluctuations on heterodyne spectroscopy in the gaussian light case. We have already remarked in section 3 that if clipping is being carried out the IFS term cannot be eliminated from the heterodyne spectrum by simply turning up the reference beam unless the mean count rate of the scattered light is small ($\bar{n}_s \ll 1$). In this situation, however, a great simplification of the error calculations occurs because assuming that $\bar{n}_c \gg \bar{n}_s$ we expect that the dominant noise will arise from the coherent field component. Thus, for example

$$\alpha^4 \text{ Var } \hat{G}^{(2)}(\tau) = \frac{1}{N} \text{Var }(n(\tau)n(0)) + \frac{2\alpha^4}{N} \sum_{n=1}^{N-1} \left(1 - \frac{k}{N}\right) \times$$

$$\left[\langle n(kT_p+\tau)I(kT_p)I(\tau)I(0)\rangle - \left|G^{(2)}(\tau)\right|^2 \right]$$

$$+ \frac{2\alpha^3}{N}\left(1 - \frac{\tau}{NT_p}\right) \langle I(2\tau)\ I(\tau)\ I(0)\rangle$$

$$\simeq \frac{1}{N}(2n_c^3 + n_c^2) + \frac{2}{N}\left(1 - \frac{\tau}{NT_p}\right) n_c^3$$

$$\simeq \frac{1}{N}(4n_c^3 + n_c^2) \tag{4.83}$$

The relative variance of the biassed estimate $g^{(2)}(\tau)$ may in a similar way be shown to be of the form[21]

$$\frac{\text{Var } \hat{g}^{(2)}(\tau)}{(g^{(2)}(\tau) - 1)^2} = \left(4N\ \bar{n}_s^2 \left| g^{(1)}(\tau) \right|^2 \cos \tau\Delta\omega \right)^{-1} \tag{4.84}$$

However, in the absence of the coherent component the analogous IFS formula is given by (4.80) (with $f(A) = 1$) in the limit $\bar{n}_s \ll 1$. The ratio of these two expressions ranges from 1/8 at $\tau = 0$ to zero as $\tau \to \infty$ indicating an advantage in heterodying of about an order of magnitude. This advantage may be evaluated quantitatively for the linewidth estimator when $\Delta\omega = 0$. In this case we obtain

$$\frac{\text{Var }\hat{\Gamma}}{\Gamma^2} = \begin{cases} 2\gamma/\bar{n}_s^{-2}N & \text{DS} \quad (4.85) \\ 21.2\gamma/\bar{n}_s^{-2}N & \text{IFS} \quad (4.86) \end{cases}$$

These formulae apply only for an ideal correlator of course, but it is possible to evaluate the clipped case for a heterodyne system. This shows, at worst, an increase in the standard deviation of Γ by a factor 1.5 over the ideal homodyne case. Since in the absence of the coherent component the optimum clipping level is zero for $\bar{n}_s \ll 1$ and we would obtain the result (4.86) it is evident that a factor of at least 7 can be gained by using Doppler rather than intensity fluctuation spectroscopy in the case of a weak gaussian light source.

5 NON-GAUSSIAN FIELDS

As has already been pointed out, the Siegert relation does not hold for non-gaussian signals. The intensity autocorrelation function of such a field in general contains more information about the scattering process and optical arrangement than in the gaussian case and is related to the spectrum in a more complicated way. Both the first and second order autocorrelation functions are therefore of interest. If analogue signal processing techniques are acceptable then a combination of heterodyne and intensity fluctuation spectroscopy can be used to investigate these quantities as discussed in section (3.1). If a fast digital system is necessary on the other hand, because of low scattered intensity for example, then a clipping correlator will be more appropriate. However, the formulae which are essential for the interpretation of measurements using such an instrument have been established only for gaussian light. Consideration of the heterodyne case has in fact already indicated that distortion of the spectral information is produced by clipping a non-gaussian field. In the first part of this section, therefore, we consider how such difficulties might be overcome by modification of the clipping correlator described in section 3.

The second part of the section is devoted to the statistical and spectral properties of light scattered by a deep random phase screen. Although this topic is not closely related to the signal processing problems discussed in the rest of the presentation it does provide an interesting example of non-gaussian light and highlights the difficulties of interpretation which are likely to arise in measurements on non-gaussian fields generally.

5.1 Signal Processing

Suppose that during a measurement of $G_k^{(2)}$, the clipping level k is varied with a distribution q_k so that the correlation function actually constructed is

$$<G_k^{(2)}(\tau)>_k = \sum_{k=0}^{\infty} q_k \, G_k^{(2)}(\tau) = \sum_{k=0}^{\infty} q_k \sum_{n=k+1}^{\infty} \sum_{m=1}^{\infty} mp(n,m;T) \quad (5.1)$$

The sums over n and k on the right hand side can be re-ordered to give

$$\sum_{k=0}^{\infty} q_k G_k^{(2)}(\tau) = \sum_{m=1}^{\infty} \sum_{n=1}^{\infty} mp(n,m;T) \sum_{k=0}^{n} q_k \quad (5.2)$$

and if $\sum_{k=0}^{n} q_k = \beta n$ (5.3)

equation (5.2) reduces to

$$<G_k^{(2)}(\tau)>_k = \beta \, G^{(2)}(\tau) \quad (5.4)$$

Thus by clipping at levels which are distributed such that (5.3) holds we can construct the true autocorrelation function. The requirement (5.3) will in fact only be satisfied exactly by an infinite, uniform distribution of levels. However, we might hope in practice to generate a uniform distribution with a cut off which is sufficiently high that only multiple events with a very low probability of occurring will be excluded from the sum (5.1). We can easily examine the effect of such a cut-off for the gaussian-case since (5.1) then reduces to

$$<G_k^{(2)}(\tau)>_k = \sum_{k=0}^{\infty} q_k \, \bar{n} \left(\frac{\bar{n}}{1+\bar{n}}\right)^{k+1} \left(1 + \frac{1+k}{1+\bar{n}} |g^{(1)}(\tau)|^2\right) \quad (5.5)$$

Substituting the distribution

$$q_k = \frac{1}{s} \quad k < s$$
$$= 0 \quad k \geqslant s \quad (5.6)$$

in this expression leads to the formula

$$<G_k^{(2)}(\tau)>_k = \frac{\bar{n}^2}{s}\left[1 - \left(\frac{\bar{n}}{1+\bar{n}}\right)^s + |g^{(1)}(\tau)|^2 \left\{1 - \frac{1+\bar{n}+s}{1+\bar{n}} \left(\frac{n}{1+\bar{n}}\right)^s\right\}\right] \quad (5.7)$$

In the limits $\bar{n} \to 0$ and $s \to \infty$ this reduces to the true autocorrelation function as expected from (5.4). If it is to differ by less than 1% from this quantity then for $\bar{n} \gtrsim 1$ we must take $s \gtrsim 10\,\bar{n}$.

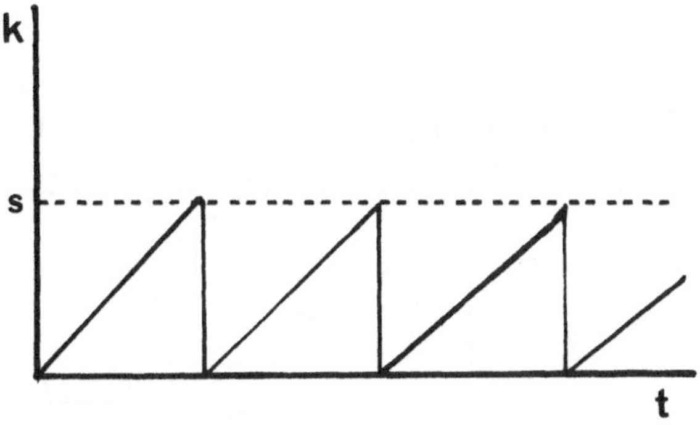

FIG 21 RAMPED CLIPPING

The distribution (5.6) may be produced by selecting clip levels using a random number generator or by a simple linear ramp method[40] (Fig 21). Care must be taken in the second mentioned case, however, since if the ramp frequency coincides with an important frequency component in the signal, correlation effects may occur. Another technique which generates a distribution approximately of the form (5.6) is scaling[41]. In this method, the true signal is correlated with a scaled version: that is, a form which has the value one upon the arrival of every s^{th} count but is otherwise zero (Fig 22). The scaling level s is set sufficiently high for the probability of registering more than one scaled count per sample time to be negligible. Since scaling may be accomplished very easily in practice, it is worth spending a little time examining the justification for such a procedure.

Assuming that only one scaled count can be recorded in the time T, after a long period $\mathcal{T}$, ms + r counts will have been detected with m recorded in the scaling channel and a remainder $r \leqslant s - 1$. If q_r is the probability distribution of the remainder, the probability of recording one or more counts in the sample interval following $\mathcal{T}$ in the scaled channel is

$$\sum_{r=0}^{s=1} q_r \sum_{n=s-r}^{\infty} p(n;T) \tag{5.8}$$

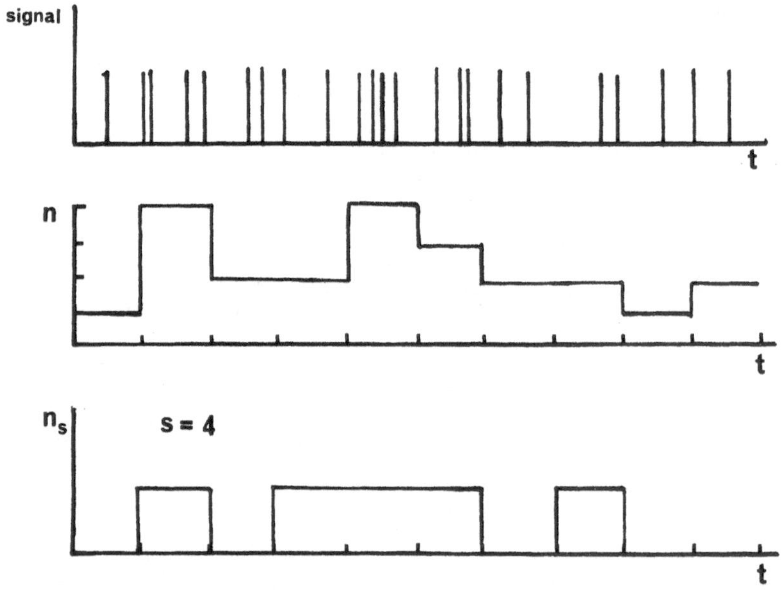

FIG 22 SCALING

The joint probability of recording one or more counts in the scaled channel and m counts in the other channel is thus

$$\sum_{r=0}^{s-1} q_r \sum_{n=s-r}^{\infty} p(n,m;T) \tag{5.9}$$

and the correlation function of photocounts scaled in one channel may then, with a little algebraic manipulation be written(42)

$$G_s^{(2)}(\tau) = \sum_{k=0}^{s-1} q_{s-k-1} \, G_k^{(2)}(\tau) \tag{5.10}$$

where $G_k^{(2)}$ is the single clipped correlation function. Equation (5.10) is identical with (5.1) and it remains to investigate the properties of q_r. For $\mathcal{T} \gg \tau_c$ the distribution of counts arriving within $\mathcal{T}$ will be approximately poissonian. The probability of finding a remainder in excess of an integral number s of scaled counts will therefore be

$$q_r = e^{-\bar{N}} \sum_{\ell=0}^{\infty} \frac{\bar{N}^{\ell s+r}}{(\ell s + r)!} \tag{5.11}$$

where $\bar{N} = \bar{n}\mathscr{T}/T$ and $\bar{n}$ is the mean number of counts per sample time as usual. For $s > 1$ equation (5.11) may be rearranged using the formula[43]

$$\sum_{n=0}^{\infty} a_{nm} z^{nm} = \frac{1}{m} \sum_{k=1}^{m} f(ze^{2\pi ik/m}) \tag{5.12}$$

where
$$f(z) = \sum_{n=0}^{\infty} a_n z^n \tag{5.13}$$

The result is[42]

$$q_r = \frac{1}{s}\left\{1+e^{-\bar{N}} \sum_{k=1}^{s=1} \exp\left(\bar{N} \cos \frac{2\pi k}{s}\right) \cos\left(\frac{2\pi k(s-r)}{s} + \bar{N} \sin \frac{2\pi k}{s}\right)\right\} \tag{5.14}$$

If $\bar{N}$ is reasonably large the sum is dominated by the first and last terms due to the form of $\cos(2\pi k/s)$ in the exponential factor. Thus a good approximation to the distribution is given by the formula ($s > 2$)

$$q_r \simeq \frac{1}{s}\left[1 + 2 \exp\left\{-\bar{N}\left(1-\cos \frac{2\pi}{s}\right)\right\} \cos\left(\frac{2\pi r}{s} - \bar{N} \sin \frac{2\pi}{s}\right)\right] \tag{5.15}$$

Note that

$$\sum_{r=0}^{s-1} q_r = 1 \tag{5.16}$$

so that (5.15) is correctly normalised. The argument of the cosine term in r on the right hand side of (5.15) changes by nearly 2π as r varies from zero to $s - 1$ so that q_r varies sinusoidally about a mean value of $1/s$ (Fig 23). For $s \leqslant 4$, q_r is in fact uniform to better than 1% for $\bar{N} > 5$ since the exponent $\bar{N}(1 - \cos 2\pi/s)$ is greater than five. In the more interesting case of large s, the cosine term in the exponent may be expanded in powers of $(2\pi/s)^2$ to show that this degree of uniformity can only be achieved for $\bar{N} > 5s^2/2\pi^2$. As we have seen already in the gaussian case for $\bar{n} \gtrsim 1$ we require $s \geqslant 10\,\bar{n}$ to avoid distortion due to the cut off in q_r at $r = s-1$. Combining these requirements we have

$$\bar{N} = \bar{n}\mathscr{T}/T > 500\,\bar{n}^2/2\pi^2 \tag{5.17}$$

$$\text{ie}\quad \mathscr{T} > 25\bar{n}T \sim \bar{n}\tau_c \tag{5.18}$$

in a typical experiment.

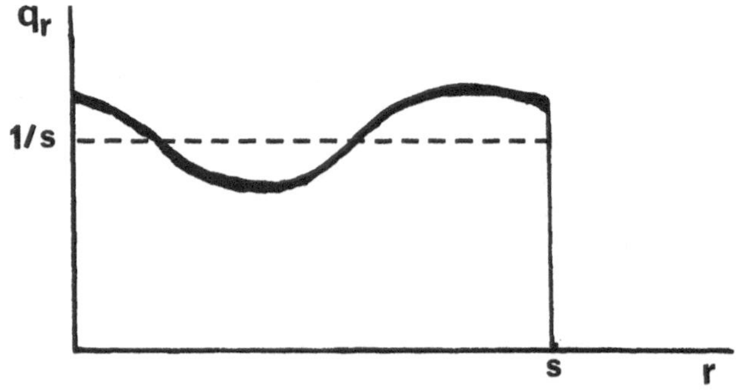

FIG 23 EFFECTIVE DISTRIBUTION OF CLIP-LEVELS DUE TO SCALING AT s

Turning now to the question of errors, it has been found using computer simulation that the statistical error in a linewidth measurement is fairly insensitive to clipping level at least for gaussian-lorentzian light [38] so that we might hope to lose little information by scanning a set of clipping levels rather than clipping at a fixed level close to the mean count rate. In the case of scaling the time $\mathscr{T}$ taken for q_r to become uniform following the occurrence of a scaled count can be regarded as roughly the time between independent samples of the scaled correlation function. For $\bar{n} \sim 1, \mathscr{T} \sim \tau_c$ which is approximately the same as the time ($\sim 1.8\ \tau_c$) between independent samples in full correlation[39]. Further, with $\bar{n} \sim 1$ and $s = 10$ we are cross-correlating signals with mean rates per sample interval of about 1 and 0.1 respectively. Calculations have shown[38] that in such a situation the expected error in linewidth should not be far from the theoretical minimum. A quantitative analysis of statistical errors can of course only be given for a field of known statistical properties but from the above considerations we might expect the accuracy of measurements in which a distribution of clipping levels is being used to be closer to that which could be achieved if full parallel correlation were possible than to that obtained by taking samples of the signal sufficiently far apart for there to be time to effect correlation between each sample without recourse to clipping procedures (batch processing).

5.2 The Deep Random Phase Screen

This is a system which simply retards the phase of an incident electromagnetic field by a randomly varying, position dependent amount typically equivalent to many wavelengths path difference. Familiar phenomena caused in this way are the twinkling of starlight and the swimming pool effect[(44)]. Examples of deep phase screens of considerable current interest are moving diffuse surfaces such as ground glass[(45)] and the dynamic scattering mode exhibited by thin layers of nematic liquid crystals[(46)]. A good deal of literature exists on the diffuse surface problem particularly,[(47)] but apparently there has been no theoretical analysis of the statistics and spectral properties of the scattered light for the non-gaussian case, when the scale of the refractive index inhomogeneities is comparable with the size of the illuminated volume.

We shall consider a simple experimental arrangement in which the light from a laser is focussed normally on to a phase screen of negligible thickness and the forward scattered radiation is detected in the far field by a photomultiplier whose axis makes an angle θ with the laser beam (Fig 24).

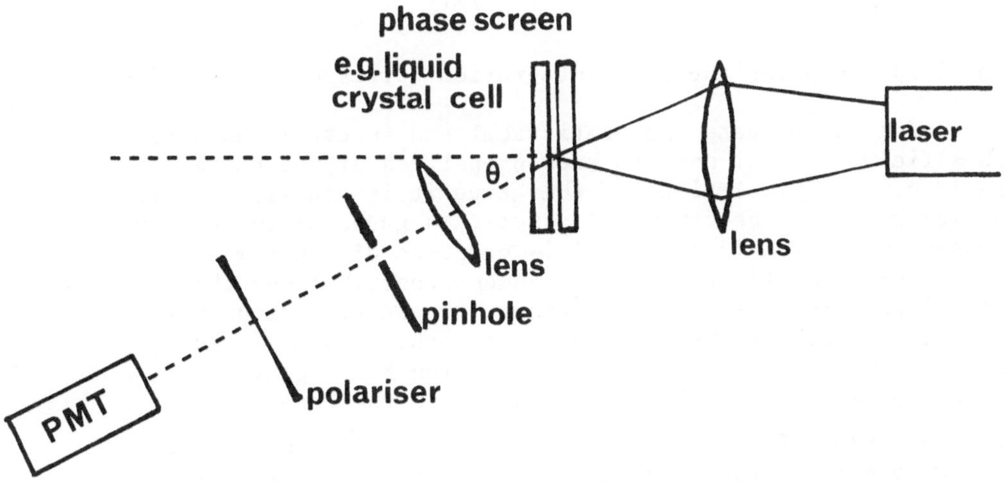

FIG 24 SCATTERING FROM A PHASE SCREEN-OPTICAL ARRANGEMENT

After passing through the phase screen the positive frequency part of the electric field may be written

$$\mathcal{E}^+(\underset{\sim}{r},0;t) = E_o \exp\{i[\phi(\underset{\sim}{r},t) - \omega_o t] - r^2/W_o^2\} \tag{5.19}$$

where $\phi(\underset{\sim}{r},t)$ is a randomly varying position dependent phase shift introduced by the screen. We have taken a cylindrical coordinate system $\underset{\sim}{R} \equiv (r,\psi,z) \equiv (\underset{\sim}{r},z)$ and located the phase screen at the $z = 0$ plane. W_o is the width of the gaussian intensity profile of the beam and therefore characterises the size of the illuminated area. According to the Helmholtz formula the field at a point $\underset{\sim}{R}$ in front of the screen is given by

$$\mathcal{E}^+(\underset{\sim}{R};t) = E_o e^{-i\omega_o t} \int_{-\infty}^{+\infty} d^2r' e^{-ik|\underset{\sim}{r}'-\underset{\sim}{R}|} e^{i\phi(\underset{\sim}{r}',t)} e^{-r'^2/W_o^2} \tag{5.20}$$

where $\underset{\sim}{k}$ is the wave vector of the light and the integral is carried out over the illuminated area. Now

$$|\underset{\sim}{r}'-\underset{\sim}{R}| = \sqrt{r^2 + z^2 + r'^2 - 2\underset{\sim}{r}.\underset{\sim}{r}'} \approx \sqrt{r^2 + z^2}\left(1 - \frac{\underset{\sim}{r}.\underset{\sim}{r}'}{r^2+z^2}\right) \tag{5.21}$$

in the far field because of the localized nature of the source. Equation (5.20) may thus be expressed in the form

$$\mathcal{E}^+(R;t) \cong E_o e^{-i\omega_o t} \int_{-\infty}^{+\infty} d^2r' \exp\{-ik(|R|(1 - |r'|\sin\theta\cos\psi'))\} \, e^{i\phi(\underset{\sim}{r}';t)} e^{-r'^2/W_o^2} \tag{5.22}$$

where ψ' is the polar angle of integration corresponding to $\underset{\sim}{r}'$.

In order to evaluate the statistical and spectral characteristics of this field we shall need to know properties of the phase variable ϕ. In calculations of this type it is usually assumed, often for want of a better model, that ϕ is gaussian in the sense discussed in section (3.1). If ϕ is composed of the sum of a large number of statistically independent contributions this is of course a justifiable assumption. Propagation through a random medium such as the atmosphere or perhaps even liquid crystal may well lead to such a distribution but a gaussian distribution is more difficult to justify in the case of scattering from diffuse surfaces[48]. A model must be assumed in order to make further progress with the analysis, however, and following previous work we shall therefore assume a joint-gaussian distribution for ϕ. It is not difficult to establish in this case the useful property

$$< \exp -i \sum_i \phi_i > = \exp -\tfrac{1}{2} <\left(\sum_i \phi_i\right)^2> \quad (5.23)$$

which we shall employ in future without further comment. In the particular situation to be considered here we shall find that the detailed structure of the spatial phase correlation function (assumed to be translationally invariant)

$$\rho(\underset{\sim}{r}) = <\phi(0;t)\phi(\underset{\sim}{r},t)>/<\phi^2> \quad (5.24)$$

will not enter the calculations because of the deep phase screen assumption

$$<\phi^2> = \overline{\phi^2} \gg 1 \quad (5.25)$$

However, we shall make the reasonable assumption that for sufficiently small $\underset{\sim}{r}$,ρ may be expanded in terms of a characteristic correlation length ξ as follows[49]

$$\rho(\underset{\sim}{r}) \cong 1 - r^2/\xi^2 \qquad r \ll \xi \quad (5.26)$$

When calculating spectral properties we shall also assume "cross-spectral purity"

$$< \phi(\underset{\sim}{r};t)\phi(\underset{\sim}{r};t')> = <\phi^2> \rho(\underset{\sim}{r}-\underset{\sim}{r}')\sigma(t-t') \quad (5.27)$$

where $\sigma(\tau) = < \phi(\underset{\sim}{r};0)\phi(\underset{\sim}{r};\tau)>/ <\phi^2>$ (5.28)

is the temporal phase correlation function for a stationary scattering process. The quantities $\overline{\phi^2}$, ξ and W_o are the three basic parameters of the model. In addition to the restriction (5.25) we shall also confine our calculations to the region

$$\overline{\phi^2} \gg \xi^2/W_o^2 \quad (5.29)$$

Because of (5.25) this inequality will be satisfied even when the scale of the phase fluctuations is comparable with the size of the illuminated area so that our results will be valid in what is expected to be a non-gaussian regime.

5.2.1 Statistics

The mean intensity of light scattered by a random phase screen is given through (5.22) and (5.23) in the far field by

$$< I(\theta;t) > = < I > = |E_o|^2 \int_{-\infty}^{+\infty}\int_{-\infty}^{+\infty} d^2r'd^2r'' \quad x$$

$$e^{ik\sin\theta(r'\cos\psi' - r''\cos\psi'')}\; e^{-\overline{\phi^2}(1-\rho(r'-r''))}\; e^{-(r'^2+r''^2)/W_o^2} \qquad (5.30)$$

Transforming to sum and difference coordinates we arrive, after integration, at the formula

$$\langle I\rangle = \pi^2\omega_o^2|E_o|^2 \int_0^{\infty} rdr\, J_o(kr\sin\theta)\exp\{-\overline{\phi^2}\,[1-\rho(r)] - r^2/2W_o^2\} \quad (5.31)$$

It may be shown similarly that the second moment of the intensity fluctuation distribution can be expressed in the form

$$<I^2> = W_o^2|E_o|^4 \int_{-\infty}^{+\infty} d^2r'd^2r''d^2r'''\exp\{2ikr''\sin\theta\cos\eta - (r'^2+r''^2+r'''^2)/W_o^2\}$$

$$\times\exp\{-\overline{\phi^2}[2-\rho(r''+r''')-\rho(r''-r''')-\rho(r'+r'')-\rho r'-r'')+ \rho(r'+r''')+\rho(r'+''')]\} \qquad (5.32)$$

where η is the polar angle corresponding to the r'' integral. (5.31) and (5.32) are difficult to evaluate in general, even for simple model correlation functions ρ, because of the exponential form of the integrand. However a "saddle point" type of approach can be used when $\overline{\phi^2} \gg 1$. Consider the function

$$f = \exp\,(\overline{\phi^2}\,\rho(r)) \qquad (5.33)$$

Bearing in mind that $\rho(r)$ will normally be a well behaved function decreasing from a value of unity at r = 0 to zero as $r \to \infty$ it is clear that f will decrease extremely rapidly from the large value $e^{\overline{\phi^2}}$ at r = 0 to unity as $r \to \infty$. The most rapid changes in f will in fact occur when r is small so that the expansion (5.26) can be used for ρ. In this region

$$f \approx e^{\overline{\phi^2}}\; e^{-\overline{\phi^2}\, r^2/\xi^2} \qquad (5.34)$$

so that f(r) is characterised by a width $\xi/\sqrt{\overline{\phi^2}}$. Considerations of this type suggest that the approximate model function[50]

$$\exp\{\overline{\phi^2}\rho(r)\} \cong 1 + (e^{\overline{\phi^2}} - 1)\exp\{-\overline{\phi^2}\, r^2/\xi^2\} \tag{5.35}$$

might be used to obtain analytical results from (5.31) and (5.32). This approximate form certainly behaves correctly in the region where f is most sensitive to $\underset{\sim}{r}$ and also has the right asymptotic value in the limit $r \to \infty$. Substituting (5.35) into (5.31) leads, after integration to

$$\langle I\rangle = \pi^2 W_o^2 |E_o|^2 \left\{ W_o^2 e^{-\overline{\phi^2}} \exp(-\tfrac{1}{2}k^2 W_o^2 \sin^2\theta) + \frac{(1-e^{-\overline{\phi^2}})}{\left(\frac{1}{W_o^2} + \frac{2\overline{\phi^2}}{\xi^2}\right)} \exp\left[-\tfrac{1}{2}k^2 \sin^2\theta \Big/ \left(\frac{1}{W_o^2} + \frac{2\overline{\phi^2}}{\xi^2}\right)\right]\right\} \tag{5.36}$$

The first term on the right is just the "aperture" diffraction pattern - in the present case the straight through laser beam profile. The second term is entirely due to the phase fluctuations. For large values of $\overline{\phi^2}$ the terms in $\exp(-\overline{\phi^2})$ may be neglected. If we also make use of the inequality (5.29) then (5.36) reduces to[51]

$$\langle I\rangle = \frac{\pi^2 W_o^2 \xi^2 |E_o|^2}{2\overline{\phi^2}} \exp(-k^2\xi^2 \sin^2\theta/4\overline{\phi^2}) \tag{5.37}$$

After a good deal of algebraic manipulation the second moment (2.14) can be evaluated, within the limits of the above approximation, to give the formula[51]

$$\frac{\langle I^2\rangle}{\langle I\rangle^2} = 2 - \frac{2\xi^2}{W_o^2} + \frac{\xi^2\overline{\phi^2}}{4W_o^2} \exp\left(\frac{k^2\xi^2 \sin^2\theta}{4\overline{\phi^2}}\right) \tag{5.38}$$

This expression reduces to the gaussian value of two in the limit $\xi \ll W_o$ corresponding to many correlation lengths (or equivalently scattering centres) within the illuminated area. However, there is clear deviation from gaussian due to the presence of the large $\overline{\phi^2}$ factor in the final term. This implies that even when $\xi > W_o$

it is possible that a second moment in excess of two will be obtained if $\overline{\phi^2}$ is large enough. Moreover, the effect may well be amplified by the exponential factor as θ is increased. This may be contrasted with the behaviour of the mean intensity (5.37) which is governed by the inverse of the angular factor present in (5.38).

The W_o and θ dependence predicted by (5.37) and (5.38) have been tested experimentally for light scattering from a thin (25 μm) layer of nematic liquid crystal undergoing electro-hydrodynamic turbulence[52]. (Figs 25 and 26) Both our own and previous work[46] have suggested that this system behaves as a deep phase screen under certain conditions. Nevertheless, a small depolarised component of the scattered light cannot be eliminated so that there are almost certainly amplitude as well as phase fluctuations present in the scattered radiation. This may explain the slight deviation from the angle dependence predicted by (5.37 and (5.38) which occurs for small values of θ. The effect of the finite thickness of the sample shows up in the measured W_o dependence. In calculating the illuminated area from the formulae of gaussian beam optics no account was taken of the spreading of the beam which occurs during passage through the sample. Consequently the values used for the experimental plot are underestimates of the true areas. This effect is negligible for large values of W_o but leads to curvature as W_o becomes comparable to the sample thickness. The relations (5.37) and (5.38) can be used in an experimental determination of the correlation length ξ and mean square phase deviation. For example, for the liquid crystal system referred to above we found $\xi = 2.6\mu m$ and $\overline{\phi^2} \sim 36$.

Extension of the above approach to higher moments becomes progressively more difficult. However, a "micro-area" model[53] enables more general results to be obtained within the framework of the above approximation. We imagine the illuminated area to be made up of N regions $\mathcal{R}$ giving statistically independent contributions to the far field. This is equivalent to neglecting terms like exp $(-\overline{\phi^2})$ in the earlier method. Thus we may write

$$\mathcal{E}^+(\theta;t) = \sum_{j=1}^{N} a_j(\theta;t)\exp(i\psi_j) \tag{5.39}$$

where the ψ_j are independent random phases, whilst the real diffraction factors are given by

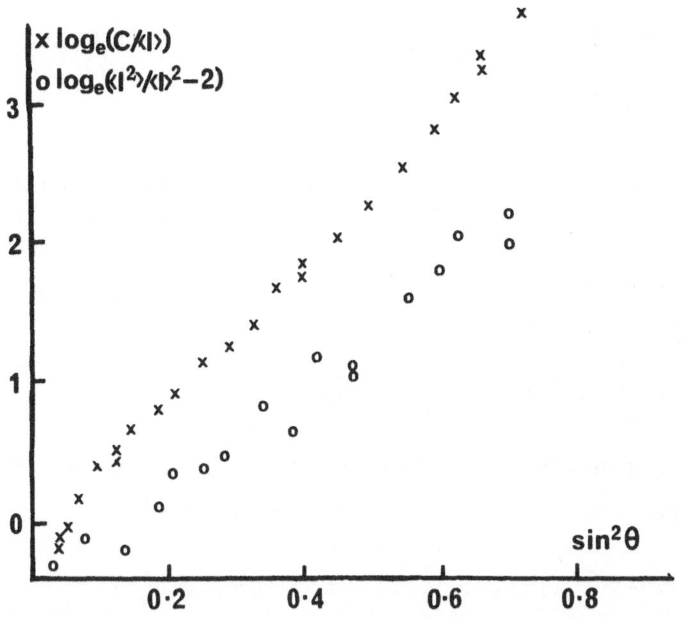

FIG 25 ANGULAR VARIATION OF THE FIRST AND SECOND MOMENTS OF THE INTENSITY DISTRIBUTION OF LIGHT SCATTERED BY A 25μm LAYER OF MBBA UNDERGOING ELECTROHYDRODYNAMIC TURBULENCE AT 20 VOLTS. THE ILLUMINATED AREA W_o IS 10.5 μm.

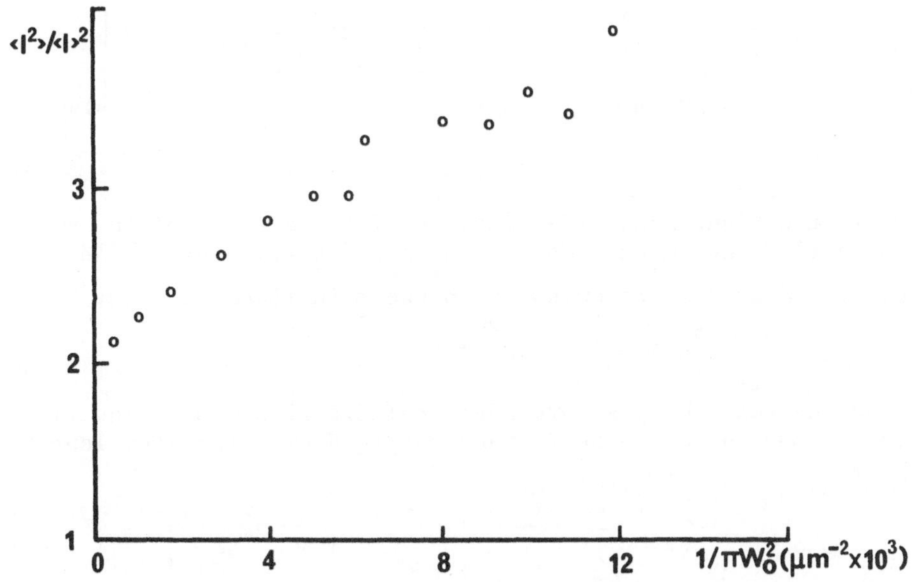

FIG 26 DEPENDENCE OF THE SECOND MOMENT ON THE SIZE OF THE ILLUMINATED AREA AT 20 VOLTS WITH $\theta = 21.6^{o}$

$$a_j^2(\theta;t) = |E_o|^2 \int_{\mathcal{R}}\int_{\mathcal{R}} \exp i\{k|\underset{\sim}{r}-\underset{\sim}{r}'|\sin\theta\cos\eta + \phi_j(\underset{\sim}{r};t)-\phi_j(\underset{\sim}{r}';t)\}d^2r\,d^2r' \tag{5.40}$$

where η is the angular variable corresponding to the polar vector $\underset{\sim}{r}-\underset{\sim}{r}'$. Expression (5.39) describes a finite random walk in the complex $\mathcal{E}$ plane with variable step lengths. This problem has been analysed by several authors[54] and the distribution relevant to the present work is quoted by Watson[55]:

$$P(I) = \frac{1}{2}\int_0^\infty u\, J_o(u\sqrt{I}) \prod_{i=1}^{N} J_o(a_i u)\,du \tag{5.41}$$

The generating function corresponding to this distribution is a confluent hypergeometric function of N variables[56]

$$\langle \exp(-\lambda I)\rangle = \psi_2(1;1,1,1,\ldots 1;-a_1^2\lambda,-a_2^2\lambda\ldots -a_N^2\lambda) \tag{5.42}$$

The result (5.41) has been used recently in the case of equal step length to evaluate the properties of light scattered from a finite number of independent particles[57]. The moments may be evaluated without difficulty in the more general case from (5.42):

$$\langle I^n\rangle = (n!)^2 \sum_{m_1=0}^{\infty}\sum_{m_2=0}^{\infty}\cdots\sum_{m_N=0}^{\infty} \prod_i a_i^{2m_i} \Big/ \Big[\prod_i m_i!\Big]^2 \Bigg|_{\Sigma m_i = n} \tag{5.43}$$

where the right hand side consists of those terms of the multiple sum satisfying $\Sigma m_i = n$. Now the a_i's are statistically independent by virtue of the random phases present in (5.40) so that

$$\langle \prod_i a_i^{2m} \rangle = \prod_i \langle a^{2m_i}\rangle \tag{5.44}$$

assuming that the a's have identical distributions. Equation (5.43) may therefore be further averaged over the step lengths to give

$$\langle I^n\rangle = (n!)^2 \sum_{m_1=0}^{\infty}\sum_{m_2=0}^{\infty}\cdots\sum_{m_N=0}^{\infty} \prod_i \langle a^{2m_i}\rangle \Big/ \Big[\prod_i m_i!\Big]^2 \Bigg|_{\Sigma m_i = n} \tag{5.45}$$

Thus, for example if $n = 1$ all the m_i must be zero except one which will be unity. There will be N such terms so that

$$\langle I \rangle = N \langle a^2 \rangle \tag{5.46}$$

If $n = 2$ on the other hand there are two types of contribution:

(1) N terms in which all the m_i are zero except one which takes the value 2 and

(2) $\frac{1}{2}N(N-1)$ terms in which all the m_i are zero except two which are equal to 1.

This gives

$$\langle I^2 \rangle = N \langle a^4 \rangle + 2N(N-1)\langle a^2 \rangle^2 \tag{5.47}$$

Similarly it may be shown that

$$\langle I^3 \rangle = N \langle a^6 \rangle + 9N(N-1)\langle a^2 \rangle \langle a^4 \rangle + 6N(N-1)(N-2)\langle a^2 \rangle^3 \tag{5.48}$$

$$\langle I^4 \rangle = N\langle a^8 \rangle + 16N(N-1)\langle a^6 \rangle\langle a^2 \rangle + 18N(N-1)\langle a^4 \rangle^2 + 72N(N-1)(N-2)\langle a^4 \rangle \langle a^2 \rangle^2 + 24N(N-1)(N-2)(N-3)\langle a^2 \rangle^4 \tag{5.49}$$

$$\begin{aligned}\langle I^5 \rangle = {} & N\langle a^{10} \rangle + 25N(N-1)\langle a^2 \rangle \langle a^8 \rangle + 100N(N-1) \langle a^6 \rangle \langle a^4 \rangle \\ & + 200N(N-1)(N-2) \langle a^6 \rangle \langle a^2 \rangle^2 + 450N(N-1)(N-2)\langle a^4 \rangle^2 \langle a^2 \rangle \\ & + 600N(N-1)(N-2)(N-3) \langle a^4 \rangle \langle a^2 \rangle^3 \\ & + 120N(N-1)(N-2)(N-3)(N-4) \langle a^2 \rangle^5\end{aligned} \tag{5.50}$$

The moments of a^2 can be evaluated from (5.40) taking advantage of the fact that the region $\mathcal{R}$ extends over the coherence length ξ of the phase. This means that we can use the approximation (5.26) after averaging over ϕ. An alternative approach which is more intuitive but mathematically equivalent is to assume that ϕ is coherent and varies linearly over the region of integration but that its gradient is gaussian distributed. To show the equivalence we observe that

$$\left\langle \exp i \sum_j \phi(r_j)-\phi(r_j') \right\rangle \cong \exp\left\{- \frac{\overline{\phi^2}}{\xi^2}\left| \sum_j (r_j-r_j')\right|^2\right\} \tag{5.51}$$

from (5.23) and (5.26). If on the other hand we assume that ϕ is linear

$$\phi = \underset{\sim}{r}.\underset{\sim}{m}/\xi \tag{5.52}$$

where m is (two dimensional) gaussian distributed[58]

$$p(m) = \frac{e^{-m^2/4\overline{\phi^2}}}{4\pi\overline{\phi^2}} \tag{5.53}$$

then

$$< \exp i\left(\sum_j \phi(r_j)-\phi(r_j')\right> = < \exp i \frac{m}{\xi} \left|\sum_j r_j-r_j'\right| \cos\theta>$$
$$= \exp\left\{- \frac{\overline{\phi^2}}{\xi^2} \left|\sum_j r_j-r_j'\right|^2\right\} \tag{5.54}$$

Whichever approach is favoured it is convenient to replace the integrals over the region $\mathcal{R}$ by infinite integrals weighted by the factor $2 \exp(-4r^2/\xi^2)$. We then obtain

$$<a^{2n}> = \frac{2(\pi\xi^2|E_o|/2)^{2n}}{n\overline{\phi^2}} \exp\left(- \frac{k^2\xi^2 \sin^2\theta}{4\overline{\phi^2}}\right) \tag{5.55}$$

Substitution of (5.55) into (5.46) and (5.47) immediately recovers the results (5.37) and (5.38) when N is interpreted as the number of phase coherence areas per illuminated area W_o^2/ξ^2.

We have thus arrived at a relatively simple model for light scattering from a random phase screen in which the phase of the emerging light is imagined to vary linearly (in two dimensions) over small independent regions of the illuminated area of mean size ξ (assumed in the present case to be larger than the wavelength of the light)(Fig 27). The typical "tilt angle" of the wave front emerging from these regions will be $\sqrt{\overline{\phi^2}}/k\xi$ which is consistent with the width predicted by equation (5.37) for the intensity distribution in the far field. When there are many "facets" on the wave front in the illuminated area structure in the intensity due to the correlation length will be averaged out. However, when the number of facets is small a "lighthouse" effect might be expected in which the incident radiation is scattered into a few well defined directions which fluctuate

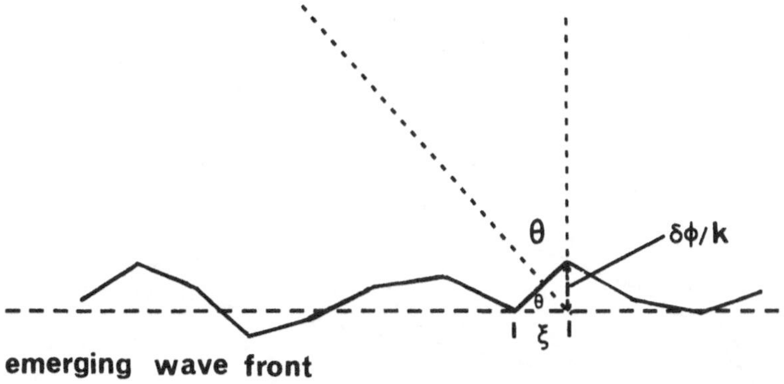

FIG 27 SIMPLE MODEL FOR LIGHT SCATTERING BY A RANDOM PHASE SCREEN

randomly with time. This effect is indeed seen experimentally in the non-gaussian regime. Interestingly in the liquid crystal system it may be observed with the naked eye even when light from a tungsten lamp is used to illuminate the sample. This is presumably because the path lengths taken by light from different parts of the very thin sample differ by only a few wavelengths which is comparable with the coherence length of white light.

Expression (5.55) can be used to predict the behaviour of the higher moments (5.48)-(5.50) and experiments with the liquid crystal system have shown reasonable agreement with the theory.[51] However, the effect of amplitude fluctuations becomes progressively more dominant as the order of the moment increases and must be taken into account. When N is reasonably large the high moments can be expressed entirely in terms of the second moment just as in the gaussian case they may be expressed in terms of the mean. In Fig 28 we show the rth moment plotted against the second moment. Good agreement can be obtained by taking amplitude fluctuations into account in a simple way. This may be done by multiplying equation (5.40) by an attenuation factor $\exp(-\eta)$ which is uniform over each micro-area. The assumption of gaussian statistics for η then leads to an additional log normal factor $\exp\{n(n-1)\overline{\eta^2}/2\}$ on the right hand side of (5.55). The broken curves in the figure are obtained by taking $\exp \overline{\eta^2} = 2$ and show good agreement with the experimental data.

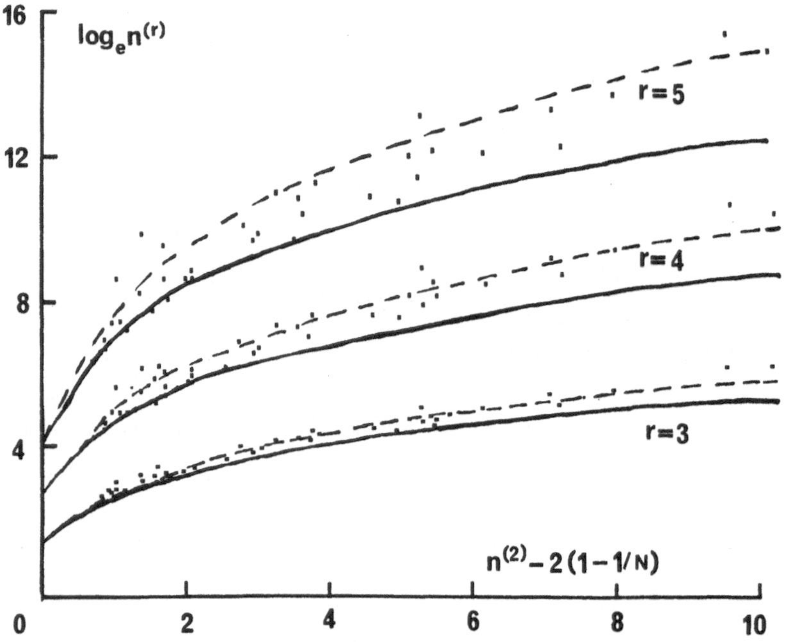

FIG 28 HIGHER ORDER STATISTICAL PROPERTIES AS A FUNCTION OF THE SECOND MOMENT : COMPARISON OF EXPERIMENT WITH THEORY.

4.2.2 Spectral Properties

Assuming "cross spectral purity" (5.27) the field spectrum of the light scattered by a random phase screen may be written

$$<\mathcal{E}^+(\theta;t)\mathcal{E}^-(\theta;t+\tau)> = \pi^2|E_o|^2W_o^2\int_o^\infty r''dr''\ J_o(kr''\sin\theta)e^{-r^2/2\omega_o^2} \times e^{-\overline{\phi^2}\,[1-\rho(r'')\sigma(\tau)]} \quad (5.56)$$

The main contribution to the integral still comes from the region $r'' \sim 0$. We shall assume that $\overline{\phi^2}$ is so large that

$$\overline{\phi^2}\,\sigma(\tau) \gg 1 \quad (5.57)$$

for all time delays of interest. (In the liquid crystal system this restricts us to the region $\sigma(\tau) \gg 0.03$ since $\overline{\phi^2} \sim 36$). This enables us to use an approximation analogous to (5.35) with a fair degree of confidence:

$$\exp(\overline{\phi^2}\sigma(\tau)\rho(r)) \doteq 1 + (e^{\overline{\phi^2}\,\sigma(\tau)}-1)e^{-\overline{\phi}^2\sigma(\tau)r^2/\xi^2} \quad (5.58)$$

Substituting (5.58) into (5.56) and assuming that

$$\overline{\phi^2}\sigma(\tau)/\xi^2 \gg 1/W_o^2 \tag{5.59}$$

by analogy with (5.29), the field spectrum may be evaluated to give

$$g^{(1)}(\theta;\tau) = \frac{\exp \overline{\phi^2}[\sigma(\tau)-1]}{\sigma(\tau)} \exp \frac{-k^2\xi^2 \sin^2\theta}{4\overline{\phi^2}}\left\{\left(\frac{1}{\sigma(\tau)} - 1\right)\right\} \tag{5.60}$$

This result has been written in normalised form using (5.37).

Since $\overline{\phi^2} \gg 1$ only the shape of the phase spectrum near $\tau = 0$ will be manifest in the behaviour of $g^{(1)}(\theta;\tau)$ by virtue of the first exponential factor. In many cases of interest

$$4\overline{\phi^2} \gg k^2\xi^2 \sin^2\theta \left(\frac{1}{\sigma(\tau)} - 1\right) \tag{5.61}$$

so that the second factor in (5.60) is of the order of unity. In this case an expansion of $\sigma(\tau)$ about $\tau = 0$ leads to

$$g^{(1)}(\theta;\tau) \cong e^{-\overline{\phi^2}\tau^2/\tau_c^2} \quad \text{for} \quad \sigma(\tau) = e^{-\tau^2/\tau_c^2} \tag{5.62}$$

$$g^{(1)}(\theta;\tau) \cong e^{-\overline{\phi^2}|\tau|/\tau_c} \quad \text{for} \quad \sigma(\tau) = e^{-|\tau|/\tau_c} \tag{5.63}$$

The shape of the phase spectrum will therefore be reflected in the spectrum of the scattered radiation for the gaussian and lorentzian cases, but its correlation time will be changed significantly by the presence of the $\overline{\phi^2}$ factor.

It should be noted that since $g^{(1)}(\theta;\tau)$ is significant only when $\tau \ll \tau_c$ the approximations (5.57) and (5.59) are justified.

Turning now to the intensity spectrum; this may be written

$$\begin{aligned}\langle I(\theta;t)I(\theta;t+\tau)\rangle = \pi W_o^2|E_o|^4 \int_{-\infty}^{+\infty} & d^2r'd^2r''d^2r''' \\ & \exp\{2ikr''\sin\theta\cos\eta-(r'^2+r''^2+r'''^2)/W_o^2\} \\ & \times \exp\{-\overline{\phi^2}[2-\rho(r''+r''')-\rho(r''-r''') \\ & - \sigma(\tau)(\rho(r'+r'')+\rho(r'-r'')-\rho(r'+r''')-\rho(r'-r'''))]\}\end{aligned} \tag{5.64}$$

Use of the approximations already mentioned (equations (5.58) and (5.59)) leads after considerable analysis to the normalised second order correlation function

$$g^{(2)}(\theta;\tau) = \left(1 - \frac{\xi^2}{W_o^2}\right)\left(1 + |g^{(1)}(\theta;\tau)|^2\right) + \frac{F(\tau)\overline{\phi^2}\,\xi^2}{W_o^2} \exp\left\{\frac{k^2\xi^2\sigma(\tau)\sin^2\theta}{2\overline{\phi^2}\,[1+\sigma(\tau)]}\right\} \quad (5.65)$$

F(0) = 1/4 so that this reduces to the second moment (5.38) when τ=0 as expected. However F(τ) is difficult to evaluate for arbitrary τ. A "micro-area" approach to the problem indicates that both the gaussian factor and the angular dependence predicted by (5.65) may be correct even for small values of σ, and a simple analytic form for F(τ) can be found by this method. An example of the resulting correlation function for a gaussian phase spectrum is shown in fig 29. Not only does the non-gaussian term distort the spectral shape, but it also increases the observed correlation time to something like the true phase fluctuation time. It is interesting that only in the non-gaussian regime can "direct"

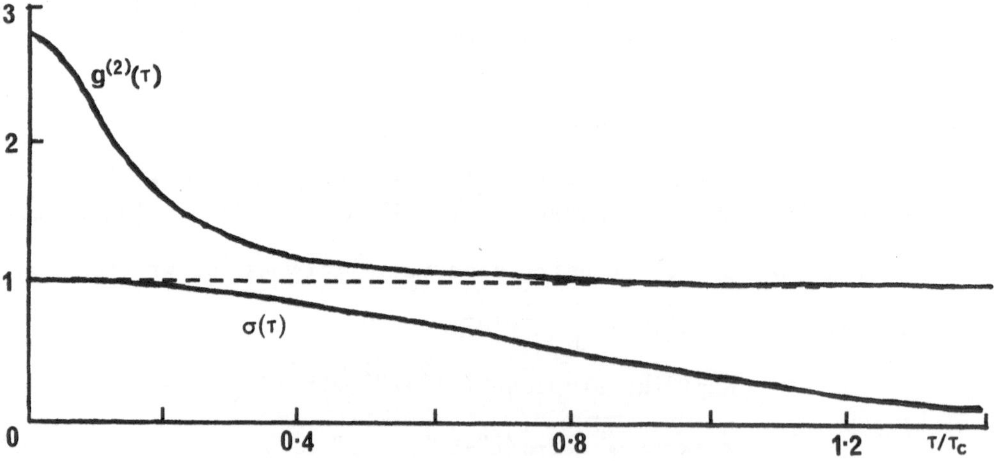

FIG 29 PREDICTED FORM OF THE INTENSITY CORRELATION FUNCTION OF LIGHT SCATTERED FROM A RANDOM PHASE SCREEN.

measurements be made through (5.65) of the phase spectrum. If we measure $g^{(1)}(\theta;\tau)$ by heterodyning or (5.65) in the gaussian limit we shall arrive at a misleading value for the fluctuation time of the scattering process itself because of the scaling factor $\overline{\phi^2}$.

In conclusion it is clear that much is to be learnt by studying simple non-gaussian systems such as the deep random phase screen. Non-gaussian statistics can prove a valuable source of information when an adequate theoretical description of such systems can be found. For example, the phase screen model should apply to a thin film of pure fluid or binary liquid mixture sufficiently close to the critical point for the range of correlation of the fluctuations to be comparable with the size of the illuminated region, provided the associated refractive index changes are large enough.

REFERENCES

1 A T Forrester, R A Gudmundsen and P O Johnson, Phys Rev 99 (1955) 1691.

2 A T Forrester, J Opt Soc Am, 51 (1961) 253.

3 F A Johnson, T P McLean and E R Pike, Proc Int Conf of the Physics of Quantum Electronics, Puerto Rico, 1965, Eds P L Kelley, B Lax and P E Tannenwald (New York: McGraw-Hill 1966).

4 E R Pike, Nuovo-Cunento 1 (1969) 277.

5 E R Pike and E Jakeman "Photon statistics and photon correlation spectroscopy" Adv in Qu Electronics 2, Ed D Goodwin (Academic Press 1973).

6 G H Heilmeier, L A Zanoni and L A Barton, Proc IEEE 56 (1968) 1162.

7 See for example W B Davenport and W L Root "An introduction to the theory of random signals and noise" (McGraw-Hill 1958) Chapters 1-8.

8 R J Glauber, Phys Rev 131 (1963) 2766.

9 H Cramer, "Mathematical methods of statistics" (Princeton University Press, N J 1946).

10 A J F Siegert, MIT Rad Lab Rep No 465 (1943).

11 G Bedard, Phys Rev, 161 (1967) 1304.

12 See also ref 8.

13 S O Rice, Bell Syst Tech J, 23 (1944) 282.

14 M Bertolotti, this volume

15 L Mandel, Proc Phys Soc, 74 (1959) 233.

16 J H Van Vleck, Harvard Univ Rad Res Lab Rep No 51 (1943)

17 J H Van Vleck and D Middleton, Proc IEEE 54 (1966) 2.

18 E Jakeman and E R Pike , J Phys A (gen phys) 2 (1969) 411.

19 E Jakeman, J Phys A (gen phys) 3 (1970) 201.

20 Ch Bendjaballah, J Phys A (math nucl gen) 6 (1973) 837.

21 E Jakeman, J Phys A (gen phys) 5 (1972) L49.

22 E Jakeman and E R Pike, J Phys A (gen phys) 2 (1969) 115.

23 See for example M Born and E Wolf, "Principles of Optics" (Pergamon press, 2nd Ed 1964).

24 L Mandel and E Wolf, Rev Mod Phys, 37 (1965)231.

25 See for example Ann Rev Astronomy and Astrophysics 6 (1968)13.

26 E Jakeman, C J Oliver and E R Pike, J Phys A (gen phys) 3 (1970) L45.

27 G Present and D B Scarl, Appl Opt, 11 (1972) 120.

28 D E Koppel, J Appl Phys, 42 (1971) 3216.

29 S H Chen and P Tartaglia, Opt Commun 6 (1972) 119.

30 F A Johnson, R Jones, T P McLean and E R Pike, Phys Rev Letts, 16 (1966) 589.

31 G Bedard, Proc Phys Soc 90 (1967) 131.

32 I DeLotto, P F Manfredi and P Principio, Energia Nucl (Milan) 11 (1964) 557.

33 E Jakeman, C J Oliver and E R Pike, J Phys A (gen phys) 4 (1971) 827.

34 E Jakeman and E R Pike, J Phys A, 1 (1968) 690.

35 Ref 7 Chapter 5

36 E Jakeman, E R Pike and S Swain, J Phys A (gen phys) 4 (1971) 517.

37 B A Saleh, J Phys A (math, nucl gen) in press.

38 C J Oliver, this volume

39 V Degiorgio and J B Lastovka, Phys Rev A 4 (1971) 2033.

40 P Tartaglia, T A Postal and S H Chen, J Phys A (math nucl gen) 6 (1973) L35.

41 P N Pusey and W I Goldberg, Phys Rev A3 (1971) 766.

42 E Jakeman, C J Oliver, E R Pike and P N Pusey, J Phys A (gen phys) 5 (1972) L93.

43 V Mangulis, "Handbook of series for scientists and engineers" (Academic press, 1965) 4.

44 L S Taylor, J Math Phys, 13 (1972) 590.

45 See for example L E Estes, L M Narducci and R A Tuft, J Opt Soc Am 61 (1971) 1301.

46 See for example Ch Deutsch and P N Keating J Appl Opt 40 (1969) 4049.

47 P Beckmann and A Spizzichino "The scattering of electromagnetic waves from rough surfaces" (Pergamon press, 1963).

48 P Beckman, IEEE trans Antennas and Propag. AP-21 (1973) 169.

49 A S Marathay, L Heiko and J L Zuckerman, Appl Opt 9 (1970) 2470.

50 M V Berry, Phil Trans Roy Soc 273 (1973) 611.

51 E Jakeman and P N Pusey, J Phys A (math nucl gen) 6 (1973) L88.

52 E Jakeman and P N Pusey, Phys Letts 44A (1973) 456.

53 See for example L H Enloe Bell Syst Tech J 46 (1967) 1479.

54 See for example Lord Rayleigh (J W Strutt) Phil Mag 6 (1919) 321.

55 G N Watson "Theory of Bessel Functions" (Cambridge University Press 1944) 419.

56 "Tables of integral transforms" Ed A Erdelyi (McGraw-Hill 1954) 385.

57 P N Pusey, D W Schaefer and D E Koppel, J Phys A (1973) to be published.

58 M S Longuet-Higgens, Phil Trans Roy Soc A249 (1956) 321.

CORRELATION TECHNIQUES

C J Oliver

Royal Radar Establishment

St Andrews Road, Great Malvern, Worcs., England

1 SOURCES AND DETECTORS

1.1 Introduction

This course on correlation techniques will consist of three parts. In the first the effect of source and detector properties, particularly their imperfections, on the measurement of correlation functions is considered. Since the technique of photon-correlation spectroscopy relies on extracting information about the scattering region from the observed correlation function we shall deduce the principle features we require for source, detector and data processing scheme. In the second part the principles of the design of a photon-counting correlation system will be considered. In the final part the choice of optimum operating conditions in photon-correlation experiments will be outlined.

A diagrammatic representation of a typical photon-correlation spectroscopy experiment is shown in figure 1. Incident light of wave vector $\underline{k}_o$ from a laser is scattered and detected with a wave vector $\underline{k}_s$. The optical field at the detector is then proportional to the spatial Fourier component of the dielectric constant fluctuations defined by the scattering angle θ. However the probability of a photodetection has been shown to be proportional to the square modulus of the positive frequency component of the optical field[1] at the detector. We can therefore define an instantaneous 'intensity' I(t) as

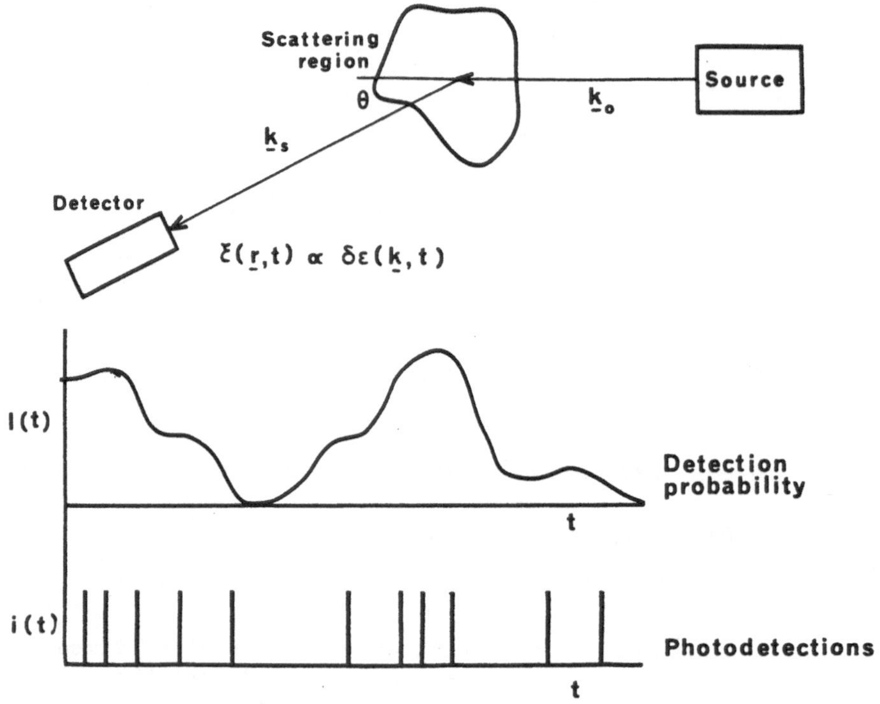

FIG 1 Block diagram for a typical laser scattering experiment. The detection probability, I(t) (ie the intensity), and the corresponding photodetections, i(t), are shown.

$$I(t) = \left|\mathcal{E}^+(t)\right|^2 . \tag{1}$$

For a scattering experiment this intensity will vary with time, as shown in figure 1. In practice any real detector has some finite response time, T', so that the detector responds to an average intensity I(t,T') given by

$$I(t,T') = \frac{1}{T'} \int_{t-\frac{T'}{2}}^{t+\frac{T'}{2}} \left|\mathcal{E}^+(t)\right|^2 dt. \tag{2}$$

However detection is a Poisson process so that the output train of photodetection pulses i(t) is a Poisson, random, train rate-modulated by the intensity fluctuations. As shown in figure 1 photodetection 'bunching' occurs in regions of high intensity. Information regarding the intensity fluctuations, and hence the scattering region itself, can be derived from the temporal distribution of these photodetections. This last statement is obviously only correct where source and detector are themselves completely free from correlations.

1.2 Data-Sampling Schemes

Let us now consider the means whereby the photomultiplier output can be converted to suitable data for processing in a correlator, or other processor. We can divide our discussion into three sections, namely: (1) the detector output (2) the sampling technique (3) the signal processor. As will be discussed later, the output of a photomultiplier, due to its mode of operation, consists of a series of pulses of different charge, q. This output can either be processed directly, including the charge variations as an additional source of noise, or it can be standardised in an amplifier and discriminator with the loss of about 2% to 5% of the original photoelectron pulses from the photocathode. The effect of the additional noise due to charge fluctuations will be considered in section 1.4.1. Let us assume here that we are concerned with a train of standardised pulses corresponding to the initial photodetections.

In order to sample such data for correlation purposes it is important that the sampling scheme itself should not introduce correlations. This implies that samples must be independent. The most efficient way of obtaining independent samples of separation T is to integrate all the incoming pulses over the time interval T, either by counting the pulses or integrating the charge. At the end of each sample the data can be transferred to the subsequent processor and the integrator reset for the next sample. In this method the incident data is all treated identically, no variable weighting is introduced, and the samples are non-overlapping and hence independent. Other, less efficient, sampling schemes can be derived in which the reset pulse is replaced by an integrating filter. The simplest such filter would be an RC integrator though filters with sharper cut-offs can be readily achieved. The samples are now made independent by the choice of the RC time constant. In order to have only a 2% residual overlap τ_{RC} is often taken to be T/4. However the effect of any such filter is to weight the incoming data so that the most recent pulses are treated with a higher weighting than those which arrived earlier. This is obviously an inefficient method of operation. To illustrate this feature of data-sampling schemes let us compare the first case, called photon-counting, with the second in more detail.

It should be made clear that the term 'photon-counting' is used here to described any scheme in which the standardised data from the photomultiplier is integrated, or counted, so that all

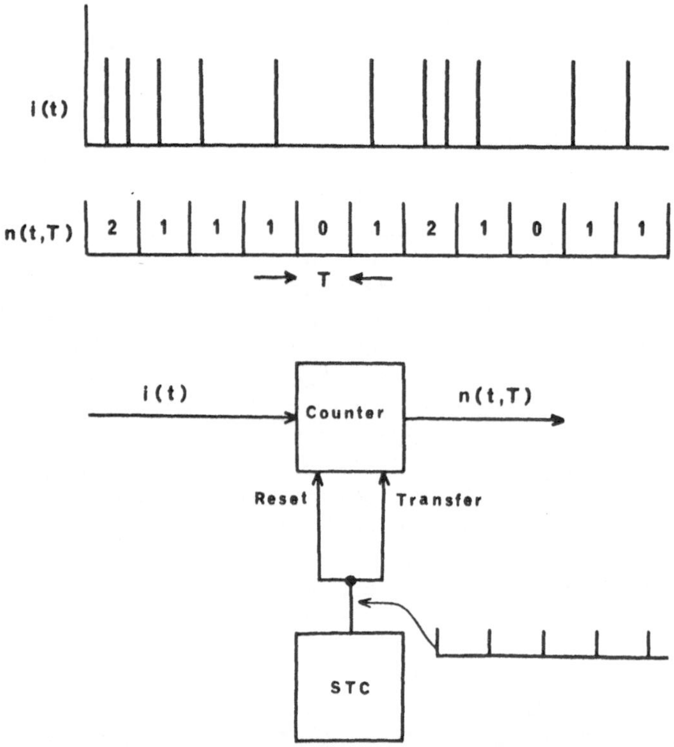

FIG 2 Block diagram for the operation of the photon-counting sampling scheme. The photodetections, i(t), and the sampled data, n(t,T), are shown together with the operation of the resettable counter.

pulses have the same weighting. Any subsequent processing may handle the data in the form of analog or digital signals provided that the resolution is adequate to preserve single photodetections throughout. For example a standard commercial correlator with an analog voltage input can be converted to a photon-counting correlator by using a scalar followed by a DAC, or by using a resettable charge integrator, as the data sampler. Alternatively it can be used with a capacitive integrator merely by using an RC combination on the input.

1.2.1 Photon-Counting. In this mode of operation the output train of photodetections is counted during a sample time of length, T. The data then takes the form n(t,T), as shown in Figure 2, denoting the counting of n photodetections in an interval T centred on time t. Experimentally this integration is achieved in a resettable counter. At the end of a sample, determined by the sample time clock (STC), the n photodetections recorded in the counter are transferred to the subsequent processing electronics, the counter is reset and the next sample begun. This process can be cycled repetitively until enough samples have been taken to give an adequate estimate of the desired information. This output data, $n(t,T)$, can be characterised both by its probability distribution $p(n,T)$ (ie the probability of detecting n counts in a period T) and by its autocorrelation function, $G^{(2)}(\tau)$. The form of the photon count probability distribution is related to the probability distribution of the intensity, $P(I)$, by[2]

$$P(n,T) = \int_0^{\infty} \frac{(\alpha I)^n}{n!} \exp(-\alpha I)\, P(I)dI. \tag{3}$$

For constant intensity, as with an ideal source, this gives a Poisson distribution

$$P(n,T) = \frac{\bar{n}^n}{n!} e^{-\bar{n}} \tag{4}$$

where $\bar{n}$ is the mean count-rate per sample time T. Intensity fluctuations would result in a non-Poisson distribution. Measurements of the photodetection statistics can therefore be used as a means of extracting spectral information[3]. Alternatively, if the detector is looking at the source directly, measurements of $P(n,T)$ can be used to reveal detector or source imperfections[4].

The autocorrelation function, as has been discussed earlier[5], is found by storing and delaying the data from the sampling counter and then multiplying this by the undelayed data. After N samples an estimate of the correlation function will be obtained defined by

$$\hat{G}^{(2)}(\tau) = \frac{1}{N} \sum_{i=1}^{N} n(iT)\, n(iT + \tau) \tag{5}$$

where τ is the delay time. The average value of this estimator for a stationary process will be

$$G^{(2)}(\tau) = \langle n(0)n(\tau) \rangle . \tag{6}$$

For a source free from all correlations, such as an ideal laser or one in which any coherence time is very short compared with the delay time, this can be factorised to give

$$G^{(2)}(\tau) = \bar{n}^2. \tag{7}$$

Any departures from this flat correlation function may be used either in measurement of the spectral properties of a scattering process[6] or, as before, to reveal source or detector imperfections. This technique offers the advantage, compared with the measurement of photodetection statistics, of revealing any characteristic correlation times associated with such processes.

In any correlation function measurement the performance can be described in terms of the accuracy with which the correlation coefficient estimator, $\hat{G}^{(2)}(\tau)$, can be measured. This is conveniently characterised by its relative variance Var $(\hat{G}^{(2)}(\tau))/\langle \hat{G}^{(2)}(\tau)\rangle^2$. The variance of $\hat{G}^{(2)}(\tau)$ for a stationary process with no correlations may be written as[7]

$$\mathrm{Var}(\hat{G}^{(2)}(\tau)) = \frac{1}{N}\ \mathrm{Var}\ (n(0)n(\tau)). \tag{8}$$

Separating the independent variables $n(0)$ and $n(\tau)$ this gives

$$\mathrm{Var}(\hat{G}^{(2)}(\tau)) = \frac{1}{N}\left[\ \langle n^2 \rangle^2 - \langle n \rangle^4\ \right] \tag{9}$$

$$= \frac{1}{N}\,(2\bar{n}^3 + \bar{n}^2) \tag{10}$$

for a Poisson distribution of n since, in this case ,

$$\langle n^2\rangle = \bar{n}^2 + \bar{n}. \tag{11}$$

The relative variance is therefore given by

$$\frac{\mathrm{Var}(\hat{G}^{(2)}(\tau))}{\langle \hat{G}^{(2)}(\tau)\rangle^2} = \frac{\mathrm{Var}(\hat{G}^{(2)}(\tau))}{\bar{n}^4} = \frac{2}{N\bar{n}} + \frac{1}{N\bar{n}^2}. \tag{12}$$

1.2.2 Capacitive Integration. In the previous section we described the photon-counting data-sampling scheme and derived the relative variance of the correlation coefficients. When using correlators with analog inputs, however, the usual approach is to use capacitive integration on the correlator input to take the place of the photon-counting sampling-unit. The correlator itself takes a short sample of the voltage across the integrating capacitor once every sample time, T. Instead of a reset operation the RC time constant τ_{RC}, selected to be short compared with T, makes the samples independent. The observed correlation function then contains a decay time of τ_{RC} together with the desired correlation time τ_c. If τ_{RC} becomes comparable with the sample time the distortion of the observed correlation function becomes severe in the first few channels. Provided this is not the case one normally ignores the effect of the integration time on the correlation function, though this must only be done after careful justification. Let us now derive the form of the correlation function using such RC integration and also the relative variance of the correlation coefficients.

The correlation function can now be described in terms of the charge Q(t) on the integrating capacitor, thus

$$G_Q^{(2)}(\tau) = \langle Q(0)Q(\tau)\rangle . \tag{13}$$

If no correlations are present at this delay then the terms can be separated as before giving

$$G_Q^{(2)}(\tau) = \langle Q\rangle^2 . \tag{14}$$

In addition the variance of the estimator $\hat{G}_Q^{(2)}(\tau)$ can be shown to be given by

$$\mathrm{Var}\,(\hat{G}_Q^{(2)}(\tau)) = \frac{1}{N}\left\{\langle Q^2\rangle^2 - \langle Q\rangle^4\right\} \tag{15}$$

which is similar to equation (9). We therefore require to calculate $\langle Q\rangle$ and $\langle Q^2\rangle$ for this process.

Suppose we consider a data-sampling scheme of a very large number of samples of duration $T'(\ll \tau_{RC})$ sufficiently small that there is a negligible probability of more than one photodetection in any sample. The presence of a photodetection in the i'th sample is denoted by the random Boolean variable Y_i which is unity when a pulse is present and zero otherise. The total charge at some point in time due to the contributions of all previous samples is therefore given by

$$Q = \sum_{i=0}^{\infty} q_o Y_i e^{-\frac{iT'}{\tau_{RC}}} , \tag{16}$$

where q_o is the constant charge per pulse. The average value of Q will therefore be given by

$$< Q > = < \sum_{i=o}^{\infty} q_o Y_i e^{-\frac{iT'}{\tau_{RC}}} >$$

$$= \bar{n}' q_o \frac{\tau_{RC}}{T'} = \bar{m}\, q_o\, \tau_{RC} \qquad (17)$$

since the terms can be separated, where $\bar{n}'$, the count rate per sample time, is given by

$$\bar{n}' = < \sum_{i=o}^{\infty} Y_i >$$

and $\bar{m}$ is the count-rate per second. Similarly the mean-square value of Q is given by

$$< Q^2 > = < \sum_{i,j=o}^{\infty} q_o^2 Y_i Y_j e^{-\frac{T'}{\tau_{RC}}(i+j)} > . \qquad (18)$$

Separating diagonal (i=j) and off-diagonal ($i \neq j$) terms we obtain

$$< Q^2 > = < \sum_{i=o}^{\infty} q_o^2 Y_i e^{-\frac{2T'i}{\tau_{RC}}} >$$

$$+ \sum_{i,j=o}^{\infty} q_o^2 < Y_i > < Y_j > < e^{-\frac{T'i}{\tau_{RC}}} > < e^{\frac{T'j}{\tau_{RC}}} >$$

$$- \sum_{i=o}^{\infty} q_o^2 < Y_i >^2 < e^{-\frac{T'i}{\tau_{RC}}} >^2$$

$$= \frac{\bar{n}' q_o^2 \tau_{RC}}{2T'} + \left(\bar{n} q_o \frac{\tau_{RC}}{T'} \right)^2 - \frac{q_o^2 \tau_{RC}}{2T'} \sum_{r=o}^{\infty} < Y_r >^2 . \qquad (19)$$

The last term can be evaluated by expanding

$$< n'^2 > = < \sum_{i,j=o}^{\infty} Y_i Y_j >$$

in a similar manner. This leads to

$$< n'^2 > = \bar{n}' + \bar{n}'^2 - \sum_{i=o}^{\infty} < Y_i >^2 . \qquad (20)$$

Replacing in eqn (19) above gives

$$< Q^2 > = \frac{\bar{n}' q_o^2 \tau_{RC}}{2T'} + \left(\frac{\bar{n} q_o \tau_{RC}}{T'} \right)^2 - q_o^2 \frac{\tau_{RC}}{2T} (\bar{n}' + \bar{n}'^2 - \overline{n'^2})$$

$$= \bar{m}\, q_o^2 \frac{\tau_{RC}}{2} + \bar{m}^2 q_o^2 \tau_{RC}^2 , \qquad (21)$$

since n' is assumed to have a Poisson distribution. The relative variance is then given by

$$\frac{\mathrm{Var}\ (\hat{G}_Q^{(2)}(\tau))}{< \hat{G}_Q^{(2)}(\tau) >^2} = \frac{1}{N\bar{m}\ \tau_{RC}} + \frac{1}{4N\ \bar{m}^2 \tau_{RC}^2} \qquad (22)$$

which should be compared with the equivalent expression for photon-counting data sampling given in eqn (12) above.

In order for the correlator input to follow the true intensity the integration time τ_{RC} must be short compared with the correlator sample time T. The value $\tau_{RC} = T/4$ gives a 2% residual in the intensity fluctuations which is not unreasonable. Replacing in eqn (22) we find that the relative variance for RC integration is therefore

$$\frac{\mathrm{Var}(\hat{G}_Q^{(2)}(\tau))}{< G_Q^{(2)}(\tau) >^2} = \frac{4}{N\bar{m}^2 T^2} + \frac{4}{N\bar{m}T} \qquad (23)$$

compared with

$$\frac{\mathrm{Var}\ (\hat{G}^{(2)}(\tau))}{< G^{(2)}(\tau) >^2} = \frac{1}{N\bar{m}^2\ T^2} + \frac{2}{N\bar{m}T} \qquad (24)$$

for photon-counting operation.

Thus for high count rates ($\bar{m}T \gg 1$) the relative variance, and hence the time taken to achieve the same accuracy, is a factor of two greater than for the photon-counting technique. For low count-rates the increase is a factor of four. One can use non-independent samples if required and ignore the first few channels which will show an exponential decay of time constant, τ_{RC}. For example if τ_{RC} is taken equal to T then the first four delays have to be ignored as they distort the apparent correlation function by more than 1%. The accuracy one obtains in this situation can no longer be described in terms of eqn (22) since the samples are not independent. In fact the correlation coefficients and the errors will be correlated. Data obtained with this sampling scheme will in fact never be as accurate as from the photon-counting one.

An example of this is shown in figure 3 where a measurement using the photon-counting sampling scheme (Malvern K7023 Correlator, Precision Devices and Systems Ltd) is compared with one using the same photodetections made with the capacitive sampling scheme (Hewlett-Packard HP3721A). The count-rate was $\bar{n} = 0.1$ with a sample time of 1 mS and an RC time constant of ~0.25 mS on the input of the analog correlator. Figure 3(a) shows the normalised correlation function

$$g^{(2)}(\tau) = \frac{G^{(2)}(\tau)}{\bar{n}^2}$$

obtained in photon-counting operation while figure 3(b) shows the same measurement made with capacitive integration. The greater accuracy of the photon-counting processing is apparent. Calculation of the relative variances yields percentage errors of 3.3% and 6.6%. The ratio of these is identical to the predicted percentage error ratio of 2.

Thus from considerations of the data-sampling schemes we have shown that photon-counting operation offers an advantage of between two and four in the total time required to achieve a desired accuracy. Where possible therefore photon-counting techniques should be employed.

1.3 Source Imperfections

Having outlined the data-sampling scheme to be used with ideal source and detector let us now investigate the departures from ideal performance of real sources and detectors and their effect on performance. Source imperfections can be divided, somewhat arbitrarily, into frequency and amplitude instabilities arising from various causes. Let us examine the effects of each in some detail.

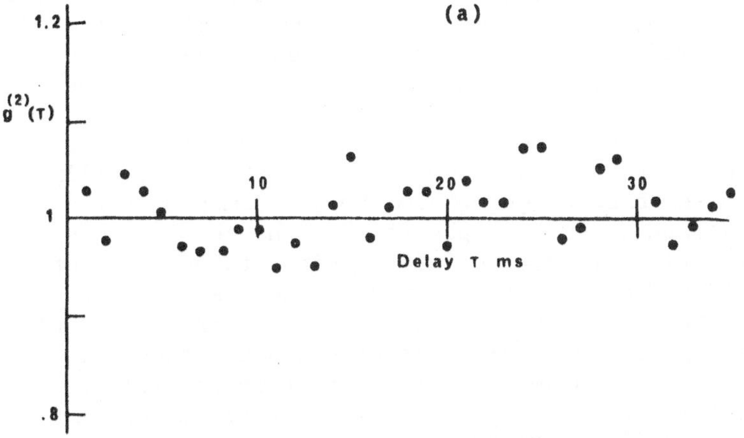

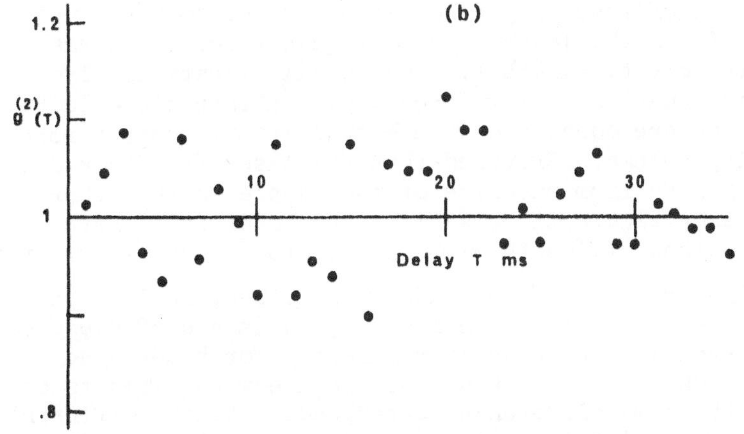

FIG 3 A comparison of photon-counting (a) and analog-integration (b) data-sampling schemes. The mean count rate per second was 100 with sample separation of 1 ms. The capacitive integration time, τ_{RC}, was ~0.25 ms.

1.3.1 Frequency Fluctuations. Since the gain of a laser is proportional to the length of the amplifying medium most lasers have a length, L, which is sufficient for many longitudinal modes of separation $c/_{2L}$ to be capable of lasing action. Typically then lasers may have a longitudinal mode separation of 100 MHz while the Doppler broadened linewidth, over which lasing may occur. is typically 2 GHz. The outputs of these lasers thus contain strong beats at the intermode frequency. Provided that one requires to measure frequency components considerably less than the mode separation these high frequency terms will be integrated out.

If transverse modes are present, however, the inter-cavity modes may have separations of only ~5 MHz between the 01 and the 00 (longitudinal) modes. In which case the effect may become apparent in correlation measurements of fairly rapid processes.

In addition to the presence of inter-mode beats any sort of correlation between the amplitudes of different modes will give additional noise terms which reduce the performance of the system below that expected. For example it is found that with an Ar^+ ion laser operating without an inter-cavity etalon the signal-to-noise ratio is a factor of two to four worse than when the same laser is operated in a single-mode after the introduction of an etalon. The higher the plasma current and hence tube gain the poorer the signal-to-noise ratio so that it is possible for the laser signal-to-noise ratio to be reduced as plasma current and hence output power is increased.

Another type of frequency fluctuation is the comparatively long-term drift of the laser output frequency due to thermal changes in the cavity-length L. Mechanical vibrations also modulate this length at a low frequency. In intensity-fluctuation spectroscopy we are concerned with mutual interference across the scattering volume. Provided that the laser frequency does not change by 1 Hz from one side of the sample to the other slower frequency fluctuations will not be important. For scattering regions of length ~100 μm therefore a stability of approximately 1 Hz in 10^{-12} seconds is required, ie 10^{12} per second drift rate. This is considerably in excess of expected drift rates even with poor quality lasers. For heterodyne measurements, on the other hand, the frequency requires to be stable over the time difference corresponding to the different light paths of the local oscillator and the scattered radiation. This normally demands an improved stability by one or two orders of magnitude. Photon-correlation spectroscopy differs from interferometric techniques in that the latter require absolute frequency stability since the absolute frequency is measured whereas the former merely require a not excessive drift rate since only frequency differences are measured.

1.3.2 Gain Fluctuations. Other source imperfections are associated with fluctuations in the gain of the amplifying medium. Any rapid fluctuations, such as plasma oscillations, will produce a departure from a flat correlation function. An example is shown in figure 4 where a sinusoidal correlation function of period 135 kHz is observed characteristic of plasma oscillations. These oscillations are only apparent when the gain is reduced below the level required for saturated operation. They can also be removed with RF excitation.

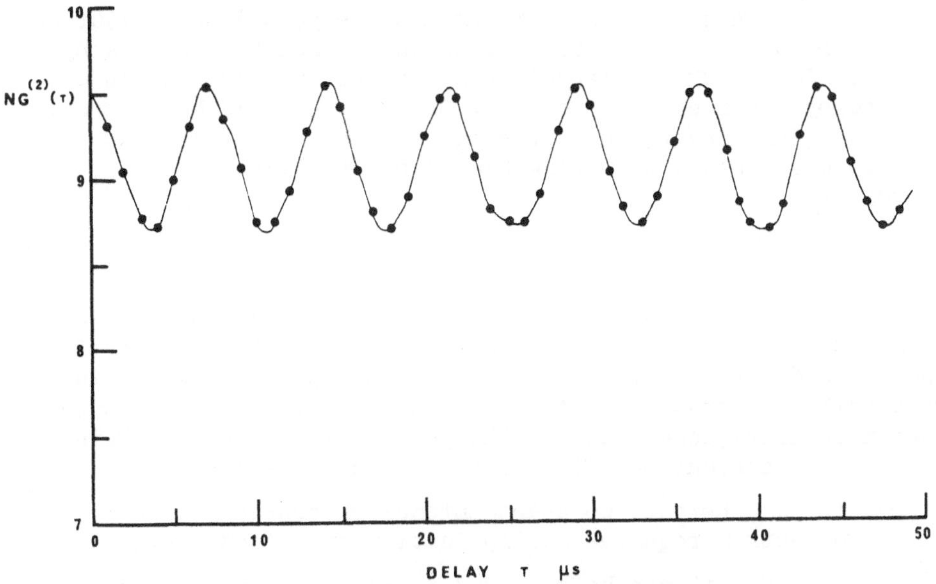

FIG 4 A measured correlation function for a HeNe laser showing plasma oscillations at a frequency of 135 kHz

While a laser may be free of such instabilities it will generally suffer from long-term amplitude drift associated with thermally-induced changes in the laser cavity alignment. Provided these fluctuations take place on a time scale long compared with the time dependent effects under study they will not give any appreciable change in the shape of the observed correlation function but will result in changes of the value of the, apparently, flat background. To illustrate this let us consider an experiment of 2N samples in which the mean count-rate was $\bar{n}$ for the first half and $\bar{n}(1+\beta)$ for the second half. The observed correlation function then takes the form

$$G^{(2)}_{obs}(\tau) = \frac{\bar{n}^2}{2}(1 + (1 + \beta)^2). \qquad (25)$$

The observed mean count rate is given by

$$\bar{n}_{obs} = \frac{\bar{n}}{2}(2 + \beta). \qquad (26)$$

The normalised correlation function is thus

$$g^{(2)}_{obs}(\tau) = 1 + \frac{\beta^2}{(2+\beta)^2} \simeq 1 + \frac{\beta^2}{4} \qquad (27)$$

for $\beta \ll 1$. Thus the departure from the expected normalisation of unity is independent of count rate and is generally speaking negligible for $\beta \ll 1$, certainly for intensity-fluctuation spectroscopy. For example for $\beta = 0.1$, which would be excessive for a reasonable laser, the error in the background normalisation is 0.0025 which would be negligible except in heterodyne spectroscopy.

1.4 Detector Imperfections

Broadly speaking, in photon-correlation spectroscopy one is more limited by the detector performance than by the source. Let us now consider the operation of a photomultiplier detector in more detail and study the effects of the imperfections encountered. As shown earlier (eqns 4 and 7) the photoemissions have a Poisson probability distribution, $P(n,T)$, and a flat correlation function, $G^{(2)}(\tau)$, for a coherent, constant-intensity, source. However each photoelectron requires multiplication by a factor of 10^7 to 10^8 before it can be detected reliably in high-speed counting circuitry. This amplification must be wide-band and must not introduce extraneous pulses. In practice the most suitable amplifier is found to be the electron-multiplier tube. Since multiplication relies on a statistical secondary-emission process this introduces a further uncertainty into the detector output beyond the photon statistics. The effect of this pulse-height distribution can be removed by photon-counting detection, in which the pulses are detected in a discriminator, with some loss arising from pulses falling below the threshold. Poor electron optics can result in a low collection efficiency for photoelectrons leaving the cathode since they are lost between the first few dynodes. More damaging than the loss of genuine pulses, from the point of view of correlation measurements however, is the introduction of spurious pulses, particularly when these are correlated with preceding pulses. Let us now consider the causes and effects of gain fluctuations and correlated after-pulses in photomultiplier tubes.

1.4.1 Gain Fluctuations. The effect of gain fluctuations in the detector is only apparent when using analog correlation techniques. Let us, therefore, generalise the analysis for capacitive integration on the input to an analog correlator, given in section 1.3.2, to include charge variation from pulse to pulse. This is conveniently done by introducing another random variable q_i to denote the charge of a pulse in the i'th interval if one were present. Equation (16) now becomes

$$Q = \sum_{i=o}^{\infty} q_i Y_i e^{-\frac{iT'}{\tau_{RC}}} \tag{28}$$

whence, using the same analysis as before,

$$< Q > = \bar{m}\,\bar{q}\,\tau_{RC} \tag{29}$$

and

$$< Q^2 > = \frac{\bar{m}}{2}\,\tau_{RC}\,\overline{q^2} + \bar{m}^2\,\tau_{RC}^2\,\bar{q}^2. \tag{30}$$

The relative variance of the correlation coefficient estimator, $G_Q^{(2)}(\tau)$, is therefore given by

$$\frac{\mathrm{Var}\,(G_Q^{(2)}(\tau))}{< G_Q^{(2)}(\tau)>^2} = \frac{< Q^2>^2 - < Q>^4}{N < Q >^4}$$

$$= \frac{1}{4N\bar{m}^2\,\tau_{RC}^2}\,\frac{\overline{q^2}^2}{\bar{q}^4} + \frac{1}{N\bar{m}\,\tau_{RC}}\,\frac{\overline{q^2}}{\bar{q}^2}$$

$$= \frac{1}{N}\left(1 + \frac{\mathrm{Var}(q)}{\bar{q}^2}\right)\left[\frac{1}{\bar{m}\tau_{RC}} + \left(1 + \frac{\mathrm{Var}(q)}{\bar{q}^2}\right)\frac{1}{4\bar{m}^2\tau_{RC}^2}\right] \tag{31}$$

where Var(q) is the variance of the charge distribution from the photomultiplier. For low count rates ($\bar{m}\tau_{RC} \ll 1$) the relative variance is greater than for standardised pulses by a factor $\left(1 + \frac{\mathrm{Var}(q)}{\bar{q}^2}\right)^2$; while for high count rates the factor is merely $\left(1 + \frac{\mathrm{Var}(q)}{\bar{q}^2}\right)$.

It is necessary, therefore, to know what the charge or pulse-height distribution for the photomultiplier output pulses is. The process of multiplication in a multi-stage secondary-emission process has been discussed by many authors, but particularly by Prescott[38] who analysed the process assuming the physical model that the secondary-emission process is Poisson, but that the effective secondary-emission ratio could vary across the dynode, due either to dynode inhomogeneities or to fluctuations in the collection efficiency from dynode to dynode as a function of position. He proposed a negative binomial (Polya) distribution for the gain of the tube such that the output charge per pulse, q, has a form

$$p(q) = \binom{n + \frac{1}{b} - 1}{n} (b\mu)^n \, (b\mu + 1)^{-\frac{1}{b} - n} \tag{32}$$

where b is a parameter describing the excess variance due to dynode inhomogeneities, μ is the mean gain per stage and n(=q/e) is the total number of electrons per pulse. Observed pulse-height distributions can be fitted to this form for a measured μ by allowing b to vary. Figures 5 and 6 show this for two photomultiplier tubes[39], an ITT FW130 and an EMI6256. The FW130 is quite well fitted by the curve for b = 0 corresponding to no

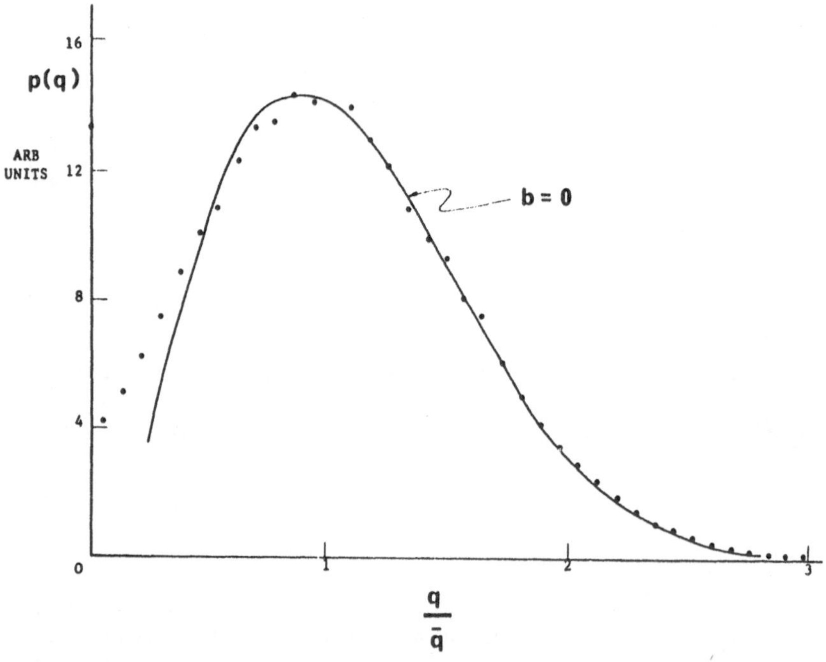

FIG 5 A measured pulse-height distribution for an ITT FW130 photomultiplier used with a gain per stage of 4.5. The continuous curve is the theoretical prediction for b = 0.

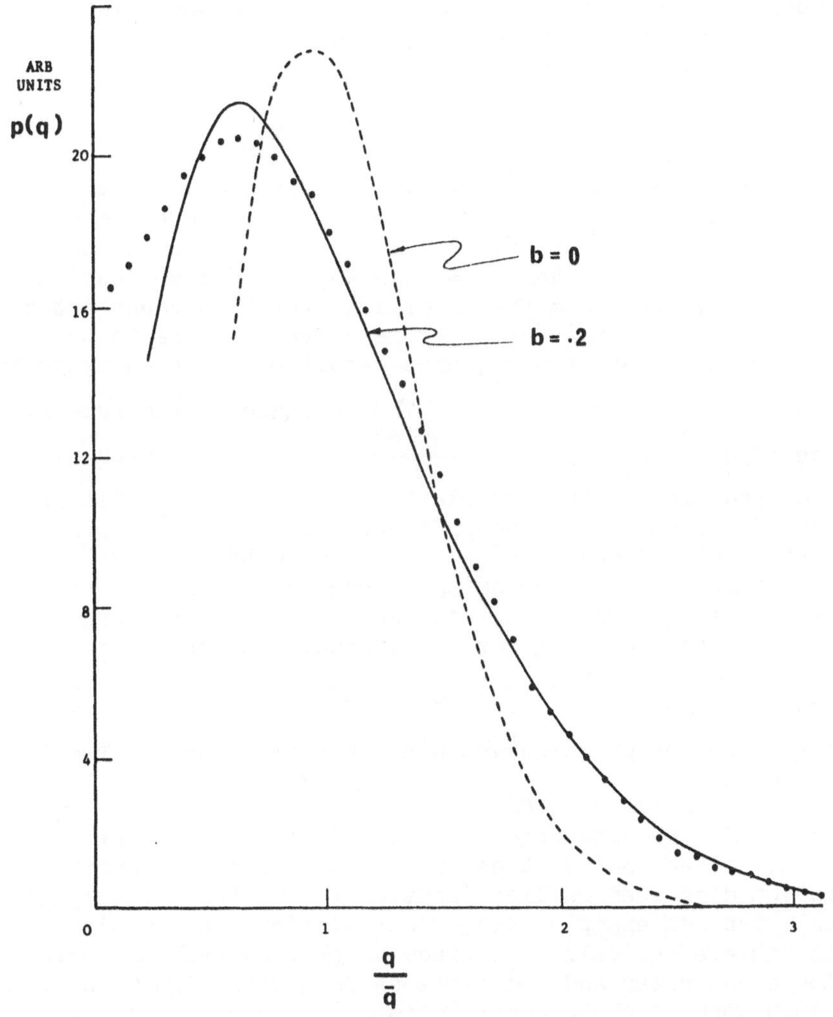

FIG 6 A measured pulse-height distribution for an EMI 6256 photomultiplier used with a gain per stage of 6.5. The theoretical predictions for b = 0 (dashed) and b = 0.2 (continuous) are included for comparison .

excess variation in the gain per stage. Multiplication is therefore a pure Poisson process. For the 6256 on the other hand b = 0.2 gives a better fit showing excess variation in the gain per stage. In the extreme case of b = 1 the pulse-height distribution would be exponential. For b ≪ 1 the relative variance of q can be approximately given by

$$\frac{\mathrm{Var}(q)}{\bar{q}^2} \simeq \frac{b\mu + 1}{\mu - 1}. \tag{33}$$

For the FW130 with b = 0 and $\mu = 4.5$ we expect the relative variance of the correlation coefficients to be increased by 1.28 for high count rates and 1.64 for low count rates. For the 6256, on the other hand, b = 0.2 and $\mu = 6.5$ so that the expected degradation is a factor of 1.42 for high count rates and 2.01 for low count rates. The relative variance in the intensity measurement, $\langle Q \rangle$, as a result of non-standardisation has been discussed elsewhere[4]. The increase in the relative variance is expected to be $1 + \frac{\mathrm{Var}(q)}{\bar{q}^2}$ which is related also to the degradation of the correlation coefficients. This is illustrated for the 6256 tube in figure 7. For standardised pulses the factor $1 + \mathrm{Var}(q)/\bar{q}^2$ is found to be 1.0, as is obviously expected, corresponding to photon-counting operation. The b = 0, Poisson multiplication, and the b = 1, exponential, expected results are shown for comparison. The observed value of $1 + \frac{\mathrm{Var}(q)}{\bar{q}^2}$ is 1.44 which is in good agreement with the value predicted from the pulse-height distribution measurement which gave b = 0.2.

To conclude the discussion of the effects of the charge distribution: if we include these with the effects of capacitive integration discussed earlier (Section 1.3.2) then, taking the 6256 as a typical example, analog correlation takes 8 times as long to achieve equivalent accuracy as photon-counting correlation with low count rates and 2.8 times as long with high count rates. Since much work in photon-correlation spectroscopy will be concerned with low count rates this implies that the use of photon-counting techniques is very important.

1.4.2 Detector Correlations. There are various feedback mechanisms in photomultipliers which can give rise to correlated after-pulses. These can be divided into internal processes, such as optical, X-ray or positive-ion feedback from the anode region induced by the presence of large pulses, or external processes associated with the HT supply or the output circuitry, such as finite recovery time following a large pulse or pulse-

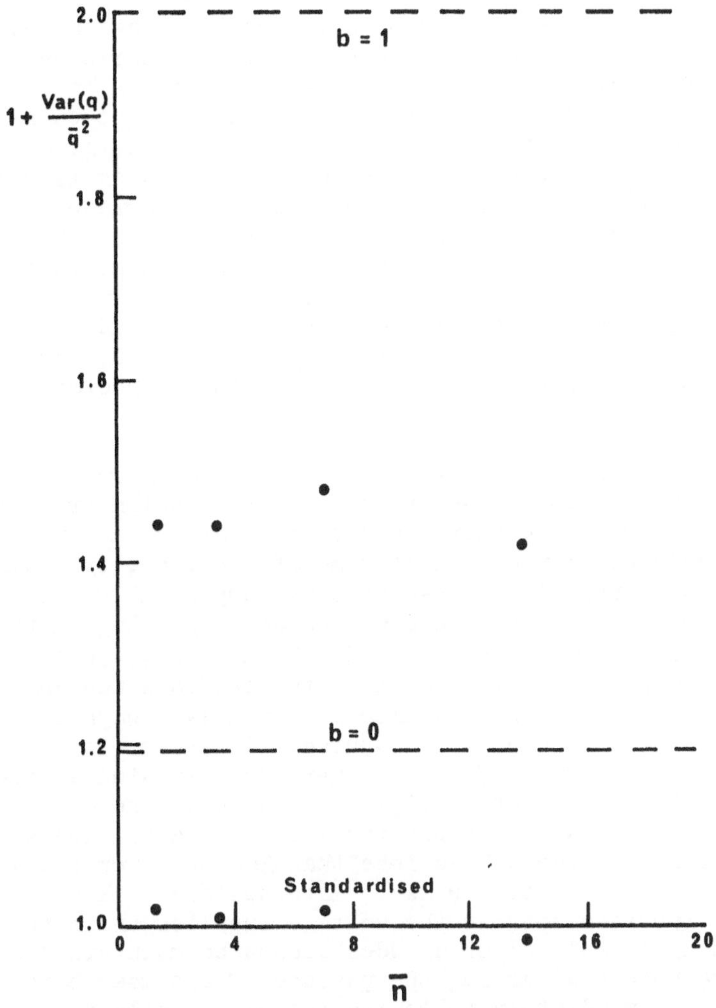

FIG 7 Measurement of the effect of the pulse height distribution, $1 + \frac{\mathrm{Var}(q)}{\bar{q}^2}$, as a function of the mean count-rate, $\bar{n}$, for the EMI6256 photomultiplier. Theoretical predictions for b = 0 and b = 1 are included.

ringing caused by output mismatching. These processes give rise to correlations, or anti-correlations, which are readily visible in correlation function measurements. Feedback effects are dependent on the amount of charge reaching the anode in a pulse. With some tubes the statistical distribution of pulse-heights due to the multiplication process may be sufficiently large to give rise to an after-pulse. More commonly, however, it requires pulses corresponding to two or more simultaneous photoemissions to give an adequately large output pulse. These can be caused by Cerenkov radiation in the glass envelope due to cosmic rays or to the presence of radioactive material such as ^{40}K in the glass. For the first type of after-pulsing every output pulse, whether it is itself an after-pulse or an original, can give rise to an after-pulse. In the second case the after-pulsing is related only to that fraction of the original pulses which corresponds to the non-statistically-large output pulses. If the feedback probability increases sufficiently the first feedback mechanism can actually give rise to oscillations whereas the second cannot. The distinction between the two mechanisms has to be made for each operating EHT voltage since increasing this greatly would probably lead to the first type of feedback with any photomultiplier tube.

Let us now consider the feedback processes in more detail. Internal feedback by light flash or soft X-ray emission from the anode region is effectively instantaneous so that the after pulse will be delayed by the transit time of the tube, typically 30-40 ns. In any particular system this may appear as a single, double-charge, pulse or as two separate pulses depending on the inherent resolution. Positive-ion feedback, however, typically takes 0.5-1 μs, depending on the mass of the feedback ion and the EHT voltage. Thus the after-pulse will nearly always be processed as a separate, correlated, pulse. An example of this kind of feedback is shown in figure 8 where positive-ion after-pulsing in an RCA PF1011 photomultiplier is shown. The clearly visible peak at 0.5μs is due to positive-ion feedback. The slight evidence of a rise towards delays less than 300 ns is probably due to the shape of the output pulse. External feedback mechanism can arise from imperfections in the voltage supplies and output circuitry. When the multiplier dynodes cannot be maintained at their operating potential during the passage of a pulse there is a reduced gain immediately following the pulse while the dynode potentials recover. This results in anticorrelations in which the photodetection probability distribution, $P(n,T)$, becomes narrower than Poisson as has been shown elsewhere[4]. If the output impedance, which should ideally be a short-circuit, gives rise to ringing then this is evinced as a damped, sinusoidal, correlation-function. An example of ringing, obtained with an EMI 9558, is shown in figure 9. The large positive tail up to $\tau = 250$ ns corresponds to the pulse shape. Both it and the characteristic ringing period of 150 ns could be integrated out using a detector dead-time of ~ 250 ns, in agreement with earlier work[4].

Having outlined some of the predominant feedback mechanisms in photomultiplier tubes let us now examine the effect these have on tube performance in the measurement of correlation functions.

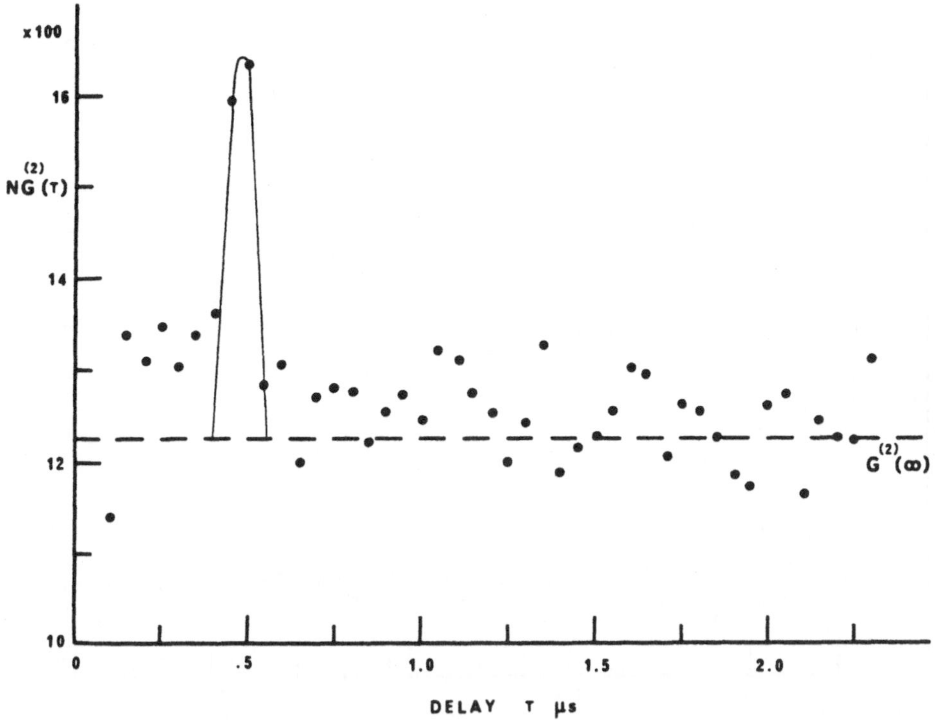

FIG 8 A measured correlation function for coherent illumination of of a RCA PF1011 photomultiplier. The sample time was 50 ns and the mean count rate $\bar{n}$ = 0.00035. The theoretical normalisation, $G^{(2)}(\infty)$, is shown as a dashed line; the continuous line indicates positive-ion after-pulsing at τ = 0.5 μs.

Let us assume that output pulses have a probability $\alpha(\tau)$ of being followed by a correlated after-pulse at delay τ. In any sample, therefore, the observed mean count rate will be given by

$$\overline{n_{obs}} = \bar{n}_o (1 + \alpha) \tag{34}$$

where $\bar{n}_o$ is the true count rate per sample and

$$\alpha = \int_o^\infty \alpha(\tau) d\tau \tag{35}$$

is the total correlated after-pulse probability. The uncorrelated part of the correlation function at delay τ can then be written as

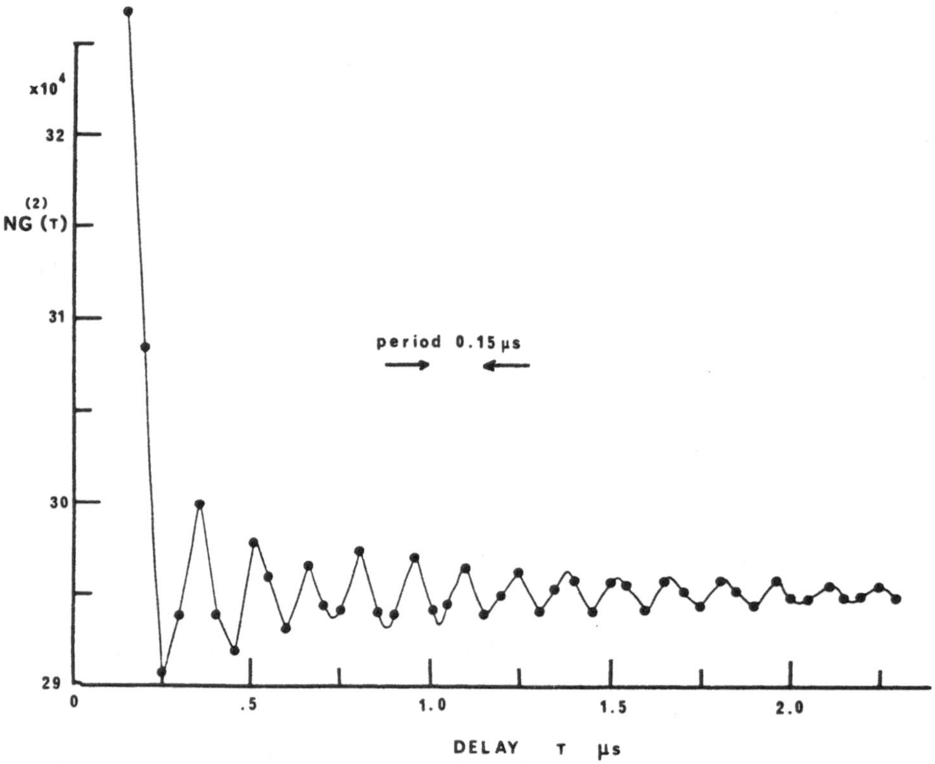

FIG 9 A measured correlation function for coherent illumination of an EMI9558 photomultiplier showing the effects of pulse shape and ringing due to mismatching on the tube output.

$$G^{(2)}_{obs}(\tau)\Big|_{uncorrelated} = \bar{n}_o^2 (1+\alpha)^2 . \qquad (36)$$

In addition there will be a contribution due to the after-pulsing which has a probability $\alpha(\tau)$. The actual contribution will depend on the after-pulsing mechanism. If all pulses can give rise to after-pulses the correlated term will be

$$G^{(2)}_{obs}(\tau)\Big|_{correlated} = \alpha(\tau)\,\overline{n_o}\,(1+\alpha) \qquad (37)$$

while for correlated after-pulses corresponding to non-statistically large original pulses this term will be

$$G^{(2)}_{obs}(\tau)\Big|_{correlated} = \alpha(\tau)\,\bar{n}_o. \tag{38}$$

The normalised correlation coefficients in the two cases are therefore

$$g^{(2)}_{obs}(\tau) = 1 + \frac{\alpha(\tau)}{\bar{n}_o\ (1+\alpha\ )} \tag{39}$$

and

$$G^{(2)}_{obs}(\tau) = 1 + \frac{\alpha(\tau)}{\bar{n}_o\ (1+\alpha)^2} \tag{40}$$

respectively. In each case the effect is most pronounced for low count rates ($\bar{n}_o \ll 1$). Therefore the presence of correlated after-pulses of this type can be very serious where weak sources and short sample times are encountered.

The second type of dependence (eqn 40) can be checked experimentally using a double pulse generator triggered by an input Poisson pulse train to simulate 100% after-pulse probability following each original pulse. For a true input count rate of $\bar{n}_o = 0.0011067$ the expected background normalisation should be 489 for 10^8 samples; the observed value was 482. The expected height of the correlation peak, chosen to lie at 400 ns, should have been 110670 + 489 = 111159; in fact the observed value was 109525. The difference here is due to the dead-time correction in the pulse generator of ~1.5%.

Using the analysis of equation (40) the correlated after-pulse probability, $\alpha(\tau)$, for the RCA PF1011 photomultiplier shown earlier in figure 8 can be calculated to be 0.2 after-pulses per 10^3 original pulses. This illustrates the importance of even small after-pulse probabilities where low count rates are encountered.

1.5 CONCLUSIONS

In this first part we have examined the causes and effects of the most common imperfections in laser sources and photomultiplier tube detectors. The advantages of digital, photon-counting, detection and storage over analog techniques has been demonstrated. For intensity-fluctuation spectroscopy we have seen that the only serious source imperfection is

	Normalised Factorial Moment	10^7 samples	
		Experiment	Theory
COHERENT SOURCE	$n^{(1)}$	1.0000	1.0000
	$n^{(2)}$	1.0002	1.0000 ± 0.0009
	$n^{(3)}$	1.0001	1.0000 ± 0.0017
	$n^{(4)}$	1.0006	1.0000 ± 0.0030
	$n^{(5)}$	1.003	1.000 ± 0.012
	$n^{(6)}$	0.994	1.000 ± 0.036

		10^5 samples		9×10^6 samples	
		Experiment	Theory	Experiment	Theory
INCOHERENT SOURCE	$n^{(1)}$	1.000 ± 0.036	1.000 ± 0.038	1.000	1.000
	$n^{(2)}$	2.00 ± 0.15	2.00 ± 0.16	2.004	2.000 ± 0.017
	$n^{(3)}$	6.05 ± 0.85	6.00 ± 0.85	6.05	6.00 ± 0.09
	$n^{(4)}$	24.7 ± 5.9	24.0 ± 5.7	24.7	24.0 ± 0.6
	$n^{(5)}$	130 ± 45	120 ± 49	130	120 ± 5
	$n^{(6)}$	850 ± 461	720 ± 510	850	720 ± 54

Table 1 Normalised factorial moments for coherent and incoherent light

instability on a time scale comparable with the scattering process under study. Consideration of detector imperfections has shown correlated after-pulses to be important where low count-rates are encountered. For heterodyne spectroscopy, in addition to the rapid source fluctuations mentioned above, the effects of long term drift may be appreciable. In addition any effects of correlated after-pulses will be exaggerated by the reduced intercept for the time-dependent part of the correlation function.

However, when both detector and source are ideal one should observe the correct probability distributions and correlation functions for both coherent and incoherent sources. Table 1 gives the example of the normalised factorial moments of the photon-counting distributions which should be given by

$$n^{(r)} = 1 \quad \text{for coherent light}$$

and

$$n^{(r)} = r! \quad \text{for incoherent light.}$$

The agreement between experiment and theory is within the experimental errors to the 6th moment. With such a combination of source and detector used for a scattering experiment one can be confident that the properties of the train of photodetections accurately reflect those of the scattering process.

2 CORRELATOR DESIGN

2.1 Introduction

In the first part we discussed the extraction of data from a given scattering experiment. This data takes the form of a rate-modulated Poisson-random train of photodetection pulses. Information about the intensity fluctuations of the scattered light and hence about the scattering process is contained in the temporal (or frequency) dependence of this data, provided that the laser source and photomultiplier detector are themselves free from correlations. The different methods by which the desired spectral information can be extracted from this data have been compared already elsewhere(5). Parallel-channel time-domain processing has been shown often to have inherent advantages in flexibility over the other techniques. I shall therefore restrict my treatment to a discussion of the different approaches adopted to the extraction of the spectral information in the time domain.

The first basic design criterion for such a system is that it should be able to operate with small delay times (T) since the maximum frequency that can be measured with such a system will be of the order 1/2T. Secondly, it should be a parallel-channel system measuring several delays simultaneously since a scanned, single-channel, system is obviously wasteful of information. Thirdly, any time lost performing signal-processing should be minimised so that no incident photodetection pulses are wasted. Fourthly, and obviously, the system requires to be as simple as possible, consistent with the previous criteria, in order to reduce its cost.

It is instructive to trace the development of temporal data-analysis techniques from single-channel delayed-coincidence measurements through single-stop time-to-amplitude-conversion (TAC) techniques and multi-stop systems to the autocorrelator. Accordingly we shall consider single start-stop systems, such as delayed coincidence and TAC techniques, in section 2.2. The extension of the methods to multi-stop techniques, in order to avoid distortion, is discussed in section 2.3. Section 2.4 then deals with the full many-bit autocorrelator while section 2.5 describes the application of clipping to such a correlator in order to increase its signal utilisation efficiency. In section 2.6 we shall consider the advantages of input signal derandomisation in terms of removing distortion due to time jitter and delay mismatch in the correlator and outline its implementation in a photon-counting correlator. Finally in section 2.7 we shall consider other modes of operation of such a photon-correlator using the same basic components and control signals.

2.2 Single-Stop Techniques

The temporal information that we require to extract from the photodetections is the correlation function

$$g^{(2)}(\tau) = \frac{\langle n(0)\, n(\tau)\rangle}{\bar{n}^2} \tag{41}$$

or, the related quantity, the conditional probability that a photon at time 0 will be followed by another at time τ , $p_c(0,\tau)$.

The simplest measurement of this conditional probability is to use a delayed coincidence method with samples, of duration sufficiently short for there to be a negligible chance of detecting two pulses per sample time, separated by the delay τ . This technique was proposed by Rebka and Pound(8) and has been used successfully in the measurement of correlation functions with short decay times(9),(10). Provided that there is only a negligible probability of detecting more than one photon in the, arbitrarily-short, sample time, T, this method yields an undistorted form for the conditional probability. Nanosecond decay times can readily be measured with this technique using a variable delay line. However the technique is a single-channel method, that is to say that information from all values of delay other than the selected τ is ignored. Where low count rates are involved this effect can lead to greatly increased experimental duration.

While it would be possible to use a bank of such delayed coincidence units to overcome this short-coming the more usual approach is to use the measurement of the single interval statistics for the separation of the photodetections. This is usually performed with a time-to-amplitude converter (TAC) with multichannel pulse-height-analyser storage. This technique, though basically a single stop method, operates effectively in parallel channels since the actual time interval between pairs of pulses is measured. This method, well known in nuclear physics, has the advantage of measuring delays down to 1 ns readily but has the attendant disadvantage of distortion due to photon pile-up. This results from the fact that the probability of detecting a photon at a large delay is reduced by the probability that a photon was already detected in the intervening time. The observed conditional probability, $P_{obs}(0,iT)$, for detection of a photon at time zero and at a delay iT is then given by

$$P_{obs}(0,iT) = P_c(0,iT)\left[1 - \sum_{j=1}^{i} P(jT)\right] \tag{42}$$

(following Coates[11]), where $P_c(0,iT)$ is the true conditional probability and $P(jT)$ is the probability of detecting a pulse in the j'th delay interval. If we take the simple case of a constant intensity so that the photon statistics are Poisson then

$$\sum_{j=1}^{i} P(jT) = 1 - e^{-\bar{m}Ti} \tag{43}$$

where $\exp(-\bar{m}Ti)$ is the probability of detecting no pulses in the time interval iT and $\bar{m}$ is the count-rate per second. Hence

$$P_{obs}(0,iT) = P_c\ (0,iT)\ e^{-\bar{m}Ti}. \tag{44}$$

Thus the observed conditional probability distribution is exponential whereas the true distribution is flat. If we have a 100 channel instrument then the last channel will be distorted by 1% when $\bar{m}T$, the count-rate per sample time, is 10^{-4}. This distortion is therefore a considerable problem and is aggravated when the signal intensity fluctuates.

Two methods have been suggested to overcome this distortion, both suffering from the fact that, as with dead-time corrections, one requires to know the true form of the conditional probability to assess the remaining distortion. The first method[11] is to correct the data by expanding the true probability, $P_c(0,iT)$, as a power series in $P_{obs}(0,iT)$,

$$P_c(0,iT) = P_{obs}(0,iT) + \alpha \left[P_{obs}(0,iT) \right]^2 + \beta \left[P_{obs}(0,iT) \right]^3 + \ldots\ldots \qquad (45)$$

Provided $P_{obs}(0,iT) \gg [P_{obs}(0,iT)]^2$ the distortion in this correction due to imperfect knowledge of α, which depends on the spectral and statistical properties of the light source will not be very great. Alternatively,one may use an inhibit technique[12] so that one takes only data in which one stop pulse is recorded in the delay range of interest. As the count-rate increases, therefore, the efficiency of this method is reduced. The distortion involved in this technique is very similar to that involved in the correction; choice between the two is basically determined by the users preference.

2.3 Multi-Stop Techniques

There are two useful techniques which effectively remove the distortion inherent in the single-stop method by increasing the number of stop pulses recorded. Suppose one starts a recording system and measures the time of arrival of the next m pulses having delays of $\tau_1, \tau_2 \ldots \tau_m$. Then the conditional probability distribution, $P_c(0,iT)$, can be measured from all the differences $\tau_i - \tau_j$ ($i \neq j = 1$ to m). This technique will only distort, as in eqn (42), when more than m photodetections occur in the delay range of interest rather than 1 photodetection as previously. Suppose we take a single-stop system, as proposed by Chopra and Mandel[13], for photon-correlation then a mean count rate, $\bar{m}T$, of 0.017 in a 100 channel system gives a distortion in the last channel of $e^{-1.7} = 0.2$ whereas with a six-stop system the distortion is only 0.01. Thus this scheme is considerably less distorted than the single-stop technique. However to avoid distortion this is still a low count-rate system with a maximum count-rate of typically about 0.3 counts per coherence time. In addition, where intensity-fluctuations are encountered, giving an increased probability of n exceeding 6 during the full delay range, the mean count-rate has to be still further reduced. The count-rate taken above gives 4% distortion with the Bose-Einstein distribution of photoncounts obtained in the short-sample-time limit with Gaussian light. Non-Gaussian statistics will require still lower count rates to avoid distortion.

A second technique, which can be described as triggered multichannel scaling, was proposed by Chen and Polonsky-Ostrowsky[14]. In this technique the multichannel scaler is triggered when the total number of photodetections, n, during a sample time, T, exceeds a preselected clip-level, k. (This clipping technique will be discussed in more detail later). Following this start pulse the multichannel scaler is stepped each sample time, T, and the incoming photodetections recorded. Since all pulses following the start pulse are recorded this system is free from distortion as the count-rate is increased. Since it uses clipping it can only be directly applied to light having Gaussian statistics though techniques are available, as will be discussed in part 3, for handling non-Gaussian light. This technique suffers from the disadvantage that it only looks at correlations between the start pulse and the subsequent stop pulses and ignores the information contained in the time intervals between stop pulses. This is not a serious difficulty with Gaussian light leading to a reduction in efficiency of typically a factor of 10[15]. However with non-Gaussian statistics the efficiency is very much more reduced.

The obvious extension is to construct a multi-start, multi-stop system which, in fact, would be a full correlator. It should be made clear that each of the schemes outlined above, together with the many-bit and clipped correlators described below, have their own particular merits and disadvantages. The choice of which system is suited to a particular experiment depends on the conditions involved.

2.4 Many-Bit Correlation

Since full correlation offers freedom from most of the problems encountered in single-stop and multi-stop systems described above, we wish, therefore, to construct correlation functions of the form

$$g^{(2)}(\tau) = \frac{G^{(2)}(\tau)}{\bar{n}^2} = \frac{<n(0,T)n(\tau,T)>}{\bar{n}^2}$$

where the input data from the sampling scheme takes the form $n(t,T)$, denoting n photodetections in a sample of duration T centred at time t. In figure 10 an ideal, many-bit correlator to perform this operation is shown. Data is collected for the sample time T in the data sampler; at the end of each sample its contents are then multiplied by the delayed version of the signal, $n(rT,T)$, contained in the r'th delay channel; the product $n(0,T)\ n(rT,T)$ is then added to the r'th storage element. By repetitive sampling the correlation coefficients are then averaged. A discussion of such a correlation system must, therefore, consider the four basic elements:- (1) the data sampler, (2) the delay elements, (3) the multipliers, (4) the storage elements - and the influence each part has on the performance of the system.

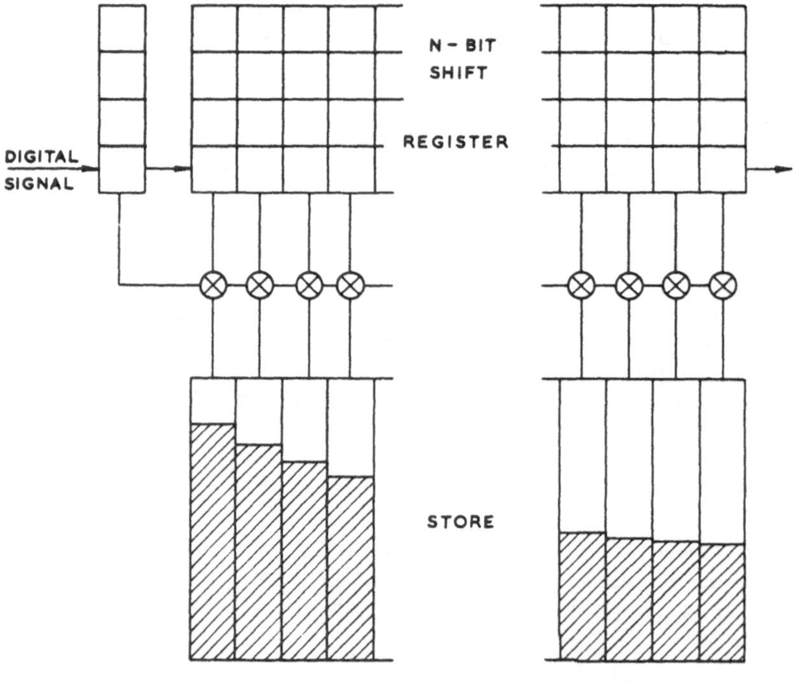

FIG 10 A block-diagram of a many-bit correlator.

The limiting components in a many-bit correlator system are generally the multiplier and storage sections. Suppose that it takes some time T_s to add a number, such as a multiplication product, to the store. Then, obviously, if the sample time T is less than T_s some of the input data will not be able to be used, even if all multiplications for each delay can be performed simultaneously. In practice multiplication of m bit numbers can be performed by a serial-parallel manipulation taking m separate multiplications and additions of separation T_s each. But such a multiplier, sufficiently complex for simultaneous multiplication of all delay channels, is expensive. Thus the multiplication is performed sequentially, channel by channel, leading to a total processing time per sample a factor of M (where M is the number of channels) larger than the time taken for one channel. The total time needed to perform the complete evaluation of the M correlation coefficients from each sample of duration T is T_p where T_p is given by

$$T_p = MmT_s \; . \tag{46}$$

One can define a signal-utilisation efficiency, β, as

$$\beta = \frac{T}{T_p} = \frac{T}{MmT_s} \tag{47}$$

which describes the performance of such a processor. Such a sampling scheme, applied to photon-correlation, is shown in Figure 11. The input signal is sampled with a clock period T given by the sample time clock (STC). After a delay of T_p, during which the processing occurs, the second sample is taken and so on.

Where the signal-utilisation efficiency falls below 100%, ie $T_p > T$, sequential processing is required. That is to say that one collects the data n(t,T) for a batch of M + 1 consecutive samples, determined by the sample-time clock (period T), to fill the delay-storage elements and the non-delayed element as shown in figure 11. One then stops data-collection and performs the requisite processing. At the end of the processing time, T_p, a further batch of M samples of delayed signal and one of undelayed signal are taken and the process repeated. Thus, from the point of view of signal-utilisation, we see that many-bit correlation will result, for the shorter sample times, in reduced data-acquisition. The effect of this inefficiency will be discussed later.

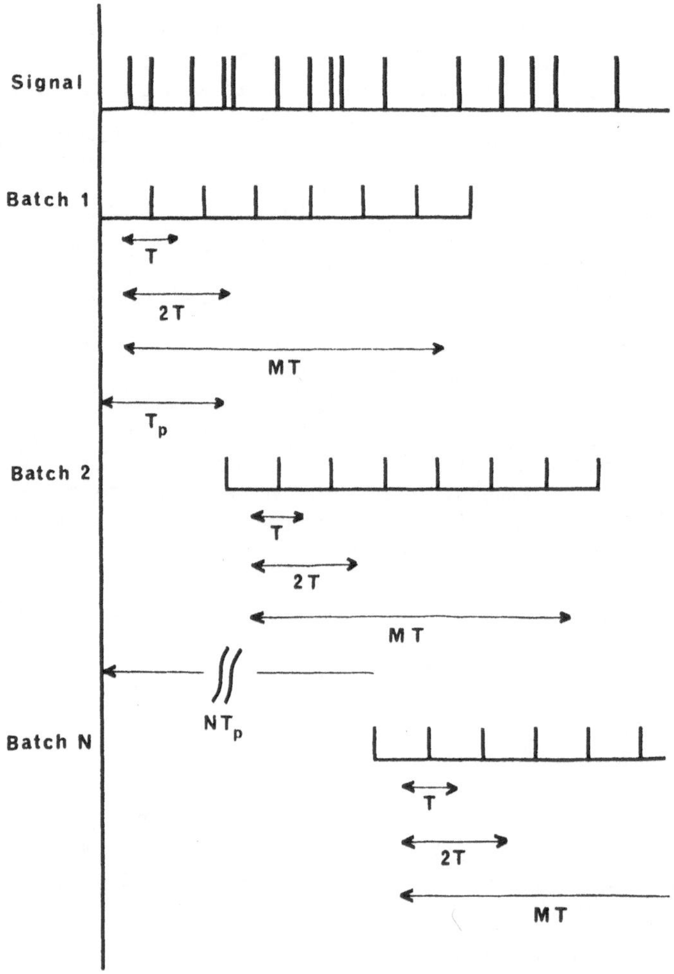

FIG 11 Schematic diagram of the photon-counting sequential-sampling scheme for a correlator of M channels with sample time T and processing time T_p.

However, to counteract this deficiency we must bear in mind that such an instrument is indeed a full correlator calculating the exact correlation function within the dynamic range of the instrument. When used for photon-counting operation, provided that the number of photodetections during a sample time T do not exceed $2^m - 1$, it gives the exact correlation function. Even where this number is exceeded, so that one has to use more than one photodetection per correlator bit, the accuracy is not reduced too drastically. There are therefore both advantages and disadvantages in the practical operation of such a many-bit correlator. It is obviously important therefore to quantify these effects.

2.4.1 Sequential Processing and Signal-Utilisation Efficiency. Let us first consider in more detail the sequential operation that will result from processing times long compared with the sample time. Obviously the important question to answer is to what extent sequential processing reduces the accuracy with which a spectral linewidth parameter can be determined. It is no longer adequate to assume that there are no correlations from channel to channel since it is these very correlations that we desire to measure. Though one might expect consecutive samples to be independent at low count-rates, where the photon noise dominates, at high count rates this will no longer be true and the error would be expected to be related to the coherence properties of the source. A detailed theoretical analysis of the accuracy obtained from the use of photon-correlation spectroscopy to determine a single linewidth parameter for Gaussian-Lorentzian light has been given by Jakeman, Pike and Swain(16) (referrred to as JPS71 hereafter). They analyse the errors involved with the sampling scheme shown in figure 11. The unbiassed estimator for the observed unnormalised correlation function is given by

$$\hat{G}^{(2)}(\tau) = \frac{1}{N} \sum_{i=1}^{N} n(iT)n(iT + \tau) , \tag{48}$$

where N is the number of samples (ie products for each delay τ). τ is in fact a negative quantity, to be consistent with figure 11, however, for stochastic stationary processes negative and positive delays are equivalent and the correlation function is independent of iT. As shown in JPS71 the variance of the estimator $G^{(2)}(\tau)$ can be written as

$$\mathrm{Var}\,(\hat{G}^{(2)}(\tau)) = \frac{1}{N}\,\mathrm{Var}\,(n(0)n(\tau)) + \frac{2\bar{n}^4}{N}\, f(\tau,T_p,T,\tau_c) , \tag{49}$$

where the two terms are given explicitly in JPS71 eqns (16) and (A3) respectively. In the limit that

$$\gamma = T/\tau_c \ll 1, \quad \tau \gg T_p \text{ and } \tau \gg \tau_c \tag{50}$$

the first term reduces to

$$\mathrm{Var}\,(n(0)n(\tau)) = \bar{n}^{4}\left(3 + \frac{4}{\bar{n}} + \frac{1}{\bar{n}^{2}}\right) \tag{51}$$

and, after many samples ($N \to \infty$), the second term becomes

$$f(\tau,T_p,T,\tau_c) = \frac{5e^{-\frac{4T_p}{\tau_c}} + 4e^{-\frac{2T_p}{\tau_c}}}{1 - e^{-\frac{4T_p}{\tau_c}}}$$

$$+ \frac{1}{\gamma}\left(1 - \frac{1 - e^{-2\gamma}}{2\gamma}\right) + \frac{1}{\bar{n}}\,. \tag{52}$$

The relative variance of the correlation function can be evaluated under these conditions. The full analysis of JPS71 assumes that $T_p = T$ and then includes a consideration of the effects of biassing on the error in the measurement of the linewidth. However a reasonable comparison of the relative errors obtained with different sequential processing times can be made by considering the relative variance of the correlation coefficients as given above. Let us therefore consider a 100 channel correlator with sample time $T = \tau_c/20$ ($\gamma = 0.05$). The delay value τ is taken to be the last channel (ie $\tau = 100\ T$). For selected signal-utilisation efficiencies β ($= T/T_p$) of 1, 0.1 and 0.01 the relative variance of the last channel can be written as

$$\frac{\mathrm{Var}(\hat{G}^{(2)}(\tau))}{\langle \hat{G}^{(2)}(\tau)\rangle^{2}} = \frac{1}{N_{\beta=1}}\left(90.1 + \frac{6}{\bar{n}} + \frac{1}{\bar{n}^{2}}\right),$$

$$= \frac{1}{N_{\beta=0.1}}\left(11.91 + \frac{6}{\bar{n}} + \frac{1}{\bar{n}^{2}}\right), \text{ or}$$

$$= \frac{1}{N_{\beta=0.01}}\left(4.93 + \frac{6}{\bar{n}} + \frac{1}{\bar{n}^{2}}\right) \tag{53}$$

respectively where

$$N_{\beta=1} = 10\ N_{\beta=0.1} = 100\ N_{\beta=0.01}. \tag{54}$$

From these results one can obtain the real time required to achieve the same accuracy with the different signal-utilisation efficiencies since this is inversely proportional to the relative variance. The ratio of these times to the time required in a full real-time processor ($\beta = 1$) is shown in figure 12 as a function of the count rate per coherence time of the source, r. Below $r = 1$ one is in the photon-noise limit so that with a sequential-processing system of efficiency β one requires to take a factor $1/\beta$ longer to achieve the accuracy of a real-time processor. This is indeed the conclusion to be expected where each sample is independent. In the large count-rate limit ($r > 100$) the processing time required is only a factor $\gamma/\beta = T_p/\tau_c$ longer for sequential processing than for real-time processing provided $T_p > \tau_c$. Thus,

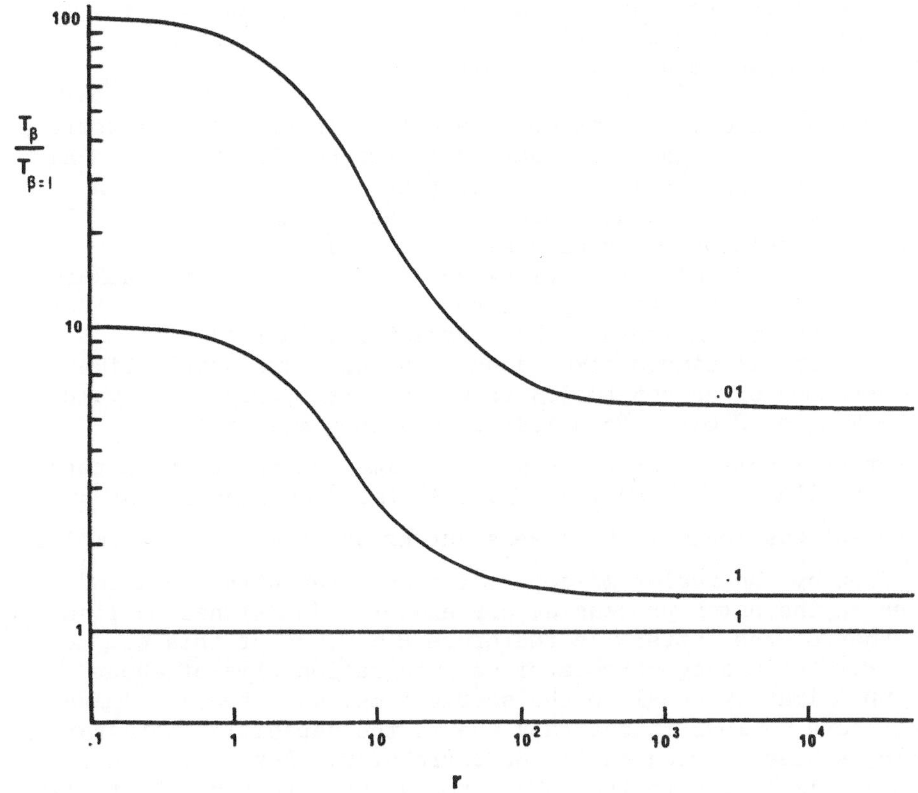

FIG 12 A comparison of the relative time taken for batch-processing correlators of signal-utilisation efficiency β, to achieve the same accuracy as a real-time correlator, $\beta = 1$. The ratio, $T_\beta/T_{\beta=1}$, of the total elapsed times is shown as a function of the count-rate per coherence time, r.

again as expected, sequential-processing only requires to take the same number of independent samples as real-time processing. In this limit an independent sample is obtained every coherence time (τ_c) of the source. The loss in accuracy of a sequential-processor compared with a real-time processor is therefore much smaller in this situation than where low count-rates are involved. Similarly for continuous analog input signals we would deduce that the effect of sequential processing is to increase the time to achieve the desired accuracy by a factor of T_p/τ_c, which could be typically about 5. Thus real-time processing is not so important in this situation as it is for photon-noise-dominated signals. In the majority of photon-correlation spectroscopy experiments a count rate of r = 10 counts per coherence time is reckoned to be exceptionally large; in practice rates down to r = 0.01 are frequently encountered. For this type of measurement it is essential to have an instrument which gives real-time processing in the frequency range of interest. This can be demonstrated by examples taken from the fields of diffusion coefficient measurement and anemometry.

2.4.2 Diffusion Coefficient Measurement. In order to illustrate the effect of the signal-utilisation efficiency light from an Ar^+ ion laser operating at 488 nm was scattered from the Brownian motion of the macromolecule haemocyanin at an angle of 130°. Two different correlators were used simultaneously to measure the intensity-fluctuation correlation-function. The Hewlett-Packard 3721A correlator is a batch-processor for sample times less than 330 μs whereas the Malvern K7023 correlator is a real-time processor for all sample times down to 50 ns. The sample-time on both instruments was set to 3.3 μs which corresponded to a value for $\gamma = T/\tau_c$ of 0.033. Both instruments then measured the correlation function for the same real time determined by a total samples setting of 8 X 1024 on the HP3721A. The duration of the measurement was found to be 4 secs during which 1.2×10^6 samples were taken by the real-time correlator compared with 8.2×10^3 samples on the batch-processing correlator. The signal-utilisation efficiency of the latter was therefore $\beta \approx 0.68\%$ at this sample time. For the analog correlator an integration time of about 3μs, approximately equal to the sample time, was chosen. There is therefore no loss in accuracy due to the capacitive integration sampling scheme adopted on the HP correlator. Any reduction in accuracy will be due to the batch processing involved. Initially a mean count rate of $\bar{n} = 0.6$ was used, equivalent to r = 18 counts per coherence time of the source. Measured correlation functions for real-time and batch-processing are shown in figure 13(a) and 13(b) respectively. The tail at the small delay end of the batch-processing result is caused by the integration time τ_{RC} being

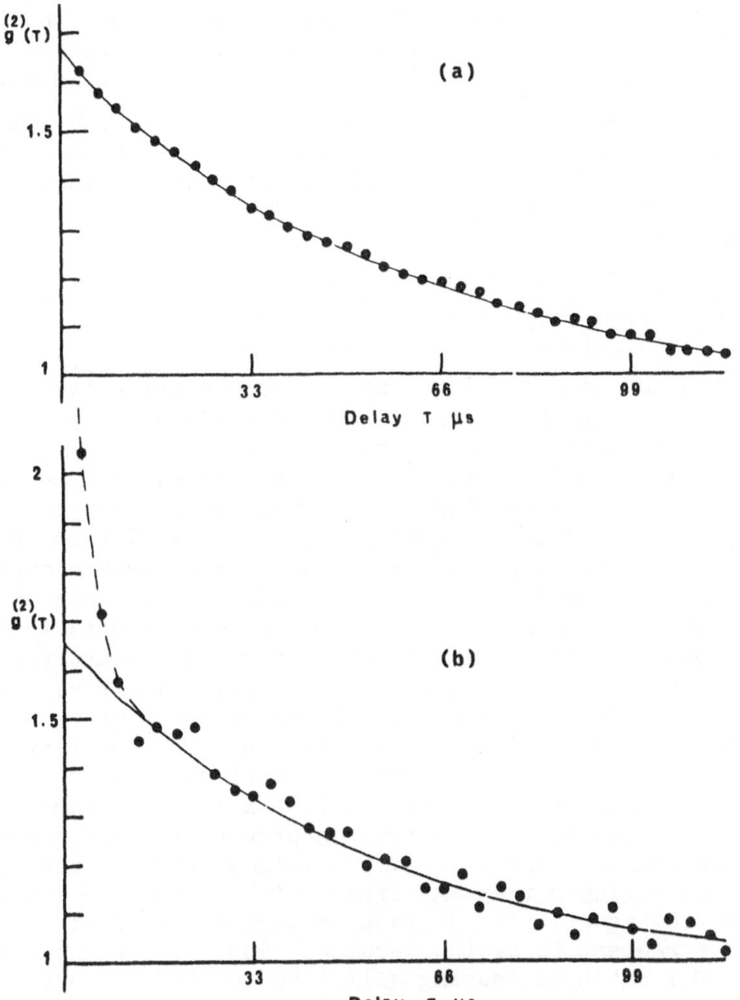

FIG 13 A comparison of real-time and batch-processing correlators in the measurement of the diffusion coefficient of haemocyanin. 1.2×10^6 samples were taken with the real-time correlator (a) in the time that 8.2×10^3 samples were taken with the batch-processing system (b). An RC integration time, τ_{RC}, of ~ 3μs was used with the batch-processing system which is responsible for the dotted tail shown at short delays. The mean count rate was 0.6 counts per sample time.

close to the sample time. In fact the observed correlation function contains both time constants which have to be unfolded to obtain the correct answer. From figure 13(a) the optical field coherence time τ_c is found to be 99 μs. The continuous line is the single-exponential fit to the data. The same-time-constant exponential is shown as a continuous line in the batch-processed result in figure 13(b); the dashed line indicates the integration time constant. The relative variance of the correlation function

$$\frac{\mathrm{Var}\ (\hat{G}^{(2)}(\tau))}{<\hat{G}^{(2)}(\tau)>^2} = \mathrm{Var}\ (\hat{g}^{(2)}(\tau)) \qquad (55)$$

was calculated for longer delay times, giving a percentage error of 1% and 3.2% in the two cases. Thus with this count rate one requires to take 10 times the elapsed real-time in order to achieve the same accuracy with the batch processor as with the real-time processor. Replacing the relevant parameters in equation (53) the predicted errors are 0.9% and 4.7% respectively. The result for the batch-processor is complicated by the capacitive integration on the input so that the agreement is as close as could be expected. The mean count rate was next reduced to $\bar{n} = 0.1$ ($r = 3$) without any other changes being made. The measured correlation functions are shown in figures 14(a) and 14(b). The real-time correlator again gives the same optical coherence time of 99 μs as shown by the exponential fit to the data. However the results on the batch-processing correlator are now so wildly varying that a computer fit was impossible. The continuous line in figure 14(b) shows an exponential of the expected coherence time for comparison. As before the relative variance of the long-delay-time coefficients was calculated giving percentage errors of 1.3% and 17% respectively. Calculation from equation (53) gives predicted values of 1.4% and 14% so that agreement is again adequate. Thus with a mean count rate of $\bar{n} = 0.1$ batch-processing will take 100 times longer than real-time processing to achieve the same accuracy. As $\bar{n}$ is reduced further the difference will tend asymptotically to the reciprocal of the signal-utilisation efficiency β, ie 150. In order to avoid the tail at short delay times a shorter integration time should be used which will have the effect of reducing the apparent value of $\bar{n}$ in the batch-processing mode and hence reducing the accuracy still further.

2.4.3 Anemometry Measurements. A second field in which photon-correlation spectroscopy is proving increasingly useful is anemometry. For a typical sinusoidal correlation-function measurement, taken in a supersonic wind tunnel by scattering light off particles moving through a set of fringes[17], the total count rate at the detector, using about 80 mW of 488 μm radiation from

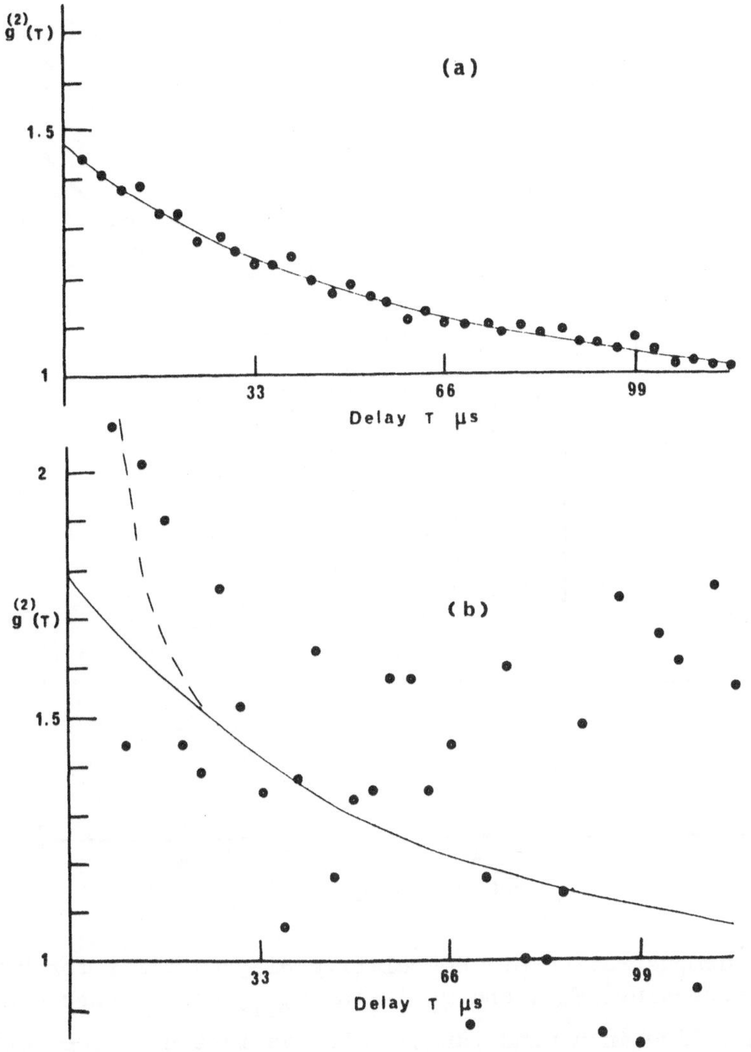

Fig 14 As for figure 13 with a mean count rate of 0.1 counts per sample time.

an Ar^+ ion laser, was about 10^4 counts per second. For the measured velocity of approximately 500 ms^{-1} the Doppler frequency was 4 MHz. Thus the count-rate was 0.002 counts per Doppler cycle. Though the counts arrived in bursts of a few photons rather than uniformly distributed the count rate is always small so that the advantage of a real-time processor with such low count rates can be expected to be considerable. Suppose we assume that this detected count rate per second is independent of velocity, and hence Doppler frequency, and that batch-

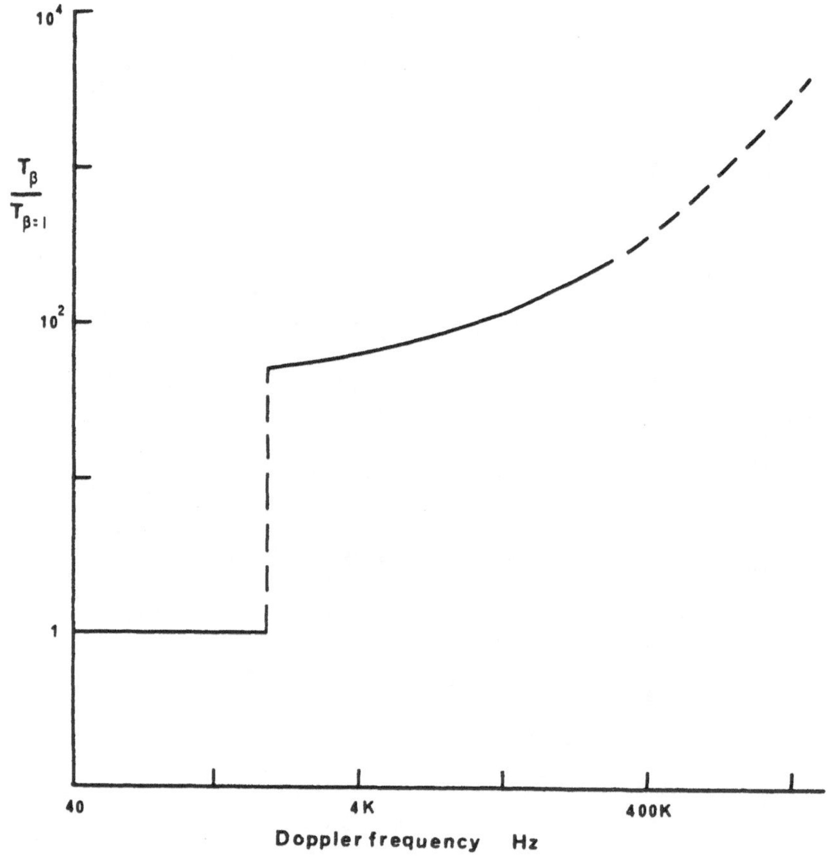

FIG 15 A demonstration of the relative elapsed times for batch-processing, T_β , and real-time $T_{\beta=1}$, correlators measuring air flow in a wind tunnel. The variation of the ratio $T_\beta / T_{\beta = 1}$ as a function of the Doppler shift is shown assuming that the scattering cross-section is independent of wind speed. A velocity of 125 m s^{-1} gives an approximate Doppler frequency of 1 MHz in the geometry used.

processing is performed in a 100 channel correlator which takes 150 μs to calculate the 100 individual coefficients, then the relative experimental duration required to achieve the same accuracy as a real-time processor is shown in figure 15 as a function of the Doppler frequency. For Doppler shifts in excess of 40 KHz, corresponding approximately to a velocity of 5m s^{-1}, the count rate per Doppler cycle is considerably less than unity so that the full disadvantage of batch-processing, T/T_p, is apparent. It is assumed that sample times down to 50 ns are possible with such an instrument. Actually Doppler frequencies in excess of ~ 400 KHz cannot be measured with most analog correlators as shown by a dashed line in figure 15. At about 1 kHz Doppler shift, corresponding to an approximate velocity of .12 m s^{-1}, the instrument becomes a real-time processor so that there is no longer any difference between the two systems. However at Doppler frequencies of 4 MHz, the advantage of real-time processing is a factor of approximately 4000 shorter experimental duration compared with the batch processor. Similar calculations can be performed for any correlator system leading to similar conclusions and illustrating the great importance of real-time correlators for use in anemometry.

2.5 One-Bit Correlation

As we have shown many-bit correlators suffer from the disadavantage of having poor signal-utilisation efficiency at sample times short compared with their processing time. If it were possible to use a one-bit shift register to contain the delayed signal then multiplication would be merely a single operation which could readily be performed simultaneously in each channel. The basic requirements in using such a system are:

1 The analytic form of the correlation function must be known so that spectral parameters can be extracted,

2 The attendent loss in accuracy resulting from one-bit quantisation is small.

Since in practice finite detectors and sample times are used the first requirement demands that any correction factor involved must be independent of delay time. In practice, therefore, it is desirable that the observed function closely resembles the full correlation function. It has been shown that the technique of single-clipping[18] satisfies the first of these requirements.

If we refer back to figure 10, then in single-clipping the delayed version of the data $n(t,T)$ is converted to a one-bit quantised form by introducing the clip counts $n_k(t)$ defined by

$$n_k(t) = 0 \quad \text{if} \quad n(t,T) \leq k$$
$$n_k(t) = 1 \quad \text{if} \quad n(t,T) > k$$

where k is the clip level. Since the delay register only contains one-bit quantised data the multiplication process has been simplified as required so that simultaneous parallel processing of all delays is possible. Construction of such a single-clipping correlator can be carried out using TTL integrated circuitry. In this case a basic time resolution of 50 ns can be achieved in which all internal processes are controlled by a 20 MHz crystal clock to give accurate timing.

For light having Gaussian statistics the normalised single-clipped correlation function has the form described by Jakeman[5], namely

$$g_k^{(2)}(\tau) = 1 + C\left|g^{(1)}(\tau)\right|^2 , \tag{56}$$

where $g^{(1)}(\tau)$, the first-order correlation function, is the Fourier transform of the optical spectrum and C is a constant depending on the operating conditions such as sample time, clip level, count-rate and detector area[19-21]. Since the full correlation function for Gaussian light is given by the Siegert relation[22],

$$g^{(2)}(\tau) = 1 + \left|g^{(1)}(\tau)\right|^2, \tag{57}$$

the distortion introduced by clipping is not important, so that the first requirement outlined above is satisfied provided that the signal statistics are Gaussian.

The loss in accuracy entailed in single-clipped correlation can be shown by measurement and computer simulation[23] to be very slight provided clipping is performed near the mean count-rate, $k = \bar{n}$. The dependence of accuracy on the operating conditions will be discussed in more detail in part 3. The second requirement of an acceptable correlator system therefore appears also to be satisfied by single-clipping. This statement must, however, be qualified by indicating that the error analysis referred to above applies only to a single spectral component.

2.5.1 Single-Clipping. The clipping condition for the input data given above can be realised using the scheme shown in figure 16. The incoming photodetection pulses, i(t), are counted in a scaler which has a preset overflow (clip level) k such that if k is exceeded the overflow pulse will set a flip-flop, acting

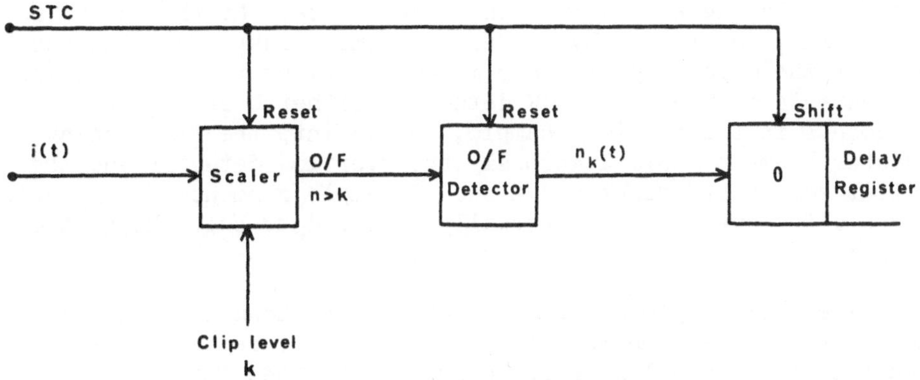

FIG 16 A block diagram of the clipping gate.

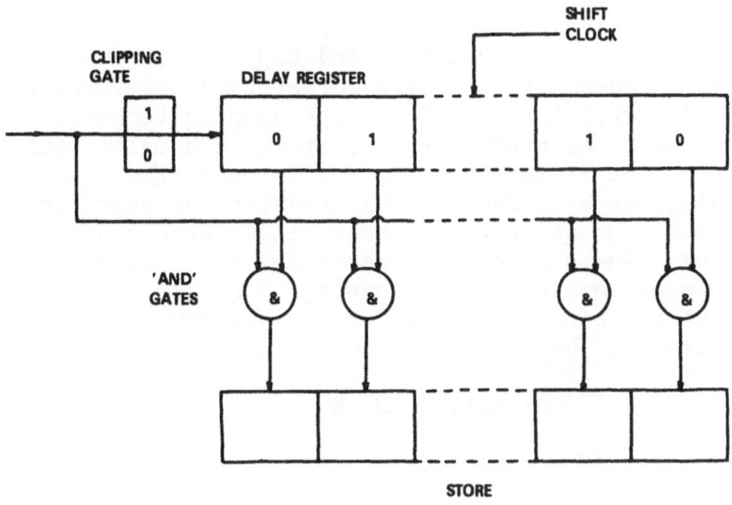

FIG 17 A block diagram of the complete single-clipping correlator.

as an overflow detector, into the 'one' state. At the end of each sample time, controlled by the sample time clock (STC), the contents of the delay register are moved one place to the right and the contents of the over-flow detector, either 0 or 1 depending on the counts in the previous sample, loaded into the first delay register element. Simultaneously the overflow detector and the initial scaler are reset and a second sample commenced. The value of the clip-level can be externally selected to suit the particular experiment.

The full single-clipped correlator is shown in figure 17. The delay register consists of a one-bit shift register carrying a series of 0's and 1's representing the clipped form of the delayed signal. Each sample-time-clock pulse shifts the contents of this delay register one place to the right and enters the new clipped counts from the clip gate as described above. To form the single-clipped correlation function this delayed, clipped, data requires to be multiplied by the unclipped input. Since the unclipped signal is available in serial form, ie the photodetections $i(t)$, the multiplication by '1' or '0' can readily be performed in an 'and' gate as shown. The simplicity of this multiplication process enables all channels to be computed simultaneously provided that each storage channel has independent access. Each storage channel therefore consists of an independent scaler able to count the incident photodetection rate.

Many laboratories now use this type of single-clipped correlator for intensity-fluctuation and Doppler spectroscopy. Instead of using independent scalers for storage one version of this type of correlator(15)(24) uses a dumping technique to enable the use of an available core store in a multichannel scaler. In practice it is found that division of the output from the 'and' gates by a factor of 16 reduces the overflow rate to the 1 per 1.3 ms, required for compatability with the store, in most situations. Where high count-rates and short sample-times give a larger overflow rate than one per 1.3 ms use of 'complementary' clipping(24), in which the clipping scheme is now the complement of before, ie

$$n_k(t) = 1 \quad \text{if} \quad n(t_1T) \leqslant k$$

$$n_k(t) = 0 \quad \text{if} \quad n(t_1T) > k\,.$$

Used with $k = 0$ this reduces the data rate to the store below that for the normal clipping at the same level whilst giving the same accuracy. With the optimum clip level $k = \bar{n}$ however there is no appreciable difference in the two schemes.

2.5.2 Double Clipping. The technique of double-clipping has been proposed[25] as a useful technique in photon-correlation spectroscopy. This is simply achieved by feeding the 'and' gates from the output of the clipping gate (fig 17). The technique can be particularly helpful where weakly scattering samples containing large quantities of spurious strong scatterers (eg dust) are encountered. The total number of counts added to the store arising from scattering off the dust particles is then restricted to 1 per sample where the actual count-rate may be much higher. Obviously the observed correlation function will be distorted by the presence of dust, but considerably less than if single-clipping or indeed full correlation were used.

The disadvantage of double-clipping lies in the complicated form of the observed correlation function. This has been given by Jakeman and Pike[18], for clipping at k = 0 only, as

$$g_{00}^{(2)}(\tau) = \left[1 + \frac{1-\bar{n}}{1+\bar{n}} \left|g^{(1)}(\tau)\right|^2\right] \Big/ \left[1 - \left(\frac{\bar{n}}{1+\bar{n}}\right)^2 \left|g^{(1)}(\tau)\right|^2\right]. \quad (58)$$

This relation only applies in the limit of short-sample times and small detector areas. Unlike the case for single-clipping the correction for finite apertures and sample times is delay-dependent so that analysis cannot really be performed with any accuracy. However it is found experimentally that when clipping at $k = \bar{n}$ the double-clipped correlation function can be reasonably well represented by a single spectral component of the same shape as for the single-clipped correlation function together with a flat background. Empirically therefore the technique has some use for obtaining approximate results with poor samples.

2.6 Input Signal Derandomisation

The correlator system described so far suffers from the disadvantage that pulses arriving close to the sample-time-clock pulse may be lost or treated differently in the various correlator channels due to fractional variation in component speed. This results in distortion of the correlation function[25]. The usually adopted remedy is to gate the input signal and impose a dead time during which all the processing, such as shifting and resetting, is performed. This is wasteful of time and will result in reduced signal-utilisation efficiency reaching a value of 50% at the maximum speed where the dead time is equal to the sample time. In addition, where short sample times (~ 50 ns) are adopted distortion is often caused by mismatch between the delays in the two correlator channels. Both these problems can be circumvented by derandomisation of the input signal, illustrated in figure 18. The input photodetection pulses, i(t), are of standard shape with a width equal to half the detector dead time which may be typically 50 ns for an ITT FW130 for example. These are fed into an edge-triggered flip-flop together with the system's basic 20 MHz clock. The incident photodetections set the flip-flop into its '1' state; the next positive edge (say) of the clock then resets the flip-flop and gives a standard (25 ns) width output pulse synchronised with the clock. This output is derandomised so that the output consists of pulses, synchronised to the basic clock, present only when a photodetection was recorded during the previous clock period. This technique allows processing in the interval between clock pulses since there will never be any change in the derandomised version of the input signal, $i_d(t)$, during this time. Thus no time will be lost in processing with such a system and all incident pulses are used. It is a true real-time processor for all sample times up to the basic clock rate of 50 ns. In addition, since the timing is all referred to the basic clock, one merely requires to match the delays to the correct clock pulse (ie $\pm$ 25 ns) to achieve the undistorted correlation function.

2.7 Other Modes of Operation

The basic configuration of shift register, gates and store, shown in figure 19, can be used in modes other than correlation with which this section has been chiefly concerned. Both photon-statistics measurements and signal-recovery, in addition to auto- and cross-correlation, are possible by rearrangement of the basic control functions. The four different inputs in this basic circuit are: A, gates input, B, register input, C, register shift-clock and D, register reset. Table 2 contains the information

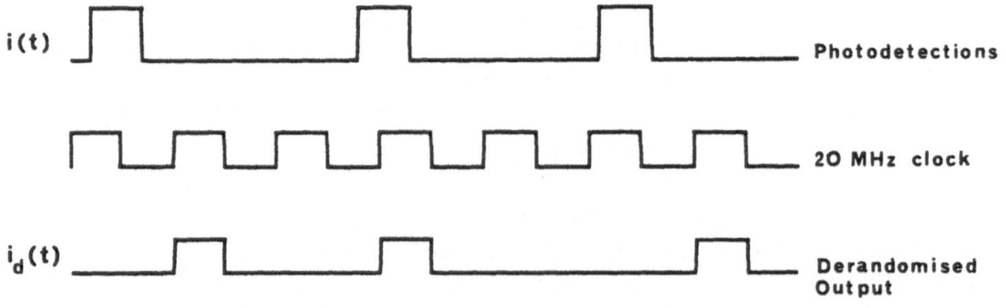

FIG 18 A wave-form diagram showing the operation of the derandomiser circuit.

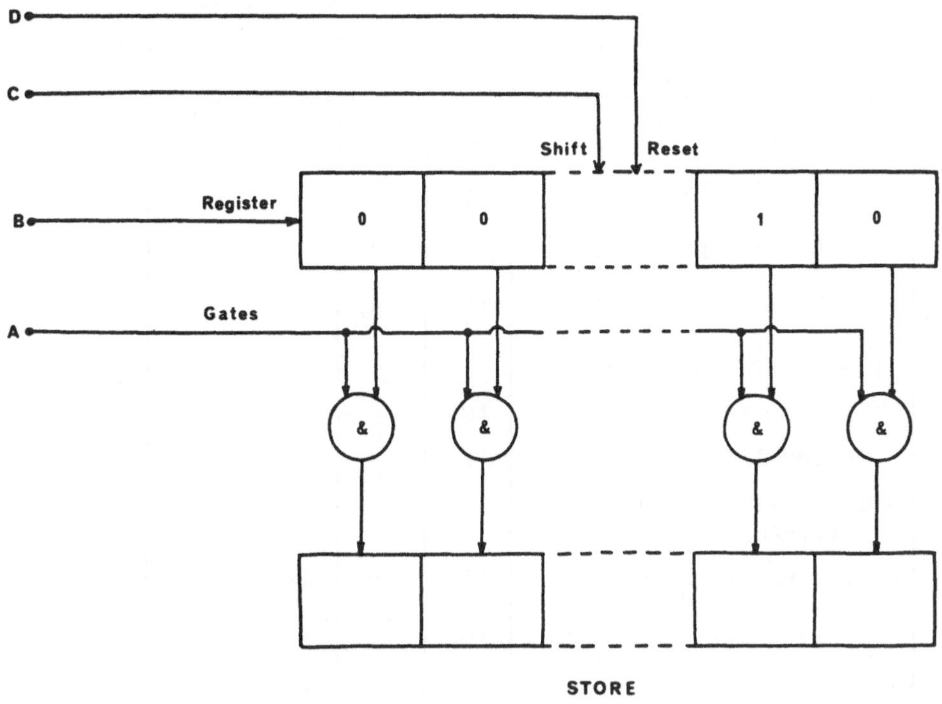

FIG 19 A block diagram of the correlator delay-register, multiplier gates and store showing the four different inputs required for operation either as a correlator, a probability analyser or a multiscaling signal-averager.

as to what control functions are required on these four inputs in order to perform the desired analysis functions.

2.7.1 Correlation. Single- and double-clipped correlation have already been described in detail. The required functions are: I_1, the derandomised input, $i_d(t)$, to channel 1, I_1^k, the output of the clip gate, $n_k(k)$, in channel 1, and STC, the sample time clock.

For correlation the register shift clock (C) is controlled by the STC and the register input (B) by I_1^k. For single-clipped correlation the gates input (A) is controlled by function I_1 giving the product $I_1(0) \times I_1^k(\tau)$ which is then stored. For double-clipped correlation the gates input (A) is controlled by function I_1^k giving the product $I_I^k(0) \times I_1^k(\tau)$. Cross-correlation can

MODE	A	B	C	D
CORRELATION				
Single-clipped	I_1	I_1^k	STC	-
Double-clipped	I_1^k	I_1^k	STC	-
Cross	I_2	I_1^k	STC	-
PROBABILITY ANALYSIS				
Differential	STC	'0'	I_1	STC
Integral	STC	'1'	I_1	STC
MULTISCALING SIGNAL-AVERAGING				
External trigger	I_1	'0'	STC	Ext trigger
Internal trigger	I_1	'0'	STC	Int trigger

Table 2 The control operations required to perform the different analysis modes.

be performed by using a second input function, I_2, on the correlator which, like I_1, has been derandomised but not clipped. I_2 is then fed to the gates input(A) to form the cross-correlation product $I_2(0) \times I_1^k(\tau)$.

2.7.2 Probability Analysis. In situations where the statistics of the optical field are required rather than the spectral properties it is useful to measure the photodetection statistics P(n,T), the probability of n photodetections in a sample of duration T. This is related to the probability distribution of the incident intensity, P(I), by the relation[2]

$$P(n,T) = \int_0^\infty \frac{(\alpha I)^n}{n!} \exp^{-\alpha I} P(I)\, dI \qquad (59)$$

where α is the detector quantum efficiency. The integral form of this density function, which can be written as

$$P(n \leqslant k,T) = \sum_{0}^{k} P(n,T) \quad , \qquad (60)$$

is also sometimes of interest.

These quantities can be measured, as shown in table 2, using the functions I_1, the derandomised input to channel 1, and STC, the sample time clock. In addition to the three inputs used for correlation the shift-register-reset input is now required. On applying a pulse to this input each delay register element is reset to '0' except the first which is set to '1'. Thus at the beginning of a sample the sample time clock (STC) resets the delay register (D). As input pulses, I_1, arrive they shift the information in the delay register to the right so that the '1' from the first channel will be in channel n + 1 when n pulses have been collected. For measurement of the probability density function, P(n,T), the register input (B) is held at '0' so that this value is transferred into the delay register on each right shift operation. At the end of a sample, therefore, in which n photodetections occurred the delay register will contain '0's except in element n + 1 which will contain a '1'. At this point the end-of-sample pulse (STC) is fed to the gates input (A) so that a '1' is added to location n + 1 of the store. By repeating this process many times the desired probability distribution will be built up. The integral distribution is constructed in the same way except that the register input (B) is held at '1'. At the end of the sample in which n photodetections occurred the delay register elements 1 to n + 1 will then contain '1' with the remainder at '0'.

2.7.3 Multiscaling Signal-Averaging. Multiscaling provides a useful technique for extracting weak repetitive signals from noise. It is equivalent to cross-correlation of the input signal with a synchronised delta-function. It is useful to have both external triggering, in which a trigger pulse derived from outside equipment starts the multiscaling process, and internal triggering, in which the multiscaling process is started in synchronism with the sample time clock at the same time giving a trigger output for peripheral equipment. The initial trigger pulse applied to D performs the register operation setting '0' into all channels except for the first which is set to '1'. This '1' controls the storage channel into which the input signal, I_1, is counted via the gates input (A). At the end of each sample the delay register contents are shifted to the right by STC pulses applied to C so that the '1' is sequentially scanned down the register. The register input (B) is maintained at '0' so that no further '1' 's are loaded into the delay register. When the '1' has swept down the complete delay register the system can then be readied for the next trigger pulse.

We have seen therefore that by simple rerouteing of the same basic signals and control functions all the required operations can be performed with the same basic components.

3 USE OF CORRELATORS

3.1 Introduction

In this part I will discuss the effect of the various operating conditions on the measurement of spectral parameters. With the exception of the dependence on clip-level these results are perfectly general and will apply to all types of correlator provided one makes allowance for any property, such as batch-processing, which might affect the result. Though detailed analytic and theoretical treatment is only given here for a particularly simple case, most of the general conclusions will also apply in more complicated situations.

We shall discuss the application of single-clipped correlation to intensity-fluctuation spectroscopy in section 3.2 and to heterodyne spectroscopy in section 3.3. In each case the fitting procedure required to analyse the data, distortion of the correlation function by various spurious processes and the attainable statistical accuracy will be discussed. Finally, in section 3.4, techniques for the analysis of non-Gaussian signals will be outlined for one-delay-bit correlators. The loss in accuracy of such systems compared with a full correlator will also be discussed.

3.2 Application Of Correlation To Intensity Fluctuation Spectroscopy

In order to discuss the application of photon-correlation to intensity-fluctuation spectroscopy (IFS) let us continue to confine our attention to the simple case of a single Lorentzian spectral component with Gaussian field statistics. Conclusions drawn from this discussion can then be applied, at least in principle, to more complicated spectra.

3.2.1 Fitting Procedure. The unnormalised single-clipped correlation function can be written as

$$G_k^{(2)}(rT) = B(1 + C \exp(-2\Gamma\, rT)) \tag{61}$$

where T is the sample time, rT is the delay, B is a normalisation constant and C is a constant depending on both spatial and temporal integration, count-rate and clip-level[20,21]. The effect of dead-time is to introduce a delay-dependent correction to C so that this effect must be reduced as far as possible. One could, therefore, perform a least-squares fitting procedure to the data using equation (61) with variables B, C and Γ. However the normalisation constant B is given in the theory as

$$B = G_k^{(2)}(\infty) = \bar{n}\,\bar{n}_k \tag{62}$$

where $\bar{n}$ and $\bar{n}_k$ are the mean unclipped and clipped count-rates per sample time respectively. Thus the data would be better fitted using a two parameter fit for C and Γ to

$$G_k^{(2)}(rT) = \bar{n}\bar{n}_k\,(1 + C\exp(-2\,\Gamma\,rT)). \tag{63}$$

In principle C can be evaluated analytically[20,21] but in practice it is best measured in the same experiment as Γ so that the identical conditions apply. A one parameter fit is therefore unsuitable.

In real experiments one does not measure the true mean values but only estimates derived over N samples. The estimates for $\bar{n}$ and $\bar{n}_k$ are then

$$\hat{n} = \frac{1}{N}\sum_{i=o}^{N} n(iT)$$

$$\hat{n}_k = \frac{1}{N}\sum_{i=o}^{N} n_k(iT), \tag{64}$$

where the summation is performed in monitor counters. Provided the total experiment duration, NT, is many coherence time, τ_c (10^4 say), the difference between the estimator and the true mean serves merely to introduce a distortion into the normalised correlation function of order $\frac{1}{N}$[16], thus

$$\frac{\hat{G}_k^{(2)}(\tau)}{\hat{n}\,\hat{n}_k} = \frac{G_k^{(2)}(\tau)}{\bar{n}\,\bar{n}_k} + O\left(\frac{1}{N}\right) \tag{65}$$

where the hats denote estimators obtained over N samples.

A two-parameter, iterative, least-squares fitting procedure to the normalised correlation function given by

$$g_k^{(2)}(rT) = \frac{G_k^{(2)}(rT)}{\bar{n}\,\bar{n}_k} = 1 + C\exp(-2\Gamma rT) \tag{66}$$

is the preferred method of analysis. However non-iterative, and therefore faster, least-squares procedures can also be derived. Suppose we introduce a variable y_r such that

$$y_r = g_k^{(2)}(rT) - 1 = C\exp(-2\Gamma\, rT). \tag{67}$$

Then by taking logs we obtain

$$\log(y_r) = -2\Gamma\, rT + \log C \tag{68}$$

which can then be fitted as a least-squares linear fit giving

$$\hat{\Gamma}T = \frac{6\sum_{r=1}^{M} r\,\ln(y_r) + 3(M+1)\sum_{r=1}^{M}\ln(y_r)}{3M^3 + 2M^2 - 3M - 3} \tag{69}$$

where M is the total number of correlator channels. This procedure suffers from the disadvantage that for large r, so that $\Gamma rT \gg \tau_c$, y_r can become negative due to the photon-statistics. Even if it does not actually become negative the log function will distort for poor statistical accuracy.

An alternative analytic least-squares fit can be obtained by eliminating C from equation (67). This leads to a solution

$$\hat{\Gamma}T = \frac{1}{2}\left[\ln\left(\sum_{r=1}^{M} y_r^2\right) - \ln\left(\sum_{r=1}^{M} y_r\, y_{r+1}\right)\right] \tag{70}$$

for $\hat{\Gamma}T$. This form has the advantage that the logs are not taken until after averaging so that the prcoedure neither blows up nor suffers from distortion as badly as that of eqn (69) when poor data is encountered. However it can only be applied to single exponential fits.

3.2.2 Model Dependence. So far we have confined our attention to extraction of a linewidth parameter for a single exponential alone. In principle one should be able to invert correlation functions to obtain the distribution of diffusion coefficients present, for example. Unfortunately,for monotonic functions such as mixtures of exponentials,this inversion is not possible as the composite correlation function is relatively insensitive to individual terms. In practice this means that analysis has to be performed in terms of some model which is assumed to describe the scattering situation correctly. For example,long molecules having both rotational and translational diffusion components can be represented as a two component system with four parameters $(C_1, \Gamma_1, C_2, \Gamma_2)$; similarly a mixture of non-interacting particles. However,in practice,extraction of more than 4 parameters is hardly feasible. In measurements of this type therefore we can only evaluate parameters of an assumed model without much evidence as to whether the model we have chosen can strictly be applied. On the other hand,where non-monotonic correlation functions are expected, as in Doppler spectroscopy for example, the data may readily be inverted to give velocity distributions. There is,therefore, no need to assume any particular model in order to perform the analysis. However in turbulence measurements it is often convenient to determine merely mean and mean-square velocities while assuming some form for the velocity distribution.

3.2.3 Distortion of the Correlation Function. In measurements of diffusion coefficients we have assumed that the scattering process can be described in terms of a given model. However if spurious scattering from 'dust' or non-specific aggregates in the sample or from the cell walls (ie flare) occurs,or if the laser intensity fluctuates during the experiment,the model will no longer apply. Analysis of the observed data in terms of the model will therefore give unreliable results.

Suppose we consider first fluctuations in the laser intensity during the course of the experiment. As was shown in eqn 27 a drift of β during the experiment would lead to an error in the normalisation of $\alpha = \frac{\beta^2}{4}$. The variable y_r in eqn (70) now takes the value of $y_r + \alpha$ so that replacing in eqn (70) we obtain

$$\hat{\Gamma}_{\alpha}T = \frac{1}{2}\left[\ell n\left(\sum_{r=1}^{M}(y_r+\alpha)^2 - \ell n\left(\sum_{r=1}^{M}(y_r+\alpha)(y_{r+1}+\alpha)\right)\right]\right. \quad (71)$$

If we assume that $\Gamma T \ll 1$ and $\alpha \ll 1$ then this can be expanded to give

$$\hat{\Gamma}_{\alpha} \simeq \Gamma(1 - 2\alpha) \quad (72)$$

in the limit $M\Gamma T \ll 1$ and

$$\hat{\Gamma}_{\alpha} \simeq \Gamma(1-4\alpha) \quad (73)$$

in the limit $M\Gamma T \gg 1$. In practice $M\Gamma T$ takes values in the region of 2 and the apparent linewidth is then given by

$$\hat{\Gamma}_{\alpha} \simeq \Gamma(1 - 3\alpha), \quad (74)$$

ie the apparent linewidth is reduced by three times the fractional misnormalisation. This is also observed experimentally. In practice the effect of the misnormalisation can be removed by taking the value of the correlation function at some delay of the order of 8 τ_c as an estimate of the normalisation. This, of course, now introduces problems in the interpretation of the results in terms of some chosen model.

Next let us consider the effect of extraneous laser light entering the detector. Provided this light enters the detector from a different direction from the scattered light no interference will take place and the intensities add. The shape of the time-dependent part of the observed correlation is not affected but the relative height of the flat background term will be increased. If the extraneous light enters satisfying the coherence condition then interference will take place and the observed correlation function will be that for heterodyne detection.

$$g^{(2)}_{het}(\tau) = 1 + \frac{2\bar{n}_s\bar{n}_o}{(\bar{n}_s + \bar{n}_o)^2} g^{(1)}(\tau) + \left(\frac{\bar{n}_s}{\bar{n}_s + \bar{n}_o}\right)^2 \left| g^{(1)}(\tau) \right|^2, \quad (75)$$

where the first term is the heterodyne term, the second term is the desired IFS term and $\bar{n}_s$, $\bar{n}_o$ are the mean signal and local-oscillator count-rate per sample time respectively. In figure 20 the effect of force fitting the observed $g^{(2)}_{het}(\tau)$ as an IFS correlation function of single linewidth $\hat{\Gamma}$ using eqn (70) is

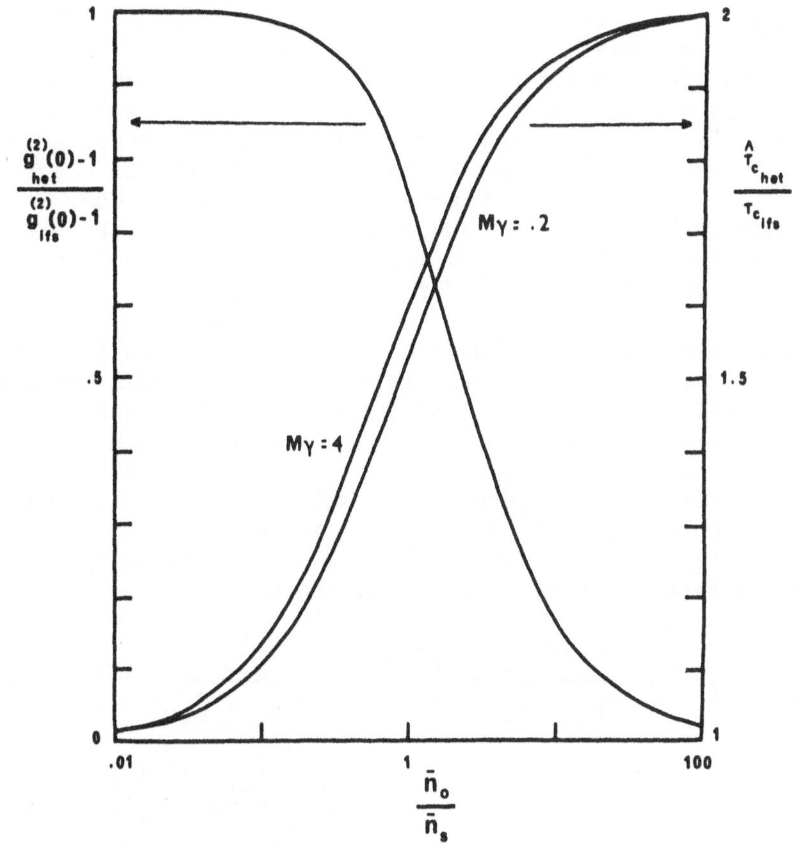

FIG 20 The dependence of the ratio of the intercepts of the correlation function, $g^{(2)}(0) - 1$, for mixed scattered and local oscillator fields (het) and for scattered light alone (IFS) on the ratio of local oscillator to scattered intensities, $\bar{n}_o/\bar{n}_s$. Also shown is the dependence of the apparent coherence time, $\tau_{c\ het}$, obtained by force-fitting the mixed fields with a single exponential. The ratio of the apparent coherence time to the true IFS one, $\tau_{c\ ifs}$, is given as a function of $\bar{n}_o/\bar{n}_s$.

investigated as function of $\bar{n}_o/\bar{n}_s$. In addition the variation of the intercept at $\tau = 0$ is evaluated from eqn (71). Supposing, for example, we have flare scattering of 10% of the signal strength ($\bar{n}_o/\bar{n}_s = 0.1$) then the intercept is reduced by only 1%

but the apparent correlation time is increased by about 12%. This makes it very difficult to demonstrate that no local oscillator is entering the detector in any given measurement. For simple situations one can investigate the dependence on angle which will probably be erratic where flare scattering is involved. However this cannot be applied in situations where one wishes to measure the angular dependence itself. When in doubt local-oscillator signal should be introduced deliberately and the two linewidths compared. Often it will be found necessary to use heterodyne spectroscopy for the measurements.

Another type of distortion mentioned above arises from the presence of spurious scatterers. In this case we shall observe self-beat terms for the sample and for the 'dust' together with a cross-term corresponding to interference between the two components. Since the time constant for 'dust' scattering is liable to be very long compared with that for the sample this cross term can be regarded as a heterodyne term with the dust scattering acting as a local oscillator. The correlation function can then be written approximately as

$$g^{(2)}_{IFS}(\tau) \simeq 1 + \frac{2\bar{n}_s \bar{n}_d}{(\bar{n}_s + \bar{n}_d)^2} g_s^{(1)}(\tau) + \left(\frac{\bar{n}_s}{\bar{n}_s + \bar{n}_d}\right)^2 \left|g_s^{(1)}(\tau)\right|^2 + \left(\frac{\bar{n}_d}{\bar{n}_s + \bar{n}_d}\right)^2 \left|g_d^{(1)}(\tau)\right|^2 \tag{76}$$

where $\left|g_d^{(1)}(\tau)\right|^2$ is the IFS term for the dust scattering. This last term can be represented by a flat addition to the theoretical background. Even if this were eliminated, by taking $G_k^{(2)}(8\tau c)$ as the background estimate, one would obtain a distortion in the apparent linewidth due to the heterodyne term which, for small fractions of spurious scattering, is of order

$$\hat{\Gamma} \simeq \Gamma \left(1 + \frac{\bar{n}_d}{\bar{n}_s}\right) \tag{77}$$

as shown in figure 20. In addition one encounters the problem of interpretation of the data in terms of a specific model because of the somewhat arbitrary normalisation chosen.

Another serious difficulty encountered with spurious scattering from dust is that the number density of the dust particles is generally so low that non-Gaussian statistics will be encountered due to concentration fluctuations in the number of scatterers[26]. Single-clipped correlation is then not applicable.

3.2.4 Accuracy. So far we have considered the effect of distortions of the observed correlation function on the spectral parameters obtained. However it is also of great importance to assess the accuracy that can be obtained in any given measurement. The parameters encountered in such a measurement will be the clip level k, count rate per coherence area per coherence time, R, the detector area, A, and the sample time, T. Let us, as before, confine our attention to the simple case of intensity-fluctuation spectroscopy measurement of a single coherence time, τ_c, for a Gaussian-Lorentzian source. We are concerned, therefore, with the accuracy with which this parameter can be determined under a given set of experimental conditions. Several authors[16,27-32] have made error analyses of this type, the most complete treatments being given by Degiorgio and Lastovka[31], Jakeman, Pike and Swain[16,30] and Kelly[32] which for simplicity will be denoted by DL71, JPS70, 71 and K71 respectively. These three treatments are not directly comparable, however, since DL71, JPS70, 71 and K71 analysed the errors obtained using three-, two-, and one-parameter fits to the data respectively. As was explained earlier the most suitable choice of fitting procedure is to use equation (76) above, namely

$$\frac{G_k^{(2)}(rT)}{\bar{n}\,\bar{n}_k} = (1 + C \exp(-2\Gamma\, rT)) = g_k^{(2)}(rT)$$

where the two variable parameters are C and Γ. This approach was adopted in JPS70, 71 who analysed the result of performing a least-squares fitting procedure on the data to this expression. They also considered the effects of weighting but found that the result was not strongly dependent on the weighting used. In fact, uniform weighting gives results little different from the optimum weighting under normal conditions. In the subsequent discussion on errors[23], therefore, we will confine our results to performing two-parameter fits using uniform weighting. Throughout the results apply to measurements taken over 10^4 coherence times, ie $NT/\tau_c = 10^4$.

3.2.4.1 Normalisation. In any real experimental situation one is analysing estimates of the correlation function obtained over N samples. The normalisation itself is also against estimates for $\langle n \rangle$ and $\langle n_k \rangle$, as described in equation (70). The larger the number of samples taken the closer these estimates come to their mean values. JPS 70, 71 have shown that the effect of using these estimates to perform the normalisation of the correlation-function estimate, $G_k^{(2)}(rT)$, is to introduce biasing terms of order $\frac{1}{N}$, thus

$$\hat{g}_k^{(2)}(rT) = 1 + C \exp(-2\,\Gamma\, rT) + O\left(\frac{1}{N}\right). \tag{78}$$

The estimate of $g_k^{(2)}(rT)$ can be derived from a single, long experiment, so that biasing is negligible, or from a series of short experiments which can each be normalised independently. Even if a series of short measurements are taken the observed data, $G_k^{(2)}(rT)$, can be normalised subsequently against the average values for n and n_k obtained over all the measurements, or normalised independently against the individual short-time estimates. An experimental and computer-simulation comparison of the accuracy attainable from the two normalisation techniques applied to the same data, referred to as subsequent (subs) and independent (ind) processing, is shown in figure 21. Agreement between computer simulation and experiment is close. The results clearly reveal the advantage obtained from using independent normalisation in spite of the bias it introduces. Experimentally-speaking, independent normalisation offers the advantage of removing some of the effects of any laser drift incurred. In addition, this type of normalisation has the property of removing some of the photon-noise from the measured function. Therefore the best experimental technique would use independent normalisation with the number of samples reduced till the bias becomes unacceptably large at, say, $N = 10^3$. Many such measurements, using one-line computer facilities, would give the greatest accuracy in the determination of Γ in a given time.

3.2.4.2 Clip Level. The accuracy of linewidth determination will also depend on the clip-level. Increasing the clip-level has the effect of increasing the magnitude of the time-dependent part of the correlation function, but also of reducing the number of clipped counts recorded and hence increasing the uncertainty in the correlation coefficients. The variance of the normalised correlation coefficients, and hence the accuracy with which Γ can be determined, will result from a combination of these effects.

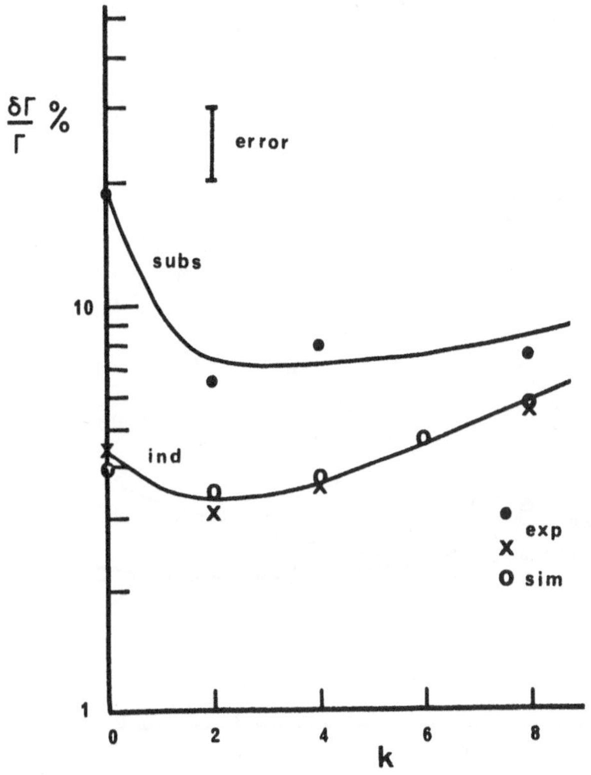

FIG 21 The dependence of the linewidth error on the normalisation technique. Single-clipping measurements (exp) and simulation (sim) are shown for $T/\tau_c = 0.1$ and $A/A_c = 0.01$.

Both subsequent (subs) and independent (ind) normalisation are used. A smooth curve is drawn through the data to aid interpretation.

Measurements and computer simulation for the dependence of accuracy on clip-level for various mean count rates is shown in figure 22. The form of the curves is such that choosing a clip level such that $k = \bar{n}$ gives the best performance. However the dependence on k is not very strong. An important feature of these results is the comparison of the computer-simulated results for full correlation with $\bar{n} = 1$ and 10, shown by a dashed line, and the minimum error achieved with single-clipped operation at $k = \bar{n}$. The errors are indistinguishable within the 20% experimental uncertainty indicating that, in the measurement of one linewidth parameter at least, no important information has been sacrified by clipping. This nullifies the advantage of many-bit correlation when applied to this simple situation.

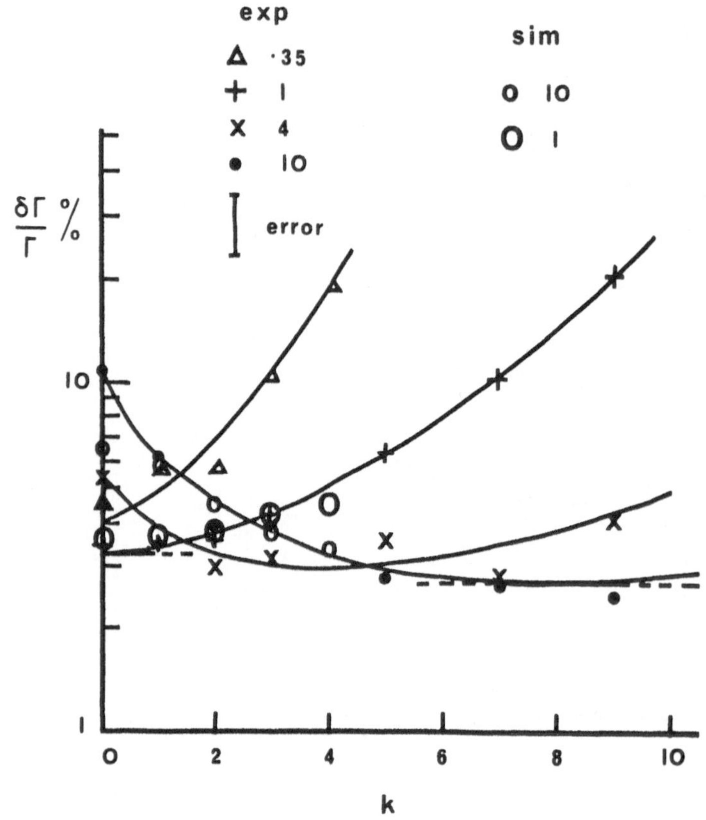

FIG 22 The dependence of linewidth error on clip level as a function of the mean count rate, $\bar{n}$. Both computer-simulated (sim) and measured (exp) data are shown. The results for simulated full correlation at $\bar{n} = 1$ and 10 are included as dashed lines.

3.2.4.3 Sample-Time, Count-Rate and Detector Area. The effect of increasing the detector area is to reduce the magnitude of the time-dependent part of the correlation function relative to the background. In addition, however, it has the effect of increasing the count-rate and thus reducing photon-noise. Reducing the source area by focussing the laser beam to a smaller waist with a shorter focal length lens will have the effect of increasing the correlation function intercept for the same count-rate. Thus good experimental practice would dictate that the count-rate per coherence area from the source should be maximised by reducing the source diameter as far as is practicable. The results of JPS71 show that the dependence of the accuracy on the count-rate per coherence time saturates at about r = 10 (figure 23). This behaviour could be surmised from the fact that at low count rates the uncertainties will be dominated by the photon-noise while at high count rates the intensity fluctuations are dominant. Thus there is no advantage to be gained by increasing r beyond this point.

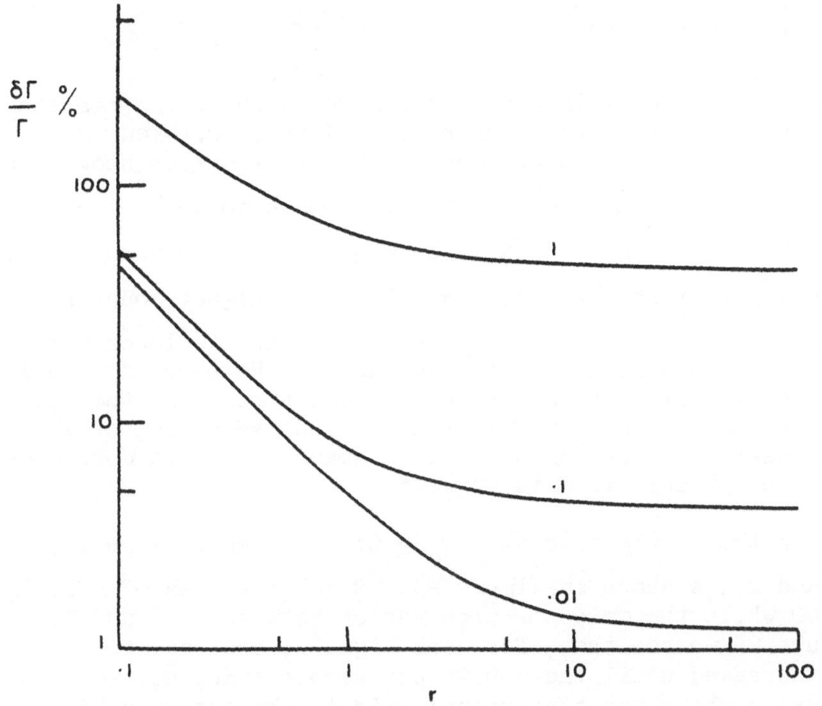

FIG 23 Linewidth error as a fraction of the counts per coherence time r (= R A/A_c) for an infinite-channel full correlator. The sample time, T, is chosen so that ΓT takes values 0.01, 0.1 and 1 as shown.

For a source of a certain scattering cross-section and linewidth one can now choose various operating settings of the sample time, T, and detector area, A, which give the optimum performance for various count-rates per coherence area, R. The error in the determination of the linewidth Γ can therefore be expressed as a function of these parameters having the form $\frac{\delta\Gamma}{\Gamma}$ $(A/A_c, T/\tau, R)$. The spatial- and temporal-integration effects enter in the form of fractions of coherence area, A_c, and time, τ_c. For convenience this error dependence on the three parameters can be expressed in terms of pairs of these parameters with the effect of the third made negligible, ie.

(1) $\frac{\delta\Gamma}{\Gamma}$ $(T/\tau_c, R)$ $\qquad A/A_c \ll 1$

(2) $\frac{\delta\Gamma}{\Gamma}$ $(A/A_c, R)$ $\qquad T/\tau_c \ll 1$

(3) $\frac{\delta\Gamma}{\Gamma}$ $(A/A_c, T/\tau_c)$ $\qquad R \ll 1.$

The first of these is shown in figure 24 where the results for a 20 channel correlator single-clipped at zero, taken from JPS 71, are compared with experiment. The four curves shown are for $R = 10, 10^2, 10^3$ and 10^4; A/A_c was chosen to be 0.01 and values of T/τ_c between .01 and .4 selected. For low count rates the optimum performance was at $T/\tau_c \simeq 0.1$. At higher count rates the result of clipping at zero moves the minimum to lower values of this ratio, in agreement with the theory. However if clipping at $k = \bar{n}$ were adopted, as it is in subsequent results, the optimum performance still occurs at $T/\tau_c \simeq 0.1$. Thus selection of T such that the observed correlation function spans about two coherence times of the optical field is optimum.

Fixing the sample time at $T/\tau_c = 0.1$ the second dependence, on A/A_c and R, is shown in figure 25. R takes values of 0.1, 1, 10 and 100 while the detector area varies between 0.03 and 3. From figure 25 we see that, for weak signals, accuracy improves as A is increased until the counts per sample time, $\bar{n}$, exceed a certain value where the statistical gain in the accuracy of each correlation coefficient is balanced by the reduction in the magnitude of the exponential term in the correlation function. The approximate positions of these minima are indicated by a line in figure 25 corresponding to count-rates per sample time of $\bar{n} = 0.05$, 0.1, 0.25 and 1.2 for R = 0.1, 1, 10 and 100

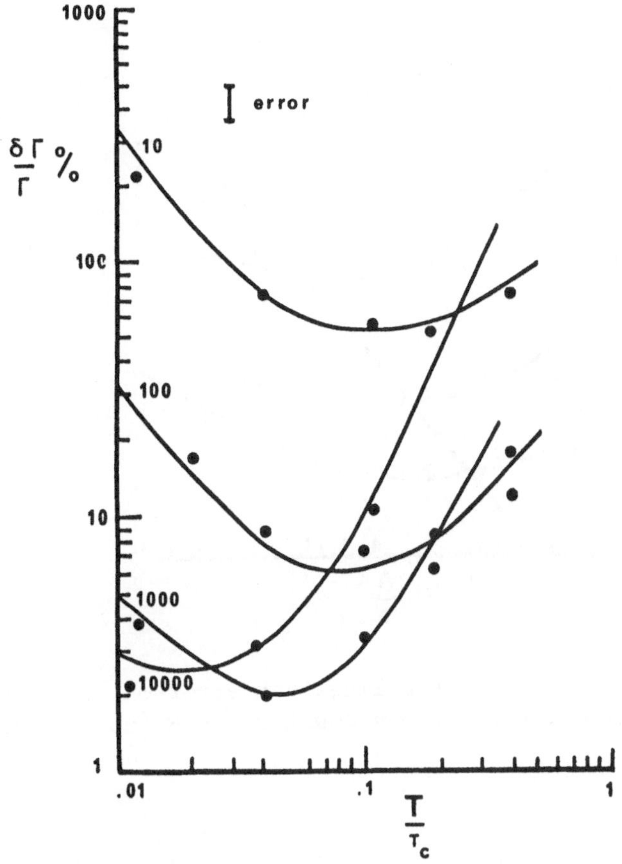

FIG 24 The dependence of the linewidth error on sample time (T/τ_c) as a function of count rate R (as labelled). Single-clipping was performed at k = 0 throughout.

respectively. In fact a reasonable compromise would be to maintain the detector area at one coherence area, ie $A/A_c = 1$.

A difference between photon-correlation spectroscopy and wave-analyser measurements should be noted at this juncture. Where both $\bar{n}$ and A/A_c are very much greater than one the error in the linewidth measurement in correlation spectroscopy is found to be proportional to $\sqrt{A}$ as shown in figure 25, in agreement with DL71. However DL71 have also shown that for a wave analyser the error in Γ reaches an asymptotic

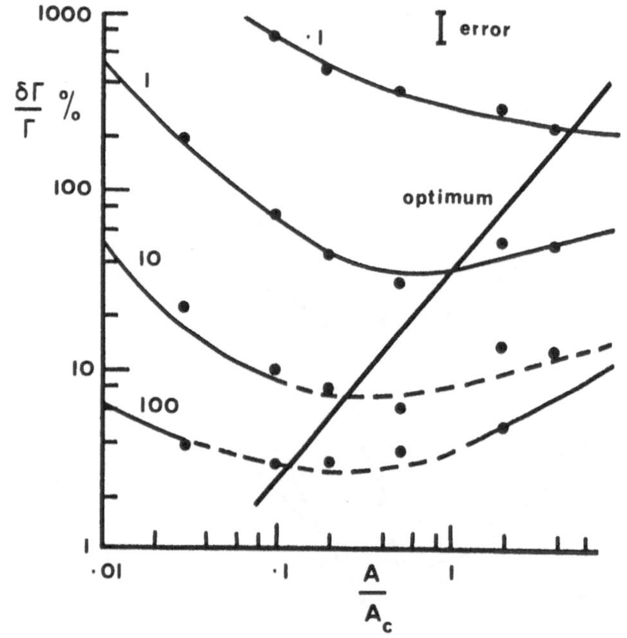

FIG 25 The dependence of the linewidth error on detector area (A/A_c) for various count-rates R (as labelled).

minimum as A increases. The difference between the two techniques lies in the different flat background terms encountered, and in the information that is discarded. In correlation the shot-noise term at $\tau = 0$, ie $< n^2 >$, is discarded and the dc term, proportional to $\bar{n}^{2}$, retained. With the spectrum analyser the dc term (f = 0) is rejected and the shot noise term ($f \to \infty$) is retained as the background. If all the information were used in each case the errors would, of course, be identical.

The dependence of the error on the detector area and the sample time $\frac{\delta\Gamma}{\Gamma}$ (A/A_c, T/τ_c) for $R \ll 1$, is shown in figure 26. Again $T/\tau \triangleq 0.1$ proves to be the optimum choice.

3.2.4.4 Operating Principles. The results of this section on the accuracy of linewidth measurements in intensity-fluctuation spectroscopy can be summarised as operating principles as follows:

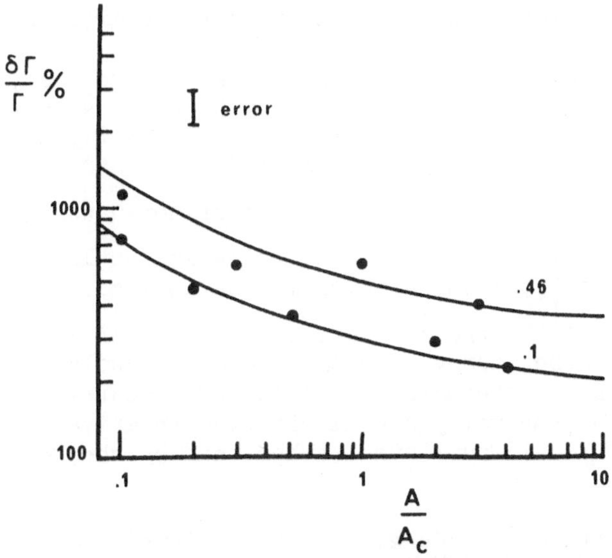

FIG 26 The dependence of the linewidth error on detector area (A/A_c) for different sample times (T/τ_c, as labelled). R was 0.1 throughout.

(1) The source area should be reduced as far as possible, consistent with having large numbers of scattering particles within the source.

(2) Many short measurements, normalised independently against their total counts, are preferable to a single long one.

(3) The clip-level should be set equal to $\bar{n}$ in which case the loss in accuracy due to clipping is insignificant.

(4) The measured correlation function should span about two optical coherence times.

(5) A detector subtending about one coherence area gives overall optimum performance.

It must be borne in mind that the above results and discussion have been applied merely to a single-component Lorentzian spectrum with Gaussian field-statistics. Where spectra containing more than one component are encountered similar operating principles will apply. The situation with non-Gaussian statistics will be discussed later.

3.3 Application of Clipped Correlation to Heterodyne Spectroscopy

Where source statistics are not known or where asymmetric spectra are encountered heterodyne, or Doppler, spectroscopy offers the advantage of giving a Doppler spectrum directly proportional to the spectrum of the incident field[19]. In addition, for weak signals ($\bar{n}_s \ll 1$, $\bar{n}_s \ll \bar{n}_c$), the effect of non-Gaussian signal statistics is dominated by the shot noise in the local oscillator so that single-clipped correlation may be applied. As has been suggested by Pike[33] the statistical accuracy of linewidth measurements can be improved using heterodyne analysis, since the information in heterodyne spectroscopy is carried by the cross-product of the noisy signal with the relatively less noisy coherent local oscillator, rather than by the square of the signal.

3.3.1 Fitting Procedure. For heterodyne spectroscopy the normalised single-clipped photon-correlation function is as given in eqn (75) above, ie

$$g^{(2)}_{het}(\tau) = 1 + \frac{2\bar{n}_s\bar{n}_o}{(\bar{n}_s + \bar{n}_o)^2}\, g^{(1)}(\tau) + \left(\frac{\bar{n}_s}{\bar{n}_s+\bar{n}_o}\right)^2 \left|g^{(1)}(\tau)\right|^2$$

From figure 20 we can estimate the amount of local oscillator required to make the heterodyne term dominate over the IFS term. For 1% distortion in the measured optical coherence time $\hat{\tau}_c$ the local oscillator should be thirty times the scattered signal. If smaller values of local oscillator are used the function will be distorted and one would need to fit to the full expression. If we generalise the correlation function to include the effects of finite detectors and sample times the clipped heterodyne correlation function can be written as

$$g^{(2)}_k(rT) = 1 + C \exp(-\Gamma\, rT) \tag{79}$$

for a Lorentzian source. Where the local oscillator is thirty times the scattered intensity the intercept at $rT = 0$ for point sources and zero sample times has a value $C = 0.06$. Typically, spatial and temporal integration will reduce this value to approximately 0.02. Since the exponential part of the correlation function is so small, fluctuations in the flat background become extremely important. Such fluctuations are often encountered in practice due to mechanical alignment variations in the local oscillator beam. From section 3.2.3 eqn 74 we see that a misnormalisation of α gives rise to approximately a 3α decrease in the apparent linewidth in IFS measurements. Since the intercept for heterodyne spectroscopy is at 0.02 instead of approximately 0.4 the normalisation requires to be a factor of 20 closer to the theoretical value. 1% distortion in the line-width therefore arises from a misnormalisation of only 0.0165%, compared with 0.33% for IFS measurements. This error in the normalisation would arise from local oscillator fluctuations of only 0.26% (eqn (27)) which, of course, are quite generally encountered. Therefore, it will be necessary to use a background normalisation obtained from the long-delay correlation coefficients ($\tau > 8\ \tau_c$), assuming that fluctuations on a time short compared with $8\tau_c$ are all due to the source under study. Once again therefore, we encounter difficulties in interpretation of data under nearly all circumstances which render heterodyne spectroscopy less attractive than it might appear to be at first sight.

3.3.2 <u>Distortion of the Correlation Function by Spurious</u> **Scattering.** In addition to the problem of normalisation outlined above we need to consider the effect of distorting features such as 'dust' or non-specific aggregates on the correlation measurement. In the discussion of IFS measurements it was indicated that scattering of this sort is generally from small numbers of scatterers so that non-Gaussian statistics are encountered due to number fluctuations. However, provided the local oscillator dominates the dust scattering, single-clipped correlation may still be applied. Thus badly contaminated samples cannot be measured at all with single-clipped correlation of the intensity fluctuations whereas they can be by heterodyne spectroscopy, provided the fluctuations of the dust scattering are slow. Provided the local oscillator dominates both dust and sample scattering the combined correlation function for heterodyne spectroscopy can be written as

$$g^{(2)}_{het}(\tau) = 1 + \frac{2\bar{n}_o}{(\bar{n}_s+\bar{n}_d+\bar{n}_o)^2}\left[\bar{n}_s g^{(1)}_s(\tau) + \bar{n}_d g^{(1)}_d(\tau)\right] \qquad (80)$$

for small apertures and sample times. As described previously, $g^{(1)}_d(\tau)$ has a coherence time which is long compared with the sample correlation function. Thus the term involving $g^{(1)}_d(\tau)$ can be treated as an additional flat background term and therefore eliminated as before, subject to the same qualifications. Thus the presence of dust will no longer distort the measured line-width for the sample provided that it has a time constant much greater than the sample. In addition one can perform clipped correlation, even though the dust-scattering statistics are not

Gaussian. Therefore heterodyne spectroscopy is very useful where the preparation of clean samples is difficult.

3.3.3 Accuracy. In addition to the advantages described above heterodyne spectroscopy also offers a gain in statistical accuracy in the ideal case where $\bar{n}_d = 0$. A comparison of the observed correlation function obtained by direct and heterodyne detection of scattering from haemocyanin is shown in figure 27[33].

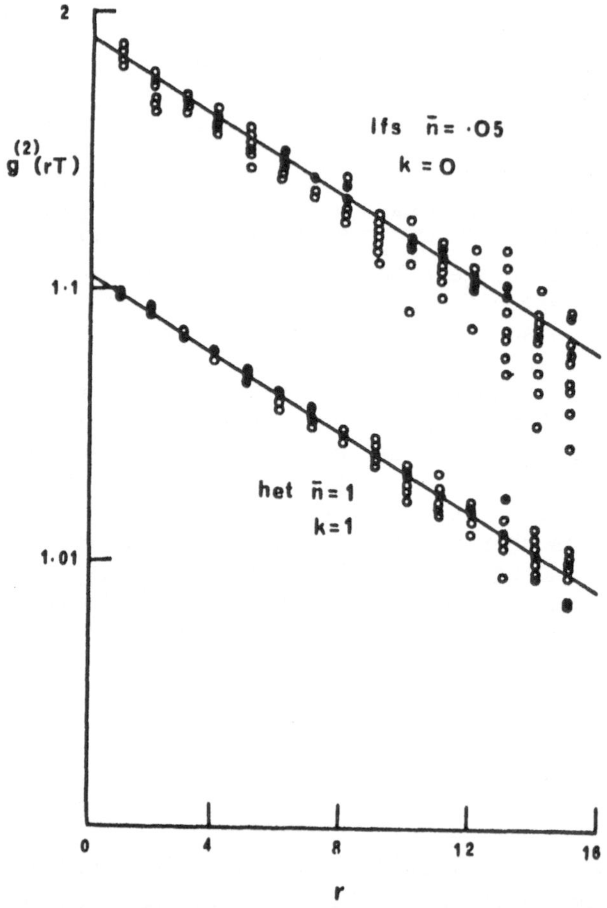

FIG 27 An experimental comparison of several IFS and heterodyne measurements of the correlation function obtained from Brownian motion of a weak sample of haemocyanin. For the IFS measurements the channel separation, T, was 21.125 μs, for the heterodyne (het) measurements it was 41.25 μs. The greater statistical accuracy of the heterodyne data is apparent.

The greater spread of the intensity-fluctuation data is apparent. A theoretical comparison of the two cases, assuming full correlation in the short sample time and small area limit, has been given by Jakeman[34] who finds that the relative variance of the linewidth estimator in the two cases is given by

$$\frac{\mathrm{Var}(\hat{\Gamma})}{<\hat{\Gamma}>^2} = \frac{2\Gamma T}{\bar{n}_s^2 N} \qquad (81)$$

for heterodyne detection and

$$\frac{\mathrm{Var}(\hat{\Gamma})}{<\hat{\Gamma}>^2} = \frac{21.2\ \Gamma T}{\bar{n}_s^2 N} \qquad (82)$$

for direct detection. Thus there is a gain of a factor of ten in the time taken to achieve a given accuracy with heterodyne detection compared with direct detection. However this advantage is reduced by the practical difficulties involved in the optical mixing of the scattered and local oscillator fields.

Thus heterodyne detection, though it offers advantages in terms of accuracy and the handling of non-Gaussian statistics and signals,including 'dust' scattering, under most circumstances can only be analysed using large delay coefficients as a background estimate. Interpretation in terms of a particular model therefore becomes non-rigorous except, in the situation where the scattered spectrum is known to have only one component.

3.4 Non-Gaussian Statistics

Situations in which non-Gaussian optical field statistics are encountered are being increasingly studied by photon-correlation. Obviously many-bit correlators have an inherent advantage in this domain since their operation is independent of the signal statistics. Single-stop delayed coincidence measurements are also independent of statistics provided the mean count-rate is low. The correction is dependent on statistics, however. Moreover, triggered multiscaling[14] and single clipping cannot be applied as they stand since the Jakeman-Pike equations[18] only apply for Gaussian light. Of course for low mean count-rates ($\bar{n} < 0.1$) the clipped correlation function is essentially undistorted so that non-Gaussian signals can always be handled in this region. However it can be shown[35] that correlation using a one-bit delay signal is possible if we have a uniform random selection of clip levels.

3.4.1 Techniques. Various experimental techniques for achieving a random selection of clip-level, and hence the full correlator function, are available. They amount to programming the clip level in various ways such as:

(1) Uniform random selection of k

(2) Sequential selection of k from a saw-tooth distribution

(3) Scaling[36,37].

Let us examine these in more detail and discuss their limitations. The one-bit clipping gate is shown in figure 28 including the programmable clip-level, k.

To achieve uniform random clip-level selection several differing pseudo-random generators can be used each controlling one bit of the variable clip level. Each sample time a new random number is generated having a uniform distribution. Since the clip level from sample to sample is completely uncorrelated this solution to random clipping is the most satisfactory. However,since one is not using the amplitude fully,it is not as efficient as a full correlator at extracting information from single events.

Ramp-clipping is somewhat simpler than uniform random clipping,since the clip level can be programmed by the output of a counter to which one is added each sample time up to some overflow where it resets. The distribution of clip levels is therefore uniform, provided many (P) such cycles are taken leaving a possible non-uniformity of order 1/P. However,since the clip level is completely correlated from sample to sample this technique is useless for single events and relies on the random nature of the signal to decorrelate the clip-level from the signal.

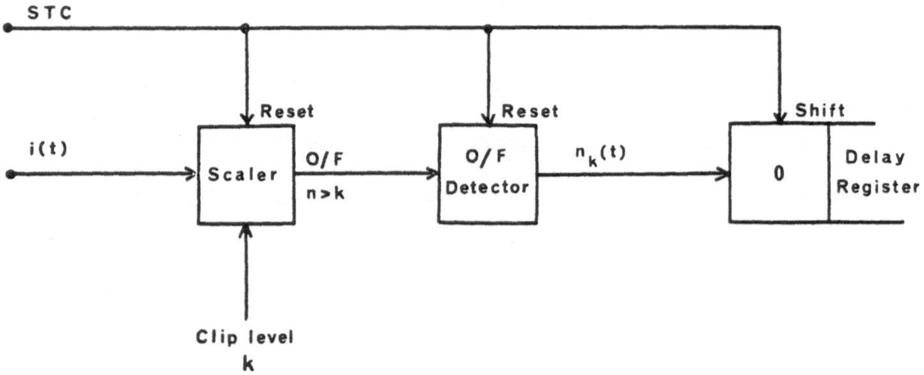

FIG 28 Block diagram of the clipping gate showing the programmable clip-level, k.

The third technique, scaling[36,37], is the simplest in terms of hardware since one merely removes the reset pulse to the clip gate in figure 28. The clip level k now determines the division ratio of the scaler. The total counts(r) left in the clip gate scaler at the end of each sample now act as the complement of a preset clip-level for the next sample. An overflow will occur in the next sample if more than k-r pulses arrive. After a comparatively small number of samples this remainder, r, is uncorrelated from the initial conditions by the signal- and photon-statistics. This technique is therefore equivalent to a random selection of clip levels with a uniform probability. The method is again unsatisfactory for single-event signals since the correlation between clip levels still partially persists. However the situation is better than with a ramp-clipping system for the same type of signal.

It must be realised that these one-bit methods, such as scaling for handling non-Gaussian signals,must be operated using the scaling level set sufficiently high for there to be a negligible possibility of two counts per sample time in the one-bit channel. It is not correct to use a subsequent clip level to further reduce this rate since one no longer has a uniform selection of clip levels in the original scaling circuit. However for Gaussian signals there is no such restriction on the clip levels chosen since different clip levels give a time-dependent term differing only in its amplitude.

Scaling can also be applied in generating the trigger pulse for the multiscaling mode of correlation[14]. However, unlike normal scaled operation,with the multiscaling technique for non-Gaussian signals one requires the apparatus to give a uniform probability of a trigger pulse across the entire delay range of interest. Otherwise, with repetitive signals, for example, one obtains synchronisation effects which distort the correlation function. The scaling level, therefore, must be sufficient for there to be a negligible probability of two scaled counts over the delay range of interest. This waste of information is due to the fact that this technique is only a single-start method,as opposed to the multistart operation of true correlation.

3.4.2 Accuracy. Having outlined techniques where non-Gaussian signals can be handled with a one-bit correlator it is obviously important to assess the loss in accuracy incurred compared with full correlation. Measurements with Gaussian sources[35] have shown that scaling is not appreciably worse than clipping at the optimum k for rates less than $\bar{n} = 2$. As the mean count-rate is increased,the width of the photodetection probability distribution increases rapidly,so that a much greater range of clip levels has to be covered. A greater fraction of the time is spent at clip levels giving poor accuracy so that the overall accuracy is reduced. As the optical field fluctuations become progressively less Gaussian giving broader and broader probability distributions this effect will become more and more important. It seems probable,therefore,that for non-Gaussian sources the use of one-bit techniques leads to an appreciable loss in accuracy. Optimum performance of such a one-bit system will result from adjusting the mean photodetection count-rate so that the contributions to the uncertainty due to the intensity-fluctuations and to the photon-statistics are comparable.

REFERENCES

1 R J Glauber,1963, Phys Rev Letts, **10**, 84.

2 L Mandel, 1959, Proc Phys Soc, **74**, 233.

3 E Jakeman, C J Oliver and E R Pike, 1968, J Phys A, **1**, 406.

4 R Foord, R Jones, C J Oliver and E R Pike, 1969, Appl Opts, **8**, 1975.

5 E Jakeman, 1973, NATO ASI, Capri.

6 R Foord, E Jakeman, C J Oliver, E R Pike, R J Blagrove, E Wood, and A R Peacocke, 1970, Nature, **227**, 242.

7 W B Davenport and W L Root, 1958, Random Signals and Noise, (New York : McGraw-Hill).

8 G A Rebka and R V Pound, 1957, Nature, **180**, 1035.

9 B L Morgan and L Mandel, 1966, Phys Rev Letts, **16**, 1012.

10 D B Scarl, 1968, Phys Rev, **175**, 1661.

11 P B Coates, 1968, J Phys E, **1**, 878.

12 C C Davis and T A King, 1970, J Phys A, **3**, 101.

13 S Chopra and L Mandel, 1972, Rev Sci Instrum, **43**, 1489.

14 S H Chen and N Polonsky-Ostrowsky, 1969, Opt Commun, **1**, 64.

15 R Nossal, S H Chen and C C Lai, 1971, Opt Commun, **4**, 35.

16 E Jakeman, E R Pike and S Swain, 1971, J Phys A, **4**, 517.

17 J B Abbiss, T W Chubb, A R G Mundell, P R Sharpe, C J Oliver and E R Pike, 1972, J Phys D, **5**, L100.

18 E Jakeman and E R Pike, 1969, J Phys A, **2**, 411.

19 E Jakeman, 1970, J Phys A, **3**, 201.

20 E Jakeman, C J Oliver and E R Pike, 1971, J Phys A, **4**, 827.

21 D E Koppel, 1971, J Appl Phys, 42, 3216.

22 A J E Siegert, 1943, MIT Rad Lab Report No 465.

23 A J Hughes, E Jakeman, C J Oliver and E R Pike, 1973, J Phys A, 6, to be published.

24 S H Chen, P Tartaglia and N Polonsky-Ostrowsky, 1972, J Phys A, 5, 1619.

25 A Fraser, 1971, Rev Sci Instrum, 42, 1539.

26 S H Chen, P Tartaglia and P N Pusey, 1973, J Phys A, 6, 490.

27 G B Benedek, 1968, Polarisation, Matiere et Rayonnement, (Paris Presses Universitaries de France)

28 H A Haus, 1969, Proc Int School of Physics, 'Enrico Fermi', Course XL11, (New York : Academic Press), p111.

29 H Z Cummins and H L Swinney, 1970, Progress in Optics, Vol VIII, ed E Wolf (Amersterdam : North-Holland).

30 E Jakeman, E R Pike and S Swain, 1970, J Phys A, 3, 255.

31 V Degiorgio and J B Lastovka, 1971, Phys Rev A, 4, 2033.

32 H C Kelly, 1971, JQE (IEEE), 7, 541.

33 E R Pike, 1970, Rev Phys Tech, 1, 180.

34 E Jakeman, 1972, J Phys A, 5, 249.

35 E Jakeman, C J Oliver, E R Pike and P N Pusey, 1972, J Phys A, 5, L93.

36 P N Pusey and W I Goldberg, 1971, Phys Rev, A3, 766.

37 D W Schaefer and B J Berne, 1972, Phys Rev Letts, 28, 475.

38 J R Prescott, 1966, Nucl. Instr.Meths., 39, 173.

39 R Jones, C J Oliver and E R Pike, 1971, Appl. Opts., 10, 1673.

LIGHT BEATING SPECTROSCOPY

H.Z. Cummins

Department of Physics, New York University

New York, New York 10003, U.S.A.

1. INTRODUCTION

During the past decade, light beating spectroscopy has developed into a major new technique for analyzing optical fields with an effective resolution orders of magnitude greater than was available with traditional spectroscopic techniques. In all the early light beating experiments, the optical field under study illuminated a photodetector whose output current was then analyzed with an audio or r.f. spectrum analyzer in order to determine the photocurrent power spectrum.

About five years ago, the digital photoelectron autocorrelation technique which was pioneered by the Malvern group came into use, and in many areas has essentially replaced spectrum analysis. The lectures at this institute are primarily concerned with various aspects of the digital correlation technique. Nevertheless, photocurrent spectrum analysis continues to be an important technique, particularly in experiments involving photocurrent frequencies higher than 1 MHz, or involving complicated spectral structure requiring resolution of 100 or more. In this seminar we will review the history of the photocurrent spectral analysis approach to light beating spectroscopy, summarize the relevant theory, discuss the instrumentation briefly, and finally compare this method to the photoelectron autocorrelation method. Two reviews of this topic appeared several years ago in which the details of the various derivations which we will only summarize in this seminar can be found.[1,2]

Light beating spectroscopy began in 1955 when Forrester, Gudmundsen and Johnson observed that when the Zeeman-split Hg 5461 Å emission line illuminated a photoelectric detector, the two strong optical components, differing in frequency by ~10^{10} Hz, beat with each other in the square-law detector producing a 10^{10} Hz component in the photocurrent.[3]

With the advent of lasers, this process immediately came into widespread use since light from two lasers can be mixed colinearly on a photodetector, and the beat note which can easily be seen on an electronic spectrum analyzer then provides a precise measurement of the stability of the lasers. Similarly, the photocurrent produced when light from a multimode laser illuminates a photodetector exhibits beats at frequencies equal to the separation between each pair of laser modes. A number of experiments in the early 1960's also demonstrated that laser light scattered from a moving system could be mixed with some unshifted light from the same laser, and the beat frequency then gives the Doppler shift directly.

In Fig. 1 (from Yeh and Cummins - Ref. 4) light which is scattered from a flowing fluid is mixed with unscattered light on a photomultiplier, and the beat frequency then gives the Doppler shift from which the flow velocity can be deduced (laser Doppler velocimetry). Application of light beating spectroscopy to problems of flow and turbulence will be discussed further in the seminar of Dr. Crosignani.

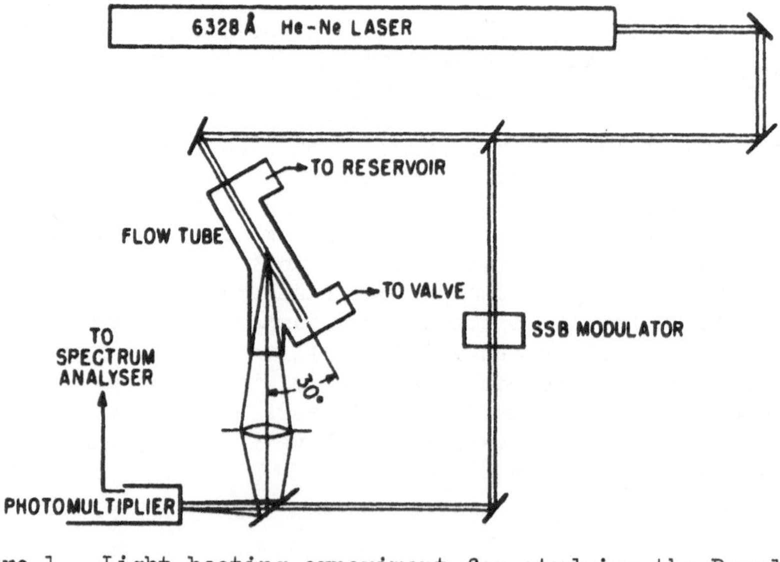

Figure 1. Light beating experiment for studying the Doppler shift in laser light scattered by a flowing fluid (Yeh and Cummins - reference 4).

These examples of light beating spectroscopy, the beating of two lasers (or beats between different modes of a multimode laser) and beating of the output of a laser with part of the output of the same laser which has been frequency shifted during scattering, can be easily understood in terms of simple semi-classical theory outlined in Dr. Jakeman's first lecture.

2. ELEMENTARY THEORY OF "BEATING"

Suppose that the positive frequency complex optical field or "analytic signal" (which we identify as Jakeman's E^+) at the detector is:

$$E(r, t) = \int_0^\infty E(r, \omega) \exp(-i\omega t)\, d\omega \tag{1}$$

The photocurrent i(t) is given by

$$i(t) = \sigma E^*(t)\, E(t) \tag{2}$$

where σ is the quantum efficiency in appropriate units and we have assumed that the photodetector is effectively a point.

For those cases where the field consists of two components with frequencies ω_1 and ω_2, Eq. (2) predicts:

$$i(t) = \sigma[|E_1|^2 + |E_2|^2 + 2\,\mathrm{Re}(E_1{}^*E_2 \exp(i(\omega_1-\omega_2)t)] \tag{3}$$

so that the photocurrent consists of a d.c. component plus a "beat" term at the difference frequency $(\omega_1-\omega_2)$. Consideration of the discrete nature of photoelectric emission also adds a shot noise term to Eq. (3), but we defer consideration of the shot noise to the next section.

Extension of this simple analysis to more complicated optical spectra was discussed by Forrester in 1961.(5) In Forrester's approach, the optical spectrum $I(\beta)$ is divided into "slices", and each pair of slices beats at the detector to produce a component in the photocurrent spectrum at the difference frequency $\omega = \beta_1 - \beta_2$, as shown in Fig. 2.

If it is assumed that all pairs of components with a common frequency difference ω will produce beat signals with random phases, then the photocurrent power spectrum $P_i{}^+(\omega)$ can be easily found by

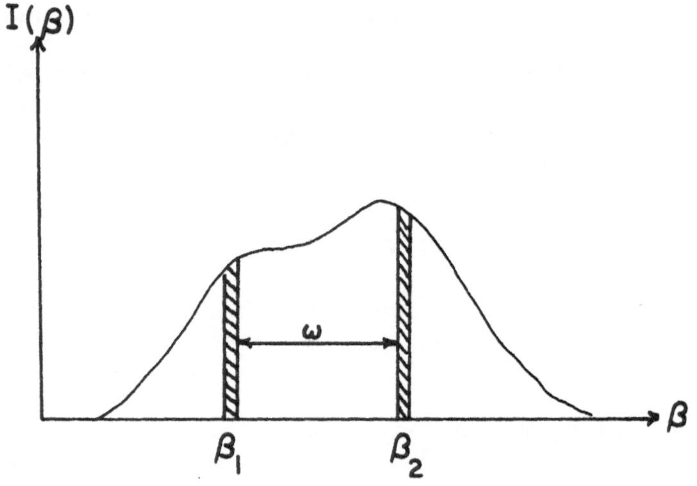

Figure 2. Forrester's approach for finding the photocurrent power spectrum from the optical spectrum $I(\beta)$. (Forrester - reference 5.)

straightforward integration:

$$P_i^+(\omega) = 2\sigma^2 \int_0^\infty I(\beta)\, I(\beta + \omega)\, d\beta \qquad (4)$$

where $P_i^+(\omega)$ indicates positive frequencies only, and the d.c. and shot noise components have been neglected.

The convolution formula [Eq. (4)] predicts the photocurrent power spectrum if the optical spectrum $I(\beta)$ is specified. For example, if the optical field is a Lorentzian centered at β_o,

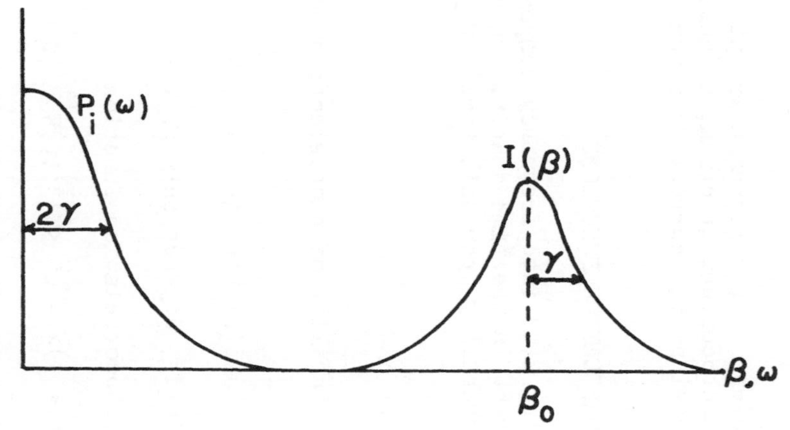

Figure 3. A Lorentzian optical spectrum $I(\beta)$ of width γ produces a self-beating photocurrent spectrum which is a Lorentzian of width 2γ.

$$I(\beta) = \langle I \rangle \frac{\gamma/\pi}{(\beta-\beta_o)^2 + (\gamma)^2} \tag{5}$$

then the photocurrent spectrum is also a Lorentzian centered at $\omega = o$, with half-width twice as great as $I(\beta)$ as illustrated in Fig. 3.

$$P_i^+(\omega) = 2\langle I \rangle^2 \sigma^2 \frac{2\gamma/\pi}{\omega^2 + (2\gamma)^2} \tag{6}$$

Similarly, Eq. (4) predicts that a rectangular optical spectrum produces a triangular photocurrent spectrum, a Gaussian optical spectrum produces a Gaussian photocurrent spectrum, etc.

3. CONNECTION WITH CORRELATION FUNCTIONS

Instead of resolving $E(t)$ into frequency components, we can work with the photocurrent in the time domain. Thus, Eq. (2) can be redefined as the probability (per unit time) of photoelectron emission:

$$W^{(1)}(t) = \sigma E^*(t)\, E(t) \tag{7}$$

Similarly, the joint probability of photoelectron emission occurring at t and $t + \tau$ is:

$$W^{(2)}\,(t,\, t + \tau) = \sigma E^*(t)\, E(t)\, E^*(t + \tau)\, E(t + \tau) \tag{8}$$

which for stationary fields, depends only on τ.

The photocurrent autocorrelation function

$$C_i(\tau) = \langle i(t)\, i(t + \tau)\rangle = e^2 \langle W^{(1)}(t) W^{(1)}(t + \tau)\rangle \tag{9}$$

consists of two contributions: one from distinct events [Eq. (8)], and another from the same photoelectron emitted at t and $t + \tau$. Thus, after some manipulation

$$C_i(\tau) = e\langle i \rangle\, \delta(\tau) + \langle i \rangle^2\, g^{(2)}(\tau) \tag{10}$$

where

$$g^{(2)}(\tau) = \frac{<E^*(t)\ E(t)\ E^*(t+\tau)\ E(t+\tau)>}{<E^*E>^2}$$

The photocurrent power spectrum $P_i(\omega)$ can be found from $C_i(\tau)$ via the Wiener-Khintchine theorem:

$$P_i(\omega) = \frac{1}{2\pi} \int_{-\infty}^{\infty} C_i(\tau) \exp(i\omega\tau)\, d\tau \tag{11}$$

Eqs. (10) and (11) show that the photoelectron correlation function [Eq. (10)] which is essentially the quantity measured on a digital correlator, and the photocurrent spectrum are a Fourier transform pair. Thus, in principle at least, the same information can be extracted from the optical signal with either technique.

If the optical field is a Gaussian random process (in the complex E space) then:

$$g^{(2)}(\tau) = 1 + |g^{(1)}(\tau)|^2 \tag{12}$$

(which is the Siegert relation discussed by Dr. Jakeman) where

$$g^{(1)}(\tau) = <E^*(t)\ E(t+\tau)>/<E^*E>$$

is the Fourier transform of the normalized optical spectrum $I(\omega)$. Eqs. (10), (11) and (12) give:

$$C_i(\tau) = e<i>\ \delta(\tau) + <i>^2\ (1 + |g^{(1)}(\tau)|^2) \tag{13}$$

$$P_i^+(\omega) = <i>^2\delta'(\omega) + \frac{e<i>}{\pi} + \frac{<i>^2}{\pi} \int_{-\infty}^{\infty} |g^{(1)}(\tau)|^2 \exp(i\omega\tau) d\tau \tag{14}$$

where positive and negative frequency terms in $P(\omega)$ have been combined. The three terms in Eq. (14) are the d.c. component, the shot-noise and the light beating spectrum. This last term is in fact equivalent to the convolution equation (4) derived from Forrester's approach, so that Forrester's assumption of random phases is actually equivalent to the assumption that $E(t)$ is a Gaussian random process - a connection which Dr. Jakeman has already discussed.

4. SELF BEATING AND HETERODYNE TECHNIQUES

There are two experimental approaches to light beating spectroscopy whose designations vary from author to author. We will arbitrarily use the terms self-beating and heterodyne.

In self-beating, the optical signal under analysis (e.g. light scattered from some preparation) illuminates the photodetector whose output then fluctuates in response to the fluctuating intensity of the optical field (hence the alternative term - intensity fluctuation spectroscopy). In the heterodyne approach, an additional optical signal - usually derived from the same laser which generates the signal field - is mixed with the signal field at the phototube and acts as a local oscillator. In practice, the local oscillator signal is frequently provided in scattering experiments by the walls of the scattering cell or by bits of opaque material intentionally placed within the scattering volume.

In the heterodyne technique, the field E(t) has two components, the signal and the local oscillator fields, $E_s(t)$ and $E_{LO}(t)$. If the local oscillator intensity is much bigger than the signal intensity, then the current correlation function is closely approximated by:

$$C_i(\tau) \simeq ei_{LO}\,\delta(\tau) + i_{LO}^2 + 2i_{LO}\langle i_s\rangle \times$$

$$\times \{\exp(i\omega_{LO}\tau)\, g_s^{(1)}(\tau) + \exp(-i\omega_{LO}\tau)\, g_s^{(1)*}(\tau)\} \tag{15}$$

If, for example,

$$g_s^{(1)}(\tau) = \exp(-i\omega_o\tau - \gamma|\tau|) \tag{16}$$

then

$$C_i(\tau) = ei_{LO}\delta(\tau) + i_{LO}^2 + 4i_{LO}\langle i_s\rangle\cos[(\omega_{LO}-\omega_o)\tau]\exp(-\gamma\tau) \tag{17}$$

$$P_i(\omega) = \frac{ei_{LO}}{\pi} + i_{LO}^2\,\delta'(\omega) + 4i_{LO}\langle i_s\rangle \frac{\gamma/\pi}{(\omega-|\omega_o-\omega_{LO}|)^2 + \gamma^2} \tag{18}$$

So the photocurrent power spectrum reproduces the optical spectrum, with the center frequency shifted from ω_o down to $\omega_o-\omega_{LO}$.

This connection between the optical spectrum and the heterodyne spectrum is in fact generally true. The signal term is just

i_{LO} $<i_s>$ times the normalized optical spectrum, shifted down in frequency to $\omega_o - \omega_{LO}$. The resemblance of this result to the operation of ordinary radios is evident in Forrester's designation "superheterodyne detection". Note that the principal result - that the photocurrent power spectrum reproduces the optical spectrum - is independent of the statistics of the optical field.

In the self-beat technique, the photocurrent autocorrelation function and power spectrum are given by Eqs. (10) and (11), and will not, in general, be simply related to the optical spectrum since the relation between $g^{(1)}$ and $g^{(2)}$ depends on the statistics of the field.

If the field is Gaussian so that Eq. (12) holds, then Eqs. (13) and (14) show that $C_i(\tau)$ and $P_i(\omega)$ are simply related to $g^{(1)}(\tau)$. For the signal field characterized by Eq. (16),

$$I(\omega) = <I> \frac{\gamma/\pi}{\gamma^2 + (\omega-\omega_o)^2} \tag{19}$$

$$C_i(\tau) = e<i> \delta(\tau) + <i>^2 + <i>^2 \exp(-2\gamma\tau) \tag{20}$$

$$P_i^+(\omega) = \frac{e<i>}{\pi} + <i>^2 \delta'(\omega) + 2<i>^2 \frac{2\gamma/\pi}{\omega^2 + (2\gamma)^2} \tag{21}$$

Eq. (21) recovers the previous convolution result [Eq. (6)] with $<i> = \sigma<I>$ and also includes the shot noise and d.c. terms. The effects of spatial coherence which we will not consider will generally reduce the third term in both Eq. (20) and (21) by a factor α of order unity.

5. APPARATUS AND SIGNAL TO NOISE CONSIDERATION

Processing of the photocurrent by spectrum analysis consists, basically, of scanning a tunable electrical filter and recording the voltage or power present as a function of frequency. Commercial spectrum analyzers are available which can cover the range from a small fraction of one Hz well up into the GHz range.

The signal to noise ratio obtainable with this technique, defined as the ratio of signal amplitude to the rms fluctuation in signal plus shot noise can be shown to be:

$$\left(\frac{S}{N}\right)_{\omega} = \left(\frac{P_s(\omega)}{P_s(\omega) + P_{sn}}\right) (1 + \delta T)^{\frac{1}{2}} \qquad (22)$$

where P_s and P_{sn} are the signal and shot-noise power respectively, δ is the bandwidth of the spectrum analyzer, and T is the time constant of the output circuit. The derivation of Eq. (22) and its evaluation for several combinations of the parameters δ and T was discussed in detail by Cummins and Swinney.[1]

6. COMPARISON WITH PHOTOELECTRON CORRELATION TECHNIQUES

The single filter photocurrent spectrum analysis technique suffers from several serious disadvantages with respect to digital correlation methods. First, since spectrum analyzers are usually analog devices, there are inevitable nonlinearities which distort the spectrum limiting the accuracy. Second, as the center frequency is swept, all the signal falling outside of the bandwidth δ is effectively lost. The efficiency of this technique is then δ/Δ_ω where Δ_ω is the total range of frequencies being scanned. As the resolution Δ_ω/δ is increased, the efficiency decreases. This problem can, in principle, be overcome with the use of multichannel or time-compression techniques, both of which have been incorporated into some commercial spectrum analyzers, but the frequency dependence of the response of these systems has generally been a serious problem.

In general, it proves to be difficult to obtain data with accuracy of better than ~3% over wide frequency ranges with analog spectrum analyzers, although for narrow-band spectra the accuracy may be significantly improved. Thus for broadband low frequency spectra, digital correlation techniques which have 100% efficiency and no analog errors have the clear advantage in both accuracy and speed. There are, however, two areas in which spectrum analyzers have significant advantages. To illustrate, suppose a digital correlator with N channels has been used to analyze an optical field. The output then provides N values of $C_i(\tau)$, for $\tau = t_o, 2t_o \ldots\ldots Nt_o$. A digital Fourier transform then gives (N/2) points of the power spectrum $P_i(f)$ at $f = (1/t_o), (2/t_o) \ldots\ldots (N/2t_o)$.

Since state-of-the-art electronics currently limit t_o to a minimum of perhaps $(1/40)^{th}$ μsec, the maximum frequency which can be studied with the digital correlator is ~20 MHz. But spectrum analysis can easily be extended to several hundred MHz with photomultiplier tubes where the transit time of the tube rather than the electronics places the upper limit on the frequency. Furthermore, by using fast photodiodes rather than photomultipliers, photocurrent spectra can be extended into the GHz range. In fact, the 1955 experiment of Forrester et al. detected light beats at ~10 GHz.

Secondly, suppose the photocurrent spectrum contains a narrow line of unknown shape of width δf centered at a non-zero frequency f_o plus other uninteresting structure. In order to measure the spectral profile of the line by Fourier transformation of the photocurrent autocorrelation function would require a correlator with at least $4(f_o/\delta f)$ channels if one wants at least four points of P(f) within the full width at half maximum of the line of interest. As f_o increases, the number of correlator channels required to resolve a given δf increases with f_o. With a spectrum analyzer, the bandwidth δf can easily be made very small, while the operation of the instrument is relatively independent of f_o.

To illustrate, we consider the light beating spectra obtained by Eden and Swinney in which Brillouin scattering from xenon near the critical point is mixed with laser light. (The spectra are shown in Fig.   of Professor Swinney's lectures where the experiments are discussed.)

As the temperature is changed, both the position and width of the Brillouin component change, and the velocity and attenuation of sound can be deduced from the spectra.

Note that the narrowest component occurs at ~11.5 MHz with half width δf ~55 KHz. This measurement could, in principle, be performed with a correlator, but would require bin times of ~$(1/25)^{th}$ microsecond and at least 800 channels in the correlator.

In conclusion, we suggest that photocurrent spectrum analysis represents an important technique for light beating spectroscopy which in some cases is more appropriate than correlation measurements. We particularly emphasize that our discussion has not included a survey of current instrumentation which has progressed significantly in recent years, and suggest that anyone designing an experiment should consult the manufacturers of both spectrum analyzers and correlators for the specifications of the most recent equipment before deciding on specific instrumentation.

6. REFERENCES

1. H.Z. Cummins and H.L. Swinney, in: Progress in Optics, Vol. VIII. Edited by Emil Wolf (North Holland Publishing Co., Amsterdam, 1970), p. 133.

2. G.B. Benedek, in: Polarisation Matiere et Rayonnement. (Presses Universitaire de France, Paris), p. 49.

3. A.T. Forrester, R.A. Gudmundsen and P.O. Johnson, Phys. Rev. 99, 1691 (1955).

4. Y. Yeh and H.Z. Cummins, Appl. Phys. Letters 4, 176 (1964).

5. A.T. Forrester, J. Opt. Soc. Am. 51, 253 (1961).

LIGHT SCATTERING BY LIQUIDS AND GASES

P. LALLEMAND

Laboratoire de Spectroscopie Hertzienne de l'Ecole

Normale Supérieure - 24 Rue Lhomond 75005 Paris France

In this chapter some of the applications of light scattering to the study of liquids and gases will be discussed. They include thermodynamic properties, transport coefficients that may be frequency dependent in the case of relaxing fluids, and rotational motions when the molecules are optically anisotropic. This wealth of possible information has attracted many people to this field which is now quite mature, and experiments and theories are very well developped now. This chapter is an introduction to this subject. In part A, we shall introduce relaxation phenomena in liquids and show how they influence the spectrum. Then we shall discuss in detail density fluctuations in low pressure gases in part B. Finally, part C will be devoted to the depolarized part of the spectrum, together with some models of molecular reorientation in liquids.

A ▪ LIGHT SCATTERING BY FLUIDS

I Generalities.

We shall first consider a pure simple fluid, and calculate the power spectral density of the scattered light using a molecular theory for light scattering. This will give us an expression for the scattering cross-section in terms of microscopic variables. As this expression requires to know much more about the fluid than one can ever think of, we shall then give a discussion of some of the physical ideas of the approximate methods which are used to calculate spectra. We shall use the method of the generalized Langevin equation which allows to tie the well known transport equations using phenomenological transport coefficients, with a microscopic description of the system. This method due to Mori rests on very sound mathematical grounds, but its derivation is much beyond the scope of this chapter. [1,2,3]

a Scattering cross-section

Let us consider a plane light wave

$$\vec{E} = \vec{E}_0 \exp(i(\vec{k}_i.\vec{r} - \omega t))$$

incident on a set of N identical atoms. Each atom has a polarizability α that gives rise to an oscillating dipole moment $\vec{P} = \alpha \vec{E}$ that radiates light in all directions. We know from classical electrodynamics that such a dipole produces at a distant point $\vec{R}$ a field

$$\vec{E}(\vec{R},t) = \vec{k}_s \times (\vec{k}_s \times \vec{E}_0) \frac{\alpha}{|\vec{R}-\vec{r}_i|} \exp\ i(\vec{k}_i.\vec{r}_i - \omega(t - \frac{(\vec{R}-\vec{r}_i)}{c}))$$

where $\vec{r}_i$ is the location of the dipole, and $\vec{k}_s$ the wave vector of a plane wave of frequency ω propagating in the direction of $\vec{R}$. This is valid only when the medium in which the radiation takes place is optically isotropic.

The field at point $\vec{R}$ is obtained by adding the contributions of all the atoms, so that

$$\vec{E}_s(\vec{R},t) = \vec{k}_s \times (\vec{k}_s \times \vec{E}_0) \frac{\alpha}{R} \exp(i(k_s R - \omega t)) \sum_i e^{i(\vec{k}_i - \vec{k}_s).\vec{r}_i}$$

where we have used the approximation $\vec{k}_s.(\vec{R}-\vec{r}_i) \approx k_s R - \vec{k}_s.\vec{r}_i$ which is valid for $|R| \gg |r_i|$.

We see that the scattered field behaves like a spherical wave and that its amplitude depends upon two factors. First, there is a term $\vec{k}_s \times (\vec{k}_s \times \vec{E}_0) = \vec{k}_s^2\ E_0 \sin\ \phi$ where ϕ is the angle between the polarization vector of the incident field and the direction of the scattered field. This term depends only upon the geometry of the experimental set-up. The second factor

$\sum_i \exp\ i(\vec{k}_i - \vec{k}_s).\ \vec{r}_i$ depends upon the positions of all the atoms in the medium and it is the physically significant term.

We have not included here any local field correction, but for practical purposes one can replace the input electric field amplitude E_0 by $E_0(n^2+2)/3$ where n is the index of refraction of the medium, if we follow the well known Lorentz-Lorenz local field correction. We shall come back to this question at the end of this chapter.

From now on, we shall use $\vec{k} = \vec{k}_i - \vec{k}_s$, which is called the scattering wave-vector, the magnitude of which is

$k = (2\pi/\lambda_0)2n \sin(\theta/2)$ where λ_0 is the wave length of the incident light beam (in vacuum), and θ the scattering angle.

As is discussed in other chapters of this book, we now calculate the correlation function of the scattered field. We put :

$$C(\tau) = \left\langle \sum_{i,j} \exp i\, \vec{k}.(\vec{r}_i(\tau) - \vec{r}_j(0)) \right\rangle \propto \langle \vec{E}(\tau).\ \vec{E}(0) \rangle$$

We can use $C(\tau)$ to determine the integrated scattered intensity which is proportional to $C(0)$, so that the scattering cross-section is given by

$$\frac{d\sigma}{d\Omega} = \frac{\omega^4}{c^4} \sin^2\phi \quad \alpha^2\ C(0)$$

This quantity is often measured through turbidity measurements in which one determines the extinction coefficient h of an otherwise transparent medium, as one has for systems where spatial correlations are short range

$$h = \frac{8\pi}{3} \frac{\omega^4}{c^4} \alpha^2 \quad C(0)$$

The expression we obtained for $C(\tau)$ involves a detailed description of the positions of all the particles. This is usually beyond our knowledge of the system that is usually described by only a very few quantities, so that we need some approximate methods to deal with this problem. We shall first derive an expression for the scattering cross-section in terms of the pair distribution function which plays a central role in the study of fluids. Let us write

$$C(\tau) = \sum_{i,j} e^{i\vec{k}.(\vec{r}_i(\tau) - \vec{r}_j(0))} =$$

$$= \sum_{i,j} \iint e^{i\vec{k}.(\vec{R}-\vec{R}')}\ \delta(\vec{R}-\vec{r}_i(\tau))\ \delta(\vec{R}'-\vec{r}_j(0))d\vec{R}\ d\vec{R}'$$

and assuming that the fluid is invariant for translations, let us change variables $\vec{R}$ and $\vec{r} = \vec{R} - \vec{R}'$. We then get

$$C(\tau) = N \int e^{i\vec{k}.\vec{r}} \sum_j \delta(\vec{r}_j(0) + \vec{r} - \vec{r}_i(\tau))\ d\vec{r}$$

We can give a physical interpretation of the microscopic quantity

$\sum_j \delta(\vec{r}_j(0)+\vec{r}-\vec{r}_i(\tau))$. We can say that it is proportional to the probability of finding a particle at location $\vec{r}_i$ and time τ, knowing that there was an other particle at location $\vec{0}$ and time 0. This probability is the Van Hove correlation function $G(\vec{r},\tau)$ and its spatial Fourier transform, which is related to $C(\tau)$, is called the intermediate scattering function, used for instance in neutron scattering. If we take $\tau = 0$, the sum over indices i and j breaks up into two parts, and one has

$$G(\vec{r},0) = \delta(\vec{r}) + \rho g(\vec{r})$$

where ρ is the mean density of the fluid and $g(\vec{r})$ the pair distribution function. Therefore we find that the scattering cross-section is proportional to the $\vec{k}$ Fourier component of the pair distribution function plus a term independent of $\vec{k}$, due to incoherent scattering. One can thus determine the pair correlation function from experimental measurements provided the magnitude of the wave-vector used is of the order of the reciprocal of typical scales of length occuring in the problem. In the case of simple fluids, typical lengths are of the order of a few atomic diameters, so that the $\vec{k}$ vectors used in light scattering experiments are much too small to show any features of $g(\vec{r})$ and $\partial\sigma/\partial\Omega$ does not depend upon the scattering angle. This is not the case when there are large correlation lengths as for instance in large biological molecules or near critical points. More will be said about it in other chapters.

Now although we have reduced the problem from an expression involving the coordinates of all particles to the pair distribution function, it turns out that we don't need so much information. In fact, the knowledge of only a very few average quantities is enough for most light scattering applications.

Let us consider the quantity $\rho = \sum_i \delta(\vec{r}-\vec{r}_i)$ whose ensemble average value is the mean density of the system. One can show that the ensemble average $C(\tau)$ is

$$C(\tau) = \sum_{i,j} e^{i\vec{k}.(\vec{r}_i(\tau)-\vec{r}_j(0))} = \int e^{i\vec{k}.\vec{r}} < \rho(\vec{r},\tau)\ \rho(\vec{0},0) > d\vec{r}$$

which now involves the autocorrelation function of the density ρ, so that we would like to study the fluctuations of the system from a macroscopic point of view.

The application of statistical methods to the calculation of thermal fluctuations is treated at length in textbooks on statistical thermodynamics. One considers a system that is in thermal equilibrium at some temperature T, and one takes a volume element V of the system that contains a very large number of molecules, but is very small compared to the wavelength of light. Assuming that spatial correlations are very short range, one can show that the probability for the volume element to exhibit fluctuations in volume, temperature, pressure and entropy (or more variables for more complicated systems) is given by

$$W(\delta V, \delta T, \delta P, \delta S) \propto \exp - \frac{\delta E}{k_B T}$$

where δE is the corresponding change in the free energy of the system. One has :

$$\delta E = \frac{1}{2}[\delta S\ \delta T - \delta V\ \delta P]$$

and after some standard manipulations, one ends up with

$$W(\delta V, \delta T) \propto \exp\left[- \frac{C_v}{2k_B T^2}(\delta T)^2 + \left(\frac{\partial P}{\partial V}\right)_T \frac{(\delta V)^2}{2k_B T}\right]$$

which shows that δV and δT are statistically independent gaussian variables, and we get their variances

$$\overline{(\delta V)^2} = k_B T V \chi_T$$

$$\overline{(\delta T)^2} = \frac{k_B T^2}{C_v}$$

$$\overline{\delta V \cdot \delta T} = 0$$

where $\chi_T = -\frac{1}{V}\left(\frac{\partial V}{\partial P}\right)_T$ is the isothermal compressibility and C_v the specific heat at constant volume.

This approach to fluctuations is quite useful, as it may be generalized to more complicated cases. Furthermore it shows that the scattered field E_S is a gaussian random variable, so that there is no particular problem about using the Siegert relation.

b Time dependence of correlation functions and generalized Langevin equation

There exist several ways to obtain approximate expressions for time correlation functions, when the amplitude of the fluctuations is small enough, so that one can neglect any nonlinear effects that could otherwise take place. We shall use as example for the discussion, the simple case of the one-dimensional brownian motion in which one big particle of mass M moves in a bath of a large number of small particles of mass $m \ll M$. In this problem that we shall treat using classical mechanics, we are interested in the correlation function $\langle V(t)\ V(0)\rangle$ where $V(t)$ is the velocity of the big particle at time t. This system can be described either phenomenologically through the equation :

$$M \frac{d\langle V\rangle}{dt} = -\xi \langle V \rangle$$

where ξ is a friction coefficient, that could be derived from the Stokes-Einstein formula, or it could be described by the coordinates of all the particles at all times which could be deduced from Hamilton's equations of motion.

First we can use the so-called Onsager hypothesis which says that the time behaviour of a correlation function is the same as that of the corresponding macroscopic variable. This means that $\langle V(t)\ V(0)\rangle$ is assumed to be equal to $\langle V(t)\rangle / \langle V(0)\rangle$ deduced from the phenomenological equation of motion for the big mass given above, the solution of which is simply

$$\frac{\langle V(t)\rangle}{\langle V(0)\rangle} = e^{-t/\tau} \qquad \text{where} \quad \tau = \frac{M}{\xi}$$

Second, we can use the general linear response theory, in which one looks for the generalized susceptibility χ of the system, when some external force is applied. If B is the generalized force corresponding to the variable A, one has generally [5]

$$\langle A(\omega)\ \overset{*}{A}(\omega)\rangle = \frac{\hbar}{2\pi} \mathfrak{Im}\ \chi_{AB}(\omega) \coth \frac{\hbar\omega}{2k_B T}$$

where $\chi_{AB}(\omega)$ is the susceptibility of the system, and where we have included quantum corrections at finite temperature. This method is used quite often in the literature, although the physical meaning of the generalized force may not always be physically transparent.

We shall use here a third method which is encountered in the more recent papers on light scattering. It is due to Mori who generalized the approach of the Langevin equation. It is quite powerful as it gives a justification for Onsager's hypothesis, so that one can use phenomenological equations to calculate correlation functions. Let us first recall what is the Langevin equation for the one-dimensional brownian motion. If we take a given set of initial conditions, we can write for the momentum of the big particle

$$M \frac{dV(t)}{dt} = \mathcal{F}(t, X_0, V, \{x_i, p_i\})$$

where $\mathcal{F}$ is the force acting on the big particle at time t due to the interactions of the big particle with all the small particles, which are characterized by $\{x_i, p_i\}$ (their initial positions and momenta at time 0). Let us break this force into two terms :

$$\mathcal{F}(t, X_0, V, \{x_i, p_i\}) = -\xi V(t) + F(t, X_0, V, \{x_i, p_i\})$$

as we expect that $\mathfrak{F}$ will exhibit a systematic part of the same kind as the phenomenological force $-\xi\langle V\rangle$ and a random part F that will depend on $\{x_i, p_i\}$. If we take now an ensemble over all initial conditions $\{x_i, p_i\}$, but for a given value of V, we shall get

$$M \frac{d\langle V(t)\rangle}{dt} = -\xi \langle V(t)\rangle + \langle F(t)\rangle$$

and if we have properly chosen ξ, then $\langle F(t)\rangle = 0$, as we expect to recover the phenomenological equation. Now we might have used some other decompositions of the force $\mathfrak{F}$, but they would be of any use only if F obeys the following conditions

- the mean value of F at any given time is zero $\langle F(t)\rangle = 0$
- the correlation time of F must be much smaller than that of V(t), so that we put here $\langle F(t)\, F(0)\rangle = 2B\,\delta(t)$, where the physical meaning of the delta-function is that $\langle F(t)\, F(0)\rangle$ vanishes for times much smaller than $\tau = M/\xi$.
- the random force F(t) is assumed to be uncorrelated with V(0) for all times, $\langle V(0)\, F(t)\rangle = 0$.

This separation of $\mathfrak{F}$ into two parts, together with the preceding requirements on F is not at all obvious, although one may think that the big particle will have to experience many collisions before its velocity may change sign, whereas the randomization of the motion of the small particles will be very fast, that is of the order of the time between collisions. This difference in time scales for the big particle coordinates and for those of the small particles leads to a definition of slow variables and of fast variables. Here the separation of variables between slow and fast variables is fairly easy, but it may be very difficult in more complicated situations.

Thus starting from a phenomenological equation of motion, we have obtained the Langevin equation where the statistical motion of the thermal bath is represented by the random driving force F(t).

$$M \frac{dV(t)}{dt} = -\xi V(t) + F(t)$$

$$\langle F(t)\rangle = 0$$

$$\langle V(0)\, F(t)\rangle = 0$$

$$\langle F(t)\, F(0)\rangle = 2B\,\delta(t)$$

We shall now use the Langevin equation to calculate the velocity autocorrelation function. Let us first take a particular set of initial conditions defining a random force F. We have to solve the equation :

$$M \frac{dV(t)}{dt} = -\xi V(t) + F(t)$$

which means that we are looking for an initial value problem. As will be done many times later, we take the Laplace transform of our equation of motion, taking advantage of the following property of the Laplace transformation : If we call the Laplace transform

$$\tilde{F}(s) = \int_0^\infty e^{-st} F(t)\, dt$$

then we get for time derivatives

$$F(t) \rightarrow \tilde{F}(s) \qquad \text{and} \qquad \dot{F}(t) \rightarrow s\tilde{F}(s) + F(0)$$

Therefore we have here

$$sM\ \tilde{V}(s) = -\xi\, \tilde{V}(s) + \tilde{F}(s) + MV(0)$$

so that

$$\tilde{V}(s) = \frac{\tilde{F}(s) + MV(0)}{sM + \xi}$$

and we obtain the correlation function

$$<\tilde{V}(s)\ V(0)> = \frac{1}{sM+\xi}\ M <V(0)\ V(0)> + \frac{1}{sM+\xi} < \tilde{F}(s)\ V(0)>$$

Now as we assumed that $<F(t)\ V(0)>=0$ for all times, we have $<\tilde{F}(s)\ V(0)> = 0$ thus

$$<\tilde{V}(s)\ V(0)> = \frac{<V(0)\ V(0)>}{s+\xi/M}$$

and taking the inverse Laplace transform, we get

$$\frac{<V(t)\ V(0)>}{<V(0)\ V(0)>} = \exp - \frac{t\xi}{M}$$

which is identical to the result we obtained using Onsager's hypothesis. This very simple discussion of the use of the Langevin equation illustrates the method that we shall use later to compute spectra.

This approach to the time evolution of statistical systems has been generalized by Mori who has given a formal solution of the many body problem of the motion of a large number of particles interacting with one another. When classical mechanics applies, this solution is obtained by solving Liouville's equation for time scales that are long compared to the correlation times of "fast" variables, but small compared to that of the "slow" variables, or macroscopic dynamical variables of the system. This allows to find an equation of motion for the dynamical variables of the system that is of the Langevin type, in which the influence of all the fast variables is described by the random force F. This equation is then identified with the phenomenological form of the Langevin equation, so that one can obtain microscopic expressions for the phenomenological transport coefficients.

For a system described by only one dynamical variable a(t), one has :

$$\frac{da(t)}{dt} = i\Omega \; a(t) - \int_0^t ds\phi(s) \; a(t-s) + F(t)$$

where Ω is the resonance frequency of the system, $\phi(s)$ a time-dependent transport coefficient that is often called a memory function and F(t) a random driving force. One has

$$< a(0) \; F(t) > = 0$$

$$< F(t) > \; = 0 \qquad \text{for all } t$$

$$i \; \Omega \langle a(0)a(0)\rangle = < a(0) \; (\frac{d}{dt} a(t)) \quad >_{t=0}$$

$$< F(0) \; F(t) > = <a(0) \quad a(0) \quad >\phi(t)$$

We shall use later a generalisation of these relations to several variables, so that Ω and $\phi(s)$ will be matrices . Ω and ϕ can either be obtained from their microscopic expressions, or one can use phenomenological quantities. Thus in the case of the one dimensional brownian motion, that we discussed in detail previously, we shall put

Ω = 0 because there is no linear restoring force and

$\phi = \frac{\xi}{M} \delta(t)$ where ξ is a phenomenological friction coefficient

Microscopic expressions for Ω and ϕ can be calculated for very dilute systems where one can consider only two body interactions. In dense systems this is not the case and the problem is very complicated. There exist now methods to deal with dense systems. One of them is to take a small number N of particles, with N∼1000, and to solve the classical equations of motion of these N particles using a large computer. One can then derive numerical values for Ω and ϕ . This method called molecular dynamics, (4) has given very good pictures of the actual molecular movements inside the fluids.

2 Spectrum of the light scattered by a simple fluid

a Use of the Langevin equation

Let us now use the Langevin equation to calculate the density autocorrelation function. One can show that for a simple fluid, the set of dynamical variables contains the following quantities :

$$\vec{\rho}(\vec{r},t) \quad , \quad T(\vec{r},t) \quad \text{and} \quad \vec{v}(\vec{r},t)$$

Among these five variables, only three play a role for a given $\vec{k}$ vector, as the transverse velocity does not couple to light waves, and as one can show that it does not couple to the other three variables in the linear regime (no vortex). So we have

$$A = \begin{pmatrix} \rho \\ T \\ v_z \end{pmatrix}$$

where v_z is the component of $\vec{v}$ parallel to $\vec{k}$.

What about M ? Following our preceding argument, M can be taken identical to the form one finds phenomenologically. So we get using the linearized hydrodynamics equations, and taking the Fourier transform for wave vector k :

$$\frac{d\vec{A}}{dt} = -\begin{pmatrix} 0 & 0 & -ik\rho_0 \\ 0 & \dfrac{\lambda\, k^2}{\rho_0 C_v} & -i\,\dfrac{k}{\beta}(\gamma-1) \\ -i\dfrac{kC_0^2}{\rho_0\gamma} & -ik\beta\,\dfrac{C_0^2}{\gamma} & \dfrac{4}{3}\eta_s\dfrac{k^2}{\rho_0}+\eta_b\dfrac{k^2}{\rho_0} \end{pmatrix}\vec{A}+\vec{F}(t)$$

Here ρ_0 is the mean density, λ the thermal conductivity, C_v the specific heat at constant volume, $\gamma=C_p/C_v$ the ratio of the specific heats, $\beta=\frac{1}{V}\left(\frac{dV}{dT}\right)_p$ the thermal expansion coefficient at constant pressure, $C_0=(\gamma/\rho_0\,\chi_T)^{1/2}$ the adiabatic velocity, χ_T the isothermal compressibility, η_s the shear viscosity and η_b the bulk viscosity.

We are now going to calculate $\langle\rho(\vec{k},t)\rho(\vec{k},0)\rangle$. As this involves an initial value problem in a linear system, it is convenient to take the Laplace transform of A. We get

$$s\,\vec{A}(s) = -\overset{\Rightarrow}{M}\,\vec{A}(s) + \vec{F}(s) + \vec{A}(0)$$

so

$$\vec{A}(s) = \left(\frac{1}{s\overset{\Rightarrow}{I}+\overset{\Rightarrow}{M}}\right)\left(\vec{A}(0)+\vec{F}(s)\right)$$

and we calculate the correlation function

$$\langle \vec{A}(s).\vec{A}(0)\rangle = \langle\vec{A}(0)\,\frac{1}{s\overset{\Rightarrow}{I}+\overset{\Rightarrow}{M}}\,\vec{A}(0)\rangle + \langle\vec{F}(s)\,\frac{1}{s\overset{\Rightarrow}{I}+\overset{\Rightarrow}{M}}\,\vec{A}(0)\rangle$$

Now our assumption that $\langle\vec{A}(0)\cdot\vec{F}(t)\rangle = 0$ leads to
$\langle\vec{A}(0).\vec{F}(s)\rangle = 0$ so we have finally

$$\langle\vec{A}(s).\,\vec{A}(0)\rangle = \langle\vec{A}(0)\,\frac{1}{s\overset{\Rightarrow}{I}+\overset{\Rightarrow}{M}}\,\vec{A}(0)\rangle$$

We find that the correlation function we look for is equal to a function of s multiplied by an ensemble average over the initial conditions, which, as we know, can be calculated from statistical thermodynamics [5].

Once we know $<\vec{A}(s).\vec{A}(0)>$, we take the inverse Laplace transform to get $<\vec{A}(t)\cdot\vec{A}(0)>$, or we can directly obtain the spectrum using

$$<\vec{A}(\omega).\vec{A}(0)> = 2\mathrm{Re}\left[<\vec{A}(s).\vec{A}(0)>\right]_{s=i\omega}$$

b <u>Calculation of the density autocorrelation function</u>

To be more specific, we shall now write explicitly the density autocorrelation function. We are interested in the matrix element $<\vec{A}(s)\vec{A}(0)>_{\rho\rho}$ which is proportional to the $\rho\,\rho$ matrix element of the inverse of the matrix $s\vec{\vec{I}} + \vec{\vec{M}}$, because the initial conditions given by statistical thermodynamics yield a diagonal matrix for $<\vec{A}(0).\vec{A}(0)>$ We obtain :

$$G(k,s) = \frac{<\rho(k,s)\,\rho(k,0)>}{<\rho(k,0)\ \rho(k,0)>} = \frac{N(s)}{D(s)}$$

$$= \frac{\begin{vmatrix} \dfrac{\lambda k^2}{\rho_0 C_v} + s & -ik\,\dfrac{(\gamma-1)}{\beta} \\ -ik\beta\,\dfrac{C_0^2}{\gamma} & \dfrac{4}{3}\eta_s\,\dfrac{k^2}{\rho_0} + \eta_b\,\dfrac{k^2}{\rho_0} + s \end{vmatrix}}{\begin{vmatrix} s & 0 & -ik\rho_0 \\ 0 & \dfrac{\lambda k^2}{\rho_0 C_v} + s & -ik\,\dfrac{(\gamma-1)}{\beta} \\ -ik\,\dfrac{C_0^2}{\rho_0\gamma} & -ik\beta\,\dfrac{C_0^2}{\gamma} & \dfrac{4}{3}\eta_s\,\dfrac{k^2}{\rho_0} + \eta_b\,\dfrac{k^2}{\rho_0} + s \end{vmatrix}}$$

If we develop both the numerator N(s) and the denominator D(s) in this expression, we obtain polynomials in s. In order to take the inverse Laplace transform, it is most convenient to express this ratio in terms of fractions involving single poles. Thus we write

$$G(k,s) = \sum_i \frac{a_i}{s - s_i}$$

As we know that the inverse Laplace transform of $a_i/(s-s_i)$ is simply given by $a_i e^{ts_i}$, we see that $G(k,t) = \sum_i a_i e^{ts_i}$.

We thus see that we first have to find the poles of D(s), that is the roots of the denominator D(s). The equation D(s)=0 is called the dispersion equation. Here we solve it for real k and complex s. Had we studied acoustically generated sound waves, we would have solved it for real ω and complex k, as we would have studied the spatial decay of monochromatic sound waves. This distinction leads to small differences in the roots when damping is large.

The dispersion equation has been solved approximately by many authors[6] so that we shall only quote the results. As D(s) is a polynomial of degree 3, we find three roots, which are given for small damping by :

$$s = -\frac{\lambda k^2}{\rho_0 C_p} = -x_0$$

$$s_{1,2} = -\Gamma\frac{k^2}{2} \pm ikC_0[1-\frac{1}{2}\Gamma^2\frac{k^2}{C_0^2}] = -x_1 \pm ix_2$$

where

$$\Gamma = [\frac{4}{3}\frac{\eta_s}{\rho_0} + \frac{\eta_b}{\rho_0} + \frac{\lambda}{\rho_0}(\frac{1}{C_v} - \frac{1}{C_p})]$$

These values are valid when both x_0 and x_1 are small compared to x_2 .One then finds as power spectral density :

$$G(k,\omega) \propto (1-\frac{1}{\gamma})\frac{x_0}{x^2+x_0^2} + \frac{x_1}{2\gamma}[\frac{1}{(\omega-x_2)^2+x_1^2} + \frac{1}{(\omega+x_2)^2+x_1^2}]$$

$$-\frac{x_1}{2\gamma x_2}\left[\frac{\omega-x_2}{(\omega-x_2)^2+x_1^2} - \frac{\omega+x_2}{(\omega+x_2)^2+x_1^2}\right]$$

c Features of the spectrum

We find 5 terms in $G(k,\omega)$.

- The first one is centered at zero frequency, and its width depends upon heat propagation. It is called the Rayleigh line.
- The other lines are centered at $\pm x_2$. Around x_2 or $-x_2$, we find a Lorentzian line, called a Brillouin line, of width x_1 , and an antisymmetric component. This antisymmetric line is usually weak, but it must be taken into account when high precision measurements are performed as it shifts the maximum of the Brillouin peak towards zero frequency. The physical origin of the antisymmetric component lies in the fact that we are studying a damped harmonic oscillator (sound wave) driven by white noise (thermal noise)[7].

We see from these results that we can measure

$\lambda/\rho_0 C_p$, $\frac{4}{3}\eta_s + \eta_b$, C_0 and γ from a measurement of the ratio of the integrated intensities of the Rayleigh and Brillouin lines, as we have $I_{Ray} / 2I_{Bri} = \gamma - 1$ (Landau-Placzek ratio)[8].Furthermore the total scattered intensity is proportional to χ_T. The location of the Brillouin peak (x_2) is related to the velocity of sound C at frequency x_2 . We see that

C is a function of k, and that it decreases with k. This phenomenon is called classical dispersion, and it takes place when damping becomes large, that is when η_s/ρ_0 is large. This can be easily achieved in gases where η_s/ρ_0 varies approximately as the reciprocal of the density. We shall discuss this problem in part B.

3 Light scattering in relaxing fluids.

We are going now to consider fluids in which sound velocity depends upon frequency. We have already mentionned classical dispersion that plays an important role in low density gases, but not in ordinary liquids as the frequencies that can be reached by light scattering are very small compared to typical time scales in liquids (collision times for instance). We consider here the case of fluids composed of polyatomic molecules that may exchange internal energy when colliding with one another. We shall start with a simple model, then discuss acoustic in such media, then describe light scattering.

a Physical model

The simplest example of a relaxing fluid is a gas of diatomic molecules. We know that the energy of each molecule can be divided into translational, vibrational and rotational energy. Each of which may be characterized by a temperature : T_{trans} , T_{vib} and T_{rot} associated to a specific heat C_{trans}, C_{vib} and C_{rot} . We consider here very light molecules in which the vibrational levels are so far apart that C_{vib} is negligible at room temperature. We are now going to send a forced sound wave in such a gas. If the sound frequency is small compared to the classical dispersion frequency, the translational temperature will adjust instantaneously to the pressure variations due to the sound wave. What about the rotational temperature ? This may not be the case, as it takes some time for the rotational degrees of freedom to reach equilibrium with the translational temperature.

Suppose that T_{rot} relaxes with a time constant τ , such that

$$\frac{dT_{rot}}{dt} = -\frac{1}{\tau}(T_{rot} - T_{trans})$$

(usually τ is defined in terms of the ratio $Z = \tau/\tau_c$ where τ_c is the mean time between kinetic collisions in the gas. Typically, Z varies from a few units in N_2 or O_2 to very large numbers for vibrational relaxation : 2×10^7 in O_2 at room temperature).

Let us now calculate T_{rot} (t) when the sound wave is present. We have $T_{trans} = T_0 + \delta T\, e^{i\omega t}$ so that

$$T_{rot} = T_0 + \delta T \frac{e^{i\omega t}}{1+ i\omega\tau}$$

We can thus define an effective specific heat at constant volume

$$C_v(\omega) = C_v(\infty) + \frac{C_{rot}}{1+ i\omega\tau}$$

where $C_v(\infty)$ is the specific heat at frequencies very large compared to τ^{-1} but small compared to the classical relaxation frequency. Likewise, we have an effective specific heat at constant pressure

$$C_p(\omega) = C_p(\infty) + \frac{C_{rot}}{1+ i\omega\tau}$$

From there, we calculate the velocity of sound, which is equal to $C = (\gamma/\rho\,\chi_T)^{1/2}$ where γ is now a function of ω. We find that C is complex. Its real part is the velocity, and the imaginary part is related to damping. So we see that the presence of a relaxation time for the internal degrees of freedom leads to a dispersion of the sound waves. We are now going to study this effect for light scattering experiments.

b Langevin equations

We shall extend the model used for simple fluids, by adding to our set of dynamical variables, a new variable ξ that represents the density of internal energy at point $\vec{r}$, and at time t. It is usually more convenient to choose this new variable in such a way that it is orthogonal at time 0 to the other variables of the set. (Several choices can be made[10][11] The determination of the new dynamical matrix is much more complicated than for the pure simple fluid case. If one studies the small frequency, small wave vector regime, one can develop the dynamical matrix in powers of k. This has been done up the second order in k by Weinberg and Oppenheim [10] who get explicit expressions of the matrix elements of M in terms of correlation functions of microscopic variables. Some of these matrix elements can be determined in well defined experiments, but one cannot get all the terms. A similar approach has been used by Desai and Kapral[11] who have calculated all the required terms in the case of a gas of polyatomic molecules which is considered as a mixture of two species : ground state and excited state, which can interact like in chemical reactions. We shall not go here into any further details about these theories. We should just say that they are developments of thermodynamic theories as were used by Mountain [12] in the case of light scattering, or earlier by Mandelstam and Leontovitch[13] for acoustics.

c Generalized hydrodynamics

Now as light scattering is usually not sensitive to the degree of excitation of the molecules, we do not need to know the value of the variable ξ . So what is commonly done is to eliminate ξ from the set of equations. We contract our set $\{\rho, T, v_z, \xi\}$ into

$\{\rho, T, v_z\}$.(We still consider systems in which transverse waves are not coupled to the dielectric constant). Doing this, we shall define a new dynamical matrix $\mathfrak{M}$ for the reduced set $\{\rho, T, v\}$. This new matrix $\mathfrak{M}$ involves now frequency dependent terms that appear when ξ is eliminated. We call them frequency dependent transport coefficients. Usually, the only important term that comes in is a Frequency dependent bulk viscosity. We thus have

$$\frac{d}{dt}\vec{A} = -\begin{pmatrix} 0 & 0 & -ik\rho_0 \\ 0 & \frac{\lambda k^2}{\rho_0 C_v} & -i\frac{k(\gamma-1)}{\beta} \\ -ik\frac{C_0^2}{\rho_0\gamma} & -ik\beta\frac{C_0^2}{\gamma} & \frac{4}{3}\eta_s\frac{k^2}{\rho_0}+\eta_b\frac{k^2}{\rho_0}+\eta_b(s)\frac{k^2}{\rho_0} \end{pmatrix}\vec{A}+\vec{F}(t)$$

where $\eta_b(s)$ is the Laplace transform of the time dependent part of the bulk viscosity. We see that this equation differs from that of a simple fluid only with respect to the term $\eta_b(s)$. It can be shown [9] that in order for this set of equations to yield the same dispersion of sound velocity as in our simple model described in section 2a, we must take

$$\eta_b(s) = (C_\infty^2 - C_0^2)\,\rho_0\tau'\,\frac{1}{1+s\tau'}$$

where $\tau' = \dfrac{(C_v - C_i)}{C_v}\tau$. Here C_0 and C_∞ are the low and high frequency limits of the sound velocity, C_v is the total specific heat at constant volume, and C_i is the internal specific heat (that is C_{rot} in the case of rotational relaxation). Zwanzig [14] has obtained the same expression for $\eta_b(s)$ using more rigorous arguments.

These equations which define generalized hydrodynamics with frequency dependent transport coefficients, lead to very good agreement with acoustical experiments for liquids. We shall see in the next part that they fail in the case of gases.

d Calculation of the density autocorrelation function

We shall follow the same derivation as for simple fluids, but the mathematics will be more involved owing to the term $\eta_b(s)$. We have as before

$$G(k,s) = \langle \rho(k,s)\ \rho(k,0)\rangle = \frac{N_I(s)}{D_I(s)}$$

where

$$N_I(s) = \begin{vmatrix} \frac{\lambda k^2}{\rho_0 C_v} + s & -ik \frac{\gamma - 1}{\beta} \\ -i\beta k \frac{C_0^2}{\gamma} & \frac{4}{3}\eta_s \frac{k^2}{\rho_0} + \eta_b \frac{k^2}{\rho_0} + \eta_b(s) \frac{k^2}{\rho_0} + s \end{vmatrix}$$

and

$$D_I(s) = sN_I(s) + k^2 \frac{C_0^2}{\gamma} \left(\frac{\lambda k^2}{\rho_0 C_v} + s \right)$$

In order to get two polynomials N(s) and D(s), we have to multiply $N_I(s)$ and $D_I(s)$ by $1 + s\tau'$. We see that N(s) is a polynomial of degree 3, and D(s) a polynomial of degree 4. Therefore, the dispersion equation will have 4 roots instead of 3. Mountain[15] has shown that the fourth root is real, and thus corresponds to a line centered at zero frequency. It is commonly called the Mountain line. Its width is of the order of τ^{-1}. Let us now show how this comes about.

We first use dimensionless quantities, putting

$$x = \frac{s}{kC_0} \quad ; \quad a = \frac{\lambda k^2}{\rho_0 C_v k C_0} \quad ; \quad b = \left(\frac{4}{3}\frac{\eta_s}{\rho_0} + \frac{\eta_b}{\rho_0}\right) \frac{k^2}{kC_0} \quad ;$$

$$b_1 = (C_\infty^2 - C_0^2) k^2 \frac{\tau'}{kC_0} \quad \text{and} \quad \theta = k\, C_0 \tau'$$

We obtain

$$D(x) = \theta x^4 + x^3[1+\theta(a+b)] + x^2[a+b+b_1 + \theta(1+ab)] + x[1+a(b+b_1) + \theta\frac{a}{\gamma}] + \frac{a}{\gamma}$$

and

$$N(x) = \theta x^3 + x^2[1+\theta(a+b)] + x[a+b+b_1 + \theta(1 - \frac{1}{\gamma} + ab)] + a(b+b_1) + 1 - \frac{1}{\gamma}$$

One can obtain approximate solutions of the dispersion equation for given conditions. We shall choose two different situations. We consider first the case of sound dispersion where $C_0 k\tau' \approx 1$ like is done when ultrasonic waves are studied, and the case of very long relaxation times of the order of the reciprocal of the Rayleigh linewidth.

α Case of ultrasonic dispersion . Mountain[15] and others[7] have studied the case where $C_0 k\, \tau' \approx 1$, that is with the new notation $\theta \approx 1$, for small classical damping, that is $a \ll 1$ and $b \ll 1$. The approximate roots of the dispersion equation are then

$$s_0 = -\frac{\lambda k^2}{\rho_0 C_p} \qquad \text{as for a simple fluid}$$

$$s_1 = - \frac{C_0^2}{v^2 \tau'}$$

and

$$s_{2,3} = - \Gamma_B \pm ivk$$

where

$$(v\, k\, \tau')^2 = \frac{1}{2} [(C_\infty k\, \tau')^2 - 1] + \frac{1}{2} [(1 - C_\infty^2 k^2 \tau'^2)^2 + 4C_0^2 k^2 \tau'^2]^{1/2}$$

and

$$2\Gamma_B = kC_0 \left[a + b_0 - \frac{C_0^2}{v^2} \frac{a}{\gamma} + \frac{b_1}{1 + (v\, k\, \tau')^2} (1 - a\, k\, C_0 \tau') \right]$$

We can interpret this result in the following way : the spectrum is composed of the Rayleigh line insensitive to the relaxation process, the Mountain line centered at zero frequency, and the Brillouin doublet. The width and the location of the Brillouin lines are now a function of $C_0 k\, \tau'$. For $C_0 k\, \tau' << 1$, we have the low frequency velocity of sound, whereas we have to high frequency velocity of sound for $C_0\, k\, \tau' >> 1$. Usually the amplitude of the Mountain line is quite small, but it can be observed [16]. Its presence shows up mainly through the shift and broadening of the Brillouin lines, thus providing an efficient technique to determine relaxation times.

These expressions for the spectrum are already quite complicated, but they become almost untractable when there are more than one relaxation time. In that case, we can take

$$\eta_b(s) = (C_\infty^2 - C_0^2)\, \rho_0 \int_0^\infty \frac{\tau\, g(\tau) d\tau}{1 + s\tau}$$

where $g(\tau)$ is the distribution of relaxation times. We define a mean relaxation time $\bar{\tau}$ as

$$\bar{\tau} = \int_0^\infty \tau g(\tau)\, d\tau$$

assuming that

$$\int_0^\infty g(\tau)\, d\tau = 1.$$

It is better then to go back to the expression $G(k,s) = N_I(s)/D_I(s)$ that directly involves $\eta_b(s)$, and to compute the spectrum taking

$$\mathrm{Re}\, (G\, (k,s)\,)_{s\, =\, i\, \omega}$$

So we have seen that a detailed study of the Brillouin spectrum in relaxing fluids allows to extend the ultrasonics techniques in order to determine shorter relaxation times.

β Long relaxation times

An other very interesting case arises when the relaxation time τ is much longer than the period of the sound wave. Such long relaxation times occur in supercooled liquids, the viscosity of which increases very fast when the temperature is lowered. In these systems, the bulk viscosity relaxes at low frequencies, involving usually quite a broad distribution of relaxation times, as will be discussed by Ostrowsky [17]. It is therefore necessary to use directly $G(k,s) = N_I(s) / D_I(s)$ as mentionned before. Nevertheless, we way give an expression for G(k,s) when $\tau \gg \rho_0 C_p / \lambda k^2$, that is when the Mountain line is the narrowest feature of the whole spectrum.

Taking the expressions for $N_I(s)$ and $D_I(s)$, and considering s of the order of $1/\tau$, the leading terms yield

$$G(k,s) \approx \frac{\eta_b(s) \frac{k^2}{\rho_0}}{s\eta_b(s) \frac{k^2}{\rho_0} + C_0^2 \frac{k^2}{\rho_0}}$$

If the dispersion is small, that is $(C_\infty^2 - C_0^2)\,\tau s \ll \frac{C_0^2}{\gamma}$, one can neglect $\eta_b(s)$ in the denominator, and then $G(k,s) \approx \eta_b(s)$. So, one can directly measure $\eta_b(s)$ from the correlation function of the scattered field. For cases of interest, the dispersion is not small, so one cannot make this approximation, but there are some doubts about the validity of the model in that case. As discussed by Ostrowsky [17] studies of the Mountain line in a very viscous fluid afford a way to extend ultrasonic measurements towards longer relaxation times.

This discussion has shown that light scattering is a useful technique to determine transport coefficients and their frequency dependence in the range from about 1 to 10^{-11} sec. It is thus used for vibrational or rotational relaxation in many fluids, but can be used to study structural relaxation only in fairly viscous liquids[7][17]. Thus there is no overlap between the experimentally measurable relaxation times, and the time scales involved in the long tails of frequency dependent transport coefficients that have recently been discovered in simple fluids like argon by molecular dynamics calculations [18] .

We shall now make a few remarks about the experimental techniques that are commonly used in this field. As the cross section for light scattering in ordinary fluids is usually quite small, less than $10^{-6}\,cm^{-1}$ per steradian, the intensity of the scattered light per coherence solid angle is so small that signal to noise considerations limit the use of light beating techniques to very small spectral widths, say less than 10^5 Hz. For larger widths, one must use high resolution Fabry-Perot interferometers. Narrow lines include the Rayleigh line at small scattering angle, the line due to concentration fluctuations and the Mountain line when the relaxation time is quite long. All these have been studied in great details thanks to the excellent resolution of light beating spectroscopy. But most of the studies have to be made with the finite resolution of interferometers, which means that details of the line shapes are often lost while taking the convolution of the theoretical line shapes and the instrumental line shape. However, one could conceivably study the stimulated counterpart of light scattering, which provides enough power to use light beating techniques. But it is likely that the use of high power lasers might create more problems than it can solve.

B ▪ LIGHT SCATTERING BY GASES

The subject of light scattering by gases has been of interest for two reasons. First gases are much simpler systems than liquids, so that one expects to have a much better understanding of their properties. Second, they exhibit time scales of interest that can be taken close to the reciprocal of the Brillouin shift. This is different from ordinary liquids where characteristic lengths or frequencies cannot be reached by light scattering, at least as long as one uses visible lasers as light sources.

In a gas, the mean time between collisions, or equivalently, the mean free path can be varied quite easily by changing the pressure. Standard gas kinetic theory [19] relates a mean time between collisions τ_c to the experimentally measured shear viscosity by

$$\eta_s = 1.27\ p\ \tau_c$$

when one considers the gas as composed of elastic spheres, if p is the pressure. This yields for argon $\tau_c = 1.7 \times 10^{-10}$ sec. at room temperature and atmospheric pressure. This corresponds to a mean free path

$$\ell_c = \tau_c \times (3k_B T/m)^{1/2} = 7.35 \times 10^{-5} \text{ cm}$$

which is close to the wavelengths of the density fluctuations studied by backward light scattering. As we shall see, the possibility of changing ℓ_c with pressure leads to line shapes that are similar to that of a liquid for $\ell_c \ll \ell_{Fluc}$, or to Doppler line shapes when $\ell_c \gg \ell_{Fluc}$, that is when the molecules scatter independently of one another. This has been studied experimentally in monoatomic gases, [20][21][22] in polyatomic gases [23][24] and in mixtures .[21][25]. One usually characterizes the situation with a parameter $y = (k\ \ell_c)^{-1}$ where k is the scattering wave vector. The results of experiments are in good agreement with the theoretical spectra that we are going to derive now.

Gases have been studied for a long time by kinetic theory in order to relate interatomic potentials to virial coefficients and to transport coefficients. Furthermore, acoustic measurements [26] have been performed on low density gases and have been compared to theoretical calculations [27] that were subsequently adapted to the case of light scattering experiments.

We shall first study the case of a pure monoatomic gas in which atoms are represented by their positions and momenta. We can use several methods to compute the spectrum of the density fluctuations, according to whether we use a macroscopic or a microscopic description of the local state of the gas. We shall start from a microscopic description.

1 Distribution function

We consider a small volume element $d\vec{r} = dx\,dy\,dz$ centered at point $\vec{r}$, and define the distribution function as

$$dN(\vec{r},\vec{u},t) = f(\vec{r},\vec{u},t)\ d\vec{r}\ d\vec{u}$$

where $dN(\vec{r},\vec{u},t)$ is the number of atoms present at time t in the volume $d\vec{r}$, the velocity of which is $\vec{u}$ within the velocity volume element $d\vec{u}$.

It is well known that for thermal equilibrium at temperature T_0, the distribution function is given by Maxwell distribution

$$f_0\ (\vec{r},\vec{u}) = N\left(\frac{m}{2\pi k_B T_0}\right)^{3/2} \exp - \frac{mu^2}{2k_B T_0}$$

where N is the total number of atoms per unit volume, and m is the mass of an atom. The knowledge of f allows to calculate all the properties of the gas that are averages over one particle coordinates. They are obtained by taking the moments of the distribution f.

2 Moments of the distribution function

We consider some physical quantity $y(\vec{r},\vec{u})$ that depends upon the position and velocity of single atoms. We shall obtain its average value at point $\vec{r}$ through the expression

$$< y(\vec{r},t) > = \int f(\vec{r},\vec{u},t)\ \ y(\vec{r},\vec{u})\ d\vec{u}$$

$< y >$ is called a moment of the distribution f for the function $y(\vec{r},\vec{u})$ by extension of the usual definition of moments when y is a power of the velocity $\vec{u}$.

Among the moments of f, those which correspond to quantities that are conserved during collisions are specially important. We know that the number of particles, the momentum and the kinetic energy are conserved when atoms collide, so that we consider

$$y(\vec{r},\vec{u}) = m$$

$$y(\vec{r},\vec{u}) = m\vec{u} = \vec{p}$$

$$y(\vec{r},\vec{u}) = \frac{1}{2}\ mu^2$$

such that

$$\rho(\vec{r},t) = \int f(\vec{r},\vec{u},t)\, d\vec{u} \quad \text{is the density}$$

$$\vec{p}(\vec{r},t) = m\int f(\vec{r},\vec{u},t)\, \vec{u}\, d\vec{u} \quad \text{is the momentum density}$$

$$Q(\vec{r},t) = k_B T_{trans} = \frac{m}{2}\int f(\vec{r},\vec{u},t)\, u^2\, d\vec{u} \quad \text{is the kinetic energy density}$$

We can now describe fluctuations in the gas from two points of view. We may first use a macroscopic description in which one writes down equations for ρ, $\vec{p}$ and T_{trans} as we did for simple fluids, or else use a microscopic description to determine $f(\vec{r},\vec{u},t)$ for given initial conditions $f(\vec{r},\vec{u},0)$ and then compute ρ, $\vec{p}$ and T_{trans}. Both methods require the use of Boltzmann equation that tells how $f(\vec{r},\vec{u},t)$ depends upon t.

3 Boltzmann equation

We recall briefly that Boltzmann equation can be derived [28] assuming that one considers a gas of particles interacting with short range two body forces, and one uses classical equations of motion for time scales that are long compared to collision times. This limits the use of this model to low density where the mean free path is large compared to the range of the interatomic forces. One finds

$$\frac{\partial}{\partial t} f(\vec{r},\vec{u},t) + \vec{u} \cdot \frac{\partial f(\vec{r},\vec{u},t)}{\partial \vec{r}} = \left.\frac{\partial f}{\partial t}\right)_{collision}$$

when no external forces are present.

The term $\left.\frac{\partial f}{\partial t}\right)_{collision}$ can be expressed in terms of collision cross-sections. If two particles of initial velocities $\vec{u}$ and $\vec{u}_1$ collide in such a way that their velocities are $\vec{u}'$ and $\vec{u}_1'$ after collision, one has for isotropic interatomic potentials

$$\left.\frac{\partial f}{\partial t}\right)_{collision} = \int d\vec{u}_1 \int d\Omega\, |\vec{u} - \vec{u}_1|\, I(|\vec{u} - \vec{u}_1|, \theta)[f(\vec{r},\vec{u}',t) f(\vec{r},\vec{u}_1',t) - f(\vec{r},\vec{u},t)\, f(\vec{r},\vec{u}_1,t)]$$

where θ is the angle between $\vec{u}$ and $\vec{u}'$, I the collision cross-section and $d\Omega$ the solid angle element of the final velocity u'.

In order to study fluctuations around equilibrium, one linearizes Boltzmann equation in terms of small departures from Maxell distribution. We put

$$f(\vec{r},\vec{u},t) = f_0(\vec{u})\, [\, 1 + h(\vec{r},\vec{u},t)\,]$$

We then get a linear equation for h

$$\frac{\partial h(\vec{r},\vec{u},t)}{\partial t} + \vec{u}.\ \frac{\partial h(\vec{r},\vec{u},t)}{\partial \vec{r}} = J\ h(\vec{r},\vec{u},t)$$

where J is the linearized collision operator given by

$$J= \int d\vec{u}_1\, f_0(\vec{u}_1) \int d\Omega\ |\ \vec{u}_1-\vec{u}\ |I(|\vec{u}_1-\vec{u}|,\theta)[\ h(\vec{r},\vec{u}_1,t) + h(\vec{r},\vec{u}'_1,t) -$$

$$h(\vec{r},\vec{u},t) - h(\vec{r},\vec{u}',t)\ ]$$

Knowing h, one can easily find the fluctuating part of ρ, $\vec{p}$ and T_{trans} by calculating the moments of m, $m\vec{u}$ and $\frac{1}{2}\ mu^2$.

$$\rho(\vec{r},t) = m \int f_0(\vec{u})\ h(\vec{r},\vec{u},t)\ d\vec{u}$$

$$p(\vec{r},t) = m\vec{v}(\vec{r},t) = m \int f_0(\vec{u})\ h(\vec{r},\vec{u},t)\ \vec{u}\ d\vec{u}$$

$$Q(\vec{r},t) = \frac{m}{2} \int f_0\ (\vec{u})\ h(\vec{r}.\vec{u},t)\ u^2 d\vec{u}$$

4 Macroscopic equations

We shall now discuss macroscopic equations for ρ, $\vec{p}$ and T_{trans} . If we take the moments of the conserved quantities m, $m\vec{u}$ and $\frac{1}{2}\ m\vec{u}^2$, we get [29]

$$\frac{\partial \rho}{\partial t} + \rho_0\ \text{div}\ \vec{v} = 0$$

$$\rho_0\ \frac{\partial v_i}{\partial t} = -\ \frac{\partial P_{i\alpha}}{\partial r_\alpha}$$

$$\frac{\partial Q}{\partial t} = C_v\ T_{trans} = -\ \text{div}\ q$$

where $P_{ij} = \rho_0 < u_i\ u_j >$ is the stress tensor, and $q_i = \frac{1}{2}\rho_0 < u_i u_\alpha u_\alpha >$

is the energy current density, and ρ_0 the equilibrium density.

We thus have to express P_{ij} and q_i in terms of ρ , v and T_{trans} . This can be done by successive approximations starting from the Maxwell distribution

$$f(\vec{r},\vec{u},t)=n(\vec{r},t)\left(\frac{m}{2\ k_B T(\vec{r},t)} \right)^{3/2} \exp\left[- \frac{m}{2k_B\ T(\vec{r},t)} (\vec{u}-\vec{v}(\vec{r},t))^2 \right]$$

where $n(\vec{r},t)$, $v(\vec{r},t)$ and $T(\vec{r},t)$ are now functions of position and time.

Chapman and Enskog have developped an iterative method to calculate $f(\vec{r},\vec{u},t)$ in terms of powers of the gradients of n, v and T. From the various approximations for f, one obtains different expressions for P and q. Here are the results for the case of longitudinal waves in the x direction.

Order 0 $P_{xx}^{(0)} = P$ hydrostatic pressure

$$q_x^{(0)} = 0$$

Order 1 $P_{xx}^{(1)} = P_{xx}^{(0)} - \frac{4}{3}\eta_s \frac{\partial v}{\partial x}$ Newton's law (η_s shear viscosity)

$q_x^{(1)} = -\lambda \frac{\partial T}{\partial x}$ Fourier's law (λ thermal conductivity)

Order 2

$$P_{xx}^{(2)} = P_{xx}^{(1)} + \frac{2}{3}(\omega_3 - \omega_2)\frac{\eta_s^2}{\rho_0 T}\frac{\partial^2 T}{\partial x^2} - \frac{2}{3}\omega_2\frac{\eta_s^2}{\rho_0^2}\frac{\partial^2 \rho}{\partial x^2}$$

$$q_x^{(2)} = q_x^{(1)} + \frac{2}{3}(\theta_4 - \theta_2)\frac{\eta_s^2}{\rho_0}\frac{\partial^2 v}{\partial x^2}$$

Order 3

$$P_{xx}^{(3)} = P_{xx}^{(2)} + \frac{2}{3}(\omega_7 + \omega_8)\frac{\eta_s^3}{\rho_0 P_0}\frac{\partial^3 v}{\partial x^3}$$

$$q_x^{(3)} = q_x^{(2)} + (\theta_6 + \theta_7)\frac{\eta_s^3}{\rho_0^2 T}\frac{\partial^3 T}{\partial x^3} + \theta_6\frac{\eta_s^3}{\rho_0^3}\frac{\partial^3 \rho}{\partial x^3}$$

This procedure yields expressions for η_0 and λ involving the interatomic potentials. The numerical factors θ's and ω's have been worked out for hard sphere and Maxwell molecules [30]. They are of the order of unity.

Using these expressions for P_{xx} and q_x, one obtains 4 hydrodynamics equations traditionally called :

Order 0	Kirchhoff
Order 1	Navier-Stokes
Order 2	Burnett
Order 3	Super-Burnett

One can then use these equations as we did in part A. This has been done for monoatomic gases, and when the results are compared either to experiments or to the more rigorous theory that will be discussed later, one finds that this series of equations is useful only for $k\ell_c < 0,3$ for Burnett, with slight improvements for the super-Burnett. This was to be expected as inpection shows that the ratios of the successive terms added to P_{xx} or q_x are of the order of $k\ell_c$. Thus this development cannot converge for large values of $k\ell_c$. To avoid this problem, one must directly solve the linearized Boltzmann equation.

5 Study of the linearized Boltzmann equation

We are going to show how one can solve the linearized Boltzmann equation for given initial conditions. We have

$$\frac{\partial h}{\partial t}(\vec{r},\vec{u},t) + \vec{u}.\frac{\partial h}{\partial \vec{r}}(\vec{r},\vec{u},t) = J\, h\, (\vec{r},\vec{u},t)$$

As we are interested in calculating the density autocorrelation function for a given wave vector $\vec{k}$, we first take a spatial. Fourier transform, then a Laplace transform, that we note $h(\vec{k},\vec{u},s)$. We have

$$sh(\vec{k},\vec{u},s) + i\vec{k}.\,\vec{u}h(\vec{k},\vec{u},s) - Jh(\vec{k},\vec{u},s) = h(\vec{k},\vec{u},0)$$

where J is the linearized collision operator that acts on velocity variables.

a Eigenvectors of the collision operator

It is very convenient, as we shall see later, to use the eigenvalues and eigenvectors of the collision operator J. The set of all functions of $\vec{u}$ is a vectorial space, and from it one defines a Hilbert space using a scalar product of two vectors. If $|\phi_i>$ and $|\phi_j>$ are elements of the Hilbert space, the scalar product $<\phi_i \mid \phi_j>$ is defined by

$$<\phi_i \mid \phi_j> = \int \phi_i(\vec{u})\, f_0(\vec{u})\, \phi_j(\vec{u})\, d\vec{u}$$

where $f_0(\vec{u})$ is the Maxwell distribution for a uniform gas at temperature T_0 . The reason for this particular choice of a scalar product is that is allows easy calculations of fluctuations. Let us consider a function $y(\vec{r},\vec{u},t)$ and a distribution function $f(\vec{r},\vec{u},t) = f_0[1 + h(\vec{r},\vec{u},t)]$ The moment of y is given by

$$\begin{aligned} < y(\vec{r},t) > &= \int f(\vec{r},\vec{u},t) \quad y(\vec{r},\vec{u},t) \; d\vec{u} \\ &= y_0(\vec{r},t) + \int y(\vec{r},\vec{u},t) \; h(\vec{r},\vec{u},t) \; f_0 \; (\vec{u}) \; d\vec{u} \\ &= y_0(\vec{r},t) + < y | \; h > \end{aligned}$$

so that the fluctuating part of $< y >$ is just $< y| \; h >$.

The eigenvectors of J are defined by

$$J | \psi_n > = \lambda_n | \psi_n >$$

where the set of the $| \psi_n >$ is chosen to be orthonormal in the sense of the scalar product defined above. One can show that the set $\{| \psi_n > \}$ is a basis of the Hilbert space, so that we write for any h

$$h(\vec{k},\vec{u},s) = \sum_n C_n (\vec{k},s) \; \psi_n \; (\vec{u})$$

We get then, assuming $\vec{k}$ to be in the z-direction :

$$\sum_n sC_n(k,s)\psi_n(\vec{u}) + iku_z \sum_n C_n(k,s)\psi_n(\vec{u}) - J \sum_n C_n(k,s)\psi_n(\vec{u}) = h(k,0)$$

In order to obtain a set of linear equations in C_n , we take the scalar product of this equation with $<\psi_m|$. Thus

$$sC_m(k,s) + ik\sum_n < \psi_m | u_z\psi_n > C_n(k,s) - \lambda_m C_m(k,s) = <\psi_m|h(k,0) >$$

as we assumed that $< \psi_m | \psi_n > = \delta_{mn}$

Now, we need to know the initial conditions h(k,0). As we are studying density fluctuations, we assume that there is one extra atom at location $\vec{r}_0$ and time 0, the velocity of which is given according to an equilibrium Maxwell distribution. Thus

$$f(\vec{r},\vec{u},0) = f_0 \; [\, 1 + \delta(\vec{r} - \vec{r}_0) \,]$$

so that $h(\vec{k},\vec{u},0) = 1$, which as we shall see is one of the eigenvectors of J.

Obviously, finding the eigenvectors of J is not an easy task, but this has been done for the particular case of Maxwell molecules (molecules that interact through a potential $V = -K/r^4$ where r is the distance between two molecules), and for hard spheres . Among the eigenvectors, one finds the functions of u that are conserved during collisions. They are for a monoatomic gas :

$$\psi_1 \; (\vec{u}) \propto 1$$

$$\psi_2 \; (\vec{u}) \propto u_x$$

$$\psi_3 \; (\vec{u}) \propto u_y$$

$$\psi_4 \; (\vec{u}) \propto u_z$$

$$\psi_5 \; (\vec{u}) \propto \frac{2}{3} u^2 - 1$$

all of which correspond to eigenvalues equal to zero. Therefore, our initial conditions are $h(k,u,0) \propto |\psi_1>$. We can now solve the set of linear equations for $C_m(k,s)$ as the knowledge of the $|\psi_m>$ and the λ_m allows to calculate the"matrix element" $<\psi_m|U_z \psi_n>$.

b Approximate solutions

It is known that the set of the eigenvectors is infinite, and that apart from the first few eigenvectors that correspond to conserved quantities, all the eigenvalues are negative. Furthermore, the set of eigenvalues is not bounded. Thus the set of equations for $C_m(k,s)$ is infinite and cannot be used for practical purposes. We thus need to make some approximations to reduce the problem to a finite number of terms. This can be done in two ways.

1 Wang-Chang Uhlenbeck method

Wang-Chang and Uhlenbeck [31] studying the application of the linearized Boltzmann equation to calculate the speed of sound in a dilute gas, used a truncation scheme. They ordered the $|\psi_m>$ for decreasing λ_m, and decided to discard all terms beyond $m > N$. Thus they reduced the problem to a system of N equations, which can be formally solved. Let us call

$$(\overset{\rightrightarrows}{M})_{mn} = <\psi_m | u_z \psi_n>$$

$$(\overset{\rightrightarrows}{L})_{mn} = \lambda_n \delta_{mn}$$

$$(\overset{\rightrightarrows}{I})_{mn} = \delta_{mn}$$

and

$$(\vec{C}(k,s))_m = C_m(k,s)$$

then

$$(s\overset{\rightrightarrows}{I} + i k \overset{\rightrightarrows}{M} - \overset{\rightrightarrows}{L})\vec{C} = \vec{C}_0$$

and as the fluctuating part of the density is $|\psi_1>$ we get :

$$G(k,s) = <\rho(k,s)\,\rho(k,0)> \propto \left(\frac{1}{s\overset{\rightrightarrows}{I} + ik\overset{\rightrightarrows}{M} - \overset{\rightrightarrows}{L}}\right)_{11}$$

This procedure is somewhat equivalent to the series of hydrodynamics equations, and it it limited to fairly small values of $k\ell_c$ although one can calculate G(k,s) for quite large numbers of functions using computers [32].

2 Models of the collision operator

In this method, one replaces the actual collision operator by a more tractable one. This was first done by Bhatnagar, Gross and Krook [33] who replaced the collision term $(\partial f/\partial t)_{collision}$ in Boltzmann equation by a single term involving a relaxation time.

Suppose that we know the first N eigenvectors and eigenvalues of J, we shall replace all the other unknown eigenvalues by a single value μ. So, instead of the true collision operator which can be expressed in terms of projection operators as

$$J = \sum_{n=1}^{\infty} \lambda_n |\psi_n\rangle\langle\psi_n|$$

we shall have

$$J' = \sum_{n=1}^{N} \lambda_n |\psi_n\rangle\langle\psi_n| + \mu \sum_{n>N} |\phi_n\rangle\langle\phi_n|$$

where μ is a constant negative number such that $|\mu| > |\lambda_n|$ for all n, and the ϕ_n are the basis of a subspace that is complementary to the subspace spanned by the N vectors $|\phi_n\rangle$. We now calculate J'h. We obtain :

$$J'|h\rangle = \sum_{n=1}^{N} \lambda_n |\psi_n\rangle\langle\psi_n|h\rangle + \mu \sum_{n>N} |\phi_n\rangle\langle\phi_n|h\rangle$$

$$= \mu|h\rangle + \sum_{n=1}^{N} (\lambda_n - \mu)|\psi_n\rangle\langle\psi_n|h\rangle$$

using the fact that

$$\sum_{n=1}^{N} |\psi_n\rangle\langle\psi_n| + \sum_{n>N} |\phi_n\rangle\langle\phi_n| = I$$

the unit operator.

We now consider the approximate linearized Boltzmann equation:

$$sh(k,s) - iku_z\, h(k,s) = J'h(k,s) + h(k,0)$$

when we replace J'h(k,s) by the expression given above, we obtain :

$$|h(k,s)\rangle = \sum_{n=1}^{N} \frac{|\psi_n\rangle(\lambda_n-\mu)}{s-iku_z-\mu}\langle\psi_n|h(k,s)\rangle + \frac{|h(k,0)\rangle}{s-iku_z-\mu}$$

This expression can be used to calculate the first N moments of f, so that

$$\langle\psi_{n'}|h(k,s)\rangle = \sum_{n=1}^{N} (\lambda_n-\mu)\langle\psi_{n'}|\frac{n}{s-iku_z-\mu}\rangle\langle\psi_n|h(k,s)\rangle +$$

$$+ h(k,0)\langle\psi_{n'}|\frac{1}{s-iku_z-\mu}\rangle$$

We have now a set of N linear equations where the unknowns are the moments $<\psi_{n'}| h(k,s)>$, the first of which is the fluctuating part of the density. If we define a matrix M by its matrix elements

$$M_{nm} = (\lambda_n - \mu) < \psi_n \mid \frac{\psi_m}{s - iku_z - \mu} >$$

we then have

$$< \rho(k,s) \; \rho(k,0) > \propto \sum_{m=1}^{N} \left(\frac{1}{I-M} \right) < \psi_m \mid \frac{1}{s - iku_z - \mu} >$$

This method leads to fairly complex numerical calculations as the matrix elements of M are themselves functions of k and $s = i\omega$, but it yields very accurate results both for large and small values of $k\ell_c$. It has been used by several authors who showed that it is sufficient to take about 20 moments [32] in order to get very good spectra for any value of $k\ell_c$.

The theoretical spectra obtained in this way have been found to be in very good agreement with experimental spectra measured by Clark [21] in Xenon. Modeling the collision operator has been done for two extreme cases as far as interatomic forces are concerned : the Maxwell molecules that interact with a r^{-4} potential, a very "soft" potential, and the hard spheres. The differences between the two cases are so small that one has little hope to deduce much information concerning the interatomic potential from light scattering experiments.

6 Generalization to polyatomic gases

One can in principle generalize the previous treatment to polyatomic gases. We must now take a distribution function that includes internal degrees of freedom, that is vibrational and rotational quantum numbers (v and j) of the molecules. Let us consider again molecules that are light enough so that only the rotational degrees of freedom can be excited at room temperature. One can define a distribution function $f(\vec{r},\vec{u},t,j)$ that we shall use for problems dealing only with scalar quantities (e.g no magnetic field is present). In the same way as one can derive the Boltzmann equation, Wang-Chang and Uhlenbeck [34] have shown that f is a solution of the following equation

$$\left(\frac{\partial}{\partial t} + \vec{u}\frac{\partial}{\partial \vec{r}}\right) f(\vec{r},\vec{u},t,j) = \sum_{k\ell m} \int d\vec{u}_1 \int d\Omega \; |\vec{u}_1 - \vec{u}| \sigma_{jk}^{\ell m}$$

$$[\; f(\vec{r},\vec{u}',t,\ell) \; f(\vec{r},\vec{u}_1',t,m) - f(\vec{r},\vec{u},t,j) \; f(\vec{r},\vec{u}_1,t,k) \;]$$

One can then deal with this equation exactly as we did for

monoatomic gases. The main difference lies in the fact that very little is known concerning the collision operator, which now includes inelastic collisions as far as the j quantum number is concerned. Desai et al [35] have calculated spectra using a model of the collision operator for 7 moments. The agreement between experimental data and theoretical spectra is quite good in H_2 HD and D_2 .

An other description of polyatomic gases uses generalized hydrodynamics, in which one takes a frequency dependent bulk viscosity as is done to interpret acoustic measurements. Calculations are exactly the same as in a liquid and lead to good agreement with experiment when the rotational relaxation time is of the order of the period of the sound waves giving rise to the Brillouin doublet. The values that have been deduced for relaxation times from light scattering spectra [24] are in good agreement with the results of other kinds of measurements [36].

On the other hand, the use of a frequency dependent bulk viscosity yields very poor results when the relaxation time is very long compared to the reciprocal of the Brillouin shift. In part A, we saw that a long relaxation time meant a sharp Mountain line. This is observed in a liquid, but not in a gas. This result has been explained by Desai et al [37]. We shall now give a very simplified version of the reason why the Mountain line does not show up in a gas.

As we did in part A, we consider 4 dynamical variables $\{\rho, T, v, \xi\}$, where ξ is the density of internal energy. ξ satisfies now the following equation

$$\frac{\partial \xi}{\partial t} = - \frac{\xi}{\tau} + D^{\star} \Delta \xi \quad + \text{coupling terms with } \rho \text{ ,v and T.}$$

We have a relaxation time as for liquids, but we added a new term $D^{\star} \Delta\xi$ which tells that excited molecules can diffuse with a self-diffusion coefficient $D^{\star}$.
If we consider now a given wave vector k, we shall have

$$\frac{\partial \xi}{\partial t} = - \left(\frac{1}{\tau} + D^{\star} k^2 \right) \xi \quad + \text{coupling terms}$$

Thus we get an equation for ξ similar to what we had for a liquid, but now it involves an "effective" relaxation time

$$\frac{1}{\tau_{eff}} = \frac{1}{\tau} + D^{\star} k^2$$

The detailed calculations involve other modifications of the dynamical matrix used in part A, but the main effect is due to changing the relaxation time.

If we consider now a gas where $\tau \propto 1/P$, but where $D^{\star} \propto 1/P$, the Mountain linewidth which is of the order of τ_{eff}^{-1} cannot become very small when the pressure decreases because of the term $D^\star k^2$ which takes over at some low pressure. A detailed study of light scattering in SF_6 shows that this model fits well with experiments.

This k-dependence of the relaxation time should be included in liquids too, but as the diffusion coefficients are much smaller than in gases, they are usually insignificant. However there seems to exist liquids [38] showing fairly long relaxation times τ that are of the order of $(D^\star k^2)^{-1}$ where $D^\star$ is the self-diffusion coefficient.

All what has been done in polyatomic gases uses approximations that are not expected to be valid for $k\ell_c$. Thus, there is still work to be done both experimentally and theoretically, as for most gases the collision number Z is so small that the significant $k\ell_c$ range for relaxation overlaps with the $k\ell_c$ range for classical relaxation.

D ■ DEPOLARIZED LIGHT SCATTERING

We are going now to study the depolarized part of the light scattered by a fluid. We shall first discuss the physical origin of the depolarization, then we shall calculate the correlation functions of the scattered field in terms of molecular coordinates. We shall discuss a few models of molecular reorientation to interpret the short and long time behaviour of rotational motions. Finally we shall briefly describe the depolarization due to translational motions, often referred to as collision induced light scattering.

1 Origin of depolarization

Until now, we considered optically isotropic molecules in which the induced dipole moment was parallel to the incident electric field vector ($\vec{P} = \alpha \vec{E}$), thus the scattering was polarized. But in general the molecular polarizability is a second rank tensor which can be shown to be symmetrical for molecules that are not optically active. Therefore the tensor $\overleftrightarrow{\alpha}$ can be diagonalized to obtain

$$\overleftrightarrow{\alpha} = \begin{pmatrix} \alpha_1 & 0 & 0 \\ 0 & \alpha_2 & 0 \\ 0 & 0 & \alpha_3 \end{pmatrix}$$

The principal axes of $\overleftrightarrow{\alpha}$ are fixed relative to those of the molecule, provided the molecule is rigid. For simplicity, we shall consider only molecules that are cylindrically symmetric, so that $\overleftrightarrow{\alpha}$ involves only two coefficients

$$\begin{pmatrix} \alpha_\perp & 0 & 0 \\ 0 & \alpha_\perp & 0 \\ 0 & 0 & \alpha_{/\!/} \end{pmatrix}$$

when we choose the z-axis parallel to the axis of revolution of the molecule. We see that if $\alpha_{/\!/} \neq \alpha_\perp$, the induced dipole will no longer be parallel to the applied electric field, so that the scattered light will no longer be completely polarized. This is the main origin of depolarization, although we shall see later that a quite different mechanism must also be considered.

2 Calculation of single particle correlation functions

Let us first assume that individual particles in the fluid scatter independently of one another, and that each of them is very much smaller than the wavelength of the incident light beam. We shall compute the correlation function of the scattered field for particular experimental situations. We take an input electric field propagating in the x-direction, polarized either in the z-direction or the y-direction, and observe the light scattered at 90° from the incident beam, that is in the y-direction. We shall find that the scattered electric field has components in the z-direction. These different situations are usually noted Vv, Vh, Hv and Hh, where V or v, and H or h mean vertical and horizontal with respect to the scattering plane. One defines polarization ratios as $\rho = I_h / I_v$ for polarized input light, and sometimes $\rho_u = I_h / I_v$ for natural input light.

Knowing $\overleftrightarrow{\alpha}$ and the angular orientation of the molecule with respect to the laboratory axes, one can easily compute the various components of the scattered field. [39] Usually one observes the scattering from a very large number of molecules, so that we have to calculate the mean scattered intensity, as we assumed that their orientations were uncorrelated. This requires the knowledge of an angular distribution function. In the absence of any strong applied field, or flow in the fluid, the molecular orientation is random, so that the angular distribution function will be uniform. This situation can best be dealt with using the formalism of irreducible tensors for the group of rotations [40].

As we have indicated above, the polarizability $\overleftrightarrow{\alpha}$ is a symmetric second rank tensor. We shall decompose it in terms of irreducible tensor components $T^{(\ell)}_q$ which, by definition, have the property to transform in rotations as spherical harmonics. This basic property is best summarized by writing down the transformation of $T^{(\ell)}_q$ in the rotation $\mathcal{R}(\alpha\ \beta\ \gamma)$ where $(\alpha\ \beta\ \gamma)$ is a set of three eulerian angles. One has

$$\mathcal{R}(\alpha\ \beta\ \gamma)\ T_q^{(\ell)}\ \mathcal{R}^{-1}(\alpha\ \beta\gamma) = \sum_{q'=-\ell}^{+\ell} D_{q'q}^{(\ell)}\ (\alpha\ \beta\ \gamma)\ T_{q'}^{(\ell)}$$

where $D_{q'q}^{(\ell)}$ ($\alpha\ \beta\ \gamma$) is a function of the rotation angles. One can show that the set of functions $D_{q'q}^{(\ell)}$ ($\alpha\ \beta\ \gamma$) is complete, so that we shall be able to express any function of $\{\ \alpha\ \beta\ \gamma\ \}$ in terms of $D_{q'q}^{(\ell)}$ ($\alpha\ \beta\ \gamma$). One finds that

$$\overleftrightarrow{\alpha} = \bar{\alpha}\ T_0^{(0)} + \beta\ T_0^{(2)}$$

where $\bar{\alpha}$ is the mean polarizability and β is proportional to $\alpha_{/\!/} - \alpha_{\perp}$.

When one takes into account the geometry of the experiment, one finds the following result :

$T_0^{(0)}$ yields zero except in the Vv geometry,

$T_0^{(2)}$ yields a contribution in all geometries, and one has

$$\frac{3}{4}\ I_{V_v} = I_{V_h} = I_{H_v} = I_{H_h}$$

This last equality is valid only when the molecules are small compared to the wavelength of light.

Let us now calculate the correlation function of $T_0^{(\ell)}$ taking into account the motions of the molecule. We do this using rotations that transform the axes of the molecule into the laboratory axes. Let $\mathcal{R}(\Omega_0)$ be this rotation for time 0. We have as induced dipole in the z-direction

$$P_z(0) \propto \mathcal{R}(\ \Omega_0)\ T_0^{(\ell)}\ \mathcal{R}^{-1}(\Omega_0)\ E_z$$

and using the fact that $T_0^{(\ell)}$ is an irreducible tensor we get

$$P_z \propto \sum_q D_{q0}^{(\ell)}(\Omega_0)\ T_q^{(\ell)}\ E_z$$

Let us now call Ω the rotation of the molecule between time 0 and time t. We then get for the same component of the induced dipole

$$P_z(t) \propto \mathcal{R}(\Omega_0)\ \mathcal{R}(\Omega)\ T_0^{(\ell)}\ \mathcal{R}^{-1}(\Omega)\mathcal{R}^{-1}(\Omega_0)\ E_z$$

$$\propto \sum_{q'q''} D_{q'0}^{(\ell)}\ (\Omega_0)\ D_{q''q'}^{(\ell)}\ (\Omega)\ T_{q''}^{(\ell)}\ E_z$$

We must now take an ensemble average over the initial orientation of the molecule, which we assumed to be uniformly distributed, thus we have to calculate

$$C(t) \propto \sum_{q,q',q''} \int d\Omega_0 \quad D^{(\ell)}_{q'q''}(\Omega) D^{(\ell)}_{q'0}(\Omega_0) D^{(\ell)*}_{q0}(\Omega_0) T^{(\ell)}_{q''} \; T^{(\ell)}_{q}$$

Now we know that the $D^{(\ell)}_{qq'}$ satisfy an orthogonality theorem :

$$\int d\Omega_0 \; D^{(\ell)}_{qq'}(\Omega_0) \; D^{(\ell')}_{jj'}(\Omega_0) = \frac{8\pi^2}{2\ell+1} \quad \delta_{qj} \quad \delta_{q'j'} \quad \delta_{\ell\ell'}$$

thus

$$C(t) \propto D^{(\ell)}_{00}(\Omega) = P_\ell(\cos\theta)$$

This shows that different orders of irreducible tensors do not mix with one another. In fact this is the reason for using these tensors.

We thus have found that the correlation function for the irreducible component $T^{(\ell)}_0$ is

$$P_\ell(\cos\theta)$$

where θ is the angle between the direction of the molecular axis at time 0 and at time t, and where P_ℓ is the ℓth order Legendre polynomial.

The totally symmetric part $T^{(0)}_0$ gives $P_0(\cos\theta)=1$ so that, as expected, molecular rotations do not affect the polarized part of the spectrum. The depolarized part depends upon

$$P_2(\cos\theta) = \frac{1}{2}(3\cos^2\theta - 1)$$

Now, many molecules exhibit a permanent dipole moment which gives rise to microwave dispersion. In that case we need to know the autocorrelation function of the permanent molecular dipole moment (which transforms as $T^{(1)}_0$) so that dielectric relaxation will provide a measurement of $P_1(\cos\theta) = \cos\theta$. As we shall see later, most of the information that may be obtained concerning rotational motions comes from comparing correlation functions of irreducible components of different orders.

Until now we have not taken into account the geometrical phase factors due to the fact that the molecules are located at point $\vec{r}_i$ in the liquid. If we do so, we find that the polarized part depends upon the isothermal compressibility of the liquid, whereas there is no change in the depolarized component if we assume that angular correlations are negligible for molecules that are separated by the wavelength of light.

This simplified discussion has neglected two points which may be quite important. First we did not consider the finite size of the molecules, which might lead to higher order multipoles when factors $\exp(-i\,\vec{k}.\vec{r})$ are expanded in powers of $\vec{k}.\vec{r}$. Second, we have neglected angular correlations between neighbouring molecules.

This is probably wrong for many liquids, in which one finds that the depolarization ratio is different from the value calculated assuming uncorrelated molecules. [42]. One can then take models for the most probable distributions of molecular orientations and compare the theoretical results to experiments. This has been done for many liquids, [43] but the results are sometimes ambiguous. Obviously a much better way to determine angular correlations is to use the k-dependence of the total cross-section for neutrons scattering, which allows to use k vectors close to the reciprocal of the intermolecular distances.

This question of angular correlations which is already very difficult to deal with for the total scattered intensity, becomes almost untractable when the spectrum is studied. We shall thus limit the discussion to uncorrelated molecules as it seems that only molecular dynamics calculations will be able to deal with the time variation of angular correlations in liquids, in any details. Until now, these calculations have been done for models of diatomic liquids like CO[44] or N_2[45] or for water [46].

We shall thus discuss the behaviour of $< P_2 (\cos \theta) >$ where $\cos \theta = \vec{u}(0).\vec{u}(t)$ if $\vec{u}$ is a unit vector pointing in the direction of the axis of revolution of the polarizability tensor of the molecule. Usually one can break up the observed spectra into two parts, that are roughly lorentzian. Some authors describe a sharp component and a broad component [47]. Here we shall discuss these features in the time domain, in which one characterizes a long and a short time behaviour, and interpret them in terms of single molecular motions.

3 Short time behaviour

The short time behaviour of the motion is best studied by developping C(t) in powers of t

$$C(\tau) = 1 + \frac{t^2}{2!}\gamma_2 + \frac{t^4}{4!}\gamma_4 + \ldots\ldots$$

where we have included only even powers of t as we assume the equations of motion to be invariant under time reversal. The coefficients γ_{2n} are thus simply the time derivatives of C(t) for t=0.

$$\gamma_{2n} = \left.\frac{d^{2n}}{dt^{2n}} C(t)\right|_{t=0}$$

Therefore as C(t) is the Fourier transform of the power spectral density $S(\omega)$ of the scattered light, we have

$$\gamma_{2n} = (-1)^n \int_{-\infty}^{\infty} \omega^{2n} S(\omega)\, d\omega$$

The coefficients γ_{2n} are related to the even moments of the power spectral density.

For very short times, where the only useful term is $\gamma_2 t^2/2$, we can assume that the molecule is free to rotate with an angular velocity distribution function similar to a Maxwell distribution. If we assume that the molecule is spherically symmetric as far as its moments of inertia are concerned (we call them I), one has

$$P(\vec{\omega})\, d\vec{\omega} = \left(\frac{I}{2\pi kT} \right)^{3/2} \exp\left(- \frac{I\omega^2}{2k_BT} \right) d\vec{\omega}$$

from which one can show that [48]

$$\gamma_2 = -6 \frac{k_BT}{I}$$

This same model applied to the case of dielectric relaxation, in which one considers $< P_1 (\cos\theta) >$, leads to

$$\gamma_2 = -2 \frac{k_BT}{I}$$

For later times, that is when the $\gamma_4 t^4$ term comes in, we must take into account the torques applied to the molecule due to its interactions with its neighbours. Gordon [49] has shown that

$$\gamma_4 = 24 \left[4 \left(\frac{k_BT}{I}\right)^2 + \frac{1}{8I^2} <(O\, V)^2> \right]$$

where $<(O\, V)^2>$ is the mean square torque acting on the molecule. Some experimental data have been interpreted this way [50] .

An other way to study the short time behaviour uses the angular velocity autocorrelation function. Let $\vec{\omega}(t)$ be the angular velocity of the molecule, and $G(t) = < \vec{\omega}(t) . \vec{\omega}(0) >$ the angular velocity autocorrelation function. The equation of motion is given by

$$\frac{d\vec{u}}{dt} = \vec{\omega} \times \vec{u}$$

We take now $C(t) = <P_2 (\cos (\vec{u}(t) . \vec{u}(0))) >$ for small times, such that

$$C(t) \simeq 1 - \frac{3}{4} <(\vec{u}(0) . \vec{u}(t))^2 >$$

We then differentiate twice

$$\frac{d^2C(t)}{dt^2} = - \frac{3}{2} \left[<(\vec{u}(0) . \ddot{\vec{u}}(t))(\vec{u}(0) . \vec{u}(t) + (\vec{u}(0) . \dot{\vec{u}}(t))^2 > \right]$$

We know that time derivatives can be changed in correlation functions, so that

$$< (\vec{u}(0) . \ddot{\vec{u}}(t)) > = <(\dot{\vec{u}}(0) . \dot{\vec{u}}(t)) >$$

Now

$$\vec{u}(0) . \dot{\vec{u}}(t) = \vec{u}(0) . (\vec{\omega}(t) \times \vec{u}(t)) \approx 0$$

for small t, and $\vec{u}(0).\ \vec{u}(t) = (\vec{\omega}(0) \times \vec{u}(0)).(\vec{\omega}(t) \times \vec{u}(t))$

$$= (\vec{\omega}(0).\ \vec{\omega}(t))(\vec{u}(0).\vec{u}(t)) -$$

$$- (\vec{\omega}(0).\ \vec{u}(t)(\vec{\omega}(t).\ \vec{u}(0))$$

For small t, $\vec{u}(0)$ and $\vec{\omega}(t)$ are uncorrelated, so finally taking

$$\vec{u}(0)\ .\ \vec{u}(t) \approx 1$$

We get

$$\frac{d^2C(t)}{dt^2} \simeq -\frac{3}{2} < \vec{\omega}(0)\ .\ \vec{\omega}(t) >$$

We thus see that the angular velocity autocorrelation function can be approximately derived from the experimental correlation function. This allows to find the approximate time for which the angular velocity becomes zero, that is when $d^2C/dt^2 = 0$. This has been studied by Dardy and Litovitz [5 1], who have compared this time to the time between collisions in many liquids. As we shall see later this analysis that can yield much information concerning almost free rotational motion is made fairly uncertain because there is a quite different mechanism that gives rise to light scattering in the far wings.

4 Long time behaviour

The long time behaviour of the correlation function of the depolarized light scattering depends upon the details of large angle molecular reorientation. We are going to describe a model which is an extension for molecular rotation of the simple model used by Frenkel [52] to interpret the temperature dependence of the viscosity. In that model, the molecule is trapped for a fairly long time by a potential well, and the probability to go from one position or orientation θ_0 to an other one θ is proportional to a Boltzmann factor $\exp(-W / k_B T)$ where W is an activation energy. The shape of the potential well is not well defined in a liquid, as it is due to the neighbours which can move themselves. (Some models [5 3] consider the ratio of the residence time of the molecule for a given orientation, to the "structural" relaxation time defined as the correlation time of the potential well. When these times are of the same order, one speaks of structure limited motion).

We shall treat here a simplified model in which the molecule has a certain probability to make a given angular motion during one rotational step. As we are interested here only in the direction of the axis of symmetry of the molecule, our task is to study the motion of a point on a sphere, knowing the probability to go from one location of the point to the next one. If the steps are very small and completely random, we shall have the equivalent of random walk in three dimensional space, for which the distribution function of the point obeys the diffusion equation. Similarly the random walk problem (brownian motion) for a point on a sphere gives rise to a probability distribution $W(\Omega,t;\Omega_0,0)$ that obeys

$$\frac{\partial W}{\partial t} = D \Delta W$$

where Δ is the angular part of the laplacian, and D is a rotational diffusion constant. If classical mechanics is applied to calculate the friction coefficient of a rotating sphere of radius a, one finds that

$$D = \frac{k_B T}{8 \pi \eta a^3}$$

where η is the viscosity of the liquid.

If the steps are not small, this equation does not apply, and we need to make a more detailed analysis of the rotational motion, as was done by Ivanov [54]. He calculates first the probability $P(\Omega,N)$ for the molecule to end up in a given orientation Ω after N rotational steps. This probability can be decomposed in sums of the functions $D^{(\ell)}_{qq'}(\alpha \beta \gamma)$ which make a complete set. Thus

$$P(\Omega,N) = \sum_{\ell mn} C^{(\ell)}_{mn}(N)\, D^{(\ell)}_{mn}(\Omega)$$

Writing down a recurrence equation for $P(\Omega,N)$ one finds that for spherically symmetric molecules

$$P(\Omega,N) = \sum_{\ell} \frac{2\ell+1}{8\pi^2} \lambda_\ell^N \operatorname{Tr}\left\{ D^{(\ell)}(\Omega_0^{-1})\, D^{(\ell)}(\Omega) \right\}$$

where Ω_0 is the angular orientation of the molecule before the motion took place, and λ_ℓ a coefficient that depends only upon the movement during one step. Therefore the use of tensorial analysis has allowed to separate the angular dependence of $P(\Omega,N)$ from the term describing single steps. The coefficients λ_ℓ are given by

$$\lambda_\ell = \frac{1}{2\ell+1} \int P(g) \frac{\sin\left[(\ell + \frac{1}{2})g\right]}{\sin(g/2)} dg$$

where P(g) is the probability to have rotation g in one step.

We now calculate $W(\Omega,t;\Omega_0,0)$ assuming that the number of steps during time t is distributed according to a Poisson distribution $W_N(t)$, such that

$$W_N(t) = \frac{1}{N!}\left(\frac{t}{\tau}\right)^N \exp\left(-\frac{t}{\tau}\right)$$

where τ is the mean time between steps. The probability of finding the molecule at angle Ω at time t knowing that it was at angle Ω_0 at time 0, is thus

$$W(\Omega,t;\Omega_0,0) = \sum_{N=0}^{\infty} w_N(t)\ P(\Omega,N)$$

$$= \sum_{\ell} \frac{2\ell+1}{8\pi^2} \exp - \frac{t}{\tau}(1-\lambda_\ell)\ \mathrm{Tr}\{ D^{(\ell)}(\Omega_0^{-1})\ D^{(\ell)}(\Omega) \}$$

Had we used the diffusion equation describing rotational brownian motion, we would have found

$$W(\Omega,t,\Omega_0,0) = \sum_{\ell} \frac{2\ell+1}{8\pi^2} \exp(^{-tD\ell(\ell+1)}) \mathrm{Tr}\{ D^{(\ell)}(\Omega_0^{-1})\ D^{(\ell)}(\Omega) \}$$

In both cases, we obtain correlation functions that are sums over irreducible tensor order. For each order, we have a term that decays exponentially in time, with a time constant equal to

$$\tau_\ell = \frac{\tau}{1 - \lambda_\ell}$$

for the general model, and

$$\tau_\ell = \frac{1}{\ell(\ell+1)\ D}$$

for brownian motion.

We can use now these results to calculate correlation functions of quantities that depend upon the rotation angle between times 0 and t. Let us consider a function $y(\Omega)$ and its correlation function

$$F(t) = \langle y\ (\Omega) \rangle$$

We have

$$F(t) = \int d\Omega d\Omega_0\ W(\Omega,t;\Omega_0,0) y(\Omega) y^\star(\Omega_0)$$

If we decompose $y(\Omega)$ in terms of the functions $D^{(\ell)}_{mn}(\Omega)$

$$y(\Omega) = \sum_{\ell mn} D^{(\ell)}_{mn}(\Omega)\ \Phi_\ell$$

we get

$$F(t) = \sum_{\ell} |\Phi_\ell|^2 \exp - \frac{t}{\tau}(1 - \lambda_\ell)$$

as the averages over angles Ω and Ω_0 are particularly simple for the functions $D^{(\ell)}_{mn}(\Omega)$. We find again the important result that components of different orders do not mix when we calculate the correlation function. Thus each order relaxes exponentially with a time constant equal to $\tau_\ell = \tau/(1-\lambda_\ell)$ or $\tau_\ell = 1/(\ell(\ell+1)D)$ depending upon which model is used for the rotation. We see that we shall learn most about rotational motions by comparing time constants τ_ℓ for various orders ℓ . Dielectric relaxation will yield τ_1 , whereas depolarized light scattering will yield τ_2 . If now we make some assumptions concerning the kind of molecular rotation that takes place in the liquid, we shall be able to compare the ratio τ_1/τ_2 of the measured τ_1 and τ_2 to the value deduced from the model, using our theoretical expression for τ_ℓ .

If we suppose that the angular steps are very small, we can get an approximate expression for λ_ℓ . We find that the ratio of the relaxation times τ_2/τ_1 is 1/3 for both expressions of τ_ℓ. If angular steps are large, one finds that τ_2/τ_1 is close to unity. Thus many measurements have been performed in view of determining the ratio τ_2/τ_1 .

One can give a qualitative explanation of these results. When we consider $\langle P_1(\cos\theta)\rangle$ and $\langle P_2(\cos\theta)\rangle$, the angle for which P_1 and P_2 are equal to 1/e are not the same. They are about 68° for P_1 and 39° for P_2 . Thus the relaxation times are different in the case of small step rotation (brownian motion), but they are about equal when angular steps are very large, so that both P_1 and P_2 decrease very much during each step.

5 Discussion

Most experimental studies have been interpreted in terms of the models that we have just described, that is one measures ratios of relaxation times and activation energies deduced from the temperature variation of the relaxation times.

These models can account only for the gross features of the experimental data. They have the advantage of being simple and to require a very small number of parameters that can be deduced from the data. This allows a relatively easy way to compare the behaviour of different liquids. But one can point out a number of drawbacks of these models.

First as mentionned before we have neglected all kinds of intermolecular correlations. This shows up in the depolarization ratio and in the correlation time of the sharp component. If there is no correlation, one can show that relaxation times derived from NMR measurements (for intramolecular relaxation processes) are equal to relaxation times deduced from the sharp depolarized component. But correlation will affect the depolarized spectrum, and not the NMR data. In the same manner, the linewidth of depolarized Raman lines (once corrected for vibrational linewidth) [55] should not depend upon angular correlations because the molecular vibrations are almost certainly uncorrelated [56] . Cases have been found where $\tau_{NMR} \simeq \tau_{Raman} \neq \tau_{Rayleigh}$, which means that angular correlations play a role.

Second we have not discussed the relationship between the local field that acts on molecules and the incident field. This will be done in the next section, but we should mention that if molecular shapes are not spherical, then the environment of the molecule may not have spherical symmetry, thus leading to a possible source of depolarization. [42].

Interpreting the sharp part of the spectrum is already quite involved, but still the physical origin of the spectrum is fairly clear. The situation is much worse for the broad component. The approach used in section 3 has been criticized by several authors, [47] but apart from the influence of collision induced scattering, it does not seem that one can give a simple physical explanation of the short time behaviour of the correlation function of the depolarized light.

One can use a mathematical description with an enlarged set of dynamical variables, some of them being coupled to the dielectric constant. This has been done mainly in order to explain the coupling between the depolarized part of the spectrum and the hydrodynamic variables [57]. This has enabled to explain the observed data, but the physical meaning of the variables that need to be used is not always very clear, nor are the values of some of the physical parameters. A complete discussion of these questions is beyond the scope of this lecture, but we shall now discuss an other kind of depolarized scattering that plays an important role in the far wings of the spectra.

6 Collision induced light scattering

Let us consider a fluid like argon in which the atoms are optically isotropic. When we apply an incident electric field polarized in some direction, the induced dipole on each atom should be polarized parallel to that same direction, and the scattered field should be polarized. This argument neglects the fact that the electric field acting on the atom is not the same as the incident field. We know that the acting field, the so-called local field, differs from the incident field in many respects. First is does not propagate with the same phase velocity (k becomes nk where n is the index of refraction), second its amplitude differs from the input one by a factor which is usually taken as $(n^2+2)/3$, the Lorentz-Lorenz factor. If we limit the differences between input and local fields to these two features, then our conclusion concerning the polarization of the scattered field will still be valid.

Let us now discuss in more details the local field at point $\vec{r}_i$. The effect of all the other molecules of the system can be divided into two parts. Those which are far away give a uniform contribution that corresponds to the change of wave vector and of amplitude. The molecules that are very close produce a contribution

$$\vec{E}'_i = \sum_{\text{neighbours}} \vec{\vec{T}}(\vec{r}_{ij})\,\vec{E}_j$$

where $\vec{r}_{ij} = \vec{r}_i - \vec{r}_j$ ($\vec{r}_j$ being the location of a neighbouring molecule), $\vec{E}_j$ is the local field at point $\vec{r}_j$, and $\vec{\vec{T}}$ the propagator for the dipolar field produced close to $\vec{r}_j$ by the dipole $\alpha_j \vec{E}_j$. This contribution depends upon the precise structure of the environment of molecule i. If the surroundings of i had perfect spherical symmetry, then the sum giving E_i' would vanish. Therefore we see that E_i' is related to the density fluctuations around point $\vec{r}_i$. The additional field E_i' gives rise to an induced dipole P_i' on molecule i that can scatter light. Summing the contributions from all atoms of the medium, one finds for an input field polarized in the z-direction

$$< E_z^2 > \propto \sum_{i \neq j} \sum_{k \neq \ell} \alpha^2 \, T^{zz}(\vec{r}_{ij}) \, T^{zz}(\vec{r}_{k\ell})$$

and

$$< E_x > \propto \sum_{i \neq j} \sum_{k \neq \ell} \alpha^2 \, T^{xz}(\vec{r}_{ij}) \, T^{xz}(\vec{r}_{k\ell})$$

where

$$T^{zz}(\vec{r}_{ij}) = \frac{1}{r_{ij}^3} \left(1 - 3 \, \frac{z_{ij} \, z_{ij}}{r_{ij}^2}\right)$$

and

$$T^{xz}(\vec{r}_{ij}) = \frac{1}{r_{ij}^3} \left(-3 \, \frac{x_{ij} \, z_{ij}}{r_{ij}^2}\right)$$

where x_{ij}, y_{ij} and z_{ij} are the cartesian coordinates of the vector $\vec{r}_{ij}$. We see that the presence of the dipolar field around molecule i breaks the symmetry of the input field, and that the scattered field has a depolarized component.

In addition ot this effect, that one may describe as being due to fluctuations in the Lorentz-Lorenz factor, there is an other mechanism of depolarization. It may occur that the wave functions of the atoms are distorted by their neighbours in such a way that the atomic polarizability is no longer spherically symmetric inside the liquid. This effect will probably take place when the distance between atoms is so small that electronic orbitals overlap (which gives rise to strong repulsive forces). This distortion is not well known, and we shall have to make some assumptions to estimate its importance. Thus there are two origins for collision induced scattering : density fluctuations and polarizability distortions.

One can show that the term due to density fluctuations leads to 3/4 as depolarization ratio, as in the case of the Rayleigh line due to molecular reorientation. This is about what is found experimentally, so it is believed that this same value applies for polarizability distortions.

Let us now give theoretical expressions for the spectra. It is most convenient here to express the correlation function of the scattered field as

$$C(t) = \langle E_s(t)\, E_s(0) \rangle \propto \langle \sum_{\substack{i \neq j \\ k \neq \ell}} T^{zz}(\vec{r}_{ij}(t))\, T^{zz}(\vec{r}_{k\ell}(0)) \rangle$$

for the z-polarization, if only density fluctuations take place. If one assumes that the polarizability distortions are pair-wise additive, and that they can be expressed in terms of a traceless symmetric tensor with respect to the vector $\vec{r}_{ij}$, then we define a quantity

$$\beta(r) = 6\, \frac{\alpha^2}{r^3} + \gamma(r)$$

where α is the mean polarizability of the isolated atom, and $\gamma(r)$ a function of r. Then, taking advantage of the particularly simple angular dependence of T^{zz} , one can show that

$$C(t) \propto \langle \sum_{i \neq j} \sum_{k \neq \ell} \beta(r_{ij}(t))\, \beta(r_{k\ell}(0))\, P_2(\hat{r}_{ij}(t) . \hat{r}_{k\ell}(0)\,) \rangle$$

where $\hat{r}_{ij}$ is a unit vector parallel to $\vec{r}_{ij}$.

In the case of low pressure gases, the problem simplifies enormously as one can neglect terms involving more than two particles. We can use a model in which only pairs of atoms play a role. One finds that the intensity of the depolarized component is proportional to the square of the density, and that one needs to take into account both density fluctuations and electronic distortions. The total intensity is given by C(0), and to calculate C(0) we can use an ensemble average over all pairs of molecules. This requires the knowledge of the pair distribution function $n_{12}\,(\vec{r}_{12})$. One knows that for a gas

$$n_{12}\,(\vec{r}_{12}\,) \propto \exp - \frac{V(r_{12})}{k_B T}$$

where $V(r_{12})$ is the interatomic potential, thus

$$I_{dep} \propto \int |\, \beta(r)\, |^2 \exp - \frac{V(r)}{k_B T}\; r^2\, dr$$

Therefore one needs to know both the potential V(r) and the pair polarizability $\beta(r)$. The most complete studies have been performed for argon for which the interatomic potential is best known.

$\gamma(r)$ has been deduced assuming some plausible analytic form. [58] [59]. The function $\gamma(r)$ cannot be determined directly, but one can calculate it using quantum chemistry methods. Accurate values have been obtained for He-He,[60] they are in qualitative agreement with the experimental results in argon-argon, in the sense that the range of $\gamma(r)$ is much larger that the range of the strong interatomic repulsive forces. Calculations have also been done [61] for Ar-Ar but as there are many more electrons than in He-He, only qualitative agreement could reached.

In gases, this depolarized component is usually very weak, and it does not affect the other kinds of scattering. This is not the case in liquids where its contribution to the far wings of the Rayleigh line is usually not negligible. As there is no way to distinguish the collision induced part of the scattering from that due to molecular reorientation, one needs to calculate the collision induced part. Fortunately liquified Ar or Xe provide a test system because we know that there is no rotational contribution in these liquids.

Suppose first that we neglect electronic distorsions, the problem of calculating the correlation functions is extremely difficult because we need to know the 2,3 and 4 particles distribution functions

$$n_{12}(\vec{r}_{12}) \ , \ n_{123}(\vec{r}_{12},\vec{r}_{13}) \ \text{and} \ n_{1234}(\vec{r}_{12},\vec{r}_{13},\vec{r}_{14})$$

Fleury et al. [62] have tried to treat the density fluctuations as a second order Raman scattering process, but the function T that comes into the expression of the scattered field varies as r^{-3} so that one needs to consider density fluctuations in the medium whose wave vectors k are of the order of the reciprocal of the intermolecular distances, and these fluctuations are very poorly defined in fluids due to the large damping. Futhermore n_{12}, n_{123} and n_{1234} are not accurately known, and some kind of superposition approximation has to be made.

Thus, one has to turn to molecular dynamics calculations to compute C(t). Alder, Weis and Strauss[63] have first calculated the depolarization ratio assuming no electronic distorsion. Their results are in qualitative agreement with the experimental data of Thibeau et al.[64] for argon, although they are off by almost an order of magnitude at liquid density. It seems probable that an average effective polarizability depending upon density ought to be used.

The situation is much better as far as the spectrum is concerned. Until now two calculations have been performed. Berne, Bishop and Rahman [65] have calculated the correlation function of the scattered light, and they have found that the most important contribution to C(t) is the term $\langle P_2(\vec{u}(t)\ \vec{u}(0)) \rangle$. Thus the exact form of $\beta(r)$ only plays a secondary role for the spectral line shape. This result is in contradiction with some models [66][67] of binary collisions in liquids in which one selects pairs of molecules owing to the fact that $\gamma(r)$ varies very fast with r ($\gamma(r)$) has been taken to vary as r^{-n} with $n\approx 15$), as this is effective only for strong two body collisions.

More recently, Alder et al [68] have performed very detailed calculations of C(t) for different interatomic potentials. They obtain results which are in very good agreement with the experimental line shapes measured by Fleury et al [69] . The theoretical calculations do not seem to indicate particular behaviour for large t as is found for the velocity autocorrelation function or the viscosity. Still it might be worthwhile to determine accurately the center of the depolarized spectrum in argon in order to look for "intercollisional interference effects"[70] One can thus say that collision induced light scattering is fairly well understood, although much work remains to be done in order to determine the absolute intensity of the spectra, which is necessary to substract the collision induced contribution to the wings of the Rayleigh line,when one looks for the short time behaviour of molecular reorientation.

REFERENCES

[1] R. ZWANZIG Ann. Rev. Phys. Chem. 16,67, 1965

[2] H. MORI Progr. Theor. Phys. (Kyoto) 33, 423, 1965

[3] R. ZWANZIG in Lectures in Theoretical Physics edited by W.E. BRITTIN (Wiley, New York) 1961

[4] B.J. BERNE and D. FORSTER Ann. Rev. Phys. Chem. 22, 563, 1971

[5] A similar approach is discussed in L. LANDAU and I. LIFSCHITZ, Statistical Physics.

[6] R.D. MOUNTAIN Rev. of Mod. Physics 38, 205 , 1966

[7] C.J. MONTROSE, V.A. SOLOVYEV and T.A. LITOVITZ J. Acoust. Soc. America 43, 117 , 1968.

[8] H.Z. CUMMINS and R.W. GAMMON J. Chem. Phys. 44, 2785, 1966

[9] K.F. HERZFELD and T.A. LITOVITZ Absorption and Dispersion of Ultrasonic Waves, Academic Press, New York, 1959

[10] M. WEINBERG and I. OPPENHEIM Physica 61, 1, 1972

[11] R.C. DESAI and R. KAPRAL Phys. Rev. A6, 2377, 1972

[12] R.D. MOUNTAIN J. Res. Natl. Bur. Std. A72,95, 1968

[13] L.I. MANDELSTAM and M.A. LEONTOVITCH Zh. Eksp. Teor. Fiz. 7, 438 , 1937

[14] R. ZWANZIG J. Chem. Phys. 43, 714, 1965

[15] R.D. MOUNTAIN J. Res. Natl. Bur. Std. A70, 207, 1966

[16] W.S. GORNALL, G.I.A. STEGEMAN, B.P. STOICHEFF, R.H. STOLEN and V. VOLTERRA Phys. Rev. Letters 17, 297, 1966.

[17] N. OSTROWSKY This conference

[18] B.J. ALDER and T.E. WAINWRIGHT Phys. Rev. Letters 18, 988, 1967.

[19] S. CHAPMAN and T.G. COWLING The Mathematical Theory of Non-Uniform Gases 3rd edition, Cambridge University Press, 1970, page 226.

[20] T.J. GREYTAK and G.B. BENEDEK Phys. Rev. Letters 17, 179, 1966.

[21] N.A. CLARK PhD Thesis, MIT, 1970, unpublished

[22] P. LALLEMAND J. Physique (Paris) 31, 551, 1970.

[23] E.H. HARA, A.D. MAY and H.F.P. KNAAP Can.J. Phys.49,420

[24] A.M. CAZABAT-LONGEQUEUE and P. LALLEMAND J. Physique (Paris) 33 - C1 - 57 , 1972

[25] W.S. CORNALL and C.S. WANG J. Physique (Paris) 33 - C1 - 51, 1972

[26] M. GREENSPAN J. Acoust. Soc. America 28, 644, 1958

[27] See a review in Studies in Statistical Mechanics, Vol. V , edited by J. DE BOER and G.E. UHLENBECK, North Holland Amsterdam, 1970.

[28] See Ref. 19 page 46

[29] See Ref. 27 page 5

[30] See Ref. 27 page 40

[31] See Ref. 27 page 57

[32] A. SUGAWARA, S. YIP and L. SIROVICH Phys. of Fluids 11, 925, 1968.

[33] P.F. BHATNAGAR, E.P. GROSS and M. KROOK Phys. Rev. 94, 511, 1954.

[34] C.S. WANG CHANG and G.E. UHLENBECK and J. DE BOER in Studies in Statistical Mechanics, Vol. II, edited by J. de BOER and G.E. UHLENBECK, North Holland, Amsterdam, 1964.

[35] C.D. BOLEY, R.C. DESAI and G. TENTI Can. J. Phys. 50, 2158, 1972.

[36] G.J. PRANGSMA, A.H. ALBERGA and J.J.M. BEENAKKER Physica 64, 278, 1973.

[37] M. WEINBERG, R. KAPRAL and R.C. DESAI Phys. Rev. A7,1413, 1973.

[38] R.T. BEYER J. Acoust. Soc. America 27, 1, 1955.

[39] I.L. FABELINSKII Molecular Scattering of Light, Plenum Press New York 1968.

[40] M.A. ROSE Elementary Theory of Angular Momentum, Wiley NY, 1957.

[41] A. BEN-REUVEN and N.D. GERSHON J. Chem. Phys. 51, 893, 1969.

[42] E. ZAMIR and A. BEN-REUVEN J. Physique (Paris) 33- C1-237, 1972.

[43] P. BOTHOREL, C. SUCH and C. CLEMENT J. Chimie Physique (Paris) 69, 1453, 1972.

[44] C.D. HARP and B.J. BERNE Phys. Rev. A2, 975, 1970.

[45] J. BAROJAS, D. LEVESQUE and B. QUENTREC Phys. Rev. A7, 1092, 1973.

[46] A. RAHMAN and F.H. STILLINGER J. Chem. Phys. 55, 3336, 1971.

[47] T. KEYES and D. KIVELSON J. Chem. Phys. 56, 1057,1972

[48] W.A. STEELE J. Chem. Phys. 38 , 2411, 1963.

[49] R.G. GORDON J. Chem. Phys. 43, 1307 , 1966.

[50] R.L. ARMSTRONG, S.M. BLUMENFELD and C.G. GRAY Can. J. Phys. 46, 1331, 1968.

[51] DARDY and T.A. LITOVITZ

[52] J. FRENKEL Kinetic Theory of Liquids, Oxford University Press, London, 1946.

[53] D.A. PINNOW, S.J. CANDAU and T.A. LITOVITZ J. Chem. Phys. 49, 347, 1968.

[54] E.N. IVANOV J.E.T.P. 18, 1041, 1964.

[55] S. BRATOS and A. MARECHAL Phys. Rev. A4, 1078,1971.

[56] S. BARTOLI and T.A. LITOVITZ J. Chem Phys. 56,404 and 413, 1972.

[57] T. KEYES and D. KIVELSON J. Physique (Paris)33-C1-231, 1972.

[58] J.P. Mc TAGUE, W.D. ELLENSON and L.H. HALL J. Physique (Paris) 33-C1-241, 1972.

[59] P. LALLEMAND J. Physique (Paris) 33-C1-257, 1972.

[60] E.F. O'BRIEN, V.P. GUTSCHICK, V. Mc KOY and J.P. Mc TAGUE Phys. Rev. A8, 690, 1973.

[61] P. LALLEMAND, D.J. DAVID and B. BIGOT to be published

[62] J.P. Mc TAGUE, P.A. FLEURY and D.B. DUPRE Phys. Rev. 188, 303, 1969.

[63] B.J. ALDER, J.J. WEIS and H.L. STRAUSS Phys. Rev. A 7, 281, 1973.

[64] M. THIBEAU, B. OKSENGORN and B. VODAR J Physique (Paris) 29, 287, 1968.

[65] B.J. BERNE, M. BISHOP and A. RAHMAN J. Chem. Phys. 58, 2696, 1973.

[66] J.A. BUCARO and T.A. LITOVITZ J. Chem. Phys. 54,3846 1971.

[67] HYUNG KYU SHIN J. Chem. Phys. 56, 2617, 1972

[68] B.J. ALDER, H.L. STRAUSS and J.J. WEIS J. Chem. Phys. 59, 1002, 1973.

[69] P.A. FLEURY, W.B. DANIELS and J.M. WORLOCK Phys. Rev. Letters. 27 , 1493, 1971.

[70] J. COURTENAY LEWIS and J. VAN KRANENDONK Can J. Phys. 50, 2902, 1972 .

A very extensive bibliography is given by P.A. FLEURY and J.P. BOON, in Advances in Chemical Physics, Vol 24, Edited by I. PRIGOGINE and S.A. RICE, John WILEY and Sons N.Y. 1973.

APPLICATIONS OF LIGHT BEATING SPECTROSCOPY TO BIOLOGY

H.Z. Cummins

Department of Physics, New York University
New York, New York 10003, U.S.A.

I. INTRODUCTION

One of the earliest applications of light beating spectroscopy was to the analysis of the spectrum of laser light scattered by small particles undergoing diffusion in solution (Brownian motion). The first experimental observation of diffusion broadening of light scattered by polystyrene latex spheres appeared in 1964.[1] That same year R. Pecora developed a general formalism based on Van Hove's theory of neutron scattering which showed that the light scattering spectrum of diffusing particles could provide both the translational and rotational diffusion constants, and that for flexible polymers one could also extract information about the internal modes.[2]

The potential application of this new technique to biology was soon recognized. Although light scattering intensity measurements had long been employed in the characterization of biological macromolecules, the additional dynamical information available in the spectrum could provide a much more thorough characterization. In 1967, Dubin, Lunacek and Benedek reported the first determination of translational diffusion constants of biological macromolecules by light beating spectroscopy,[3] and Bergé, Volochine, Billard and Hamelin observed the dramatic spectral broadening of light scattered by motile microorganisms.[4]

Since then, the field has grown rapidly, particularly since the introduction of digital correlation techniques which permit higher precision and shorter measurement times than are possible with the analog spectrum analyzers used in the early experiments. Much of the later development has been an extension of the early work on diffusion to include such complications as polydispersity, concentration dependence, number fluctuation and monomer-dimer equilibria, although several workers have extended the technique to the study of more complex systems such as contractile muscle tissue, cell membranes, chemotactic response of motile bacteria and electrophoretic mobilities.

The application of light beating spectroscopy to biology has now reached a crucial point in its development. On one hand, the determination of translational diffusion constants of biological macromolecules, viruses, etc. from digital correlation measurements is rapidly developing into a standard and highly reliable laboratory technique. On the other, the study of more complex systems has been beset with numerous difficulties which suggest that if the technique is to eventually become widely utilized for measurements other than diffusion constant determinations, considerable refinement of both theory and experimental technique will be required.

In these lectures, I plan to present a survey of the theoretical and experimental developments of the past nine years, and to present a simple development of the theory which can serve as a basis for the detailed expositions of specific investigations in the seminars. The survey will not include all the published work in the field (which is rapidly growing to unmanageable dimensions). However, there have been several recent reviews with excellent bibliographies to which the interested reader can refer.[5-9] Similarly, in discussing the theory, I will not give all derivations in detail but will provide references to textbooks and journal articles in which detailed derivations can be found.

The most general version of the problem we will consider is illustrated in Fig. 1. Incident light (assumed to be a uniform monochromatic plane-polarized plane wave) traverses the scattering volume which contains the scattering elements (j). These elements may be distinct objects such as macromolecules, cells, viruses, etc., or subunits of larger structures.

The scattered field $\vec{E}_s$ is observed at the distant point R_o. We will ignore effects associated with multiple scattering and assume that each scattering element interacts independently with the incident field. In this approximation, the scattered field can be taken as a simple sum of contributions from the individual scatterers. The scattering elements are, however, capable of interacting with each other.

The scattered field will, in general, have two components. One will be polarized perpendicular to the scattering plane (the plane containing $\vec{k}_o$ and $\vec{k}_s$) and is shown as $\vec{E}_s$ in Fig. 1. The other component is polarized <u>in</u> the scattering plane and is termed the depolarized field. It will be designated by $\vec{E}_s^H$.

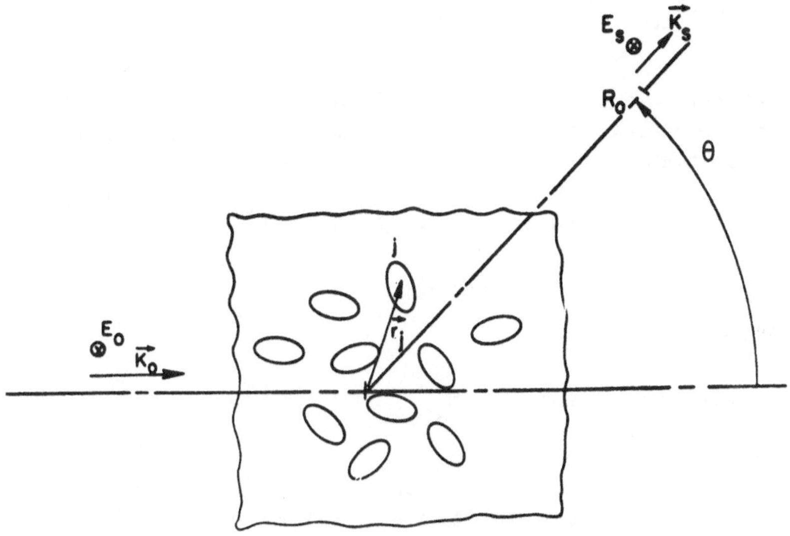

Figure 1. Geometry of a light scattering experiment.

The polarized and depolarized scattered fields can be written as

$$E_s = \sum_j E_j = \sum_j A_j \, e^{i\phi_j} \, e^{-i\omega_o t} \tag{1}$$

$$E_s^H = \sum_j E_j^H = \sum A_j^H \, e^{i\phi_j} \, e^{-i\omega_o t}$$

where A_j and A_j^H are the amplitudes of the polarized and depolarized scattered fields from the jth scattering element and ϕ_j is the phase. If we set $\phi = 0$ for a scatterer at the origin, then as Dr. Pike showed in his first lecture,

$$\phi_j = (\vec{k}_o - \vec{k}_s)\cdot\vec{r}_j = \vec{q}\cdot\vec{r}_j \tag{2}$$

where we have defined the scattering vector $\vec{q} = \vec{k}_o - \vec{k}_s$. Since the frequencies of the incident and scattered light will usually be nearly equal, $|k_o| \simeq |k_s|$ so that

$$|q| = |k_o - k_s| \simeq 2|k_o| \sin(\theta/2) = (4\pi n_o/\lambda_o) \sin(\theta/2) \tag{3}$$

where λ_o is the vacuum wavelength of the incident light and n_o is the refractive index of the medium filling the scattering volume (e.g., the solvent).

The polarized field is given by Eqs. (1) and (2)

$$E_s = \sum_j A_j(t)\, e^{i\vec{q}\cdot\vec{r}_j(t)}\, e^{-i\omega_o t} \tag{4}$$

with a similar experssion for E_s^H.

The scattering amplitude of the jth scattering element, A_j, can change in time if the scattering element changes its structure or if it is anisotropic and its orientation changes with time. The phase factors exp $[i\vec{q}\cdot\vec{r}_j]$ will change in time with the motion of the scatterer's center of mass, $\vec{r}_j(t)$.

Starting with Eq. (4), we can define various functions of the scattered field which are accessible to experimental measurement:

1. The average intensity

$$I_s = \langle|E_s|^2\rangle = \langle\sum_j \sum_{j'} A_j A_{j'}\, e^{i\vec{q}\cdot(\vec{r}_j-\vec{r}_{j'})}\rangle \tag{5}$$

2. The field autocorrelation function

$$G^{(1)}(\tau) = \langle E_s^*(t)\, E_s(t+\tau)\rangle \tag{6}$$

3. The optical spectrum

$$I(\omega) = \frac{1}{2\pi} \int G^{(1)}(\tau)\, e^{i\omega t}\, d\tau \tag{7}$$

4. The intensity autocorrelation function

$$G^{(2)}(\tau) = \langle E_s^*(t)\, E_s(t)\, E_s^*(t+\tau)\, E_s(t+\tau)\rangle \tag{8}$$

In self-beat spectroscopy, the intensity correlation function $G^{(2)}(\tau)$ or its Fourier transform $P_i(\omega)$ is measured with a correlator or spectrum analyzer, while in heterodyne spectroscopy one measures $G^{(1)}(\tau)$ or $I(\omega)$.

In most (although not all) of the problems we will consider, the scattered field will be a stationary Gaussian random process so that the normalized forms of $G^{(1)}$ and $G^{(2)}$ will be related by the Siegert relation

$$g^{(2)}(\tau) = |g^{(1)}(\tau)|^2 + 1 \tag{9}$$

where

$$g^{(1)}(\tau) = G^{(1)}(\tau)/G^{(1)}(o)$$

$$g^{(2)}(\tau) = G^{(2)}(\tau)/[G^{(1)}(o)]^2$$

Our program in these lectures will be to evaluate the quantities in Eqs. (5) - (8) with various models for the amplitudes and phase factors in Eq. (4). The models will be chosen to represent various situations of biological interest. At this point, the theory is perfectly general with the exception that we have excluded multiple scattering effects.

2. INTENSITY

Before beginning our discussion of the dynamical problem we will briefly review the subject of light scattering intensity measurements. Since intensity measurements have long been utilized in the characterization of biological materials, there is a substantial literature on the subject so that we only present a few relevant results which will be useful in our later discussion. Detailed expositions of the material can be found in textbooks[10-12] and in various review articles, e.g. Oster.[13,14] Many of the classic papers in this field have been reprinted in a bound collection by McIntyre and Gornick.[15] Some aspects of light scattering intensity measurements will also be covered at this Institute in the seminar of Professor Eisenberg.

Equation (5) for the time-averaged scattered intensity I_s can be applied to structured systems by taking over the formalism of X-ray diffraction. Thus in a periodic structure such as a muscle fiber, the intensity distribution will exhibit characteristic Bragg peaks whose spacing is related to the periodicity of the structure by the usual Laue equations. (See the seminar of Carlson and Fraser in this volume). However, if the scattering volume contains a dilute solution of identical scatterers whose positions are uncorrelated, then the cross terms in Eq. (5) average to zero and one has simply

$$I_s = N <A^2> \tag{10}$$

For small isotropic particles in vacuum, I_s was first shown by Rayleigh to be given by:

$$I_s = \frac{NI_o (2\pi)^4}{\lambda^4 r^2} \alpha^2 \sin^2\theta_1 \tag{11}$$

where α is the polarizability of the particle and θ_1 is the angle between E_o of K_s. (For the geometry of Fig. 1, $\theta_1 \equiv 90°$.) For particles in vacuum α may be related to the refractive index by $n^2-1 = 4\pi N\alpha$, while $n \simeq 1 + (dn/dc)c$ where N is the particle number density and c is the concentration in gm/cm^3.

If the scatterers are immersed in a solvent of refractive index n_o, then Eq. (11) reduces to:

$$I_s = \frac{I_o v\, 4\pi^2 \sin^2\theta_1 n_o^2 (dn/dc)^2 Mc}{N_o \lambda^4 r^2} \tag{12}$$

where N_o is Avagadro's number, M is the molecular weight of the scatterers in Daltons and v is the scattering volume.[12]

The Rayleigh ratio R is defined as the intensity scattered into a unit solid angle per unit volume of solution divided by $2 \sin^2\theta_1$:

$$R = \frac{I_s}{I_o v} \frac{r^2}{2 \sin^2\theta_1} \tag{13}$$

so that

$$R = K\, Mc \quad cm^{-1}$$

$$K = \frac{2\pi^2 n_o^2 (dn/dc)^2}{N_o \lambda^4} \quad (cm/g)^2 \tag{14}$$

For biological molecules, $(dn/dc) \simeq 0.19\ cm^3/g$ so that (for $\lambda = 6328$) $K \simeq 1.3 \times 10^{-7}\ (cm/g)^2$. As an example, for lysozyme ($M = 14{,}000$) at 1% concentration (.01 g/cm^3), $R \simeq 10^{-5}\ cm^{-1}$ which is about 20 times larger than R for pure water.[16]

Equation (14) provides the basic connection between scattered intensity and molecular weight. It is only valid, however, in the limit of very small particles and very low concentrations.

a. Size and Shape Effects

For scatterers whose dimensions are $\lesssim \lambda/20$, the electric field of the incident light is essentially constant over the volume of the scatterer and the dipole approximation represented by Eqs. (11) - (14) is adequate. For larger objects, the scattered intensity will generally be reduced because of interference effects. The effects of interference can formally be incorporated into the previous results by including a multiplicative angle dependent form factor $P(\theta)$.[12]

Evaluation of the form factor $P(\theta)$ for many geometries can be found in the literature. (c.f. Refs. 10, 11, 13, 14.) The most precise method of evaluation of $P(\theta)$, the Mie method, requires solution of the electromagnetic boundary value problem which is usually done by numerical approximation methods. (Mie solutions are frequently given as numerical tabulations.) If the scatterer is sufficiently small to satisfy the Rayleigh-Gans criterion[11,13] $(n-n_o)L << \lambda/4\pi$, then $P(\theta)$ is given by:

$$P(\theta) = \frac{1}{v} \int_v e^{iq\cdot r}\, dv \tag{15}$$

where v is the volume of the scatterer, and the evaluation involves averaging over orientations of the scatterer. Note that in all cases $P(\theta)$ goes to 1 in the limit $\theta \to 0$.

$P(\theta)$ as a function of $u = qL/2$ for spheres, discs, rods and random coils is shown in Fig. 2, from Oster.[14]

For any shape of scatterer it can be shown that the small-angle limit of $P(\theta)$ is given by:

$$P(\theta) \underset{qL<<1}{\rightarrow} 1 - q^2 R^2_G/3 \tag{16}$$

where R_G is the particle's radius of gyration.

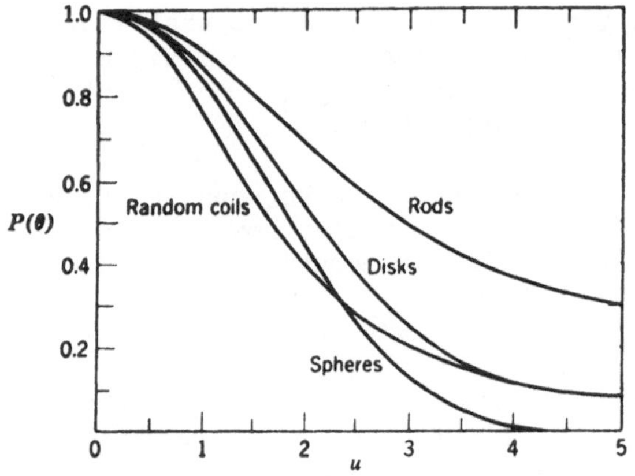

Figure 2. Form factor P(θ) as a function of $u = qL/2$ for homogeneous rods, discs, spheres and random coils [from G. Oster, Ref. 14]

b. Particle Interactions

The validity of Eq. (10) requires that the positions of the scatterers be uncorrelated, or equivalently that they be sufficiently far apart to prevent interparticle interactions.

At finite concentrations, however, some interactions will always occur so that factorization of the sum in Eq. (5) is no longer valid. In this case, it is usually more convenient to express the scattered intensity in terms of the concentration fluctuations which, in turn, depend on the chemical potential μ. (An alternative approach based on the pair correlation function will be discussed by Dr. Pusey in his seminar).

The equation of state of a non-ideal gas can be expressed by the virial expansion in the pressure P:

$$PV = RT\ (1 + BP + CP^2 + \ldots\ldots)$$

where the virial coefficients B, C ..., which would all be zero for an ideal gas, are a measure of the interaction between particles.

A similar virial expansion can be used to express the excess chemical potential of a solution with solute concentration c:

$$\mu \propto c\left(\frac{1}{M} + Bc + Cc^2 + \ldots\right), \quad \frac{\partial\mu}{\partial c} \propto \left(\frac{1}{M} + 2Bc + \ldots\right).$$

This leads to the following expression for the Rayleigh ratio:

$$R = \frac{Kc}{(1/M + 2Bc + 3Cc^2 + \ldots)} = \frac{KMc}{(1 + 2MBc + \ldots)} \tag{17}$$

where K is defined by Eq. (14). Note that as $c \to o$, Eq. (17) reduces to Eq. (14).

The magnitude of the virial coefficients depends on the nature of the interparticle interactions. For uncharged scatterers where the interaction is mainly mechanical, the primary contribution to B is from the excluded volume effect. For spheres the excluded volume value for B is $B = 4\ v/M$ where v is the specific volume (cm^3/g), so that:

$$R \simeq \frac{KMc}{1 + 8vc + \ldots} \tag{18}$$

Discussion of the excluded volume value of B for various shaped particles can be found, for example, in Tanford.[12]

c. Zimm Plots

To complete our discussion of intensity measurements, we combine Eq. (17) with the form factor $P(\theta)$ to obtain

$$R_\theta = \frac{Kc\ P(\theta)}{\frac{1}{M} + 2Bc + \ldots}$$

$$\frac{Kc}{R_\theta} = \frac{\frac{1}{M} + 2Bc + \ldots}{P(\theta)} \tag{19}$$

If experimental values of Kc/R_θ are plotted against $\sin^2(\theta/2) + Ac$ where A is an arbitrary constant, one obtains a "Zimm Plot" as illustrated in Fig. 3.

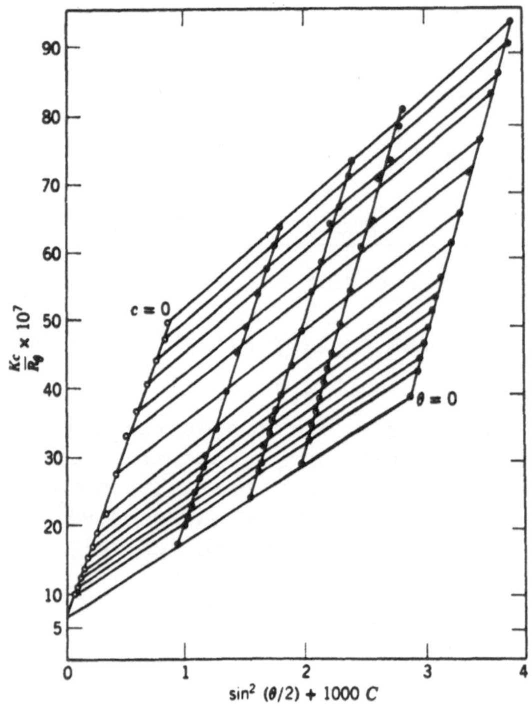

Figure 3. Zimm Plot for cellulose nitrate in acetone [from G. Oster, Ref. 14].

Extrapolating all points at constant θ to c = o (infinite dilution) gives a line which is

$$\frac{Kc}{R_\theta} = \frac{1}{MP(\theta)} \underset{\theta \to o}{\to} \frac{1}{M(1 - q^2 R_G{}^2/3)} \simeq \frac{1}{M}\,(1 + q^2 R_G{}^2/3) \qquad (20)$$

where we have made use of Eq. (16).

Similarly, extrapolating all points at constant c to θ = 0 gives

$$\frac{Kc}{R_\theta} = \frac{1}{M} + 2Bc \qquad (21)$$

So the two limiting loci are straight lines, both with intercepts at $\frac{1}{M}$. From the slopes one can find the second virial coefficient B and the radius of gyration R_G.

3. IDEAL DIFFUSION

In this section we begin our discussion of dynamics with the problem of a dilute solution of identical scatterers undergoing simple diffusion. Other aspects of the diffusion problem will be discussed in the following section. Several important aspects of diffusion measurements which will be discussed in rather general terms in this section will be further developed and illustrated with the results of some recent experiments by Dr. Pusey in his seminar.

a. Spherical Scatterers

We first consider the simplest model problem in some detail, both because of its intrinsic interest and because it provides a starting point for the analysis of more complicated problems.[2]

Let the scattering volume contain N identical spherical scatterers, each with time-independent scattering amplitude A which will, in general, depend on the scattering angle θ. Then, assuming stationarity, the field autocorrelation function is:

$$G^{(1)}(\tau) = \langle \sum_{j=1}^{N} Ae^{-iq\cdot r_j(o)} \sum_{j'=1}^{N} Ae^{iq\cdot r_{j'}(\tau)} \rangle \, e^{-i\omega_o\tau}. \tag{22}$$

For sufficiently dilute solutions the positions of the different scatterers will be uncorrelated so that the cross terms vanish, whence (dropping the now unnecessary subscripts j and j')

$$G^{(1)}(\tau) = N|A|^2 \, \langle e^{i\vec{q}\cdot(\vec{r}(\tau)-\vec{r}(o))} \rangle \, e^{-i\omega_o\tau}. \tag{23}$$

Now the ensemble average in Eq. (23) can be replaced by:

$$\langle e^{iq\cdot(r(\tau)-r(o)} \rangle = \int G_s(\vec{R}, \tau) \, e^{i\vec{q}\cdot\vec{R}} \, d^3R \tag{24}$$

where $G_s(\vec{R}, \tau)$ is the conditional probability that a particle located at the origin at time 0 will be at the position $\vec{R}$ at time τ. (Formally, G_s is the "self" part of the van Hove space-time correlation function.)

For particles undergoing free isotropic diffusion, G_s obeys the diffusion equation

$$\frac{\partial G_s}{\partial t} = D_T \nabla^2 G_s \tag{25}$$

which, together with Eqs. (23) and (24) gives:

$$G^{(1)}(\tau) = N|A|^2 \; e^{-D_T q^2 \tau} \; e^{-i\omega_o \tau} \tag{26}$$

From Eq. (7) the optical spectrum is then:

$$I(\omega) = N|A|^2 \left\{ \frac{D_T q^2/\pi}{(\omega-\omega_o)^2 + (D_T q^2)^2} \right\} \tag{27}$$

where the quantity in brackets is a normalized Lorentzian centered at ω_o with half width at half maximum of

$$\Delta\omega_{1/2} = D_T q^2. \tag{28}$$

Combining Eq. (28) with the Stokes-Einstein equation for D_T

$$D_T = \frac{kT}{6\pi\eta \; r_h} \tag{29}$$

where η is the viscosity and r_h is the hydrodynamic radius, we then have

$$\Delta\omega_{1/2} = \frac{kTq^2}{6\pi\eta \; r_h} . \tag{30}$$

For spheres in water at 20°C with 6328 Å light, for example, Eq. (30) reduces to:

$$\Delta\omega_{1/2} = \frac{49\pi \;\; \sin^2 (\theta/2)}{r_h^\mu} \tag{31}$$

with r_h^μ the hydrodynamic radius in microns. This linewidth is so small that diffusion broadening cannot be observed by convential spectroscopic techniques.

Finally, if the number of particles in the scattering volume is sufficiently large so that the central limit theorem guarantees that the scattered field is a Gaussian random process, then the normalized second order correlation function defined in Eq. (9) becomes:

$$g^{(2)}(\tau) = 1 + e^{-2D_T q^2 \tau} \tag{32}$$

In digital correlation experiments, the quantity usually determined is $\hat{g}^{(2)}(\tau) = 1 + ae^{-2D_T q^2 \tau}$ where a is a constant depending on spatial coherence, etc. as discussed in previous lectures by Oliver and Jakeman. Linear fits of $\ln(\hat{g}^{(2)}(\tau) - 1)$ vs τ thus give the slope as $2D_T q^2$ from which the diffusion constant D_T can be found.

Although we have specifically considered spherical scatterers, the results of this section also apply to scatterers of arbitrary shape which are optically isotropic and small enough so that the scattering amplitude A is independent of orientation. If the diffusion constant is analyzed in terms of Eq. (29), the quantity r_h will be an "equivalent hydrodynamic radius" and will not necessarily be equal to any dimension of the scatterer. Even for spherical scatterers the hydrodynamic radius will usually be larger than the physical size of the dry scatterer due to the presence of some solvent which moves with the scatterer. In fact, as we shall discuss later, such measurements can provide a value for the degree of solvation.

b. Non-Spherical Scatterers

Next, we consider a dilute solution of identical, optically isotropic non-spherical scatterers whose scattering amplitudes vary with orientation so that:

$$A(t) = A_o + A_1(t) \tag{33}$$

where $A_o = \langle A(t) \rangle$ is the mean scattering amplitude and $A_1(t)$ is the fluctuation in A(t), with $\langle A_1(t) \rangle = 0$. If we assume that the positions of the different scatterers are uncorrelated, and also that the position and orientation of a single scatterer are uncorrelated, the field autocorrelation function (Eq. 10) becomes:

$$G^{(1)}(\tau) = N \langle A(0)\, A(\tau) \rangle \ \langle e^{iq\cdot[r(\tau) - r(0)]} \rangle \ e^{-i\omega_o \tau} \tag{34A}$$

Combining Eqs. (33) and (34), and assuming that the center of mass motion of each particle is governed by isotropic translational diffusion, we then have

$$G^{(1)}(\tau) = \left[NA_o^2 \, e^{-D_T q^2 \tau} + N\langle A_1(0)\, A_1(\tau)\rangle \, e^{-D_T q^2 \tau} \right] e^{-i\omega_o \tau} \qquad (34B)$$

The first term in Eq. (34B) is just the pure translation term of the previous section, while the second term includes translational diffusion and orientational fluctuations as well through the term $\langle A_1(0)\, A_1(\tau)\rangle$. (Note that Eqs. (34A) and (34B) can be used for any problem in which the scattering amplitude changes with time, and are not limited to the case of orientation dependence.)

The amplitude autocorrelation function $\langle A(0)\, A(\tau)\rangle$ is often governed by rotational diffusion. For a scatterer having rotational symmetry (e.g. rods or ellipsoids of revolution) the amplitude autocorrelation function can be shown to have the form

$$\langle A(0)\, A(\tau)\rangle = A^2_{(q=o)} \sum_{\substack{\ell=o \\ \text{even}}}^{\infty} B_\ell e^{-\ell(\ell+1) D_R \tau}$$

so that

$$G^{(1)}(\tau) = NA^2_{(q=o)} \left[B_o e^{-D_T q^2 \tau} + B_2 e^{-(D_T q^2 + 6D_R)\tau} + B_4 e^{-(D_T q^2 + 20D_R)\tau} + \ldots \right] \quad (35)$$

as was shown by Pecora in 1964.[2] (The rotational diffusion problem is also considered by Professor Lallemand in his lectures.)

The coefficients B_ℓ appearing in Eq. (35) can be evaluated for various specific models. For rigid rod-shaped scatters, the analysis within the Rayleigh-Gans Approximation was discussed in the original 1964 paper of Pecora [2] and tabulations and plots of these coefficients have been given by Pecora [17] and by Cummins et al. [18] in conjunction with an analysis of the rod-shaped virus TMV. A similar derivation for flexible linear chains was given by Fujime. [19]

Figure 4 (from Pecora [17]) shows the size of the B_ℓ coefficients for rigid rods of length L as a function of qL. S_o is the pure translation term (B_o), S_1 is the leading translation-rotation term (B_2), and S_h is the sum of all higher-order terms ($B_4 + B_6 + \ldots$). The curve S is the total itensity [P(θ)] and is equal to the sum of the other three curves. Note that S_1 and S_h approach zero as $qL \to 0$ so that only the pure translation term S_o contributes to the small-angle spectrum.

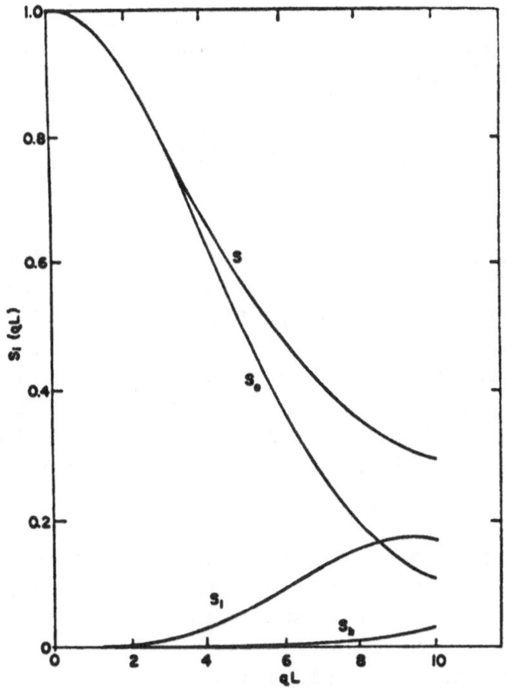

Figure 4. Intensities of the different terms in the spectrum of light scattered by optically isotropic rigid rods undergoing translational and rotational diffusion [from R. Pecora, Ref. 17].

For qL<10, the sum in Eq. (23) can be truncated after the B_2 term, so that the optical spectrum is given by:

$$I(\omega)\simeq NA^2_{q=o}\left[\frac{B_o^2\,(D_T q^2/\pi)}{(\omega-\omega_o)^2+(D_T q^2)^2}+\frac{B_2\,(D_T q^2+6D_R)/\pi}{(\omega-\omega_o)^2+(D_T q^2+6D_R)^2}\right] \quad (36)$$

There are thus two Lorentzians, one of half-width $D_T q^2 = \Gamma_o$ and the other of half width $D_T q^2 + 6D_R = \Gamma_2$.

The normalized second-order correlation function is[18]

$$g^{(2)}(\tau) = 1 + \frac{B_o e^{-2\Gamma_o\tau} + 2B_o B_2 e^{-(\Gamma_o+\Gamma_2)\tau} + B_2^2 e^{-2\Gamma_2\tau}}{(B_o + B_2)^2} \quad (37)$$

Note that the self-beat correlation function minus the background gives three exponentials, but by taking the square root one recovers the two exponentials Γ_o and Γ_2, since

$$[g^{(2)}(\tau) - 1]^{\frac{1}{2}} = [B_o\ e^{-\Gamma_o\tau} + B_2\ e^{-\Gamma_2\tau}]/[B_o + B_2]\ .$$

Finally, we emphasize that the approximation of uncorrelated orientation and position is not completely correct, and that in general there will be some coupling between translational and rotational diffusion. The effect of such coupling (which introduces additional terms into the spectrum) was first discussed by Maeda and Saito,[20] was considered in conjunction with the spectrum of TMV by Fujime[21] and by Schaefer _et al._,[22] and discussed again recently by Chow.[23] But it is not yet completely clear how the effects of the coupling can be extracted from the spectra.

In any case, the effect of rotational motion modifies the simple Lorentzian spectrum (or exponential correlation function) and one must analyze two or more superimposed Lorentzians or exponentials which turns out to be a far more difficult procedure than the single Lorentzian or exponential analysis appropriate for spherical scatterers.

c. Depolarized Scattering

If the individual scatterers have anisotropic polarizabilities, then there will in general be a depolarized component of the scattered light whose amplitude will be fully modulated by the orientational fluctuations since A_j^H must have zero mean by symmetry. Consequently, the depolarized scattered field autocorrelation function [from Eqs. (1), (6), (33) and (34)] is:

$$G_H^{(1)}(\tau) = N\langle A_1^H(o)\ A_1^H(\tau)\rangle\ e^{-D_T q^2\tau} \qquad (38)$$

Note that there is _no_ pure translation term in the depolarized spectrum, and that since A_1^H depends on the optical anisotropy of the scatterer rather than its shape, the intensity associated with Eq. (38) is relatively insensitive to angle. (Unlike B_2 ... B_4 ... in Eq. (36) which all go to zero as $\theta\to 0$.)

For small rigid anisotropic molecules with axial symmetry, the depolarized spectrum is dominated by the leading term[24]:

$$(\alpha_{zz} - \alpha_{xx})^2 \quad \cdot \quad \frac{(D_T q^2 + 6D_R)}{(\omega-\omega_o)^2 + (D_T q^2 + 6D_R)^2} \tag{39}$$

Thus by measuring the depolarized spectrum at small angles where $6D_R >> D_T q^2$, one can determine the rotational diffusion constant separately. Wada et al.[25] and Dubin et al.[26] have applied this method to determine the rotational diffusion constants of TMV and lysozyme, respectively.

4. DIFFUSION II

In this section, we first examine some complications which may occur in actual diffusion experiments and consider how they can be included in the data analysis. We then consider the connection between light beating measurements and sedimentation and intensity measurements. Finally, we present a survey of diffusion constants which have been determined from light beating experiments.

a. Polydispersity

In our discussion of the diffusion problem in the preceding section, we assumed that the scattering volume contains N identical scatterers. In practice, the scatterers may not all be identical. Instead, one may find various distributions of sizes (e.g. monomers and dimers, rods and rod fragments, viruses and dust, etc.).

The effect of polydispersity on the scattered intensity $I(\theta)$ has been considered extensively in the literature (cf. Kerker, Ref. 11, Sec. 8.3). The infinite dilution line on the Zimm plot (Fig. 3) is in general curved when polydispersity is significant, and the zero angle intercept gives

$$R_\theta \underset{\substack{c\to 0 \\ \theta\to 0}}{\to} Kc\, M_w \tag{40}$$

where M_w is the weight-average molecular weight. Various analytical procedures have been proposed for deducing the size distribution from $R(\theta)$, but they all require data of extraordinary precision to give results of any reliability.

The effects of polydispersity on the light beating spectrum have been considered by a number of authors, including Pecora,[27] Dubin,[16] Benbasat,[28] Koppel,[29] and Pusey.[30]

Dubin[16] considered the spectrum that would be produced by a mixture of two molecular species. He considered a mixture with equal number concentrations of two types of molecules, one having diffusion constant D_1 and the other $D_2 = 1.5\,D_1$, representing a possible mixture of native and denatured lysozyme. Although the resulting theoretical self beat spectrum consists of three Lorentzians, he found that a single Lorentzian fit would be almost indistinguishable from the exact spectrum with an rms error of only 0.19% as shown in Fig. 5. The error resulting from forcing the single Lorentzian fit increases with increasing D_2/D_1, but unless $D_2/D_1 \gtrsim 2.5$, the precision of experimental data must be considerably better than 1% if the presence of two species is to be detected from fitting errors alone.

Pecora and Tagami[27] considered the spectrum of polydisperse rods and coils. For the length distribution function f(L) they chose the Schulz distribution

$$f(L) = \frac{1}{Z!}\;\frac{(Z+1)^{Z+1}}{\langle L\rangle}\;L^{Z}\;e^{-(Z+1)L/\langle L\rangle}$$

for which they derived both the intensity and the spectrum of the pure translation term. A related analysis for a Schulz-Zimm distribution of spheres was given by Benbasat and Bloomfield.[28]

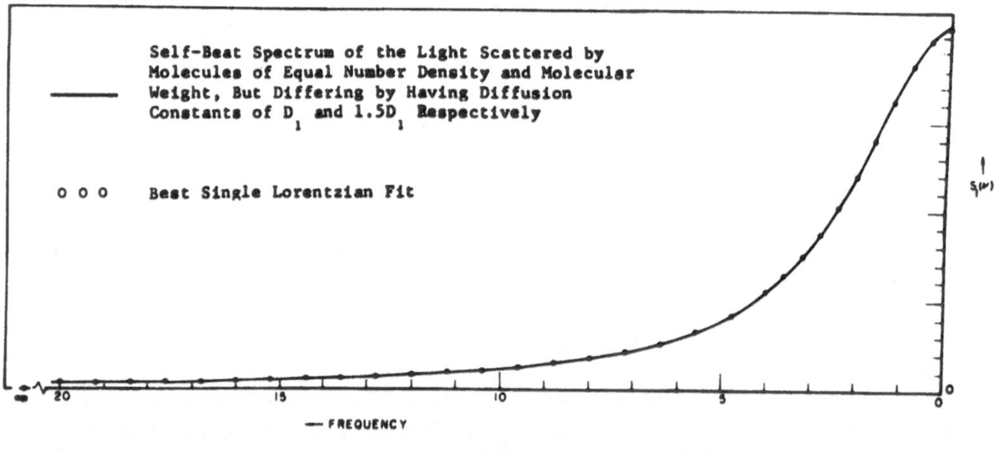

Figure 5. Self-beat spectrum of the light scattered by a two-component mixture compared to the best single-Lorentzian fit [from S. Dubin, Ref. 16].

This procedure permits Z and <L> to be determined from parametric fits of the data, but is only useful if the form of the distribution is known in advance.

A more general approach to the effect of polydispersity on the light beating spectrum has been discussed recently by Koppel[29] and by Pusey.[30] The approaches of these two authors are formally identical, although the notation employed by Pusey is somewhat simpler.

The exponential field correlation function for monodisperse spheres $|g^{(1)}(\tau)| = e^{-\Gamma\tau}$ with $\Gamma = D_T q^2$ will in general be modified by polydispersity to give a superposition of exponentials:

$$|g^{(1)}(\tau)| = \int_0^\infty F(\Gamma) e^{-\Gamma\tau} \, d\Gamma \tag{41}$$

The distribution of decay rates $F(\Gamma)$ is so defined that $F(\Gamma)d\Gamma$ is the fraction of the total scattered intensity due to molecules for which $D_T q^2$ lies between Γ and $\Gamma + d\Gamma$, and $\int_0^\infty F(\Gamma)d\Gamma = 1$. In Koppel's approach, $K_m(\Gamma)$, the cumulants of $F(\Gamma)$, are computed from the data. The first cumulant then is found to give the "Z-average" diffusion constant, the second cumulant is a measure of the width of the distribution, the third cumulant is a measure of skewness, etc. The cumulants also serve as a sensitive test of departures from monodispersity.

Pusey[30] has also discussed this approach to polydispersity and has applied it to an analysis of data obtained with samples of polystyrene in cyclohexane (which will be discussed further in his seminar). In Pusey's notation one starts from an expansion of $e^{-\Gamma\tau}$ about the mean decay rate $\langle\Gamma\rangle$, where

$$\langle\Gamma\rangle = \int \Gamma F(\Gamma) \, d\Gamma \tag{42}$$

$$e^{-\Gamma\tau} = e^{-\langle\Gamma\rangle\tau} e^{-(\Gamma-\langle\Gamma\rangle)\tau} = e^{-\langle\Gamma\rangle\tau}\left[1-(\Gamma-\langle\Gamma\rangle)\tau+\frac{(\Gamma-\langle\Gamma\rangle)^2\tau^2}{2!} + \ldots\right] \tag{43}$$

which, together with Eq. (41), gives:

$$\left.\begin{aligned}|g^{(1)}(\tau)| &= \int_0^\infty F(\Gamma)e^{-<\Gamma>\tau}[1-(\Gamma-<\Gamma>)\tau + \frac{(\Gamma-<\Gamma>)^2\tau^2}{2!} + \ldots]d\Gamma \\ &= e^{-<\Gamma>\tau}(1-<\Gamma>\tau + <\Gamma>\tau + (\tau^2/2!)\int(\Gamma-<\Gamma>)^2 F(\Gamma)d\Gamma + \ldots\end{aligned}\right\} \quad (44)$$

or finally, since $\ln(1+x) = x - \frac{1}{2}x^2 + \frac{1}{3}x^3 + \ldots$,

$$\ell n\ a|g^{(1)}(\tau)| = \ell na - <\Gamma>\tau + \frac{1}{2!}(\mu_2/<\Gamma>^2)(<\Gamma>\tau)^2 - \frac{1}{3!}(\mu_3/<\Gamma>^3)(<\Gamma>\tau)^3 + \ldots \quad (45)$$

where

$$\mu_n/<\Gamma>^n = \frac{1}{<\Gamma>^n}\int(\Gamma-<\Gamma>)^n F(\Gamma)\, d\Gamma$$

is the nth normalized moment of $F(\Gamma)$ about $<\Gamma>$.

If one truncates Eq. (45) after the first three terms, there remains a quadratic function of τ against which $\ell n|\hat{g}^{(1)}(\tau)|$ data can be fit in place of the usual linear function and the fit then determines both the mean $\bar{\Gamma}$ and the variance of $F(\Gamma)$.

One form of polydispersity which is frequently encountered is the presence of dust particles in biological preparations. Since a few large dust particles will produce considerable intensity which fluctuates very slowly, they will produce a second exponential which decays so slowly that it forms an essentially constant background.

One possible approach to separating out the effect of dust is illustrated in Fig. 6[31,32] for DNA extracted from the virus fd. Several channels of the digital correlator are shifted out in time to a point where the molecular exponential would have decayed to the background level. The residual level above the theoretical background can then be subtracted from all data points to determine the diffusion constant. Alternatively, if there are enough dust particles in the scattering volume to guarantee Gaussian statistics, then the theoretical background can first be subtracted from all the data points, and the square root of the remainder can be fit to a constant plus a single exponential.

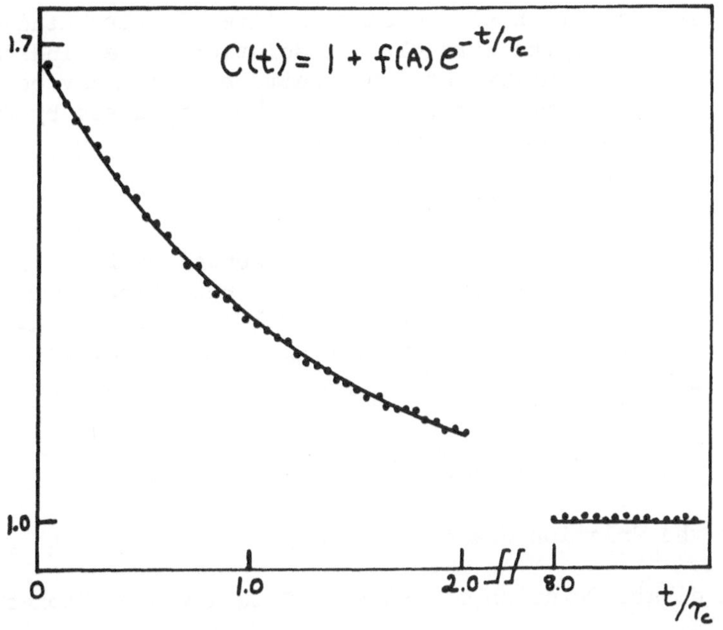

Figure 6. Detecting the effect of dust on the self-beat spectrum of DNA by the use of delayed channels [from J. Newman, Ref. 31].

b. Number Fluctuations

Schaefer and Berne[33] have considered the problem of scattering from a system containing so few scatterers that N fluctuates significantly about its mean so that the scattered field is non-Gaussian. They show that although the fluctuations in occupation number N do not effect the heterodyne spectrum, the self-beat spectrum is modified by the presence of an additional term in $G^{(2)}(\tau)$:

$$G^{(2)}(\tau) = \langle N \rangle^2 A^4 (1+e^{-2Dq^2\tau}) + A^4 \langle \delta N(o) \delta N(\tau) \rangle \tag{46}$$

This extra term decays with a characteristic time τ_N which is the time for a particle to traverse the scattering volume, and will therefore generally add a long lived "tail" onto the

$e^{-2Dq^2\tau}$ term whose decay time $\tau_D = 1/2Dq^2$ is normally much shorter that τ_N. Schaefer[34] has recently shown how this additional fluctuation can be exploited to study the dynamics of motile micro-organisms.

A similar situation also occurs in laser Doppler studies of turbulent flow where the particles in solution act as indicaters of the local fluid density, and the scattered field may be non-Gaussian.[35] This problem will be further discussed by Professors Bertolotti and Crosignani in their lectures.

c. Concentration Dependence

When the correlation functions represented by Eqs. (26) or (32) are measured in real experiments, the resulting decay rate may depend on concentration. The Einstein relation for the diffusion constant together with the virial expansion for μ, discussed in Section 2, gives:

$$D = \frac{kT}{f} \; [1 + 2MBc + \ldots] \tag{47}$$

where f is the friction coefficient.

For a dilute solution of spheres, f is given by Stokes' law:

$$f_o = 6\pi\eta r_h \tag{48}$$

which, in the limit $c \to 0$, recovers the usual Stokes-Einstein expression for D of Eq. (29).

The effect of concentration on the friction coefficient can also be represented by a virial expansion:[36]

$$f(c) = f_o \; (1 + B'c + \ldots). \tag{49}$$

Combining Eqs. (47) and (49),

$$D(c) = \frac{kT}{f_o} \; \frac{[1+2MBc+..]}{[1+B'c+..\,]} \approx \frac{kT}{f_o} \; [1+(2MB-B')c+..]=D_o[1+B_Dc+..] \tag{50}$$

The tendency of the μ-virial (2MB) and the friction virial (B') to offset each other in the diffusion virial (B_D) is discussed by Tanford (Ref. 12, page 372).

Several light beating experiments have been interpreted in terms of the virial expansion for D(c) of Eq. (50). In their study of the muscle protein myosin, Herbert and Carlson[37,38] found that B_D was essentially zero, while for the virus R17, Pusey <u>et al.</u>[39] found that it could be positive, negative or zero depending on the salt concentration of the solvent, as shown in Fig. 7. (This work will be discussed by Dr. Pusey in his seminar.)

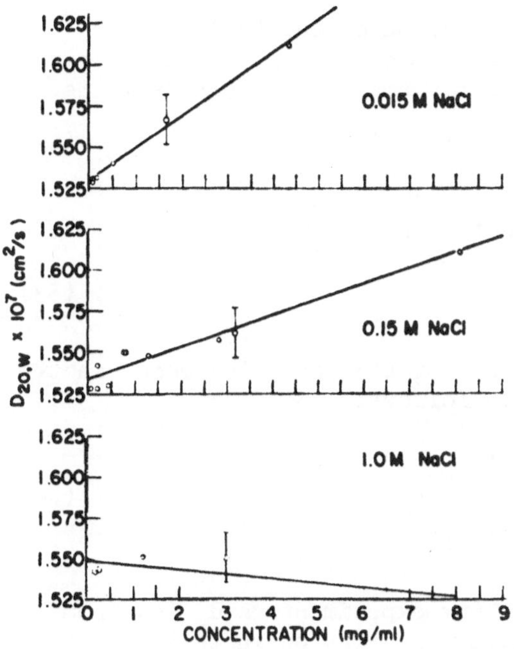

Figure 7. Diffusion coefficient $D_{20,W}$ of R17 as a function of virus concentration in 1 M, 0.15 M and 0.015 M NaCl [from Pusey *et al.*, Ref. 39].

Equation (50) can be evaluated theoretically for the simple case of rigid spheres with no long-range interactions. The μ-virial 2MB is then given by the excluded volume[12]

$$2MB = 8v \tag{51}$$

where v is the partial specific volume of the solute. The effect of concentration on the friction factor is more complicated. Pyun and Fixman[36] found B' = 7.157 so that:

$$D(c) \simeq \frac{kT}{f_o} \cdot \frac{(1 + 8vc)}{(1 + 7.157\ vc)} = D_o\ (1 + 0.843\ vc). \tag{52A}$$

Batchelor[40] has reconsidered Pyun and Fixman's analysis and finds B' = 6.55 so that:

$$D(c) = D_o\ (1 + 1.45\ vc) \tag{52B}$$

Recently, Altenberger and Deutch[41] have reanalyzed the diffusion problem starting with a full N-body diffusion equation, and have shown the origin of the factorization of Eq. (22) to give Eq. (23). Their treatment also yields a diffusion virial for hard spheres directly which for sufficiently small radii ($qr << 1$) is:

$$D(c) = D_o(1 + 2vc) \tag{52C}$$

which implies that the friction factor is:

$$f(c) = f_o(1 + 6vc) \tag{53}$$

Note that the virial expansions for D(c) and f(c) occur in two different forms which we rewrite as follows:

$$\begin{aligned} D(c) &= D_o(1 + B_D c) = D_o(1 + K_D vc) \\ f(c) &= f_o(1 + B_f c) = f_o(1 + K_f vc) \end{aligned} \tag{54}$$

Values of D(c) from light beating experiments and f(c) from sedimentation experiments are usually interpreted in terms of the B coefficients. With c the weight concentration in (gm/ml), the units of B are (ml/gm). Theoretical predictions such as Eqs. (52) and (53), however, are for K_D and K_f which are dimensionless.

In making the conversion, a problem arises with regard to the partial specific volume v. Since $vc = \phi$, where ϕ is the volume fraction of the solute, the value of v must be appropriate for the actual diffusing particles in solution. If v of the dry solute is used, the resultant K values will generally be much too large.

To illustrate, we consider a recent experiment by Newman <u>et al.</u>[32] on DNA from the virus fd. Experiments were performed at sufficiently high salt concentration (0.15M NaCl) to insure electrostatic screening.

The approximate experimental values found for the virials were:

$$B_D \simeq 51.3 \text{ (ml/gm)}$$

$$B_f \simeq 230.8 \text{ (ml/gm)}$$

If the partial specific volume of the dry solute, $v = 0.513$ (ml/gm) is used, then the K virials are:

$$K_D \sim 100$$

$$K_f \sim 450$$

However, the diffusion measurement also gives a value for the hydrodynamic radius, $r_h = 316$ Å which is 4.35 times larger than the radius of a tightly packed sphere of a dry DNA molecule, $r \sim 72.6$ Å. The volume fraction of the hydrated DNA molecules in solution is thus $(4.35)^3 = 82$ times larger than the volume fraction of the dry solute, and the appropriate K virials are therefore:

$$K_D \sim 1.20$$

$$K_f \sim 5.41$$

in reasonable agreement with the theoretical predictions.

The difference between the dry radius and the hydrodynamic radius is particularly large in this case since the molecule takes the form of a loosely coiled strand. For rigid objects (such as the virus R17), the difference between the two radii is less dramatic.[42]

Finally, it should be noted that the entire theory we have outlined here rests on the assumption that the individual scatterers interact independently with the optical field so that the total scattered field, $E_s(t)$, is a simple sum of contributions from the various scatterers. At finite concentrations multiple scattering can become significant, and the theory can become invalid. Unfortunately, the effects of multiple scattering on the spectrum have only been investigated in a few special situations[43,44] and there is a clear need for further work in this area.

d. Friction Coefficients

The friction coefficient f occurring in Eq. (47) for the diffusion constant has been studied both theoretically and experimentally by many investigators, and we comment briefly on this aspect of the diffusion problem.

Stokes' solution [Eq. (48)] for the viscous force on a sphere of radius r was extended to ellipsoids by Perrin in 1936.[45] Perrin's solutions for prolate and oblate ellipsoids of revolution averaged over orientation can be summarized by the Perrin factor f/f_o which gives the ratio of the friction constant for an ellipsoid of axial ratio a/b to the friction constant of a sphere of equal volume. A plot of the Perrin factor can be found in Tanford[12] along with the appropriate analytic expressions.

Since f/f_o is a monotonically increasing function of a/b, the diffusion constant of an ellipsoid will always be less than that of a sphere of equal volume which provides an additional method of determining the shape of non-spherical scatterers. Perrin also derived the rotational friction constant of ellipsoids which enters into the rotational diffusion constant D_R.

Additional calculations for the diffusion constants of shapes other than ellipsoids have been performed and applied to biological objects such as the rod-shaped TMV [cf Bloomfield et al.][46] Other aspects of the calculation of f(c) will be discussed by Dr. Pusey.

Finally, the friction constants of objects whose shapes are too complex to permit analytical solutions can be determined by working with scale models. Douthart and Bloomfield[47] constructed scale models of T2 bacteriophage which were immersed in a rotating oil bath, and used the measured torque to find the rotational friction constants.

e. Charge Effects

The diffusion constant of a charged macromolecule in solution may differ considerably from a neutral one of the same size and mass both because the charged macromolecule tends to pull along a cloud of counterions and because of the contribution to the chemical potential of electrostatic interactions between macromolecules, which was also considered in Dr. Eisenberg's seminar in connection with the scattered intensity.

M. Stephen[48] considered the problem of the diffusion of charged macromolecules in solution. For small molecules in the low-density limit he finds that the diffusion constant is increased by a factor

$$D_{eff} = D_o \left(1 + \frac{S_-^2}{q^2 + S_+^2} \right) \tag{55}$$

where S_+ is the Debye screening wave vector $(4\pi\rho e_+/\varepsilon kT)^{\frac{1}{2}}$ in which e_+ is the counterion charge, while S_- is similarly defined in terms of e_-, the charge on the macromolecule. Stephen's result predicts both an increase in D due to the counterion cloud and a departure from q^2 dependence in the decay rate of the intensity correlation function. These predictions have not yet been fully tested, although Pusey et al.[39] have studied the dependence of the apparent correlation function on salt concentration for the virus R17. This problem will be discussed in considerably greater detail by Dr. Pusey.

f. Connection with Sedimentation and Intensity

The results of intensity correlation measurements can be combined with sedimentation and intensity measurements to find the molecular weights, degree of solvation, molecular shape and the virials.

Sedimentation measurements in the ultracentrifuge determine the sedimentation coefficient

$$S = \frac{u}{\omega^2 r} = \frac{M}{N_o f}(1 - v\rho) \tag{56}$$

where u is the sedimentation velocity, v is the partial specific volume of the solute and ρ is the solvent density.[12]

Combining Eq. (56) with Eq. (47) for D, both in the limit of infinite dilution, gives:

$$M = \frac{S_o}{D_o} \frac{RT}{(1 - v\rho)} \tag{57}$$

The "Svedberg equation" (57) permits light scattering determinations of D_o to be combined with independent measurements of the sedimentation constant S_o and the partial specific volume v to find the molecular weight M. This has been done, for example, by Pusey *et al.*[39,42] for R17, PM2, T7 and BSV, by Dubin *et al.*[49] for T4, T5 and T7, and by Newman *et al.*[31,32] for viral DNA. (It should be noted that a similar combination of measurements has been employed in the past for molecular weight determinations using diffusion constants found by classical techniques.[12]

The diffusion constant for spheres determines the hydrodynamic radius r_h which describes the hydrodynamic object consisting of the dry scatterer plus internal solvation, together with a shell of externally associated solvent. Angular intensity (light or X-ray) measurements, on the other hand, give the physical radius r_o of the scatterer without the externally associated solvent. From the difference between r_o and r_h, that part of the degree of solvation δ (g solvent/g dry molecule) associated with the external shell of solvent can be deduced.[39,42] Similarly, the diffusion constant determined from intensity correlation measurements for non-spheres (which depend parametrically on the axial ratio for ellipsoids) can be combined with the molecular weight, partial specific volume and the radius of gyration determined from intensity measurements to estimate the molecular shape.

Finally, as Pusey et al.[39] discussed for the case of R17, the friction virial B' in Eq. (49) can be found from concentration-dependent sedimentation measurements, and, when combined with the effective virial for the diffusion constant [Eq. (50)] can be used to evaluate the osmotic virial B. (In their analysis, the friction virial was taken to be independent of salt concentration with a value of 9.32 cm^3/g.) In principle, the self-consistency of this procedure can be further checked by comparing the resulting B with the value determined by angular intensity measurements as discussed in Section 1. Similarly, the Zimm plot value of the molecular weight can be compared with the result deduced from combining diffusion constant, specific volume and sedimentation data.

g. Survey of Diffusion Experiments

To conclude our discussion of diffusion, we present a brief survey of diffusion measurements obtained by light beating spectroscopy since 1964.

TABLE I

SURVEY OF LIGHT BEATING DETERMINATIONS OF DIFFUSION CONSTANTS

1964 - 1973

<u>Prehistory</u>: Polystyrene latex spheres [Cummins 64,(1) Arecchi 67](50)

<u>First Biological Materials</u>: Bovine Serum Albumin, Ovalbumin, Lysozyme, TMV, Calf Thymus DNA [Dubin 67](3)

<u>Biomolecules</u>: Poly-α-Amino Acid [Ford 69],(51) RNase [Rimai 70](52) Myosin [Herbert 71],(37,38) Bovine Milk Casein Micelles [Lin 72, Dewan 73](53,54)

<u>Viruses</u>: T4, T5, T7 and λ [Dubin 70],(49) R17 [Pusey 72 - includes concentration and salt dependence],(39) PM2, T7, BSV and R17 [Pusey 73](42)

<u>Lysozyme</u>: (Translation and Rotation) [Dubin 71](26)

<u>Viral DNA</u>: Single-stranded DNA from fd virus [Newman 73](31,32)

TABLE I (Continued)

TMV: Six experiments to measure D_T and D_R of TMV:
(*Indicates depolarized scattering measurement)

	$D_T \times 10^7$	D_R
Cummins[18]	0.28	320
Wada*[25]		350
Fujime[21]		
c=0	0.45	390
c=0.1 mg/ml	0.37	
Wada[55]	0.33	
Schaefer[22]	0.39	420
Schurr*[56]		~340

RRE: Partial list of biological materials studied at RRE-Malvern[57,58]

	$D_T \times 10^7$
Lysozyme[59]	10.6 ± 0.2
BSA[59]	5.76 ± 0.05
Tamm-Horsfall glycoprotein[60]	0.220 ± 0.002
Histone 2 A 1	0.83 ± 0.01
α-crystalline	2.36 ± 0.02 2.20 ± 0.02
Haemocyanins[59,61]	
Helix pomatia α	1.04 ± 0.01
Helix pomatia β	1.05 ± 0.01
Archacatina marginata	1.00 ± 0.01
Pila leopoldvillensis	1.04 ± 0.01
Murex trunculus	1.03 ± 0.01
Sheep colonic mucosa glycoprotein[59]	0.238 ± 0.006
Adenovirus	
Myxovirus (influenza)	
Thyretin	
RNA polymerase	2.70 ± 0.15
γ-globulin	3.8 ± 0.2
Aldelase	3.8 ± 0.2
Catalese	3.6 ± 0.2
Aspartate amino transferase (enzyme)	
T 3 bacteriophage	
Turnip yellow mosaic virus	
Fibrinogen	2.4 ± 0.1
Keyhole limpet haemocyanin	1.03 ± 0.01
Haemoglobin	
Algae (alive and dead)	

5. OTHER TOPICS

In this final section, we will briefly summarize a number of other biological phenomena which have been investigated by light scattering spectroscopy during the last few years. The starting point in each case will be Eq. (23) or (34). We first consider two closely related problems which are also connected with the fluid flow problem discussed by Dr. Crosignani.

a. Electrophoresis

If particles of net charge q in solution are subjected to an electric field E, they acquire a drift velocity $v = Eq/f$ where f is the friction coefficient.

The electrophoretic mobility U is defined as[12]

$$U = v/E \tag{58}$$

In an experiment in which light scattered from particles all moving with uniform velocity v is mixed with unscattered light which serves as the local oscillator, the heterodyne spectrum is

$$I(\omega) = \delta(\omega-\omega_o + \vec{q}\cdot\vec{v}) \tag{59}$$

This result follows from Eq. (23) or (34A) with $\vec{r}(\tau) - \vec{r}(o) = \vec{v}\tau$ or equivalently from Eq. (24) with $G_s(R, \tau) = \delta(\vec{R} - \vec{v}\tau)$. If the particles are also undergoing translational diffusion, then the spectral line centered at $\omega = \omega_o - \vec{q}\cdot\vec{v}$ will have the usual diffusive width $D_T q^2$. If there are different molecular species present in the scattering volume, then the spectrum and correlation function of the optical field are given by:

$$I(\omega) \propto \sum_j N_j |A_j|^2 \frac{D_{Tj}q^2/\pi}{(\omega-\omega_o + \vec{q}\cdot\vec{v}_j)^2 + (D_{Tj}q^2)^2} \tag{60}$$

$$G^{(1)}(\tau) \propto e^{-i\omega_o\tau} \sum_j N_j |A_j|^2 e^{iq\cdot v_j\tau} e^{-D_{Tj}q^2} \tag{61}$$

where $v_j = U_j E$ from Eq. (58).

Thus a light beating measurement in the presence of an electric field offers two major advantages over a zero-field experiment: first, the electrophoretic mobilities and diffusion constants are determined simultaneously in a single experiment; second, if several species of scatterers are present, each species gives a contribution to the spectrum which is a Lorentzian of half-width D_Tq^2 centered about a frequency determined by the eletrophoretic mobility of that species, in contrast to the situation at zero field where the various Lorentzians are all superimposed.

In 1971, Ware and Flygare[62] studied the correlation function of light scattered from BSA in an electric field of ~100 V/cm and observed the presence of two peaks in the spectrum indicating the presence of two species of slightly different mobilities. In a subsequent study they reported measurements on mixtures of BSA and fibrinogen. Uzgiris[63] has also reported measurements using this technique on erythrocytes, staphylococcus epidermidis and polystyrene spheres.

The electrophoresis technique seems to offer great promise and is currently being investigated in a number of laboratories. The major problem to be overcome is the heating caused by conduction in the electrolytic solvents required for many biological materials.

b. Motile Microorganisms

In Section 2-a we saw that for scatterers undergoing translational diffusion, the decay rate Γ of $g^{(1)}$, or the half-width $\Delta\omega_{\frac{1}{2}}$ of $I(\omega)$ is D_Tq^2, which is the number of scattering lengths $(1/q)^2$ crossed in one second by the moving scatterer.

In 1967, Bergé, Volochine, Billard and Hamelin[4] studied the spectrum of light scattered from spermatazoa in solution. They reasoned that since motile microorganisms move further in a given time than do passive ones, the spectrum should show broadening as a consequence of the motility.

Figure 8 [from Bergé et al. - Ref. 4] shows two light beating spectra obtained with rabbit sperm. Curve I shows the spectrum obtained immediately after dilution in Locke's solution when microscopic observation showed a high motility level, while curve II shows the spectrum obtained after 30 minutes when most of the spermatozoa were dead. The second spectrum, which did not change further with time, represents ordinary diffusion of the non-motile sperm.

Nossal and Chen and their coworkers have investigated the light scattering spectrum of motile bacteria during the last few years,[64-68] and we briefly summarize their work here.

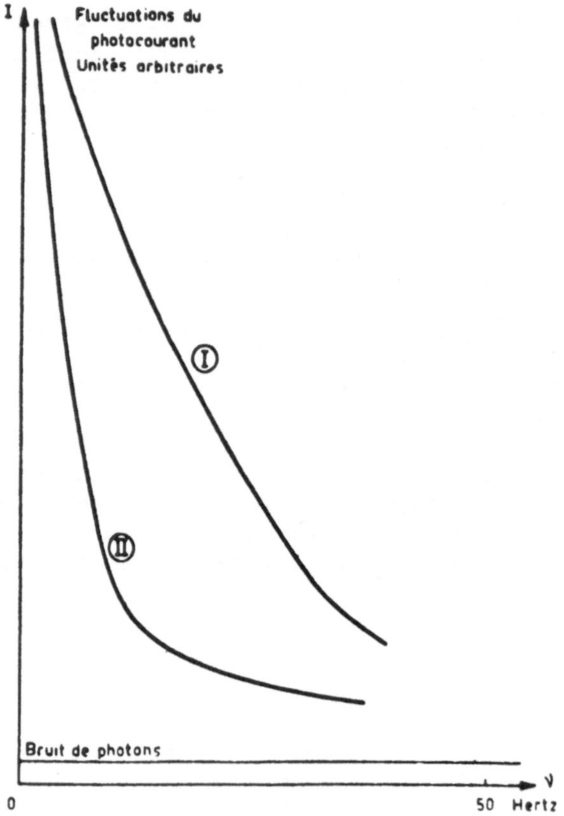

Figure 8. Self-beat spectra of rabbit sperm. Curve I - high motility, Curve II - low motility [from Bergé et al., Ref. 4].

Returning to Eq. (23) for the field correlation function, for uncorrelated scatterers assuming stationarity:

$$G^{(1)}(\tau) = A^2 e^{-i\omega_o \tau} \langle \sum_j e^{iq\cdot(\vec{r}_j(\tau)-r_j(o))} \rangle \tag{62}$$

where we are neglecting the possibility of time-dependent scattering amplitudes. If, as in the electrophoresis problem, we assume that each scatterer moves with constant velocity $\vec{v}_j$ so that $\vec{r}_j(\tau) - \vec{r}_j(o) = \vec{v}_j\tau$, and then let $P(\vec{v})$ be the normalized velocity distribution function, then:

$$|G^{(1)}(\tau)| = NA^2 \int_{\vec{v}} e^{i\vec{q}\cdot\vec{v}\tau} \; P(\vec{v}) \; d^3v \tag{63}$$

For the simplest case of isotropic geometry where $P(\vec{v})$ is independent of direction, Eq. (63) can be reduced, with the introduction of the speed distribution function,

$$P_s(v) = 4\pi v^2 P(\vec{v}) \tag{64}$$

to give:[65]

$$|g^{(1)}(\tau)| = \int \frac{\sin vq\tau}{vq\tau} P_s(v)\, dv \tag{65}$$

Since $|g^{(1)}(\tau)|$ depends on $q\tau$ only through the combination $q\tau = x$, Eq. (65) can be recast as:

$$|g^{(1)}(x)| = \int_v \frac{\sin (vx)}{vx} P_s(v)\, dv \tag{66}$$

The Fourier sine transform of Eq. (66) is then:

$$P_s(v) = \frac{2v}{\pi} \int_0^\infty x|g^{(1)}(x)| \sin(xv)dx \tag{67}$$

Fig. 9, from Nossal, Chen and Lai,[65] shows plots of $|g^{(1)}(x)|^2$ vs x for three scattering angles for light scattered from motile E. Coli bacteria. The different sets of data scale when plotted as a function of $q\tau$ as predicted. (Data for diffusing scatterers scale similarly when plotted as a function of $q^2\tau$). The swimming speed distribution $P_s(v)$ determined by Nossal, Chen and Lai from the data of Fig. 9 is shown in Fig. 10.

Nossal and Chen have also observed the light scattered from motile E. Coli bacteria placed in an oxygen gradient and have been able to observe the anisotropy of $P_s(v)$ representing the response of the bacteria to the gradient (chemotaxis).[67]

There have been several informal discussions at the Institute concerning the extraction of the swimming speed distribution $P_s(v)$ from the correlation function $|g^{(1)}(x)|$. Equations (66) and (67) show that the two functions which form the Fourier transform pair are $x|g^{(1)}(x)|$ and $P_s(v)/v$. Since the correlation data is always truncated at some point, the Fourier inversion will inevitably introduce spurious oscillations into $P_s(v)$.

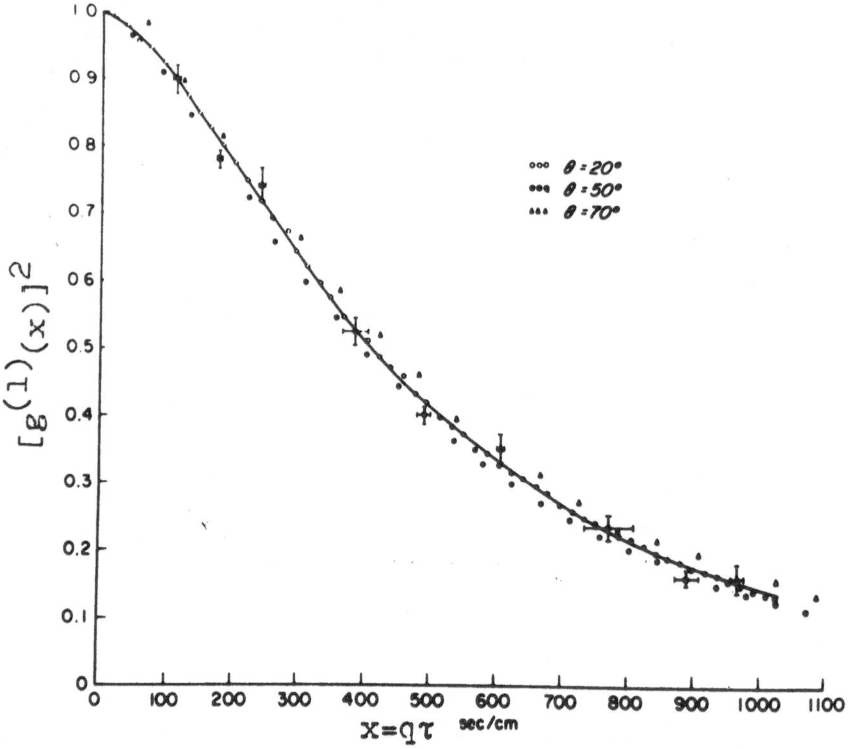

Figure 9. Autocorrelation data for motile E. Coli bacteria at three scattering angles vs x = qτ showing scaling behavior [from Nossal et al., Ref. 65].

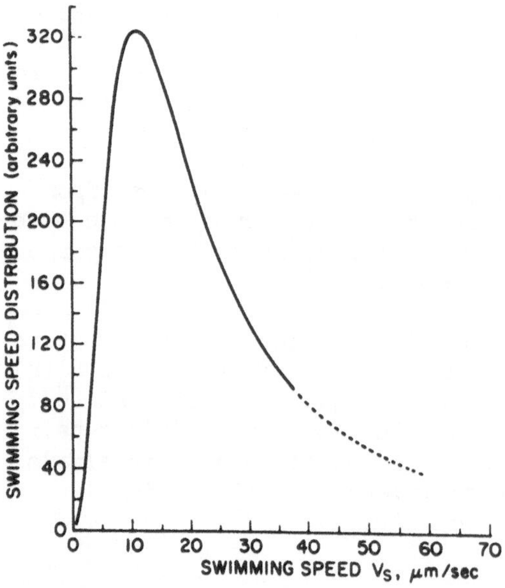

Figure 10. The swimming speed distribution $P_s(v)$ derived according to Eq. (67) from the 20° data of Fig. 9 [from Nossal et al., Ref. 65].

G. Stock and J.P. Boon suggest extrapolating $x|g^{(1)}(x)|$ by either polynomial or multi-Gaussian models far enough to suppress the wiggles, or else taking the envelope of the oscillating $P_s(v)$ as the correct distribution. If, on the other hand, one has some model for $P_s(v)/v$, then these procedures can be circumvented by finding the moments of $x|g^{(1)}(x)|$ which are related to the moments of $P_s(v)/v$, and adjusting the model parameters to optimize the agreement.

The assumption of constant scattering amplitude is dubious since bacteria are comparable in size to the wavelength of light, and they tend to wiggle when swimming. Very recently, Boon, Nossal and Chen[68] have examined the consequences of wiggling motions on the spectrum of light scattered from motile bacteria. Their preliminary analysis leads to the conclusion that for vigorously motile strains the wiggling motion has little effect on the spectra and can be neglected, but this question is not yet completely resolved.

In closing, we mention two recent experiments involving other aspects of motility.

Schaefer[34] has studied the spectra of E. Coli in a small scattering volume. The number fluctuation component of the correlation function which we discussed in Section 3b provides a measure of the time required for a single bacterium to cross the scattering volume. This experimental approach could be particularly interesting in exploring the cooperative phenomena apparently involved in the appearance of travelling bands of chemotactic bacteria as discussed, for example, by Keller and Segel,[69] which should cause enhanced number fluctuations.

Finally, we mention a recent light beating study by Fujime, Maruyama and Asakura of isolated bacterial flagella.[70] This experiment provides an estimate of the flexural rigidity of the flagella which may be useful in analyzing mechanisms of propulsion responsible for observed motilities.

c. Chemical Reactions

The problem of light scattering from systems undergoing chemical reactions has been considered theoretically by a number of authors (cf. Berne and Frisch,[71] Blum and Salzburg,[72] Bloomfield and Benbasat.[73,74] The basic conclusion of these analyses is that if the scattering amplitude depends on the combined concentration changes produced by a shift towards either side of the reaction equation, then the spontaneous fluctuations about equilibrium will modulate the scattering amplitude and the spectrum should contain an additional component with an angle-independent width about zero proportional to the relaxation rate τ_c^{-1}.

In 1969, Yeh and Keeler[75] reported that the light scattered from $ZnSO_4$ and $MnSO_4$ solutions in water exhibited such a broadening effect due to the ionization equilibrium. Their widths, measured with a Fabry-Perot interferometer indicated a relaxation time $\tau \sim 10^{-8}$ sec, but their results have not been substantiated although several other groups have repeated the experiment.

A number of experiments involving chemical reactions have been performed with various muscle proteins. Carlson and Herbert have studied myosin solutions, and interpret their spectra in terms of a concentration dependence of the relative amounts of monomers and dimers present in the equilibrium mixture which may be significant in the process of self-assembly of the large myosin subunits in muscle fibers.[71,72] Fujime and his coworkers have reported studies of F-actin-heavy meromyosin and F-actin-tropomyosin solutions by light beating spectroscopy[76-81] from which they have extracted information about the internal degrees of freedom to which we will return in the next section. These and other experiments on muscle proteins and muscle fibers will be further discussed in the seminar of Carlson and Fraser.

In most cases, the fluctuation in refractive index produced by the small fluctuations about equilibrium will be very small, and it appears that the effect of chemical reactions on the spectrum will be extremely difficult to see.

Recently, however, Berne and Gininger[82] have shown that if a chemically reacting system is studied in the presence of an electric field, then the individual Lorentzians of Eq. (60) will each be broadened by the rate at which the species producing that component reacts. This technique promises to provide a much more sensitive probe of chemical reactions than the zero-field method employed in previous studies.

To conclude this section, we mention a recent experiment by Magde, Elson and Webb in which the relaxation time for a chemical reaction was determined from the intensity correlation function of fluorescence rather than of scattered light.[83] In their experiment, a solution of DNA and the dye ethidium bromide was illuminated with 5145 Å light from an argon laser. Since the DNA and dye complex is strongly fluorescent, the reversible binding reaction DNA + Dye $\rightleftarrows$ complex gives rise to fluctuations in the fluorescence intensity, and these fluctuations were studied with an autocorrelator. (This technique will be discussed further in an informal seminar by Dr. Herbert.)

d. Conformational Changes

Macromolecules in solution frequently have different internal configurations of nearly equal energy between which transitions may occur.

There are two closely related classes of conformational transitions which have been explored by light beating spectroscopy. The simpler class involves a basically static situation in which the conformational change is induced by some change in the environment. Thus, for example, Ford, Lee and Karasz found that the diffusion constant of poly-α-amino acid solutions exhibits a change with acid concentration reflecting the helix-coil transition,[51] Rimai et al. found that the spectrum of RNase solutions exhibited changes in D_T reflecting conformational changes during both chemical and thermal denaturation,[52] and Pike reports on changes in the correlation function of BSA reflecting the dependence of conformation on solvent PH.[84]

The second class of problems involves rapid changes between configurations which can be described either as internal modes of motion or as dynamic conformational transitions. Pecora and his coworkers have considered a number of such cases, including flexible polymers,[85] flexible coil macromolecules,[86] dynamic helix-coil transitions[87] and rods with a break in the middle which acts as a "universal joint"[88] (also see Fujime - Ref. 89).

The basis of these calculations is an assumption that the amplitude correlation function for the scattered field can be represented by

$$\langle A(o)\ A(\tau)\rangle = A_o^2 + \sum_j A_j^2\ e^{-\tau/\tau_j}$$

where the τ_j are the relaxation rates for the various (uncorrelated) internal fluctuations.

Substitution into Eq. (34A) then gives the following expression for the optical spectrum:

$$I(\omega) = A_o^2 \frac{D_T q^2/\pi}{(\omega-\omega_o)^2+(D_T q^2)^2} + \sum_j A_j^2 \frac{(D_T q^2+\tau_j^{-1})/\pi}{(\omega-\omega_o)^2+(D_T q^2+\tau_j^{-1})^2} \qquad (68)$$

where the first term in Eq. (68) is the usual translational diffusion component, and the terms in the j-sum represent the contributions from the internal modes.

If the self-beating spectrum is computed from Eq. (68) with the approximation $A_j << A_o$, then the leading terms in $|g^{(1)}(\tau)|^2$ [or $P_i(\omega)$] beyond the pure translation term are exponentials with decay rates (or Lorentzians centered at zero with half-widths):

$$\Gamma_j = 2D_T q^2 + \tau_j^{-1} \tag{69}$$

which arise from beating between the pure translation term and the j-sum in Eq. (68).

Thus, low angle scattering (where $D_T q^2 \to o$) can give the dominant relaxation time. Marshall and Pecora have discussed a similar approach to helix-coil transition rates.[87]

A number of experiments based on these ideas have been reported during the last three years. Yeh[90] reports small angle measurements on the copolymer dAT (alternating adenine-thymine nucleotide chain) from which he deduces a rate for helix coil interconversion of $\sim 2.3 \times 10^3$ sec^{-1}. Fujime and coworkers have analyzed light scattering spectra of muscle protein complexes as discussed in the previous section. Their experiments are generally analyzed by fitting the observed spectra to single Lorentzians Γ_T, and plotting Γ_T vs q^2. This procedure, as shown by Eq. (69), gives the diffusion constant from the slope and τ^{-1} from the $q \to o$ intercept if the pure translational term is neglected.[91] These experiments are also controversial, however, as will be discussed in the seminar of Carlson and Fraser.

e. Bounded Diffusion and Gels

Several studies have been reported recently in which light-beating spectroscopy has been applied to systems undergoing gel transitions. French, Angus and Walton[92] studied collagen and gelatin solutions, B. Aiello et al. have investigated the gel transition in agarose-water,[9] and Fraser and Carlson have investigated the gelling of the muscle protein actomyosin[93] which will be discussed in their seminar.

A relatively simple model for the spectrum of a system undergoing a gel transition can be extracted from our discussion of diffusion in Section 3a by assuming that each particle continues to diffuse as the gel forms, but that its motion is restricted to a volume which decreases as the gel sets up.

The problem can then be handled by replacing the space-time correlation function for a free particle, G_s (R, τ), in Eq. (24) by a solution for a bounded or constrained particle.

Solutions to the diffusion equation for particles bound by springs, walls, etc. tend to be rather complex, but we can write an approximate solution based on the papers of Kac and Uhlenbeck and Ornstein: [94]

$$|g^{(1)}(\tau)| = e^{-Dq^2(1-e^{-2\gamma\tau})/2\gamma} \tag{70}$$

where γ is the average velocity/box size, or (free transit time)$^{-1}$. For short times, Eq. (70) is identical to the free diffusion result, but for long times $|g^{(1)}(\tau)|$ falls only to $e^{-Dq^2/2\gamma}$ rather than zero. This result shows that as the distance over which the particles are free to diffuse decreases, the fluctuations in the intensity will also decrease.

If the intensity correlation function is normalized to 1 at $\tau = 0$, then for free particles it should decrease to ~0.5 for long times, whereas for particles constrained in small volumes the long-time value will remain closer to 1.

Intensity correlation data for actomyosin undergoing gelling were recently obtained by A. Fraser et al.[93] The general form of the correlation functions is in agreement with the model we have just discussed, although the correctness of the model has not yet been fully established.

f. Cells, Muscles, Membranes, Nerves ....?

Beyond the study of macromolecules, viruses and the other problems of scatterers in solution which we have analyzed in some detail lies a whole range of phenomena involving larger biological structures on which the technique of photon correlation spectroscopy promises to have tremendous impact. So far, only a few preliminary experiments have been reported which we will summarize briefly in this section.

Living cells are known to be characterized by many internal dynamical processes, and it should be possible to analyze many of these processes by light scattering if the changes occurring during the process can couple to the light.

In 1972, Bargeron et al. reported that the light beating spectrum of erythrocyte ghosts (human red blood cells from which the hemoglobin has been removed) exhibited an inelastic light scattering component at ~170 Hz which they identified with a putative "metabolic pump" which supposedly sustains the transport of ions across the cell membrane.[95] Unfortunately, it later developed that the signal they had seen resulted from convective flow set up by the laser beam and was not connected with the internal dynamics of the cells.[96] This experiment generated considerable excitement since the prospect of following such processes dynamically would represent a substantial breakthrough.

Maeda and Fujime[97] have performed a light beating experiment with a system employing a microscope so that the scattering volume is very small; they say that motions within a cell should be accessible with this apparatus, and that a study of muscle fibers was in progress in 1972.

Carlson, Bonner and Fraser[98] have studied the light scattering spectrum of muscle fibers at rest and in various states of contraction and from the correlation functions they have inferred some characteristics of the dynamics of the contraction process. This work will be discussed further in the seminar of Carlson and Fraser in this Institute.

A study of blood flow within the eye using light beating spectroscopy has recently been reported by C. Riva and coworkers.[99] Shaw and Newby[100] studied the light scattering spectrum of locust brains and observed a marked increase in intensity fluctuations when potassium was added to the surrounding fluid, resembling the spectra of Fig. 8. They conjectured that the scattering was primarily due to presynaptic vesicles within the ganglia, and that these vesicles are free to move extensively only when the nerve endings are depolarized.

Recently, R.W. Piddington has observed a similar effect in locust ganglia which have been repetitively shocked for 20 minutes and has shown that the effect can be blocked with the nerve poison tetrodotoxin.[101] He has also studied scattering by the giant plant cell Spirellum nitella and was able to observe the effect of internal streaming as well as the arrest of streaming by electric shock. (Dr. Piddington will present an informal report on his work in this Institute.)

Finally, we should note that in studying complex biological structures, the validity of the assumption of independent scattering from different scattering elements may no longer hold, and that multiple scattering will become significant. This is one area where additional theoretical work is clearly needed.

In closing, we emphasize that this survey of biologically relevant light beating experiments is representative of the current status of the field, but not complete. Many new results are being reported which indicate a rapid growth of interest and activity in this field, and it now seems that light beating spectroscopy should soon become a technique of widespread use in biological research.

6. REFERENCES

1. H. Z. Cummins, N. Knable, and Y. Yeh, Phys. Rev. Letters 12, 150 (1964)

2. R. Pecora, J. Chem. Phys. 40, 1604 (1964).

3. S. B. Dubin, J. H. Lunacek, and G. B. Benedek, Proc. Natl. Acad. Sci. U.S. 57, 1164 (1967).

4. P. Bergé, B. Volochine, R. Billard, and A. Hamelin, C. R. Acad. Sci. Paris D265, 889 (1967).

5. B. Chu, Ann. Rev. Phys. Chem. 21, 145 (1970).

6. R. Pecora, Ann. Rev. Biophysics & Bioengineering 1, 259 (1972).

7. N. C. Ford Jr., Chemica Scripta 2, 193 (1972).

8. N. C. Ford Jr., R. Gabler, and F. E. Karasz, Adv. Chem. (to be published).

9. E. R. Pike and E. Jakeman, in: Advances in Quantum Electronics - Vol 2, D. Goodwin, editor (Academic Press, London - to be published).

10. H. C. Van De Hulst, Light Scattering By Small Particles (John Wiley & Sons, N.Y., 1962).

11. M. Kerker, The Scattering of Light and Other Electromagnetic Radiation (Academic Press, N.Y., 1969).

12. C. Tanford, Physical Chemistry of Macromolecules (John Wiley, N.Y., 1961).

13. G. Oster, Chem. Rev. 43, 319 (1948); J. Gen. Physiol. 33, 215 (1950).

14. G. Oster, in: Physical Methods of Chemistry, Part III-A (edited by A. Weissberger and B. W. Rossiter, Wiley, N.Y., 1972) p. 75.

15. Light Scattering from Dilute Polymer Solutions, D. McIntyre and F. Gornick, editors (Reprint Collection - Gordon & Breach, N.Y., 1964).

16. S. B. Dubin, Ph.D. Thesis, M.I.T. - 1970 (unpublished).

17. R. Pecora, J. Chem. Phys. 48, 4126 (1968).

18. H. Z. Cummins, F. D. Carlson, T. J. Herbert, and G. Woods, Biophys. J. 9, 518 (1969).

19. S. Fujime and M. Maruyama, Macromolecules 6, 237 (1973).

20. T. Maeda and N. Saito, J. Phys. Soc. Japan 27, 984 (1969).

21. S. Fujime, J. Phys. Soc. Japan 29, 416 (1970).

22. D. W. Schaefer, G. B. Benedek, P. Schofield, and E. Bradford, J. Chem. Phys. 55, 3884 (1971).

23. T. S. Chow, Phys. Fluids 16, 31 (1973).

24. R. Pecora, J. Chem. Phys. 49, 1036 (1968).

25. A. Wada, N. Suda, T. Tsuda, and K. Soda, J. Chem. Phys. 50, 31 (1969).

26. S. B. Dubin, N. A. Clark, and G. B. Benedek, J. Chem. Phys. 54, 5158 (1971).

27. R. Pecora and Y. Tagami, J. Chem. Phys. 51, 3293, 3298 (1969).

28. J. A. Benbasat and V. A. Bloomfield, J. Polymer Sci. 10, 2475 (1972).

29. D. E. Koppel, J. Chem. Phys. 57, 4814 (1972).

30. P. N. Pusey, in: Industrial Polymers: Characterization by Molecular Weight, editors J. H. S. Green and R. Dietz (Transcripta Books, London, to be published). See also J. C. Brown, R. Dietz and P. N. Pusey, (to be published).

31. J. Newman, H. L. Swinney, H. Z. Cummins, S. A. Berkowitz, and L. Day, Bull. Am. Phys. Soc. 18, 671 (1973).

32. J. Newman, L. Day, S. Berkowitz, and H. L. Swinney (to be published).

33. D. W. Schaefer and B. J. Berne, Phys. Rev. Letters 28, 475 (1972).

34. D. W. Schaefer, Science 180, 1293 (1973).

35. P. J. Bourke, et al., J. Phys. A3, 216 (1970).

36. C. W. Pyun and M. Fixman, J. Chem. Phys. 41, 937 (1964).

37. T. J. Herbert and F. D. Carlson, Biopolymers 10, 2231 (1971).

38. F. D. Carlson and T. J. Herbert, J. de Phys. 33-C1, 157 (1972).

39. P. N. Pusey, D. W. Schaefer, D. E. Koppel, R. D. Camerini-Otero, and R. M. Franklin, J. de Phys. 33-C1, 163 (1972).

40. G. K. Batchelor, J. Fluid Mech. 52, 245 (1972).

41. A. R. Altenberger and J. M. Deutch, J. Chem. Phys. 59, 894 (1973).

42. P. N. Pusey, D. E. Koppel, D. W. Schaefer, R. D. Camerini-Otero, and S. H. Koenig, Biochemistry (to be published); R. D. Camerini-Otero, P. N. Pusey, D. E. Koppel D. W. Schaefer, and R. M. Franklin, Biochemistry (to be published).

43. R. A. Ferrell, Phys. Rev. 169, 199 (1968).

44. H. C. Kelley, J. Phys. A 6, 353 (1973).

45. F. Perrin, J. de Phys. 5, 33 (1934); 7, 1 (1936).

46. V. A. Bloomfield, K. E. Van Holde, and W. O. Dalton, Biopolymers 5, 135, 149 (1967).

47. R. J. Douthart and V. A. Bloomfield, Biochemistry 7, 3912 (1968).

48. M. J. Stephen, J. Chem. Phys. 55, 3878 (1971).

49. S. B. Dubin, G. B. Benedek, F. C. Bancroft, and D. Freifelder, J. Mol. Biol. 54, 547 (1970).

50. F. T. Arecchi, M. Giglio, and U. Tartari, Phys. Rev. 163, 186 (1967).

51. N. C. Ford Jr., W. Lee, and F. E. Karasz, J. Chem. Phys. 50, 3098 (1969).

52. L. Rimai, J. T. Hickmott Jr., T. Cole, and E. B. Carew, Biophys. J. 10, 20 (1970).

53. S. H. C. Lin, S. I. Leong, R. K. Dewan, V. A. Bloomfield, and C. V. Morr, Biochemistry 11, 1818 (1972); 10, 4788 (1971).

54. R. J. Dewan and V. A. Bloomfield, J. Dairy Sci. 56, 66 (1973).

55. A. Wada, N. C. Ford Jr., and F. E. Karasz, J. Chem. Phys. 55, 1798 (1971).

56. J. M. Schurr and K. S. Schmitz, Biopolymers 12, 1021 (1973). Another experiment on TMV appeared too late to include in this survey: T. A. King, A. Knox, and J. D. G. McAdam, Biopolymers 12, 1917 (1973).

57. E. R. Pike, J. de Phys. 33-C1, 177 (1972).

58. C. J. Oliver - private communication.

59. R. Foord, E. Jakeman, C. J. Oliver, E. R. Pike, R. J. Blagrove, E. J. Wood, and A. R. Peacocke, Nature 227, 242 (1970).

60. C. J. Oliver, E. R. Pike, A. J. Cleave, and A. R. Peacocke, Biopolymers 10, 1731 (1971).

61. E. J. Wood, W. H. Bannister, C. J. Oliver, R. Lontie, and R. Witters, Comp. Biochem. Physiol. 40B, 19-24 (1971).

62. B. R. Ware and W. H. Flygare, Chem. Phys. Letters 12, 81 (1971); J. Coll. Interf. Sci. 39, 670 (1972).

63. E. E. Uzgiris, Optics Commun. 6, 55 (1972).

64. R. Nossal, Biophys. J. 11, 341 (1971).

65. R. Nossal, S. H. Chen, and C. C. Lai, Optics Commun. 4, 35 (1971).

66. R. Nosal and S. H. Chen, J. de Phys. 33-C1, 171 (1972).

67. R. Nossal and S. H. Chen, Optics Commun. 5, 117 (1972).

68. J. P. Boon, R. Nossal, and S. H. Chen (to be published); J. P. Boon (to be published); also see D. W. Schaefer (to be published).

69. E. F. Keller and L. A. Segel, J. Theor. Biol. 26, 399 (1970); 30, 225, 235 (1971).

70. S. Fujime, M. Maruyama, and S. Asakura, J. Mol. Biol. 68, 347 (1972); S. Fujime and S. Hatano, J. Mechanochem. Cell Motility 1, 81 (1972).

71. B. J. Berne and H. L. Frisch, J. Chem. Phys. 47, 3675 (1967).

72. L. Blum and Z. W. Salzburg, J. Chem. Phys. 48, 2292 (1968).

73. V. A. Bloomfield and J. A. Benbasat, Macromolecules 4, 609 (1971).

74. J. A. Benbasat and V. A. Bloomfield, Macromolecules 6, 593 (1973).

75. Y. Yeh and R. N. Keeler, J. Chem. Phys. 51, 1120 (1969).

76. S. Fujime and S. Ishiwata, J. Phys. Soc. Japan 29, 1651 (1970).

77. S. Fujime, S. Ishiwata, and T. Maeda, Biochim. Biophys. Acta 283, 351 (1972).

78. S. Fujime and S. Ishiwata, J. Mol. Biol. 62, 251 (1971).

79. S. Ishiwata and S. Fujime, J. Phys. Soc. Japan 31, 1601 (1971).

80. S. Ishiwata and S. Fujime, J. Phys. Soc. Japan 30, 302, 303 (1971).

81. S. Ishiwata and S. Fujime, J. Mol. Biol. 68, 511 (1972).

82. B. J. Berne and R. Gininger, Biopolymers 12, 1161 (1973).

83. D. Magde, E. Elson, and W. W. Webb, Phys. Rev. Letters 29, 705 (1972).

84. E. R. Pike, in: Quantum Optics, S. M. Kay and A. Maitland, editors (Academic Press, N. Y. 1970).

85. R. Pecora, J. Chem. Phys. 43, 1562 (1965); also see N. Saito and S. Ito, J. Phys. Soc. Japan 25, 1446 (1968).

86. R. Pecora, J. Chem. Phys. 49, 1032 (1968).

87. A. G. Marshall and R. Pecora, J. Chem. Phys. 55, 1245 (1971).

88. R. Pecora, Macromolecules 2, 31 (1969).

89. S. Fujime, J. Phys. Soc. Japan 31 , 1805 (1971).

90. Y. Yeh, J. Chem. Phys. 52, 6218 (1970).

91. S. Fujime, J. Phys. Soc. Japan 29, 751 (1970).

92. M. J. French, J. C. Angus, and A. G. Walton, Biochim. Biophys. Acta 251, 320 (1971).

93. A. B. Fraser, E. Eisenberg, W. W. Keilley, and F. D. Carlson, Biophysical Society Abstracts 13, 81a (1973).

94. Selected Papers on Noise and Stochastic Processes, edited by N. Wax (Dover Publications, New York, 1954).

95. C. B. Bargeron, R. L. McCally, P. E. R. Tatham, S. M. Cannon, and R. W. Hart, Phys. Rev. Letters 28, 1105 (1972).

96. C. B. Bargeron, R. L. McCally, S. M. Cannon, and R. W. Hart, Phys. Rev. Letters 30, 205 (1973).

97. T. Maeda and S. Fujime, Rev. Sci. Instr. 43, 566 (1972).

98. F. D. Carlson, R. Bonner, and A. Fraser, Cold Spring Harbor Symposium on Quantitative Biology 37, 389 (1972).

99. C. Riva, et al., Investigative Opthamology 11, 936 (1972).

100. T. I. Shaw and B. J. Newby, Biochim. Biophys. Acta 255, 411 (1972).

101. R. W. Piddington - (to be published).

CRITICAL PHENOMENA IN FLUIDS*

Harry L. Swinney†

Department of Physics, New York University

New York, New York 10003

In this paper we consider the behavior of fluids near the critical point and discuss in particular the use of light scattering as a tool for probing the dynamics of fluids near the critical point. In this volume Pike[1] and Lallemand[2] have presented extensive developments of the theory of light scattering by fluids; hence in the present paper we are able to concentrate primarily on describing the phenomena which are peculiar to the critical region and on the experiments which have explored those phenomena.

After an introduction to critical phenomena in Section 1, we consider in Section 2 the behavior of the scattering intensity near the critical point. The general features of the light scattering spectrum are discussed in Section 3, and in Sections 4 and 5 we consider in greater detail the Rayleigh and Brillouin components near the critical point.

1. CRITICAL PHENOMENA

1.1 Description of Critical Behavior

We begin by considering the phase diagram of a simple fluid, as shown in Fig. 1. The general behavior of the isotherms of a fluid, as shown in Fig. 1, was discovered in 1869 by Andrews[3] who reported measurements of the pressure and volume of carbon dioxide at different temperatures. Above a well-defined temperature, which Andrews called the "critical temperature," it was found that there existed only a single phase which was gas-like at low densities and liquid-like at high densities, but there was no abrupt change from a gas to a liquid as the volume of the container was decreased.

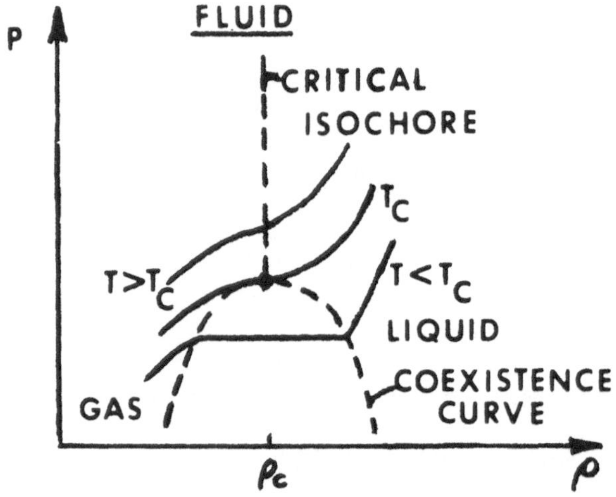

Figure 1. Isotherms of a simple fluid.

On the other hand, below T_c there was a range of mean densities for which two distinct phases coexisted, and the difference between the densities of the two phases was found to approach zero as the critical point was approached.

Andrews noted that at the critical point the pressure-density isotherm was very flat; that is, a large change in sample volume resulted in only a very small change in pressure, so the fluid is very soft near the critical point. Thus the isothermal compressibility, $\kappa_T \equiv [\rho(\partial P/\partial\rho)_T]^{-1}$, which is inversely proportional to the slopes of the P-ρ isotherms in Fig. 1, is quite large in the vicinity of the critical point.

It soon became clear, as Andrews himself had suggested, that critical points occur in all fluids, not just CO_2. To this day CO_2 remains one of the most thoroughly studied fluids near the critical point because it is easy to obtain and it has a convenient critical temperature [T_c = 30.1°C].

The general behavior of fluids in the vicinity of the critical point was explained qualitatively in 1873 by van der Waals,[4] whose now famous equation of state predicts the existence of a gas-liquid equilibrium and a critical temperature above which only a single phase exists.

Critical points also occur in many other systems. That is, by varying the temperature or some other parameter, two distinct phases can be made more and more similar until ultimately at a certain critical point, all differences vanish, and beyond that point only one homogeneous phase can exist. For a fluid the difference between the liquid and gas densities $\rho_L - \rho_G$, is the parameter that describes the order of the system. In a ferromagnet the order parameter is the spontaneous magnetization M, which goes continuously to zero as the temperature is increased to the critical temperature. The similarity in the critical behavior of fluids and ferromagnets is clear if we compare the isotherms of a simple fluid with those of a magnet, as shown in Fig. 2.

In the past ten years it has become increasingly apparent that there are quantitative as well as qualitative similarities in the behavior of fluids and magnets near critical points. That is, the asymptotic laws which describe the behavior of corresponding physical properties have been found to be essentially of the same form. For example, the order parameters are described by the simple exponential law

$$M \sim (\rho_L - \rho_G) \sim \varepsilon^{\beta},$$

where

$$\varepsilon \equiv |T - T_c|/T_c.$$

Experimentally $\beta \approx 1/3$, while the "classical" models such as the van der Waals gas and the Weiss ferromagnet invariably yield

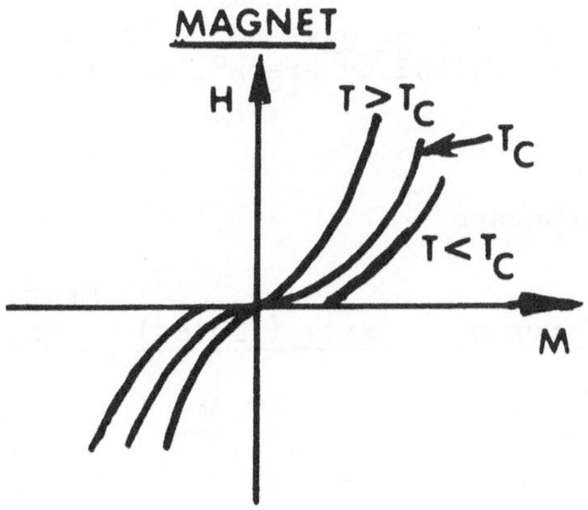

Figure 2. Isotherms of a ferromagnet.

$\beta = 1/2$ (see Table I). Similarly, the susceptibilities of fluids (κ_T) and ferromagnets [$\chi_M = (\partial H/\partial M)^{-1}{}_{T,H=0}$] are also described by a simple exponential law,

$$\chi_M \sim \kappa_T \sim \varepsilon^{-\gamma},$$

where experimentally $\gamma \simeq 1.25$, while the classical models yield $\gamma = 1$. The specific heat (at constant field or volume) is described by

$$c_H \sim c_V \sim \varepsilon^{-\alpha},$$

where experimentally α is small ($\alpha \approx 1/8$), while the classical models predict a discontinuity in the specific heat ($\alpha = 0$). [The exponents γ and α characterize the behavior of κ_T and c_v along a particular thermodynamic path, the critical isochore ($\rho = \rho_c$).]

Experiments have shown that there exists a variety of other types of systems near critical points, including, e.g., antiferromagnets, binary metallic alloys, systems with order-disorder phase transitions, and binary liquid mixtures, which are described by essentially the same exponential laws. Thus the critical exponents which characterize the behavior of corresponding properties of a wide range of systems are very nearly the same. A common feature of systems near critical points is that the range ξ over which the fluctuations in the order parameter are correlated diverges ($\xi = \xi_o\varepsilon^{-\nu}$, with $\nu \approx 2/3$), and the details of the interactions then play a secondary role, while the cooperative behavior common to all systems near critical points is of decisive importance. The gradual recognition of the universality in critical behavior has motivated intense theoretical and experimental research on the subject of critical phenomena in the past few years. (See the reviews by Kadanoff, et al.[5] and by Fisher[6], the book by Stanley[7], the volume edited by Green[8], and the reprint volume edited by Stanley[9].)

Table I. Critical exponents

Quantity	Exponent	Expt. (fluids)	Mean field theory	3-dim. Ising model
Order parameter	β	0.33-0.36	½	0.31
Susceptibility	γ	1.18-1.27	1	1.25
Specific heat	α	Small	0	0.08-0.12
Correlation length	ν	0.59-0.67	½	0.64

Let us consider now the critical point of a mixture of two components, A and B. Above the critical temperature the two components are miscible in all proportions, but if the temperature is lowered below T_c, the fluid may undergo a phase separation into an A-rich phase and a B-rich phase. If x is the mole-fraction concentration of one of the components, then the coexistence curve is described by the order parameter $x - x_c \sim \varepsilon^\beta$ with $\beta \simeq 1/3$. The parameter analagous to the pressure (or chemical potential) of a simple fluid is $\Delta = \mu_A - \mu_B$, the difference between the chemical potentials (per unit mass) of the two species; the susceptibility of the mixture is given by $(\partial x/\partial \Delta)_{P,T}$. (Some binary mixtures have a lower rather than upper critical point, with the two components miscible below T_c and immiscible above T_c, and some mixtures have both upper and lower critical points.)

In these lectures we will consider the critical behavior of simple fluids and binary liquid mixtures, which are the two types of critical systems which have been investigated using the light beating and photon correlation spectroscopic techniques, and the word "fluid" will be used to mean either a binary mixture or a one-component fluid.

1.2 Theoretical Models of Phase Transitions

The various "classical" models of critical systems, such as the Landau theory of phase transitions, the Weiss ferromagnet, the van der Waals gas, and the many refinements of these models, invariably predict $\beta = \frac{1}{2}$, $\gamma = 1$, and a discontinuity in the specific heat, predictions which are clearly in disagreement with the experimental results (see Table I).[5-9] The crucial incorrect assumption implicit in the classical equations of state is that the free energy is analytic at the critical point, and hence can be expanded in a Taylor series about the critical point. Therefore, all exponents are multiples of $\frac{1}{2}$.

The modern era in the theory of phase transitions began in 1944 when Onsager solved the two-dimensional Ising model of a ferromagnet and found that for this model the free energy is <u>not</u> analytic at the critical point. In this model there is associated with each lattice site a variable s_i that can have only two values, +1 (spin up) or -1 (spin down). The Hamiltonian for this system for a given configuration $\{s_i\}$ of the spins is[6,7]

$$\mathcal{H}_I = -J \sum_{i,j} s_i s_j - H \sum_i s_i ,$$

where the first sum, the interaction between spins, is carried out only over nearest neighbors, and the second term represents the interaction of the spins with the magnetic field H. The Ising model is isomorphic to a simple model for a fluid, a "lattice gas":

in that case the number associated with each site (1 or 0) corresponds to the presence or absence of an atom on a site. Onsager was able to calculate the partition function $Z = \Sigma \exp(-\mathcal{H}/k_BT)$, where the sum is over all configurations $\{s_i\}$ of the spins, but the three-dimensional model has not been solved in closed form. However, in the last decade numerical techniques have been extensively used to obtain the properties of the three dimensional Ising model, particularly the critical exponents, and the results for this rather crude model of a real physical system are in good agreement with the results of experiments (see Table I).

There are as many exponents as there are properties of systems near the critical point; therefore, it is natural to ask if there exist relations between exponents, and how many fundamental exponents are there. One answer to this question is provided by the <u>static scaling</u> hypothesis developed by Widom,[10] who proposed (without physical justification) that the free energy of a system near the critical point is a homogeneous function. Thus for a magnet the free energy is given by

$$G(\varepsilon,H) = \lambda G(\lambda^a\varepsilon,\lambda^b H),$$

where λ is arbitrary and a and b are unspecified constants. It is then straightforward to show from the definitions of the critical exponents and the thermodynamic relations for the various physical properties in terms of derivatives of G, that the critical exponents are all given in terms of the parameters a and b; these relations between the critical point exponents, known as the <u>static scaling laws</u>, are in reasonable agreement with experiments and with the results for the Ising model. Moreover, several equations of state which satisfy the homogeneity hypothesis have been proposed and have been found to describe accurately the available data for fluids and magnets (see Vicentini-Missoni, <u>et al</u>.[11] and Schofield, Litster, and Ho[12]).

A recent important development in the theory of critical phenomena is the "renormalization group" method developed by Wilson.[13-15] **Wilson has greatly expanded and** developed a heuristic idea which Kadanoff[16] used in a plausible derivation of the homogeneity hypothesis. Kadanoff's derivation was based on the known cooperative nature of critical behavior. Far above the critical point the spins are flipping randomly, but as the critical point is approached, droplets of correlated spins form, and the size of these droplets is given by the correlation length. (In a fluid it is the increasing size of the droplets, of "liquid" rather than correlated spins, which is responsible for the intense scattering known as critical opalescence.)

Kadanoff's basic idea was that the Ising problem for spins

separated by a lattice space "a" could be reformulated as a system of cells of length "aL" where a <<aL <<ξ.[16] Kadanoff argued that since L <<(ξ/a), all the spins in a cell point in the same direction, either up or down. This assumption is clearly not correct--such complete alignment occurs only as T→0; however, the rigorous procedure developed by Wilson is based on Kadanoff's cell construction idea, and Wilson's method yields Kadanoff's results as a special case.[13-15] It is assumed that if the original Hamiltonian contains only nearest neighbor interactions, then the Hamiltonian for the cell problem also has only nearest neighbor interactions. Thus the Hamiltonian for the site problem,

$$\frac{\mathcal{H}}{k_B T} = - \frac{J}{k_B T} \sum_{i,j} s_i s_j - \frac{H}{k_B T} \sum_i s_i ,$$

and the Hamiltonian for the cell problem,

$$\frac{\mathcal{H}_L}{k_B T} = - \frac{J_L}{k_B T} \sum_{i,j} s_i^L \, s_j^L - \frac{H_L}{k_B T} \sum_i s_i^L$$

where the sums are over cells, are formally the same. Furthermore, since the site and cell pictures both describe the same physical problem, the free energies of the two systems must be equal. The dependence of the new coupling constants J_L and H_L can, at least in principle, be calculated in terms of the old coupling constants J and H. The transformations between the two sets of coupling constants,

$$\frac{J_L}{k_B T} = u_L \left(\frac{J}{k_B T}, \frac{H}{k_B T} \right) ,$$

$$\frac{H_L}{k_B T} = \frac{H}{k_B T} \, v_L \left(\frac{J}{k_B T}, \frac{H}{k_B T} \right) ,$$

are the "renormalization group" transformations.

Kadanoff made plausible guesses for u_L and v_L which yielded the homogeneity hypothesis. Wilson, on the other hand, has derived differential equations for these transformations and has shown that the critical point corresponds to a fixed point of the transformations [e.g., $J^* = u(J^*)$]. Moreover, the values of the critical exponents can be obtained from a knowledge of the properties of the transformations and their derivatives at the special fixed point.

The importance of the Wilson method lies in the fact that the transformations u_L and v_L are analytic functions of their arguments even at the critical point. Hence pertubation methods can be used in studying the solutions of the differential equations in u and v, whereas perturbation methods cannot be applied to study the free energy at the critical point since it is not analytic.

The Wilson method also contributes to our understanding of universality. Although the Ising model results agree with experiments, the numerical methods used to investigate this model provide no physical insight into the origin of universality. In the Wilson method, however, it is clear that the functional forms of the renormalization group transformations are determined by the operator structure of the Hamiltonian and are independent of the numerical values of the coupling constants; hence many systems can have the same critical point exponents.

In this paper we will be interested primarily in the dynamic properties of critical systems. The renormalization group method provides insight and numerical results as well, and it seems clear that this approach will be applied extensively in the future to the study of the dynamics of critical behavior. Indeed, one of the techniques used by Wilson[17] for the calculation of the static critical exponents has already been applied by Halperin, Hohenberg, and Ma[18] to study the dynamic critical behavior of a simple system, the time-dependent Ginzburg-Landau model.

2. THE SCATTERING INTENSITY NEAR THE CRITICAL POINT

Now let us consider the scattering of light by fluids near the critical point. Einstein showed in 1910 that the scattering intensity is proportional to the mean square fluctuation in the dielectric constant ε. (See the papers by Pike[1] and Lallemand[2] in this volume.) In a simple fluid the scattering arises primarily from density fluctuations, and Einstein showed, using thermodynamic fluctuation theory, that

$$\langle(\Delta\rho)^2\rangle = k_B T \rho^2 \kappa_T/V .$$

Hence

$$I \propto \langle(\Delta\varepsilon)^2\rangle = (\partial\varepsilon/\partial\rho)^2\langle(\Delta\rho)^2\rangle \propto \kappa_T. \qquad (1)$$

Thus the scattering intensity should exhibit the same strong divergence as the isothermal compressibility. It is this intense scattering (critical opalescence) which makes the critical region particularly amenable to study by light scattering techniques.

Einstein treated the scattering system as a continuum in his derivation of the scattering intensity. If the system is treated as a system of particles with density $\rho(\vec{r}) = \Sigma_j \delta(\vec{r}-\vec{r}_j)$, where $\vec{r}_j$ is the position of the jth particle, then the light scattering cross section per unit volume is given by[2,19]

$$\frac{1}{V}\frac{d\sigma}{d\Omega} \quad \frac{\pi^2}{\lambda_o^4}\left(\frac{\partial\varepsilon}{\partial\rho}\right)^2 \rho \; S(q) \; , \tag{2}$$

where the static structure factor S(q) is given by a Fourier transform of the density-density correlation function,

$$S(q) = 1+\rho\int d^3r \; e^{i\vec{q}\cdot\vec{r}} <\Delta\rho\;(\vec{r}\,')\;\Delta\rho\;(\vec{r}\,' + \vec{r})>/\rho^2 \; , \tag{3}$$

where

$$\Delta\rho(\vec{r}) \equiv \rho(\vec{r}) - <\rho> \; .$$

Now the density-density correlation function is related to the <u>pair correlation function</u> g(r) by

$$g(r)-1 = <\Delta\rho(\vec{r}\,')\;\Delta\rho(\vec{r}\,'+\vec{r})>/\rho^2 \; , \tag{4}$$

where g(r) is equal to the probability (normalized to unity at infinity) of finding a particle at r <u>given that</u> there is a molecule at the origin. Rewriting Eq. (3), we have

$$S(q) = 1 + \rho\int d^3r \; e^{i\vec{q}\cdot\vec{r}} \;\; [g(r) - 1] \; . \tag{5}$$

For normal fluids g(r) - 1 decays to zero within distance equal to a few times the interparticle spacing, as shown in Fig. 3, so we can make the approximation

$$e^{i\vec{q}\cdot\vec{r}} \simeq 1$$

in the integrand and obtain

$$S(q) = 1 + \rho\int d^3 \, r[g(r) - 1] \; . \tag{6}$$

The compressibility theorem of statistical mechanics then yields[2,19]

$$S(q) = \rho \; k_B T \; \kappa_T \; , \tag{7}$$

which recovers Einstein's result.

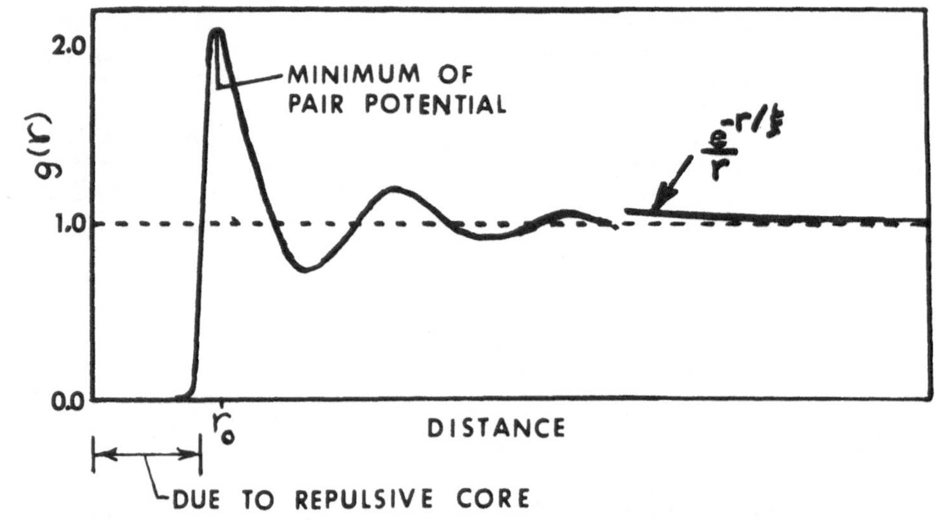

Figure 3. The pair distribution function for a fluid. The long-range tail which develops near the critical point is shown on the right.

Near the critical point the correlations become very long-ranged and the above result for S(q) is modified due to the long tail of g(r). In 1914 Ornstein and Zernike considered the effect of correlations, which were neglected in the Einstein theory. Through a plausible argument which is yet to be established on firm theoretical grounds, they obtained a form for the asymptotic behavior of the net correlation function $G(r) \equiv g(r) - 1$,

$$G_{OZ}(r) = \frac{1}{4\pi\rho R^2} \frac{\exp[-r/\xi]}{r}, \quad r \to \infty \tag{8}$$

where the long-range correlation length ξ is defined by this equation, and the direct correlation range R is assumed to be a parameter which is constant or at most slowly varying in the vicinity of the critical point. [A typical value for the coefficient in G is $(4\pi\rho R^2)^{-1} = 0.5$ Å.]

Substituting G_{OZ} into S(q), we obtain

$$S(q) \simeq \frac{1}{R^2} \frac{\xi^2}{1+q^2\xi^2}, \tag{9}$$

which away from the critical point ($q\xi \ll 1$) must yield the Einstein result; hence we have

$$S(q) = \frac{\rho\, k_B T\, \kappa_T}{1 + q^2\xi^2} \tag{10}$$

and

$$\xi^2 = \rho\, k_B T\, R^2\, \kappa_T \quad . \tag{11}$$

For a normal fluid the hole in $[g(r) - 1]$ due to the repulsive core (see Fig. 3) dominates the integral for $S(q)$ and in fact for this case the integral is negative. Near the critical point $g(r)$ for small r is no different from a normal fluid, but the long-range tail dominates the integral for $S(q)$. Note now the mathematical meaning of the droplet picture used earlier to describe critical behavior: the probability of having a particle at distance r from another particle is very slightly greater than it would be in absence of correlations. For example, for a fluid at $\varepsilon = 10^{-5}$, we have $\xi \simeq 1000\text{Å}$; hence at a distance $r = 1000\text{Å}$, $g(1000\text{Å}) = 1.00016$, while in the absence of correlations, $g = 1$. Nevertheless, it is this cooperative behavior over macroscopic distances which is responsible for the singularities in the integrals for the dynamic as well as static properties of systems near the critical point.

The correlation length is a parameter which enters all of the theoretical expressions which have been developed to describe the dynamical behavior of fluids near the critical point. Therefore, accurate values for the correlation length are crucial if the theoretical predictions for the light scattering spectrum are to be tested near the critical point. Accurate measurements are quite difficult, however, because $q\xi \ll 1$ for the values of q accessible in light scattering experiments, except for measurements very near T_c. The most accurate measurements of ξ have been performed in an experiment by Lunacek and Cannell.[20] They used a null technique to measure the ratio of the forward to backscattering intensities [see Eq. (10)] and they were able to determine ξ for ε as large as 3×10^{-2} where $(q\xi)^2 \simeq 0.002$ for backscattering with a laser wavelength $\lambda_0 = 6328\text{Å}$.

3. LIGHT SCATTERING SPECTRUM

The dynamics of the density and concentration fluctuations of a fluid near the critical point can be deduced from measurements of the light scattering spectrum. If a fluid is sufficiently far from the critical point for the inequality $q\xi \ll 1$ to apply, then the dynamics of the long wavelength fluctuations are governed by the classical equations of hydrodynamics, which were discussed by Lallemand.[2] The hydrodynamic expressions for the principal components in the spectrum of a one-component fluid, the Rayleigh line (the unshifted component which arises from the diffusive decay of the density fluctuations) and the Brillouin doublet (which arises from scattering from propagating sound waves) are given in Table II, which includes a comparison of the spectrum of a fluid (xenon) near the critical point with that of a fluid (ethyl ether[21]) far from the critical point. The qualitative behavior of the spectrum near the critical point is as follows:

(1) The scattering intensity diverges strongly, as was discussed in the previous section (critical opalescence).

(2) Most of the increase in the total scattering intensity is due to the increased intensity of the Rayleigh line; for a normal fluid the Rayleigh and Brillouin intensities are comparable, but near the critical point the intensity of the Rayleigh component, which is proportional to $c_p - c_v$, is orders of magnitude larger than the intensity of the Brillouin components, which is proportional to c_v.

(3) The Rayleigh line is extremely narrow near the critical point because of the strong divergence in c_p, which behaves in the same way as κ_T.

(4) The Brillouin frequency shift is proportional to the sound velocity, which is given by

$$v = (c_p/\rho\, c_v\, \kappa_T)^{\frac{1}{2}} \sim c_v^{-\frac{1}{2}} \sim \varepsilon^{\alpha}; \qquad (12)$$

hence the Brillouin shift should go to zero as the critical point is approached.

(5) The Brillouin linewidth, proportional to the sound attenuation, increases rapidly as the critical point is approached due to the strong divergence in the bulk viscosity ζ.

Table II. - The light scattering spectrum of a one-component fluid in the hydrodynamic region. The q-dependent quantities are calculated for $\lambda_o = 4880\text{Å}$ and $\Theta = 90°$.

Quantity	Formula from hydrodynamics	Normal fluid ethyl ether (Ref. 21) (1 atm.; 20°C)	Fluid near the critical point Xe ($\rho_c, \Delta T = 1°C$)
Total cross section (per unit volume)	$\frac{1}{V}\left(\frac{d\sigma}{d\Omega}\right)_{TOT} = \frac{\pi^2}{\lambda_o^4}\left(\rho\frac{\partial\varepsilon}{\partial\rho}\right)_T^2 k_B T \kappa_T$	1.05×10^{-5} cm^{-1}	1.0×10^{-2} cm^{-1}
Landau-Placzek ratio	$\frac{I_R}{2I_B} = \frac{c_p}{c_v} - 1$	0.43	140
Brillouin cross section	$\frac{1}{V}\left(\frac{d\sigma}{d\Omega}\right)_B = \frac{c_v}{2c_p}\left[\frac{1}{V}\left(\frac{d\sigma}{d\Omega}\right)_{TOT}\right]$	1.0×10^{-5} cm^{-1}	3.6×10^{-5} cm^{-1}
Rayleigh linewidth[a]	$\frac{\Gamma_R}{2\pi} = \left(\frac{1}{2\pi}\right)\frac{\lambda}{\rho c_p} q^2$	8.2 MHz	0.040 MHz
Brillouin shift	$\frac{\omega_B}{2\pi} = \frac{vq}{2\pi}$	3000 MHz	300 MHz
Brillouin linewidth[a]	$\frac{\Gamma_B}{2\pi} = \frac{q^2}{2\pi}\left[\frac{\frac{4}{3}\eta_s+\zeta}{\rho} + \frac{\lambda}{\rho c_v} - \frac{\lambda}{\rho c_p}\right]$	64 MHz	21 MHz

[a]Here ρ is the mass density, while in earlier sections ρ was the number density.

The spectrum for a binary mixture contains in addition to the Brillouin components a complicated central component due to the coupled density - concentration fluctuations (see Mountain and Deutch[22] and Fleury and Boon[23]). Near the critical point, however, the concentration fluctuations are dominant and the central component is essentially a single Lorentzian of width $\Gamma_R = Dq^2$, where D is the binary diffusion coefficient, which approaches zero as the critical point is approached.

Near the critical point the laws of hydrodynamics must be supplanted by a more general dynamical theory. The transition from hydrodynamic to nonhydrodynamic behavior, where $q\xi \gtrsim 1$, occurs at $\varepsilon \sim 10^{-4}$ for light scattering experiments at a 90° scattering angle. On the other hand, as we shall see, the region of strong dispersion in the sound velocity and rapidly increasing sound attenuation occurs much further from the critical point, at $\varepsilon \sim 10^{-2}$.

The dynamical behavior of fluids near the critical point has been examined theoretically by Kawasaki and Ferrell and others, and in the following sections we will compare the predictions of these theories with the results obtained in light scattering experiments.

4. THE RAYLEIGH LINEWIDTH

The width of the central component of either a one-component fluid or a binary mixture is essentially equal to the decay rate of the order parameter fluctuations, which is given (for scattering vector $\vec{q}$) by

$$\Gamma = (L/X)q^2 ,$$

where L is an Onsager kinetic coefficient (the thermal conductivity for a one-component fluid and the concentration conductivity for a binary mixture) and X is a generalized susceptibility. In 1954 Van Hove[24] pointed out that since X diverges strongly as the critical point is approached while L was presumed to be constant near the critical point, the decay rate Γ should therefore go to zero as the critical point is approached. This explained the "critical slowing down" that had been frequently reported by experimentalists: as the critical point is approached, systems require increasingly longer times to reach equilibrium.

4.1 Historical Background

The first measurements of the Rayleigh linewidth in a fluid near the critical point were reported in 1965 by Alpert, <u>et al.</u>,[25] who measured the linewidth for the binary mixture aniline-cyclohexane at the critical concentration. Alpert and coworkers used a heterodyne spectrometer that was developed by Cummins, Knable, and Yeh;[26] in this system the scattered light from the sample was mixed with laser light that had been shifted 30 MHz by a Bragg tank modulator. The measured linewidths were quite narrow, of the order of 100 Hz. Shortly thereafter Ford and Benedek developed the "self-beat" spectrometer, in which the scattered light was mixed with itself in the photodetector, and they investigated the Rayleigh linewidth for sulfur hexafluoride near the critical point.[27] In these and numerous subsequent experiments the linewidth was observed to approach zero as the critical point was approached, as expected from the Van Hove theory. Since it was known that many static properties of systems in the critical region exhibit simple power law behavior it was natural to fit the experimental results for the linewidth to the simple exponential law

$$\Gamma/q^2 = D = D_o \, \varepsilon^{\gamma-\psi} , \qquad (13)$$

where γ and ψ are, respectively, the exponents which characterize the divergences in X and L.

The results from one of the early experiments is shown in Fig. 4.[28] At the time this experiment was performed (1968), it was thought that the thermal conductivity was well-behaved or at most weakly divergent at the critical point. Hence $\psi \simeq 0$, while $\gamma \simeq 1.2$; therefore, $\gamma-\psi \simeq 1.2$. However, the exponent $\gamma-\psi$, the slope of the curve in Fig. 4, is 0.73. The conclusion was that there must be a strong divergence in the thermal conductivity.

Figure 4 includes thermodynamic data which were obtained by Sengers and Michels at the van der Waals laboratory, as well as the linewidth data.[28] Thermal conductivity measurements by conventional thermodynamic techniques are extremely difficult near the critical point, because the probability of convection diverges strongly as the critical point is approached; nevertheless, the results for the thermal diffusivity in the critical region obtained by linewidth measurements and by conventional thermodynamic techniques are in excellent agreement.

The linewidths measured in binary mixtures were found to be described in the hydrodynamic region by the simple exponential law with $\phi \equiv \gamma-\psi \simeq 0.63$, but the exponents that were obtained for the simple fluids CO_2 and xenon were somewhat higher, $\phi \simeq 0.74$, and for SF_6 the exponent was markedly different, $\phi \simeq 1.26$, contrary to the expected "universality" in critical behavior.

The larger exponents observed for the pure fluids was a matter of serious concern because the concept of universality was well established from numerous measurements of the static properties of many systems near the critical point. It was suggested that perhaps the measurements were affected by impurities in the samples, but extensive systematic studies by Bak and Goldburg,[29] (on phenol-water with hypophosphorous acid as an impurity) and by Bak, Goldburg, and Pusey[30] (on bromobenzene-water with acetone as an impurity) showed that even for fairly high impurity concentrations the critical behavior is unchanged except for a change in T_c.

In 1970 Sengers suggested that the apparently higher exponents observed for the pure fluids could be explained by taking into consideration the nondivergent background contribution to the thermal conductivity.[31] Because of the large contribution of nonanomalous background terms, the behavior of systems in the temperature region readily accessible to experiment may be very different from the true asymptotic behavior which is presumably describable by the simple exponential laws. Thus the nonsingular background contributions must first be subtracted if the data are to be analyzed over extended temperature ranges.

Generalizing the hydrodynamic result $\Gamma = (\lambda/\rho c_p)q^2$ to include a q dependence in λ and c_p, and separating λ and c_p into background and critical parts, we have

$$\Gamma = [\lambda^b/\rho(c_p)_q]q^2 + \Gamma^c\ [(c_p)_q^c/(c_p)_q]\ , \tag{14}$$

where

$$\Gamma^c \equiv [\lambda_q^c/\rho(c_p)_q^c]q^2$$

is the critical part of the linewidth. If the Ornstein-Zernike form is assumed for the q dependence of $(c_p)_q^c$ and $(c_p)_q$, then Eq. (14) becomes:

$$\Gamma = (\lambda^b/\rho c_p)q^2\ (1+q^2\xi^2) + \Gamma^c\ [(c_p)^c/c_p]\ , \tag{15}$$

where the absence of a subscript q on c_p indicates the $q = 0$ (thermodynamic) quantity.

The partition of a transport coefficient into background and critical parts is clearly a crucial part of the data analysis in any experimental investigation of the dynamics of a system near the critical point. In such experiments, which include, for example, measurements of the spin diffusion rate in magnets and the sound velocity and attenuation in fluids as well as measurements of the viscosity, conductivity and diffusivity, meaningful comparison between theory and experiment can be made only if a systematic procedure can be developed for estimating the bare transport or Onsager kinetic coefficient. For the thermal conductivity and shear viscosity of a pure fluid Sengers and Keyes have developed a method for estimating λ^b and η_s^b using data obtained far from the

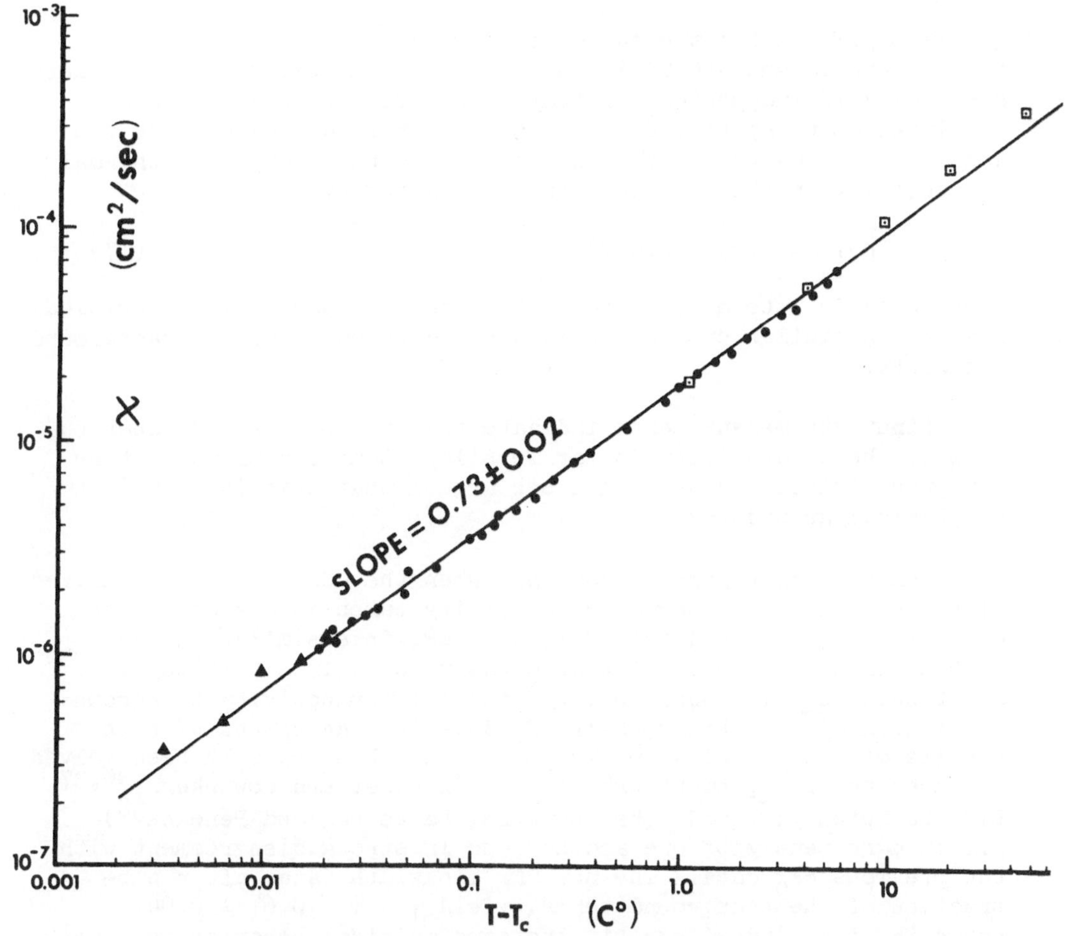

Figure 4. The thermal diffusivity of CO_2 along the critical isochore: ●,Swinney and Cummins;▲,Siegel and Wilcox (linewidth measurements);⊡,thermodynamic data (Sengers). (Figure from Ref. 28.)

critical point.[32] The procedure is based on the empirical result, frequently used in the engineering literature, that the "excess" thermal conductivity,

$$\tilde{\lambda}(\rho) = \lambda(\rho,T) - \lambda(0,T) , \tag{16}$$

[where $\lambda(\rho,T)$ is the thermal conductivity at a density ρ and temperature T, and $\lambda(0,T)$ is the thermal conductivity in the dilute gas limit] is independent of temperature for temperatures and densities up to approximately twice the critical temperature and density. The Sengers-Keyes ansatz is that the background thermal conductivity in the critical region is given by

$$\lambda^b\,(\rho,T) = \tilde{\lambda}(\rho) + \lambda(0,T) , \tag{17}$$

where $\tilde{\lambda}(\rho)$ is determined using data obtained away from the critical region. A similar expression is assumed to hold for the background viscosity.

Linewidth measurements indicate that the background contribution to the conductivity is far less important for mixtures than for pure fluids, and de Gennes has argued that this is plausible on physical grounds.[33]

Sengers and Keyes[32] found that when the CO_2 data were analyzed with the background thermal conductivity taken into account, the exponent ϕ was reduced from 0.73 to 0.62; in a similar analysis of the xenon data we found that ϕ was reduced from 0.75 to 0.64.[34] But Benedek _et al_. found that the thermal conductivity background correction did not bring their SF_6 data into agreement with the results of other fluids.[35] The SF_6 puzzle has recently been solved by three new independent experiments (Langley and coworkers;[36,37] Lim and Swinney;[37] and Feke, Hawkins, Lastovka, and Benedek[38]) all in agreement with one another and in strong disagreement with the previous SF_6 data. The new SF_6 linewidth data, after subtraction of the background terms, yield $\gamma - \psi = 0.61 \pm 0.04$; hence SF_6 does indeed exhibit the same critical behavior as other fluids. In 1968 Kadanoff and Swift[39] extended the mode-mode coupling approach of Kawasaki[40] and predicted that D should exhibit the same critical behavior as the inverse correlation length: $D \sim \xi^{-1}$. Hence the prediction was that the exponent $\gamma - \psi$ which describes the critical behavior of the linewidth should be equal to the exponent ν. The value of $\gamma - \psi$ determined in linewidth measurements agrees well with the value 0.63 obtained from measurements of the angular dependence of the intensity of the scattered light. Thus both theory and experiment indicate that $\gamma - \psi \simeq 0.63$, which together with the accepted value of $\gamma(\gamma \simeq 1.23)$, yields the exponent characterizing the divergence in the thermal conductivity of concentration conductivity, $\gamma \simeq 0.60$. This strong divergence in the conductivity was the first important new result obtained from linewidth measurements near the critical point.

There is a simple plausibility argument suggested by Arcovito et al.[41] for the expected behavior of the diffusivity of a fluid in the critical region. The density is correlated over a range ξ, so we can picture droplets of radius of the order of ξ diffusing in the fluid. Of course, the droplets are fuzzy and atoms are continuously "evaporating" and "condensing" on the droplet surface. Now by Stokes' law a sphere of radius ξ has a mobility given by $(6\pi\eta_s\xi)^{-1}$; therefore, the Einstein relation between the diffusivity and the mobility, $D = (k_BT) \times (\text{mobility})$, yields

$$D = \frac{k_BT}{6\pi\eta_s\xi} \quad .$$

As we shall see (Section 4.4), the latest theoretical mode-mode coupling and decoupled-mode predictions for the critical part of the diffusivity in the hydrodynamic region differ from the above equation only by a constant multiplicative factor which is 1.05.

So far we have considered only the hydrodynamic region where $q\xi \ll 1$. The first experiment in which linewidth measurements were extended from the hydrodynamic region into the extreme nonhydrodynamic region ($q\xi \gg 1$) was reported in 1969 by Berge, Calmettes, Laj, and Volochine, who found that for the mixture aniline-cyclohexane near T_c there was a transition in the behavior of Γ/q^2 from a q-independent, temperature-dependent behavior to a q-dependent, temperature-independent behavior.[42] Similar results have subsequently been obtained for many other fluids, as shown for example for xenon[43-45] in Fig. 5.

4.2 Experimental Considerations

Before discussing recent experimental and theoretical developments, let us consider briefly some of the experimental problems in light scattering measurements near the critical point.

The choice of a particular fluid system is not important because presumably there is a universality in behavior. Most systems that have been studied were chosen because they have a convenient T_c, near room temperature, and because thermodynamic data helpful in the interpretation of the results are available. Inert gases are the best choice for pure fluids because the results are not affected by the internal molecular degrees of freedom. The experiments of Bak, Goldburg, and Pusey[29,30] show that impurities change T_c, but unless the impurity concentration is extremely high, the behavior with respect to the measured T_c is unchanged.

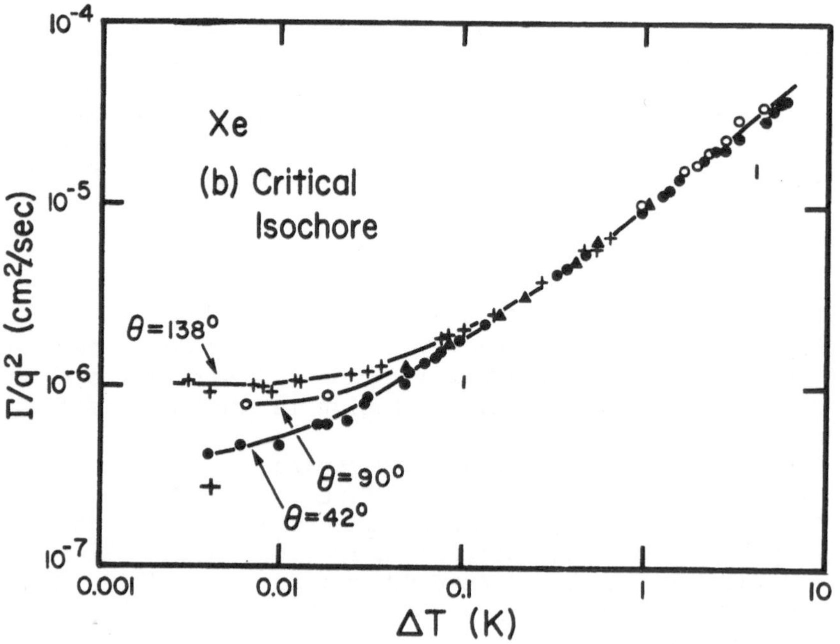

Figure 5. Γ/q^2 as a function of ΔT for xenon on the critical isochore. The data at $\Theta=42°$ (closed circles), 90° (closed triangles), and 138° (crosses) were obtained by Henry using an analog spectrum analyzer.[43,44] The open circles are data for $\Theta = 90°$ obtained by Lim with a pulse correlator.[44,45] The error bars represent the uncertainty in Γ/q^2 and in ΔT at extreme and intermediate values of ΔT.

Long-term temperature regulation to better than 1 mK is necessary if the nonhydrodynamic region within a few millidegrees of $T_c(\varepsilon \sim 10^{-5})$ is to be explored. This level of temperature control can be achieved with a double thermostat system with 0.01 K control in the outer thermostat, and temperature gradients in the inner bath can be eliminated by vigorous stirring (with a synchronous motor stirrer).

Relative temperatures can be straightforwardly measured with 0.2 mK accuracy with a thermistor and a Wheatstone bridge. The critical temperature can be determined by slowly (over periods of days) lowering the temperature and observing the temperature at which the meniscus is formed.

Equilibration times are quite long near T_c as a consequence of the "critical slowing down"; hence accurate determination of T_c, which is of course crucial for these experiments, requires patience and good long-term temperature control.

In intensity fluctuation spectroscopy it is important to avoic parasitic scattering ("flare") from the walls of the sample cell and from dust, as Oliver[46] has discussed. In our experiments we have often found that a significant amount of cell wall scattering is present. As the level of the coherent local oscillator signal from elastic scattering increases from zero to a level much greater than the scattering from the sample, the photocurrent correlation function decay rate decreases from 2Γ to Γ (see Fig. 20 of Oliver[46]). This effect is illustrated in Fig. 6 for scattering from a sample of xenon. In principle the presence of flare could be determined by testing the goodness of fit of the correlation function to a single exponential; however, this is a rather insensitive test, as Oliver has discussed. We have found that a better test in our experiments is to examine the correlation function for sample volumes near and away from the cell walls, as in Fig. 6, and then discard those data for which there is no plateau (indicated by the dashed line).

A problem peculiar to the study of fluids near the critical point is the presence of a large, gravitationally-produced density gradient:

$$(\partial\rho/\partial z) = g\rho(\partial\rho/\partial P)_T \; .$$

The derivative $(\partial\rho/\partial P)_T$ diverges as $\varepsilon^{-\gamma}$ in both pure fluids and binary mixtures.[47] The linewidth minimum as a function of the height of the beam in the sample cell occurs very near to the height corresponding to the critical density (or concentration) but the actual location of the height corresponding to the critical density (or concentration) must be determined from measurements of the intensity, which varies symmetrically as a function of height about the critical isochore (see Swinney and Henry[44] and Lim[45]).

Multiple scattering is always a concern in light scattering experiments near the critical point. Some of the binary mixtures which have been studied by light scattering have been chosen because the refractive indices of the two components are very nearly equal, so the scattering is weak and multiple scattering is not so serious. Lim has looked for the effect of multiple scattering

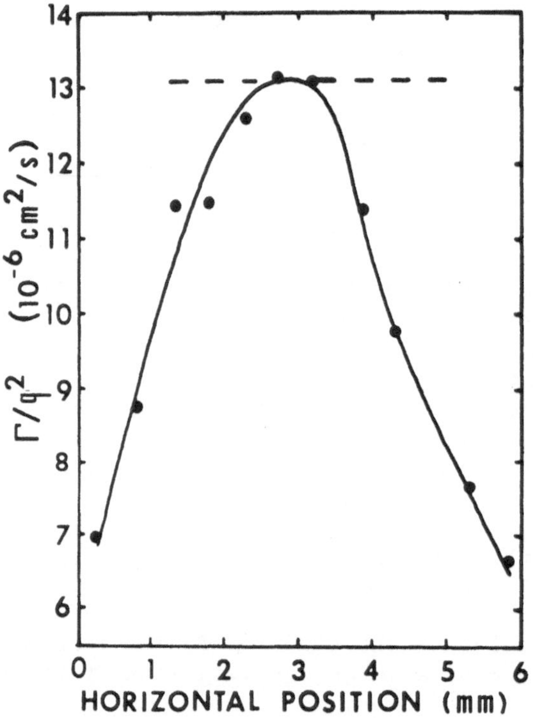

Figure 6. Γ/q^2 as a function of the horizontal position of the xenon sample cell for Θ=30.4° and ΔT=1.5K. The linewidth decreases sharply for scattering volumes near the cell walls (at the horizontal positions 0 and 6 mm) as a consequence of the intense scattering from the walls. The horizontal dashed line shows the value of Γ/q^2 measured at the same temperature for Θ=90°. (From Ref. 44.)

on the Rayleigh linewidth of fluids near the critical point by examining the linewidth for effective sample volumes of different sizes.[45] Apertures were used to collect light from volumes ranging from the primary scattering volume (the laser beam) to sample volumes which were ten times larger, and no variation of the linewidth with sample volume was detected even for $\varepsilon \simeq 10^{-5}$. Volochine and Berge[48] have argued that the effect of multiple-scattering on light beating measurements of the linewidth can be unimportant even when the effect on the scattering intensity is large because the coherence area of the multiply-scattered light can (depending on choice of apertures) be small.

4.3 Mode-Mode Coupling Theory

In the mode-mode coupling theory developed by Kawasaki[40,49] and Kadanoff and Swift[39] the hydrodynamic equations are first solved to obtain the macroscopic normal modes. These are the form

$$A_q^j(t) = A_q^j(0) \exp(-S_j t),$$

where the different j-modes are:

(1) Heat flow $\quad S_T = (\lambda/\rho c_p)q^2$

(2) Viscous flow $\quad S_\eta = (\eta_s/\rho)q^2$

(3) Sound $\quad S_s = \pm ivq + \Gamma_B.$

These modes serve as the phenomenological "bare propagators" of the theory, and the form of the coupling constants is found by a projection operator technique.

Kadanoff and Swift found that the dominant contribution to the anomaly in λ comes from intermediate states involving one heat mode and one viscous flow mode,[39]

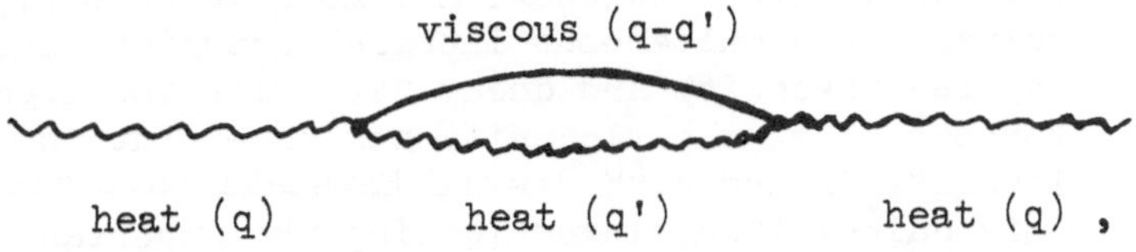

and they found that $\lambda/\rho c_p \sim \xi^{-1}$. Subsequently, Kawasaki extended this calculation and obtained an explicit integral expression for the decay rate near the critical point:[49]

$$\Gamma^c = \frac{\lambda(q)q^2}{\rho c_p(q)} = \frac{k_B T}{(2\pi)^3 \eta_s^*} \int d\vec{k} \left[\left(\frac{q}{k}\right)^2 - \left(\frac{\vec{q}\cdot\vec{k}}{k^2}\right)^2 \right] \frac{\hat{G}(\vec{q}-\vec{k})}{\hat{G}(\vec{q})} , \qquad (18)$$

where $\hat{G}(\vec{q}) = \int d\vec{k} G(\vec{r}) e^{i\vec{q}\cdot\vec{r}}$, and $G(\vec{r})$ is the density-density (or concentration-concentration) correlation function, η_s^* is the "high frequency" shear viscosity and k_B is Boltzmann's constant. Pike[1] (in this volume) has derived Eq. (18) from a generalized Green's function approach [see Eq. (126) of Pike]. Kawasaki evaluated the integral in Eq. (18) using the Ornstein-Zernike form for the correlation function,

$$\hat{G}_{OZ}(\vec{q}) \propto [\xi^{-2} + q^2]^{-1} , \qquad (19)$$

obtaining[49]

$$\Gamma^c = [k_B T / 6\pi\eta_s^* \xi^3] \; K_o(q\xi), \qquad (20a)$$

where

$$K_o(x) = \tfrac{3}{4} [1 + x^2 + (x^3 - x^{-1}) \arctan x]. \qquad (20b)$$

[In (19) and in subsequent expressions for the correlation function we omit proportionality factors independent of $\vec{q}$ or $\vec{r}$, since they cancel in (18).]

Since 1970 there have been three significant modifications to the original Kawasaki result [Eqs. (18) and (20)]:

(1) The viscosity in Eq. (18) is ambiguous since it represents an effective weighted average over all viscous modes appearing in intermediate states. The correct interpretation of η_s^* requires a self-consistent evaluation of both the viscosity and the decay rate. Kawasaki and Lo have recently solved the simultaneous integral equations involving the viscosity and decay rate with the frequency dependence of the viscosity neglected but the non-locality included.[50] Lo and Kawasaki have also extended this calculation, investigating the importance of the frequency dependence or memory effects, and they deduced an expression which relates η_s^* to the macroscopic shear viscosity η_s, thus removing the ambiguity in the shear viscosity.[51] They found

$$\eta^{*-1} = R(q\xi)\eta_s^{-1} , \tag{21a}$$

where

$$R(x) = [K(x)+\Delta K(x)]/K_o(x). \tag{21b}$$

The term $K(q\xi)/K_o(q\xi)$, which describes the effect of nonlocality on the viscosity ($K/K_o = 1$ if nonlocal effects are neglected), is given numerically in Fig. 3 of Kawasaki and Lo,[50] and $\Delta K(q\xi)/K_o(q\xi)$, which describes the effect of the frequency dependence, is given in Table I of Lo and Kawasaki.[51] The viscosity correction factor $R(q\xi)$ is shown in Fig. 7, curve (a). Note that η_s^* differs from η_s even far from T_c; in that region $\eta_s^* = \eta_s/1.063$.

(2) In Kawasaki's analysis of the order parameter fluctuations in a fluid, Dyson-type self consistent equations for the time correlations of the critical fluctuations were derived and Eq. (18) was then obtained by evaluating the contributions of the two lowest order terms to the decay rate. Recently Lo and Kawasaki[52] have investigated the contributions of the four next higher order terms and have found that the inclusion of these "vertex correction" terms reduced Eq. (20) by 2.44% for $q\xi \ll 1$ and increases (20) by 0.40% for $q\xi \gg 1$. The vertex correction $V(q\xi)$, the ratio of the corrected to the uncorrected decay rate, is shown by curve (b) in Fig. 7, which was obtained by connecting the limiting values of $V(q\xi)$, $V(\infty)$, and $V(0)$ by a smooth curve. [A calculation of this small modification to the theory for intermediate values of $q\xi$ would require the evaluation of a complicated integral expression - Eq. (2.10) in Ref. 52]. The vertex correction to the decay rate is frequency dependent, with the above values for $q\xi \ll 1$ and $q\xi \gg 1$ applying only in the zero frequency limit. Because of the frequency dependence of the vertex correction, the observed spectral line will in principle deviate from the Lorentzian lineshape, but the correction is so small that the predicted departures from the Lorentzian shape would be very difficult to observe.

(3) The integral expression [Eq. (18)] for the decay rate was evaluated by Kawasaki using the Ornstein-Zernike form for the correlation function, leading to the result, Eq. (20). But scattering experiments and the theoretical investigations of the Ising model by Fisher and

Burford[53] have shown that there are small departures from Ornstein-Zernike behavior near the critical point. The correct asymptotic form for the correlation function at the critical point is expected to be $r^{-(1+\eta)}$ with $\eta \simeq 0.05$ to 0.1, while for the Ornstein-Zernike theory $\eta = 0$.

Fisher and Burford found that correlations in the Ising model are accurately described by:[53]

$$\hat{G}_{FB} \propto (\xi^{-2} + \phi^2 q^2)^{\eta/2}/[\xi^{-2} + (1+\phi^2\eta/2)\, q^2], \qquad (22)$$

where $\phi = 0.15 \pm 0.01$, independent of the type of lattice. Swinney and Saleh[54] have evaluated the decay rate integral using $\hat{G}_{FB}$ and the result for the decay rate ratio,

$$C(q\xi) = \Gamma^c\,(\hat{G}_{FB}, q\xi)/\Gamma^c\,(\hat{G}_{OZ}, q\xi)\ , \qquad (23)$$

is given by curve (c) in Fig. 7 for $\eta = 0.1$. (The decay rate integral was also evaluated by Swinney and Saleh[54] and by Chang et al.[55] for other forms of the correlation function which have been used in the analysis of data from scattering experiments; however, $\hat{G}_{FB}$ is more satisfactory theoretically since, as explained in Ref. 53, it leads to the correct asymptotic behavior at large r both at the critical point and away from the critical point.)

With the viscosity, vertex and correlation function modifications included, the linewidth expression, Eq. (20a), in the mode-mode coupling theory becomes

$$\Gamma^c = (k_BT/6\pi\eta_s\xi^3)\ K_o\,(q\xi)\ H(q\xi)\ , \qquad (24)$$

where η_s is the macroscopic shear viscosity, and the correction factor $H(q\xi)$, which most analyses of linewidth data have heretofore assumed to be unity, is given by

$$H(q\xi) \equiv R(q\xi)\ V(q\xi)\ C(q\xi) \qquad (25)$$

and is plotted as curve (d) in Fig. 7.

The final result of the mode-mode coupling theory, Eq. (24), can be rewritten as

$$\Gamma^* = \left(\frac{1}{q\xi}\right)^3 K_o\,(q\xi)\ H\,(q\xi)\ , \qquad (26a)$$

where the "scaled" linewidth Γ^* is defined as

$$\Gamma^* = \left(\frac{6\pi\eta_s}{k_B T}\right)\left(\frac{\Gamma^c}{q^3}\right) \quad . \tag{26b}$$

Thus the theory predicts that the experimental data for Γ^* [Eq. (26b)] for different temperatures and scattering angles, obtained for various simple fluids and binary mixtures, should all fall on a single universal curve [Eq. (26a)] when the dimensionless quantity Γ^* is plotted as a function of $q\xi$. This single curve is predicted to describe the critical behavior not only along the critical isochore and the coexistence curve, but also along any other thermodynamic path in the critical region. In Section 4.5 we test the mode-mode coupling prediction for all fluids for which ξ and η_s have been independently determined.

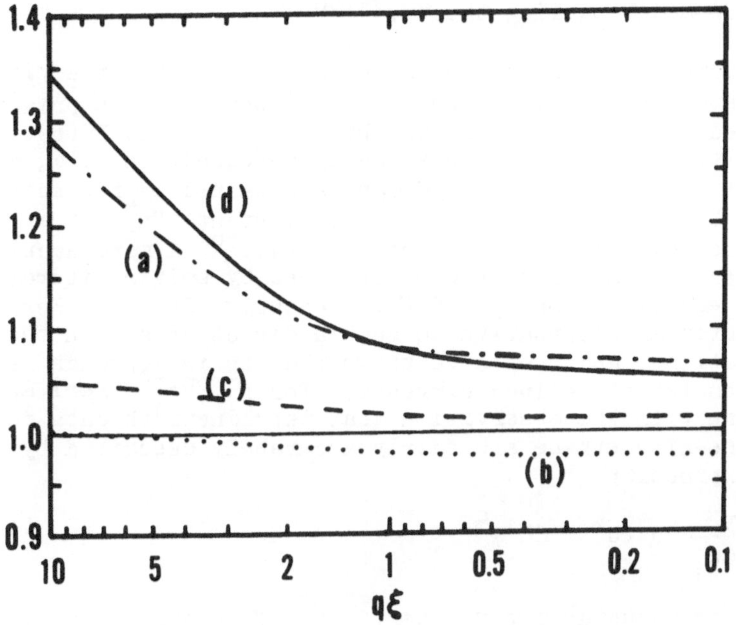

Figure 7. Modifications to the Kawasaki theory which are discussed in the text. (a) - self consistent solution for the viscosity ratio η_s/η_s^*;[50,51] (b) - lowest-order vertex corrections;[52] (c) - correlation function correction using the Fisher-Burford form for the correlation function.[54] Curve (d) includes all three corrections and gives the ratio of the modified decay rate to that of Eq. (20). (From Ref. 44.)

4.4 Decoupled-mode Theory

Ferrell has calculated the critical behavior of transport properties by factoring the currents J(t) in the current correlation functions in the Kubo formulas for the transport coefficients,[56] and then the Kubo formulas were evaluated directly, a procedure which, as Ferrell has pointed out, is equivalent to the mode-mode coupling theory without vertex corrections because the absence of internal lines (the vertex corrections) between two intermediate state propagators allows them to be factored within the Kubo integral.[57] Ferrell obtained for the critical part of the decay rate[56]

$$\Gamma^c = \frac{k_B T\, q^2}{8\pi\eta_s^* \hat{G}(\vec{q})} \int d\vec{r} \left[\frac{1}{r} + \frac{(\vec{q}\cdot\vec{r})^2}{q^2 r^3}\right] G(\vec{r}) e^{i\vec{q}\cdot\vec{r}} \quad , \tag{27}$$

where η_s^* is a constant, wavenumber-independent viscosity. If the Ornstein-Zernike form is used for the correlation function, then Eq. (27) yields the same result [Eq. (20)] that Kawasaki obtained with the mode-mode coupling theory.[54,56]

The fluctuation-dissipation or Kubo formulas for the viscosity and decay rate are a pair of coupled equations which in principle can be solved self-consistently to obtain $\eta_s(q,\omega)$ and $\Gamma^c(q)$. However, since the viscosity is only weakly dependent on q, ω, and ε, a good first approximation for Γ^c can be obtained by replacing $\eta_s(q,\omega)$ in the decay rate integral by a constant, "η_s^*"; this was the procedure followed in obtaining Eq. (24).[56] A more accurate expression for Γ^c can of course be obtained by solving iteratively the coupled equations for η_s and Γ^c. Recently, Perl and Ferrell have considered an alternative to such a direct attack on the coupled equations, and they have shown that their approach leads to a self-consistent refined expression for Γ^c.[58,59] Perl and Ferrell began with the observation that the linewidth data for 3 methylpentane-nitroethane are fairly accurately described by the empirical expression

$$\Gamma^c = \frac{k_B T}{16\bar{\eta}_s} q^2(q^2 + \xi^{-2})^{\frac{1}{2}} \quad , \tag{28}$$

where $\bar{\eta}_s$ is an adjustable parameter.[58,59] This expression for Γ^c was then used in evaluating the Kubo integral for $\eta_s(q,\omega)$, which in the hydrodynamic limit was found to have a critical part given by

$$\eta_s^c \equiv \eta_s^c(q=0,\ \omega=0) = \frac{8\bar{\eta}_s}{15\pi^2} \ln(q_D\xi) \quad , \tag{29}$$

where q_D is a free parameter to be determined by fitting the macroscopic shear viscosity data to Eq. (29). Finally, Perl and

Ferrell used their result for $\eta_s(q,\omega)$ to evaluate the decay rate integral, obtaining the following refined expression for Γ^c:

$$\Gamma^c = \frac{k_B T}{6\pi\eta_s^{eff}\ \xi^3}\ K_o(q\xi)\ , \tag{30a}$$

where

$$\eta_s^{eff} = \eta_s^b \left\{ 1 + \frac{\bar{\eta}_s}{\eta_s^b}\left(\frac{8}{15\pi^2}\right)\left[\ell n\left(\frac{q_D\xi}{(1+q^2\xi^2)^{\frac{1}{2}}}\right) + \gamma(q\xi)\right]\right\}, \tag{30b}$$

and $\gamma(q\xi)$ is a function given numerically.[59] [Some values of the function γ, which increases monotonically with increasing $q\xi$, are $\gamma(0)\simeq\gamma(0.1) = -0.492, \gamma(1) = -0.357, \gamma(2) = 0.189$ and $\gamma(\infty)\simeq\gamma(100) = = 0.090$.]

The comparison between the mode-mode coupling and decoupled-mode theories and between theory and experiment is facilitated by considering again the scaled linewidth Γ^*, Eq. (26b), which in the decoupled-mode theory can be written

$$\Gamma^* = C(q\xi)\left[\frac{K_o(q\xi)}{(q\xi)^3}\right]\left\{1 + \left(\frac{\bar{\eta}_s A}{\eta_s^b}\right)[\tfrac{1}{2}\ell n(1+q^2\xi^2) - \gamma(q\xi)] + \right.$$

$$\left. - \left(\frac{\bar{\eta}_s A}{\eta_s^b}\right)^2 [\ell n(q_D\xi) - \tfrac{1}{2}\ell n(1+q^2\xi^2) + \gamma(q\xi)]\ [\tfrac{1}{2}\ell n(1+q^2\xi^2) - \gamma(q\xi)] + \ldots\right\}, \tag{31}$$

where

$$A = 8/15\pi^2. \tag{32}$$

It is clear that Γ^* is a function only of $q\xi$ if the second and higher order terms in A can be neglected. In the hydrodynamic region the contribution of the higher order terms is less than 0.5% for $\varepsilon<3\times10^{-2}$, while in the nonhydrodynamic region the contribution of the higher order terms is larger but still typically less than 3% for the data which we analyze in this paper. The decoupled-mode expression [Eq. (31)] for Γ^* with the second and higher terms in A neglected is compared with the mode-mode coupling theory [Eq. 26a)] in Fig. 8. The upper and lower curves shown for the decoupled-mode theory correspond, respectively, to the values of $\bar{\eta}_s/\eta_s^b$ obtained for nitroethane-3 methylpentane ($\bar{\eta}_s/\eta_s^b = 1.30$) and 2, 6 lutidine-water ($\bar{\eta}_s/\eta_s^b = 0.99$); for the other fluids considered here the ratio $\bar{\eta}_s/\eta_s^b$ falls between 1.30 and 0.99.

In the hydrodynamic region the mode-mode coupling theory (with vertex corrections included) yields

$$\Gamma^* = 1.053/q\xi\ , \tag{33}$$

while the decoupled-mode theory (including higher order terms in A) yields

$$\Gamma^* = (1.050 \pm 0.003)/q\xi \ , \tag{34}$$

where the ± 0.003 in Eq. (34) is due to the small, $q\xi$-dependent contribution of the higher order terms. Thus the two theories are in excellent agreement in the hydrodynamic region. However, as we have noted, the decoupling approximation is equivalent to the neglect of vertex corrections; if the comparison of the decoupled-mode theory with the mode-mode coupling theory is made with the vertex corrections omitted from the latter theory, then the mode-mode coupling values for Γ^* are 2.7% instead of 0.3% higher than the decoupled-mode values for Γ^* in the hydrodynamic region.

The difference between the mode-mode coupling and decoupled-mode theories is much larger in the extreme nonhydrodynamic region; e.g., at $q\xi = 10$ the difference between the two theories (with the higher order terms in the decoupled-mode theory neglected) is 8.3% for $\bar{\eta}_s/\eta_s^b = 1.30$ and 12.5% for $\bar{\eta}_s/\eta_s^b = 0.99$. In the extreme nonhydrodynamic region the omission of vertex corrections would lower the mode-mode coupling theory values for Γ^* by only 0.4%; thus the vertex corrections are far too small to explain the difference between the two theories in this region.

4.5 Comparison of Theory and Experiment

A detailed comparison of the predictions of the mode-mode coupling and decoupled-mode theories has recently been reported by Swinney and Henry.[44] Here we summarize briefly the results of that comparison.

Recent independent measurements of the shear viscosity and correlation length for several systems now permit a direct comparison between the measured linewidths and the theoretical predictions, using no adjustable parameters. In particular, it is now possible to test the mode-mode coupling prediction that the scaled linewidth, $\Gamma^* \equiv 6\pi\eta_s \Gamma^c/k_BTq^3$, where Γ^c is the critical part of the linewidth, should be described by the <u>same</u> universal function of $q\xi$ for all fluids.

The experimental results for Γ^*, calculated using linewidth, viscosity, and correlation length data obtained in many laboratories over the past few years, are compared with the prediction of the mode-mode coupling theory, Eq. (26a), in Fig. 9. The decoupled-mode theory result for Γ^* differs from the mode-mode coupling theory prediction only in the nonhydrodynamic region; a comparison of the experimental results obtained in that region with the two theories is shown in Fig. 10. Separate curves are necessary for

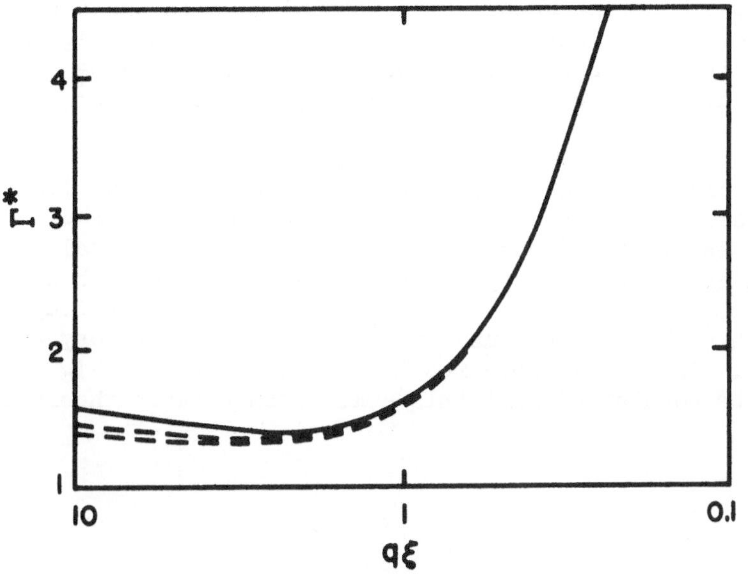

Figure 8. The theoretical predictions for the scaled linewidth $\Gamma^* = 6\pi\eta_s\Gamma^c/k_BTq^3$, in the mode-mode coupling and the decoupled-mode theories. The mode-mode coupling calculation of Lo and Kawasaki yields a single, universal curve for Γ^*, which is shown by the upper curve.[50-52] Perl and Ferrell found that the decoupled-mode approach leads to a result for Γ^* which is slightly different for different fluids, depending on the ratio $\bar{\eta}_s/\eta_s^b$.[58-59] However, for the fluids we consider in this paper the ratio $\bar{\eta}_s/\eta_s^b$ falls within the range between 1.30 and 0.99, which are the values used in drawing the upper and lower dashed curves, respectively. The curves for the decoupled-mode theory represent the theory to lowest order in $A = 8/15\pi^2$; if higher order terms in A were included, the theoretical values for Γ^* would be changed typically by 3% or less, but would depend not only on $q\xi$, but also on q_D [see Eq. (31)]. The curves for both the mode-mode and decoupled-mode theories include the correlation function modification $C(q\xi)$. (From Ref. 44.)

different fluids since the decoupled-mode theory prediction depends on the ratio $\bar{\eta}_s/\eta_s^b$, which is slightly different for different fluids. Although the amount of data in the nonhydrodynamic region is limited, the available data can be seen (Fig. 10) to be somewhat better agreement with the decoupled-mode theory than with the mode-mode coupling theory. Additional data in the nonhydrodynamic region are clearly needed; however, this is a region in which definitive experiments are quite difficult because of the density gradient, concentration gradient, and temperature control problems.

In conclusion, the experimental results for the scaled linewidth for all fluids that have been studied are described within the experimental uncertainty by the same universal function of $q\xi$, independent of the particular thermodynamic path or fluid system, and moreover the data agree with the mode-mode coupling theory within the combined uncertainties (~10%) of theory and experiment. This remarkable result is illustrated in Fig. 9. It should be emphasized that this comparison between theory and

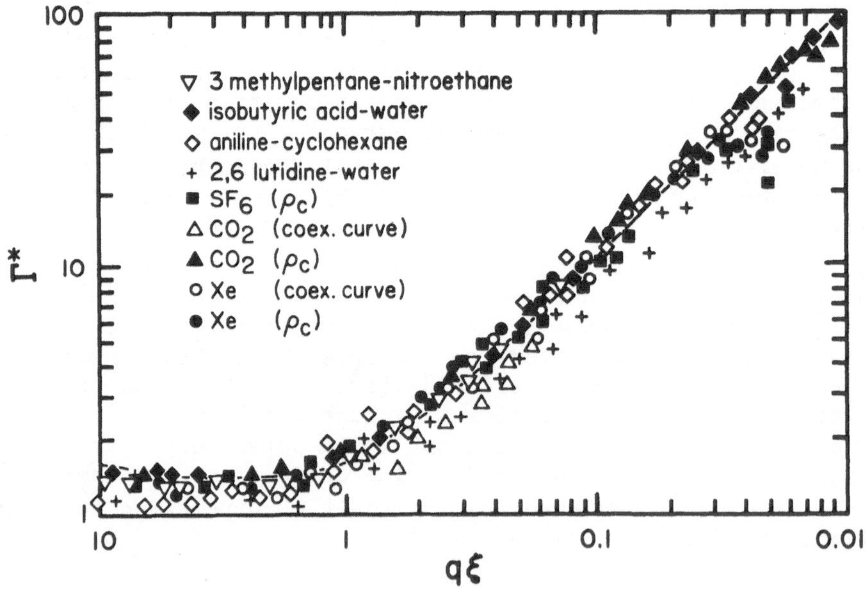

Figure 9. Comparison of experimental data for different systems and thermodynamic paths "scaled" according to Eq. (26b) with the prediction of the modified Kawasaki theory, Eq. (26a). The comparison is shown with no adjustable parameters. (From Ref. 44.)

experiment involves no adjustable parameters. More accurate measurements of the linewidth and the parameters which enter the theories must be performed before it will be possible to distinguish clearly between the different predictions of the mode-mode and decoupled-mode theories in the nonhydrodynamic region and before the importance of various corrections to the theories can be evaluated.

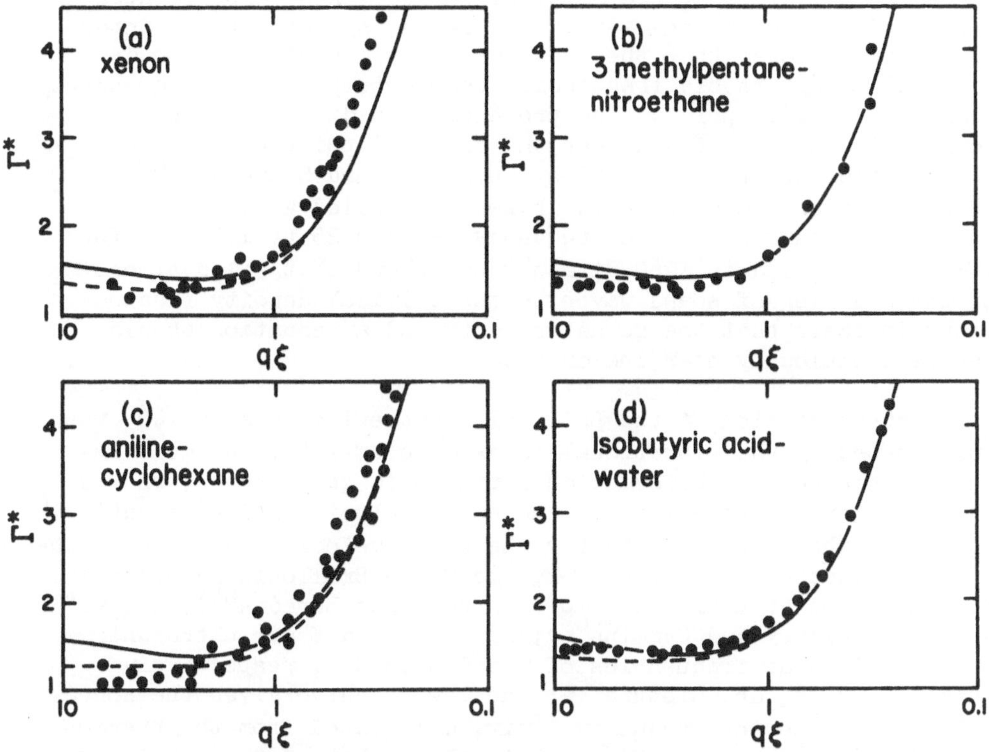

Figure 10. A comparison of the experimental results for Γ^* with the decoupled-mode and mode-mode coupling theories. (a) xenon, (b) 3 methylpentane-nitroethane, (c) aniline-cyclohexane, (d) isobutyric acid-water. For each fluid the solid curve represents the mode-mode coupling theory[49,50] with viscosity, vertex, and correlation function modifications included, and the dashed curve represents the decoupled-mode theory[56-59] with the viscosity and correlation function modifications included. In the decoupled-mode theory there is a small variation in Γ^* at a given value of $q\xi$ due to the higher order terms in Eq. (32); however, this variation is typically less than 2% for the range of q and ξ investigated in the experiments considered here, which is too small to show on the scale of these graphs. (From Ref. 44.)

5. BRILLOUIN COMPONENTS

5.1 Historical Background

As we have seen in Section 3, in the hydrodynamic region the Brillouin frequency shift should decrease and the linewidth should increase rapidly as the critical point is approached. Significant departures from the predicted hydrodynamic behavior were observed in the first Brillouin scattering experiments performed on a fluid near the critical point. These experiments, performed (in 1967[60] and 1968[61]) on CO_2 using Fabry-Perot interferometers, showed that the sound velocity and attenuation determined at hypersonic frequencies become temperature-independent near the critical point. The presence of a dispersion in the sound velocity of a fluid near the critical point had in fact already been observed in 1952 by Chynoweth and Schneider,[62] whose ultrasonic measurements in xenon revealed a small but definite dispersion in the frequency range 0.25 to 1.25 MHz for $(T-T_c) \lesssim 2°C$, and in 1960 Fixman[63] had shown that calculations of the coupling of sound waves to the critical density fluctuations indicate that the sound velocity and attenuation should behave anomalously near the critical point.

Interpretation of the Brillouin scattering data for CO_2 was complicated by the vibrational relaxation effects, but measurements subsequently performed on a monatomic gas, xenon, showed the same general behavior for the sound velocity and attenuation.[64-66] Figure 11 shows the results for the sound velocity in xenon on the critical isochore. Curves A, B, and C are Brillouin scattering data obtained at angles of 170° (Cannell and Benedek[64]), and 90° and 40° (Swinney and Cummins[65]); curves D and E are ultrasonic data obtained at frequencies of t and 0.55 MHz, respectively (Garland, Eden, and Mistura[66]); the lowest curve gives the sound velocity in the low frequency limit, calculated from the thermodynamic data [see Eq. (12)]. It is clear that as the critical temperature is approached there is an increasing difference between the Brillouin scattering velocities and the sound velocity calculated from the thermodynamic data. Very near T_c even the 0.55 MHz ultrasonic data depart from the curve described by the thermodynamic data.

The dispersion in the sound velocity near the critical point is explained qualitatively by the Kadanoff and Swift mode-mode coupling theory, which predicts that the strong critical point divergences will occur only for measurements at frequencies well below a relaxation frequency,[39]

$$\omega_R = \lambda/\rho c_p \xi^2 \quad , \tag{35}$$

which goes to zero rapidly as the critical point is approached. Thus the Kadanoff-Swift prediction is that the sound velocity will reflect the divergence in c_V if $\omega \ll \omega_R$, where ω is the measurement frequency; however, as $T \to T_c$, the strongly temperature-dependent

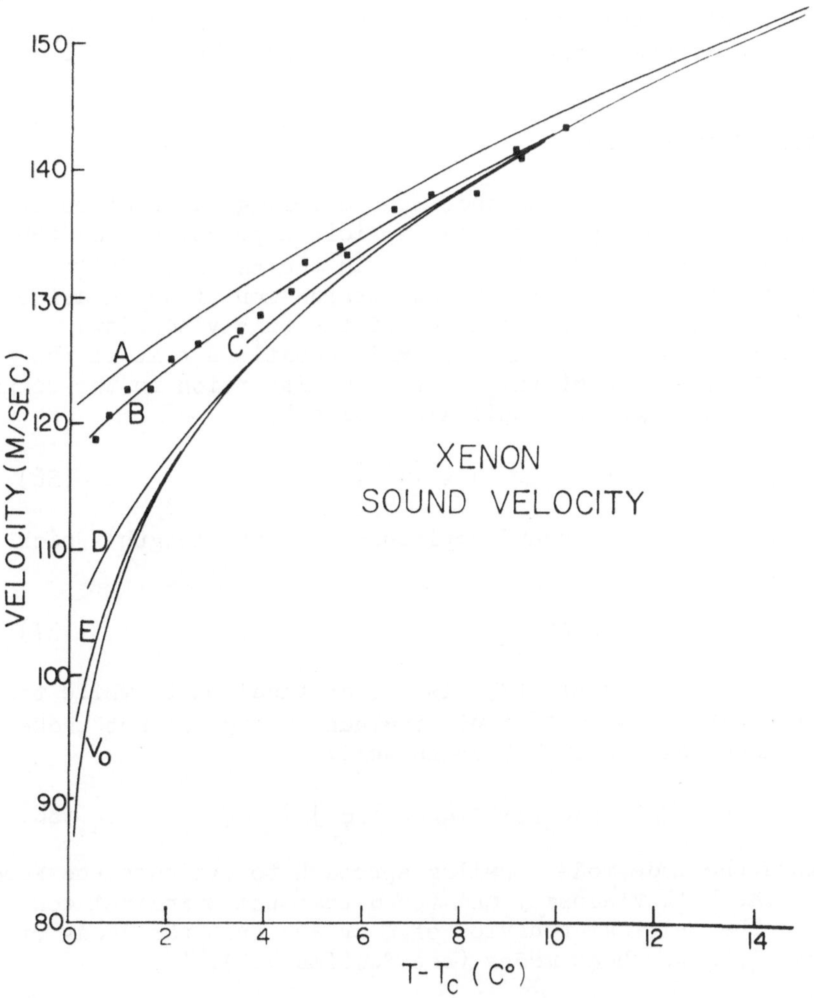

Figure 11. The velocity of sound in xenon on the critical isochore. Brillouin data: A, $\Theta = 170°$ (Ref. 64); B, $\Theta = 90°$, and C, $\Theta = 40°$ (Ref. 65). Ultrasonic data: D, 5MHz, and E, 0.55 MHz (Ref. 66). The static velocity v_o was computed from thermodynamic data. The data points shown are for the $\Theta = 90°$. (From Ref. 65.)

frequency ω_R will cross ω and the sound velocity will exhibit dispersion.

The relaxation frequency ω_R has the following physical meaning: at this frequency one period of the sound wave is equal to the time it takes heat to diffuse a distance of one correlation length. As the critical point is approached, the rate of heat diffusion, $\lambda/\rho c_p$, decreases rapidly, while the correlation length ξ is rapidly increasing; hence for measurements at any fixed frequency the temperature is ultimately reached at which heat no longer has time to diffuse across correlation length during the period of the sound wave.

5.2 Kawasaki-Mistura Theory

Kawasaki has extended the mode-mode coupling calculation of Kadanoff and Swift and has obtained explicit expressions for the sound dispersion and attenuation.[67] In the Kadanoff-Swift-Kawasaki approach, the dispersion and attenuation of sound arise primarily from the anomalous behavior of the bulk viscosity. It follows directly from the hydrodynamic equations that if the bulk viscosity $\zeta(\omega)$ is complex, there is a dispersion in the sound velocity which is given for small dispersion by

$$[v(\omega)-v_o]/v_o = (\omega/2\rho v_o^2) \text{ Im } \zeta^c(\omega) , \tag{36}$$

and an attenuation of the sound amplitude (per wavelength) which is given by

$$\alpha_\lambda^c(\omega) = (\pi\omega/\rho v_o^2) \text{ Re } \zeta^c(\omega) . \tag{37}$$

The attenuation given by Eq. (37) is the critical part, while the measured attenuation, $\alpha_\lambda = \alpha_\lambda^b + \alpha_\lambda^c$, includes a significant background or "classical" contribution as well:

$$\alpha_\lambda^b(\omega) = (\pi\omega/\rho v_o^2) [4\eta_s/3+\zeta^b+\lambda/c_v-\lambda/c_p] . \tag{38}$$

Kawasaki used the mode-mode coupling approach to evaluate the Kubo formula for the bulk viscosity and found that the principal contribution to the critical behavior of ζ arises from the decay of a sound mode into two heat modes (see Section 4.3).[67]

Unfortunately, the approach of Ferrell and Perl to the evaluation of the Kubo formulas for the transport coefficients near the critical point cannot be used for the bulk viscosity because the current-current correlation function contains the density or concentration at four space-time points, and in this case the decoupling approximation, that is, the factorization of the correlation function, is a poor approximation.[68]

Results for the sound dispersion and attenuation which are the same as the expressions obtained by Kawasaki have been derived by Mistura and coworkers,[66,69] who developed Fixman's idea that the coupling of the sound waves to the density fluctuations leads to an excess complex specific heat $\Delta(\omega)$. The dispersion and attenuation are then given by[69]

$$[v(\omega)-v_o]/v_o = -[(c_p/c_V-1)/2c_p] \operatorname{Re} \Delta(\omega) , \quad (39)$$

$$\alpha_\lambda^c(\omega) = [(c_p/c_V-1)\pi/c_p] \operatorname{Im} \Delta(\omega) . \quad (40)$$

The explicit expressions for the attenuation and dispersion derived by Mistura _et al._ from a consideration of the behavior of the complex specific heat are identical to Kawasaki's results [Eqs. (42) and (46)] obtained from an analysis of the bulk viscosity.

In the Kawasaki-Mistura theory the critical behavior of the sound mode is determined primarily by a single variable,

$$\omega^* \equiv \omega/2\omega_R . \quad (41)$$

The theoretical result for the sound dispersion can be written[67,69]

$$[v(\omega)-v_o]/v_o = F(T)\cdot J(\omega^*,\lambda^b/\lambda) , \quad (42)$$

where the weakly temperature-dependent combination of quantities $F(T)$ is given by

$$F(T) = \frac{k_B T^2}{2\pi\rho c_V \xi}\left(1 - \frac{c_V}{c_p}\right)\left(\frac{\partial\xi^{-1}}{\partial T}\right)_s^2 , \quad (43)$$

and

$$J(\omega^*,\lambda^b/\lambda) = \int_0^\infty \frac{x^2 dx}{(1+x^2)^2[1+(2\Gamma_R/\omega)^2]} , \quad (44)$$

with

$$2\Gamma_R/\omega = \{K_o(x) + (\lambda^b/\lambda)[x^2(1+x^2)-K_o(x)]\}/\omega^* . \quad (45)$$

[$K_o(x)$ is given by Eq. (20b).]

The theoretical expression for the critical part of the sound attenuation can be written[67,69]

$$\alpha_\lambda^c(\omega) = 2\pi F(T)\cdot I(\omega^*,\lambda^b/\lambda) , \quad (46)$$

where

$$I(\omega^*,\lambda^b/\lambda) = \int_0^\infty \frac{(2\Gamma_R/\omega)\ x^2\ dx}{(1+x^2)^2[1+(2\Gamma_R/\omega)^2]} \quad . \tag{47}$$

The comparison of experimental results for the dispersion and attenuation is facilitated by defining new quantities, the reduced dispersion,[65]

$$D \equiv [v(\omega)-v_o]\ /v_oF \ , \tag{48}$$

and the reduced attenuation,[70]

$$A \equiv \alpha_\lambda^c/2\pi F \ . \tag{49}$$

Then the Kawasaki-Mistura theory predicts that the experimentally determined quantities D and A should be given by

$$D = J(\omega^*,\lambda^b/\lambda) \tag{50}$$

and

$$A = I(\omega^*,\lambda^b/\lambda) \ . \tag{51}$$

For measurements very close to the critical point the theory predicts $D = J(\omega^*,0)$ and $A = I(\omega^*,0)$; hence in this limit the reduced dispersion and attenuation should each be described by a single universal curve as a function of ω^*, independent of the particular thermodynamic path or fluid system.[65,66,69,70]

In practice the background thermal conductivity is not negligible in the temperature range for which the experiments are performed. Hence D and A depend on λ^b/λ as well as on ω^*, but the dependence on λ^b/λ is much weaker than the dependence on ω^*, as Fig. 12 illustrates for the reduced dispersion. The predicted value of D for a measurement at a particular value of λ^b/λ will fall between the upper curve, for which $\lambda^b = 0$ (i.e., $\lambda = \lambda^c$), and the lower curve, for which $\lambda^c = 0$ (i.e., $\lambda = \lambda^b$); the two curves differ by at most 20%. All the reported analyses of the sound dispersion data in terms of the Kawasaki-Mistura theory have been based on Eq.(5.1) of Kawasaki[67] (or the equivalent equation of Mistura) which assumes $\lambda^b = 0$; however, it is clear that the coupling of the sound mode to the heat mode depends on the full decay rate Γ_R of the heat diffusion mode [see Eq. (3.9) of Kawasaki[67] or Eq. (18) of Mistura[69]], so a complete analysis of the dispersion and attenuation must include the effect of the background.

Another refinement of the theory for the sound dispersion and attenuation is the assumption that the dispersion is small, an assumption that was used by Kawasaki and Mistura in deriving Eqs. (42) and (46). We find that the modification of the theory if this assumption is relaxed is negligible for $\omega/\omega_R \lesssim 10$, as Fig. 13 illustrates for the sound dispersion. For $\omega/\omega_R \gtrsim 10$ the removal of the small dispersion assumption does result in a significant modification in the Kawasaki and Mistura theories, as Fig. 13 illustrates; however, for $\omega/\omega_R \gtrsim 10$ there are also other assumptions which break down (e.g., for large ω/ω_R, most of the contribution to the integrals I and J occurs for $q\xi>>1$ where the assumed Ornstein-Zernike form for the correlation function is known to be inaccurate.)[71] Therefore, it appears that the small dispersion assumption is a reasonably accurate one in the region in which the theory is applicable, although Eden, Garland, and Thoen[70] have argued to the contrary. A more important consideration in the analysis of experimental data is the choice of the values of the thermodynamic parameters which enter ω_R and F(T); the theoretical predictions for the dispersion and attenuation are particularly sensitive to the correlation length, which has been accurately determined only for a few fluids.

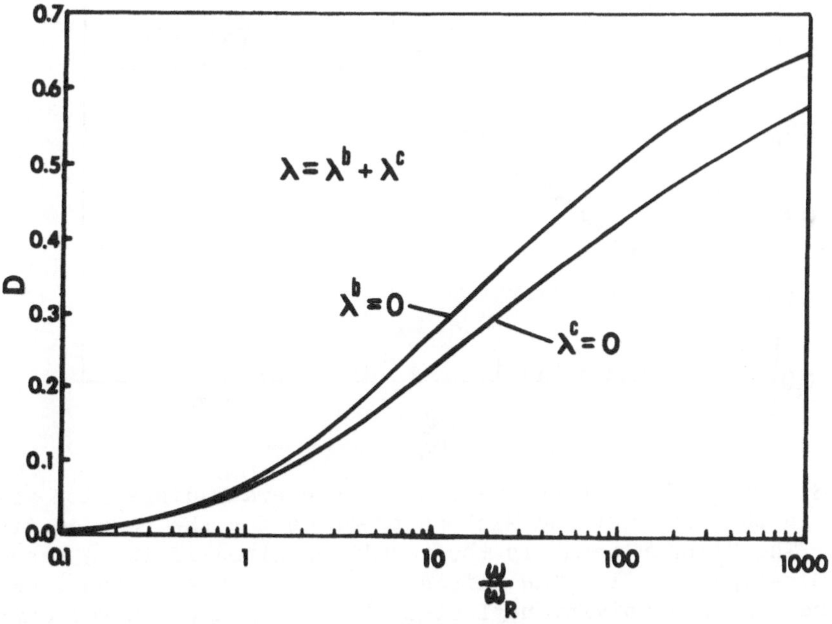

Figure 12. The effect of the thermal conductivity background λ^b on the reduced dispersion D. Very near the critical point we have $\lambda^b<<\lambda^c$, and the reduced dispersion in the mode-mode coupling theory is given by the upper curve; far from the critical point we have $\lambda^b>>\lambda^c$, and the reduced dispersion is given by the lower curve.

The theoretical expressions for the critical behavior of acoustic mode that we have presented were derived for single component fluids; however, there is also an anomalous acoustic behavior in binary mixtures, as first discussed by Fixman[63]. We will not discuss the problem of sound progation in fluids of two or more components near critical points, but instead refer the reader to the recent theoretical and experimental papers on this problem by Mistura, Tartaglia, and D'Arrigo.[72]

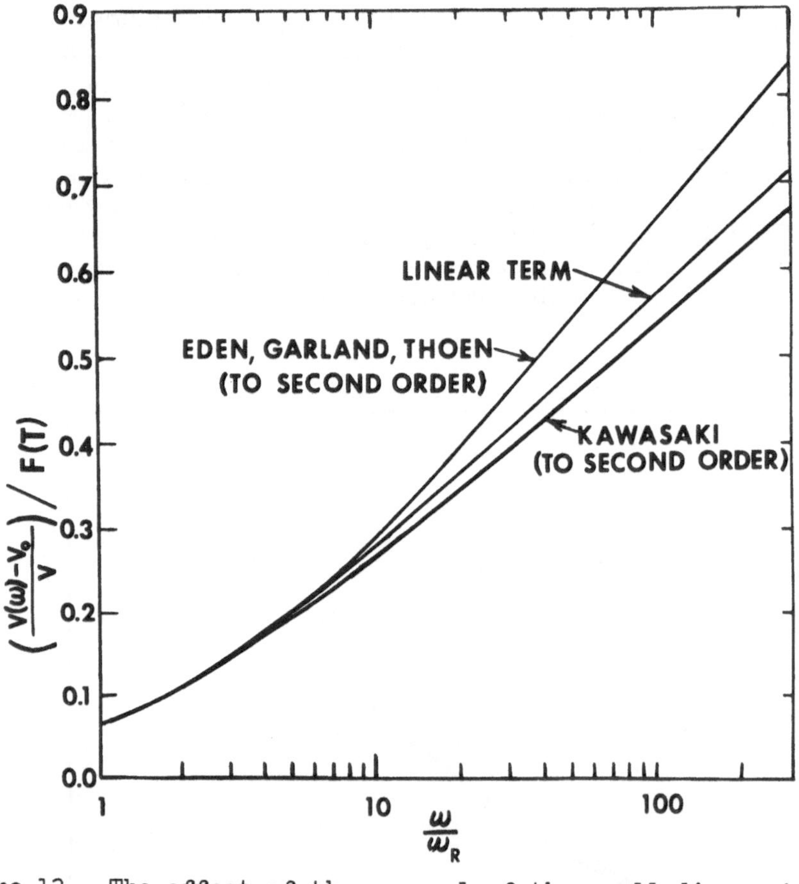

Figure 13. The effect of the removal of the small dispersion assumption in the Kawasaki and Mistura theories for the reduced dispersion. The "linear term" is the result obtained if it is assumed that $[v(\omega) - v_0]/v_0 \ll 1$. The linear term, shown with the background thermal conductivity neglected ($\lambda^b=0$), is a universal curve, independent of the fluid system, measurement frequency, or thermodynamic path, while the modified curves, which are shown for Brillouin scattering in xenon on the critical isochore at $\Theta = 90°$ and $\lambda_0=4880$ Å, depend on these parameters.

5.3 Brillouin Scattering Experiments

5.3.1 Interoferometric experiments. The primary experimental problem in obtaining interferometric Brillouin scattering spectra near the critical point is the very intense central component, which has instrumental wings which distort the Brillouin components. The contrast of an interferometric system can be greatly enhanced, however, by using two interferometers in tandem or by passing the light two or more times through the same interferometer; these techniques are discussed by Vaughan[73] and Sandercock.[74] Cannell and Benedek[64] have used two spherical interferometers in tandem to obtain Brillouin spectra for xenon near the critical point, as shown in Fig. 14. The tandem system yields a very high contrast; however, it is extremely difficult to keep the transmission maxima of two different interferometers superimposed for any period of time, especially if they have narrow linewidths. The multipass technique, which has been used extensively with plane-parallel plate interferometers in the past few years,[73,74] is easier to use, and recently Cannell, Lunacek, and Dubin[75] have constructed a double-pass spherical interferometer.

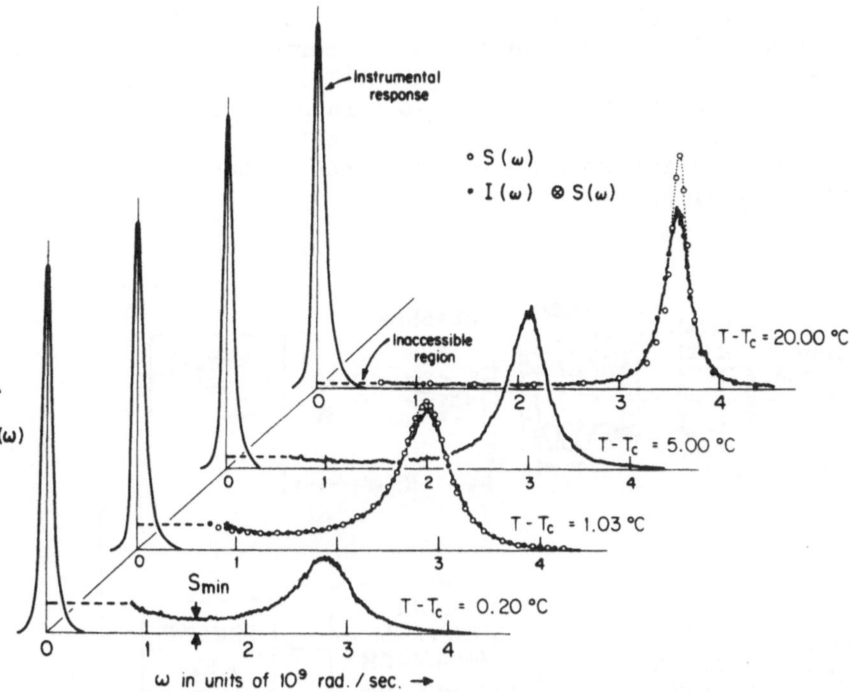

Figure 14. Brillouin scattering spectra obtained by Cannell and Benedek[64] with a tandem spherical interferometer system for xenon on the critical isochore($\Theta = 170°$). Note the additional mode between the Rayleigh and Brillouin components, which Cannell and Benedek have analyzed as a "Mountain-mode" arising from a structural relaxation at a frequency $\omega_s = v\xi^{-1}$ (see Lallemand[2] and Kadanoff and Swift[39]).

5.3.2 Optical heterodyne experiments. There have been many Brillouin scattering studies of sound propagation performed with interferometers, but in practice the problems of laser frequency stability and interferometer resolution limit these measurements to frequency shifts of the order of one hundred MHz or larger. In 1965 Lastovka and Benedek recognized that the light beating technique offers a way of extending Brillouin scattering measurements to much lower frequencies, and they were able to obtain a Brillouin scattering spectrum for toluene with a frequency shift of 30 MHz, corresponding to a scattering angle of approximately 0.5.[76,77] In that experiment the sample cell was placed inside the laser cavity and the scattered light was mixed at the photocathode with a beam of light from the laser. Although the instrumental broadening was much greater than the intrinsic Brillouin linewidth and the scattering angle was not independently measured, this experiment did demonstrate the feasibility of Brillouin scattering measurements by the light beating technique; however, no further optical heterodyne Brillouin scattering experiments were reported until the 1973 experiment described below, primarily because of the difficulties associated with the optical alignment.

Eden and Swinney[78] have recently reported an optical heterodyne Brillouin scattering experiment which was performed with an optical system that is fairly simple to align and for which an accurate measurement of the scattering angle is straightforward. This apparatus, shown in Fig. 15, was used to investigate the behavior of the sound velocity and attenuation for xenon in the vapor phase below the critical point.

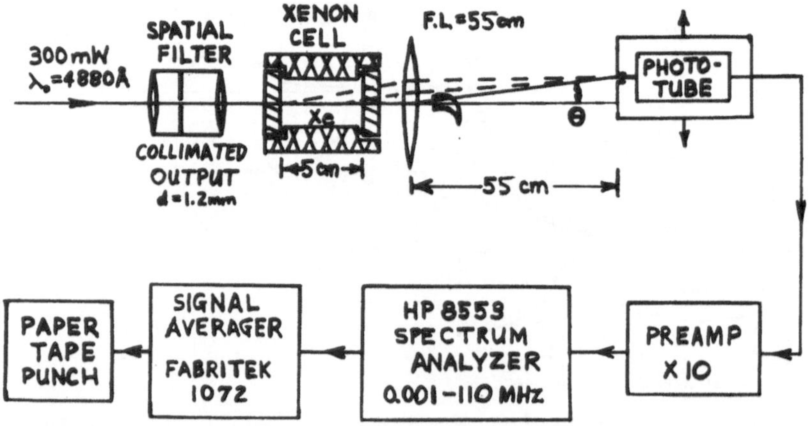

Figure 15. Optical heterodyne spectrometer used in Brillouin scattering experiment.

The light beating technique, which is much less sensitive than interferometry (Vaughan[73]), can only be used to study signals of high spectral density. Since the Brillouin linewidth Γ_B is proportional to q^2 while the scattering cross section is essentially independent of angle, high spectral density can be achieved at small scattering angles. The apparatus diagrammed in Fig. 15 is designed for measurements at scattering angles ranging from $\sim 0.1°$ to $\sim 10°$. The laser beam is spatially filtered and collimated by a diffraction-limited lens to produce a 1.2mm-diameter beam which is incident on the sample cell. The xenon cell is a 5 cm long stainless steel cylinder with 1.25 cm-thick polished fused quartz windows on the ends. The laser beam, which is incident normally to the cell windows, establishes the optic axis of the spectrometer. The collection optics consist of a lens which is on the optic axis and a small slit which is in the focal plane of the lens. The light scattered at an angle Θ with respect to the optic axis is focused into a circle (in the focal plane of the lens) of radius $f \tan \Theta$, where f is the focal length of the lens. The slit and phototube assembly are mounted on a precision translation stage which is perpendicular to the optic axis; hence the scattering angle can be varied continuously. The accuracy of the measurement of the scattering angle is limited only by the accuracy with which the focal length of the lens is known and by the diffraction of the incident laser beam; the latter can, of course, be reduced by increasing the diameter of the laser beam.

In the system in Fig. 15 the elastically scattered light from the cell windows serves as a coherent optical field which mixes with the signal of interest at the photocathode of the photomultiplier. The total average photocurrent is the sum of that due to the scattering from the sample, $\langle i_s \rangle$, and that due to the elastically scattered coherent field, i_{LO}. The power spectrum of the photocurrent centered at Γ_B is given by[79]

$$P(\omega)=\frac{e}{\pi}\left(\langle i_s\rangle+i_{LO}\right) + \frac{2\langle i_s\rangle c_v}{\pi\, c_p}\left(\frac{c_p-c_v}{c_p}\,\beta_1\langle i_s\rangle+\beta_2 i_{LO}\right)\frac{\Gamma_B}{(\omega-\omega_B)^2+\Gamma_B{}^2}, \quad (52)$$

where β_1 and β_2 are heterodyne efficiency coefficients, which are of order unity. The frequency-independent term in Eq. (52) is the shot noise, while the component of width Γ_B centered at ω_B arises from mixing of the Brillouin components with the Rayleigh component (the term proportional to $\langle i_s \rangle^2$) and with the local oscillator (the term proportional to $\langle i_s \rangle\, i_{LO}$). [The term arising from the mixing of the Rayleigh and Brillouin components actually has a width $\Gamma_R + \Gamma_B$, but $\Gamma_R \ll \Gamma_B$, so this term can be written as in Eq. (52)]. Note that the form of the spectrum given by Eq. (52) is independent of the relative size of $\langle i_s \rangle$ and i_{LO}.

A typical spectrum obtained in the optical heterodyne experiment on xenon is shown in Fig. 16. The ratio of the Brillouin signal

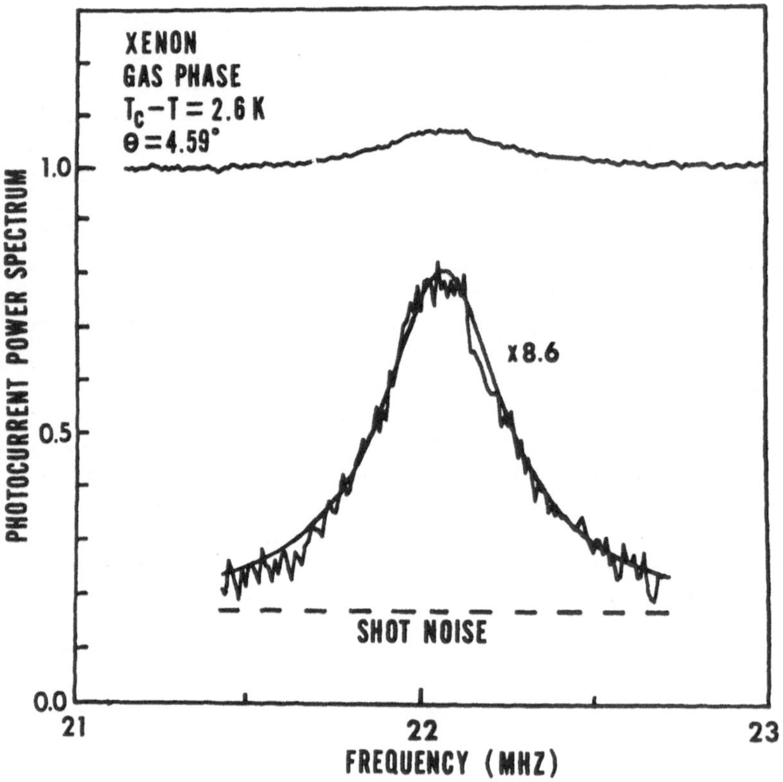

Figure 16. Optical heterodyne Brillouin spectrum for xenon on the gas side of the coexistence curve below the critical point. The Brillouin shift is 22.05 MHz and the half-width at half-maximum is 0.220kHz. The smooth curve shown with the enlarged inset is a computer-fit Lorentzian.

power at ω_B to the shot noise power, which is 0.07 for the spectrum in Fig. 16, ranged from 0.05 to 0.4 for the xenon measurements; however, the important quantity, the size of the signal relative to the fluctuations in the background, was typically 40 or more. Figure 17 shows spectra obtained for three temperatures at a scattering angle of approximately 2.3° in xenon; for a fixed external angle, the scattering angle in xenon varies slowly with temperature due to variation of the refractive index, which is well known along the coexistence curve of xenon. The critical behavior of the sound velocity is qualitatively clear from Fig. 17: the Brillouin shift decreases as the critical point is approached, reflecting the decrease in the sound velocity, while the linewidth, $\Gamma_B = \alpha_\lambda \omega_B / 2\pi$, increases rapidly, reflecting the increase in the acoustic attenuation.

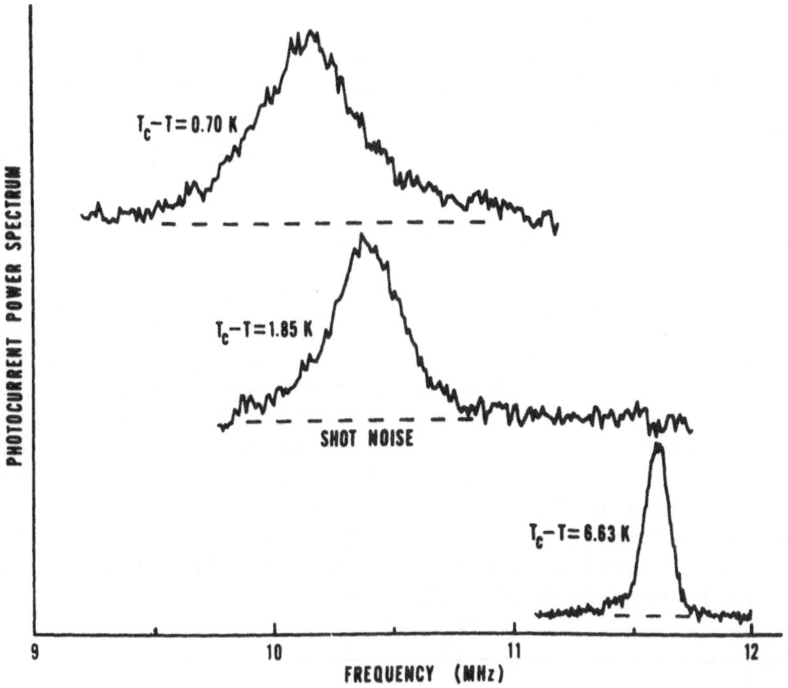

Figure 17. Brillouin spectra for xenon at three temperatures in the vapor on the coexistence curve for a scattering angle of 2.50° in air.[78] The spectra have been normalized to the same amplitude; the peak intensity to shot noise ratio decreases as the critical point is approached due to the decrease in spectral density.

Although the line broadening due to the size of the slits in front of the phototube and from the spectrum analyzer bandwidth was negligible, there was a significant contribution to the observed linewidths due to the divergence of the laser beam incident on the sample. This instrumental broadening, which is independent of angle (except for the small dispersion in v), was studied in measurements at angles sufficiently small such that the intrinsic broadening, proportional to q^2, was negligible. For the 1.2mm-diameter laser beam the instrumental line profile was well approximated by a Gaussian with a half-width at half-maximum of 46 kHz; therefore, an observed spectral line was given by the convolution of the Gaussian instrumental profile with the Lorentzian Brillouin line. The values for the intrinsic Brillouin linewidth Γ_B were obtained by finding the width of a Lorentzian which when convoluted with a Gaussian of 46 kHz half-width yielded the observed spectral line.

The optical heterodyne and interferometric techniques are complementary. The frequency instability of currently available lasers limits the useful range for interferometers to frequencies $\gtrsim$ 100 MHz, while the relatively low sensitivity of the optical heterodyne method limits this technique in practice to frequencies $\lesssim$ 100 MHz, since signals at higher frequencies usually have a spectral density which is too low to be studied by optical heterodyning.

5.4 Comparison of Theory and Experiment

An absolute comparison of the Brillouin scattering data with the predictions of the Kawasaki-Mistura theory can be made only if the parameters which enter F and ω_R are known from other experiments. The parameters are known fairly accurately along the critical isochore of xenon, so the experimental results for the reduced dispersion obtained along this path can be directly compared with theory, as shown in Fig. 18. The uncertainty in the sound velocity data in Fig. 18 is one percent or more, while the dispersion is only of the order of a few percent; hence the uncertainty in the dispersion is quite large, but within this uncertainty the data are in agreement with the theory.

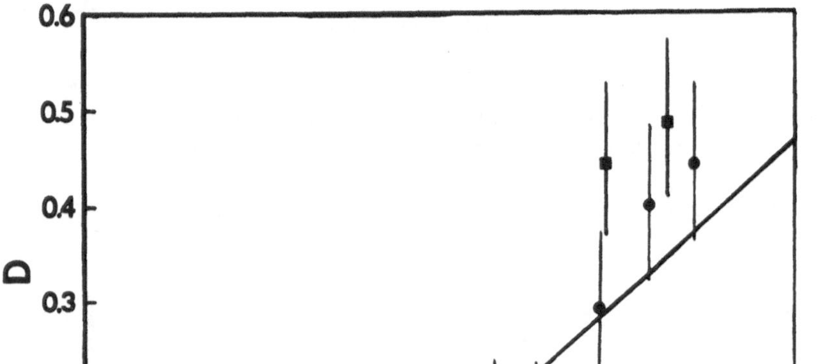

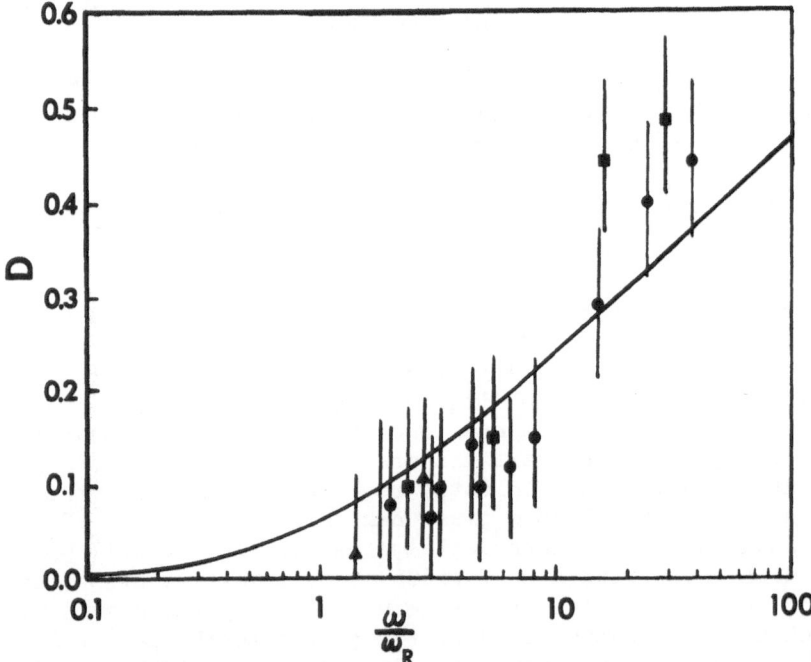

Figure 18. Comparison of the reduced dispersion [Eq. (48)] determined for xenon on the critical isochore with the Kawasaki-Mistura theory [Eq. (50)]. The reduced dispersion was deduced from the Brillouin scattering sound velocity data shown in Fig. 11: Θ=170° (squares), Cannell and Benedek;[64] Θ=90° (circles) and Θ=40° (triangles), Cummins and Swinney.[65] The parameters used to evaluate F, ω_R, and α_λ^b were taken from Ref. 44.

The results for the reduced dispersion and the reduced attenuation deduced from the optical heterodyne experiment performed for xenon on the vapor side of the coexistence curve are compared with the theoretical predictions in Fig. 19. (The dispersion graph also includes the 90° interferometric data of Cummins.[80]) Along the coexistence curve the parameters needed in order to evaluate F, ω_R, and α_λ^b are not so well known as along the critical isochore. The greatest uncertainty is in the term $(\partial\xi^{-1}/\partial T)_s$ which enters F [Eq. (43)]; this term was calculated from the Ornstein-Zernike relation $\xi^2 = \rho_n k_B T R^2 \kappa_T$ with R = 4.9Å[81] and with κ_T determined from the Schofield-Litster-Ho equation of state[12] using the parameters of Hohenberg and Barmatz.[82] The sound dispersion was computed using the zero frequency sound velocity values computed by Thoen and Garland[83] from the Schofield-Litster-Ho equation of state, and the parameters used in evaluating α_λ^b and ω_R were taken from the tabulation by Swinney and Henry.[44]

The theoretical curves in Fig. 19 were drawn with the effect of the background thermal conductivity neglected (see Fig. 12); however, for the data shown in Fig. 19 the inclusion of the background correction changes the theoretical curves by at most 10%, which is less than the uncertainty in the parameters which enter F and ω_R. The experimental results for D and A, obtained at different temperatures and frequencies, are described reasonably well by the single variable ω^*; moreover, the results for D and A agree within the experimental uncertainty with the theoretical predictions. It should be emphasized that this is an absolute comparison of theory and experiment, involving no adjustable parameters; however, there is a fairly large uncertainty in the parameters which enter the theory, particularly for F far from T_c.

In conclusion, the Brillouin scattering data that have been reported for simple fluids near the critical point agree within the (rather large) experimental uncertainty with the prediction of the Kawasaki-Mistura theory that the reduced dispersion D and the reduced attenuation A should depend primarily only on the single variable, $\omega^* \equiv \omega/2\omega_R$. However, a definitive test of the specific predictions for D and A, Eqs. (50) and (51), must await more accurate measurements of the sound velocity and attenuation and the parameters which enter the theory. Such measurements are now underway in several laboratories, and these experiments, which extend beyond the relaxation frequency ω_R into the region of the high frequency relaxation, $\omega_s = v\xi^{-1}$, should provide new information on the critical behavior of the sound propagation mode.

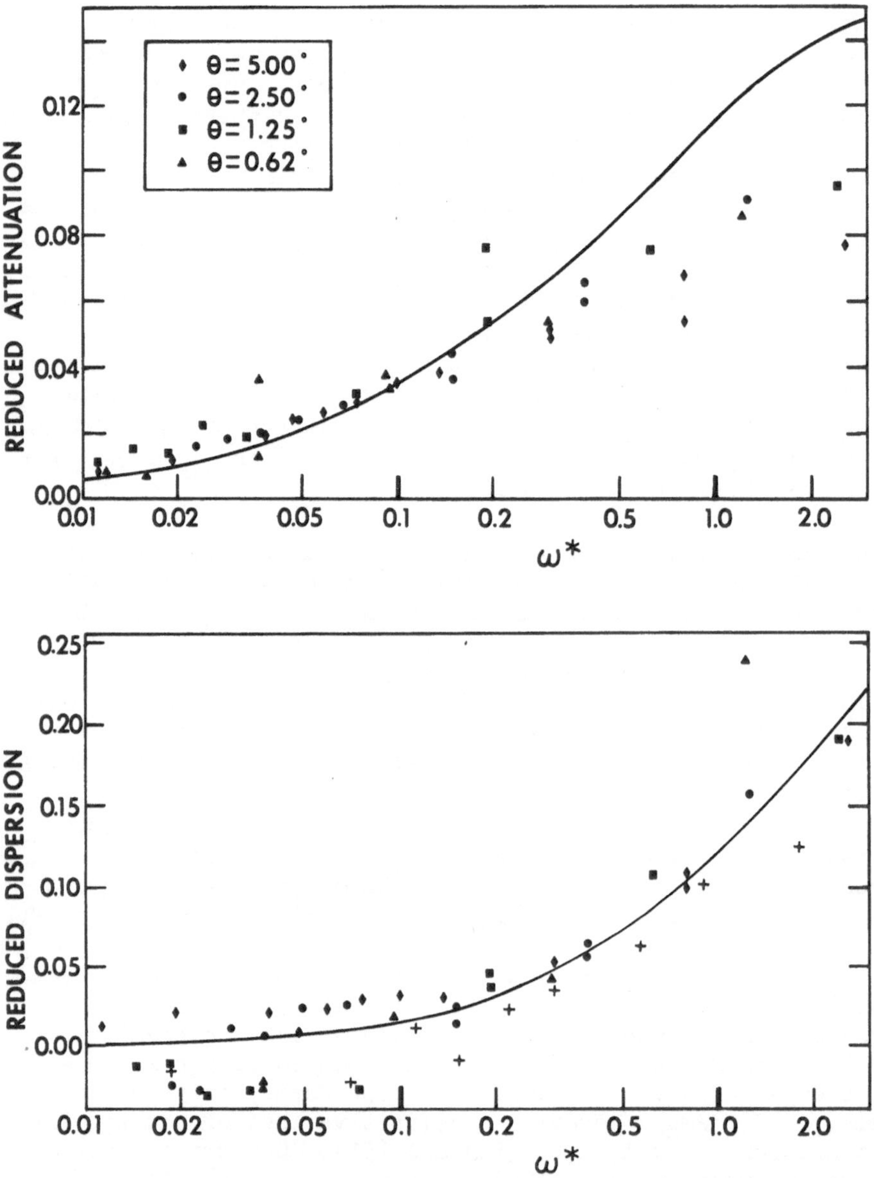

Figure 19. A comparison of the predictions of the Kawasaki-Mistura theory with Brillouin scattering measurements of Eden and Swinney of the reduced attenuation and dispersion for xenon in the vapor phase on the coexistence curve. The indicated angles are the external angles; the actual scattering angles in xenon were calculated from Snell's law. Included in the graph of the reduced dispersion are sound velocity data (+) obtained by Cummins in Brillouin scattering measurements at $\Theta = 90°$. (Figure from Ref. 78.)

6. ACKNOWLEDGMENTS

Many of the examples presented here were taken from experiments performed by present and former members of the light scattering group at New York University (now at the City College of the City University of New York), including D. L. Henry, T. K. Lim, D. Eden, and H. Z. Cummins. In particular, I would like to acknowledge that the present discussion of the Rayleigh linewidth is condensed from a paper (Ref. 44) co-authored with D. L. Henry, and the discussion of the optical heterodyne Brillouin scattering experiment is drawn largely from a paper (Ref. 78) co-authored with D. Eden. Finally, I am especially indebted to H. Z. Cummins, who has continuously stimulated and supported this research program on fluids near the critical point.

7. REFERENCES

* Research supported by the National Science Foundation.

† Present address: The City College of New York, Physics Department, Convent Avenue and 138 Street, New York, N.Y. 10031

1. E.R. Pike, this volume.

2. P. Lallemand, this volume.

3. T. Andrews, Phil. Trans. Roy. Soc. (London) 159, 575 (1869). This paper is reprinted in Ref. 9.

4. J.D. van der **Waals**, Doctoral Dissertation, Leiden, 1873.

5. L.P. Kadanoff et al., Rev. Mod. Phys. 39, 395 (1967).

6. M.E. Fisher, Rept. Prog. Phys. 30, 615 (1967).

7. H.E. Stanley, Phase Transitions and Critical Phenomena (Oxford University Press, New York, 1971).

8. M.S. Green, editor, Critical Phenomena (Academic Press, New York, 1971).

9. H.E. Stanley, editor, Cooperative Phenomena near Phase Transitions (M.I.T. Press, Cambridge, Mass., 1973).

10. B. Widom, J. Chem. Phys. 43, 3892, 3898 (1965); see also C. Domb and D. Hunter, Proc. Phys. Soc. 86, 1147 (1965).

11. M. Vicentini-Missoni, R.I. Joseph, M.S. Green, and J.M.H. Levelt Sengers, Phys. Rev. B 1, 2312 (1970).

12. P. Schofield, J.D. Litster, and J.T. Ho, Phys. Rev. Lett. 23, 1098 (1969).

13. K.G. Wilson, Phys. Rev. B 4, 3174, 3184 (1971). Refs. 14 and 15 are also helpful in understanding the renormalization group method.

14. S. Kovesi-Domokos, "Critical Phenomena and Renormalization Group for Outsiders," Technical Report COO-3285-9 (Physics Dept., Johns Hopkins Univ., Baltimore, 1972).

15. S.K. Ma, Rev. Mod. Phys. 45, 589 (1973).

16. L.P. Kadanoff, Physics 2, 263 (1966). See also Ref. 5.

17. K.G. Wilson, Phys. Rev. Lett. 28, 548 (1972); K.G. Wilson and M.E. Fisher, Phys. Rev. Lett. 28, 240 (1972).

18. B.I. Halperin, P.C. Hohenberg, and S.K. Ma, Phys. Rev. Lett. 29, 1548 (1972).

19. H.Z. Cummins, Ref. 8, p. 379.

20. J.H. Lunacek and D.S. Cannell, Phys. Rev. Lett. 27, 841 (1971).

21. S. Gewurtz, W.S. Gornall, and B.P. Stoicheff, J. Acoust. Soc. Am. 49, 994 (1971).

22. R.D. Mountain and J.M. Deutch, J. Chem. Phys. 50, 1103 (1969).

23. P.A. Fleury and J.P. Boon, Adv. Chem. Phys. 29, 1 (1973).

24. L. Van Hove, Phys. Rev. 95, 1374 (1954).

25. S.S. Alpert, Y. Yeh, and E. Lipsworth, Phys. Rev. Lett. 14, 486 (1965).

26. H.Z. Cummins, N. Knable, and Y. Yeh, Phys. Rev. Lett. 12, 150 (1964).

27. N.C. Ford, Jr., and G.B. Benedek, Phys. Rev. Lett. 15, 649 (1965).

28. H.L. Swinney and H.Z. Cummins, Phys. Rev. 171, 152 (1968).

29. C.S. Bak and W.I. Goldburg, Phys. Rev. Lett. 23, 1218 (1969).

30. C.S. Bak, W.I. Goldburg, and P.N. Pusey, Phys. Rev. Lett. 25, 1420 (1970).

31. J.V. Sengers, Ref. 8, p. 445.

32. J.V. Sengers and P.H. Keyes, Phys. Rev. Lett. 26, 70 (1971).

33. P. de Gennes, unpublished lecture, Second International Conference on Light Scattering in Solids, Paris, 1971.

34. H.L. Swinney, D.L. Henry, and H.Z. Cummins, J. Phys. (Paris) 33, C1-181 (1972).

35. G.B. Benedek, J.B. Lastovka, M. Giglio, and D. Cannell, in Critical Phenomena, ed. by R.E. Mills, E. Aschner, and R.I. Jaffey (McGraw-Hill, New York, 1971), p. 503.

36. R. Mohr and K.H. Langley, J. Phys. (Paris) 33, C1-97 (1972).

37. T.K. Lim, H.L. Swinney, K.H. Langley, and T.A. Kachnowski, Phys. Rev. Lett. 27, 1776 (1971).

38. G.T. Feke, G.A. Hawkins, J.B. Lastovka, and G.B. Benedek, Phys. Rev. Lett. 27, 1780 (1971).

39. L.P. Kadanoff and J. Swift, Phys. Rev. 166, 89 (1968).

40. K. Kawasaki, Phys. Rev. 150, 291 (1966).

41. G. Arcovito, C. Faloci, M. Roberti, and L. Mistura, Phys. Rev. Lett. 22, 1040 (1969).

42. P. Bergé, P. Calmettes, C. Laj., and B. Volochine, Phys. Rev. Lett. 23, 693 (1969).

43. D.L. Henry, H.L. Swinney, and H.Z. Cummins, Phys. Rev. Lett. 25, 1170 (1970); D.L. Henry, Ph.D. Thesis, Johns Hopkins University, 1970 (unpublished).

44. H.L. Swinney and D.L. Henry, Phys. Rev. A 8, 2586 (1973).

45. T.K. Lim, Ph.D. Thesis, Johns Hopkins University, 1973 (unpublished).

46. C.J. Oliver, this volume.

47. L. Mistura, J. Chem. Phys. 55, 2375 (1971).

48. B. Volochine and P. Bergé, Phys. Rev. Lett. 23, 693 (1969).

49. K. Kawasaki, Ann.Phys. (N.Y.) 61, 1 (1970).

50. K. Kawasaki and S.M. Lo, Phys. Rev. Lett. 29, 48 (1972).

51. S.M. Lo and K. Kawasaki, Phys. Rev. A 8,2176 (1973).

52 S.M. Lo and K. Kawasaki, Phys. Rev. A5, 421 (1972).

53. M.E. Fisher and R.J. Burford, Phys. Rev. 156, 583 (1967).

54. H.L. Swinney and B.E.A. Saleh, Phys. Rev. A 7, 747 (1973)

55. R.F. Chang, P.H. Keyes, J.V. Sengers, and C.O. Alley, Ber. Bunsen - Ges. Physik. Chem. 76, 260 (1972).

56. R.A. Ferrell, Phys. Rev. Lett. 24, 1169 (1970).

57. The comparison between the mode-mode coupling and decoupled-mode theories is discussed by R.A. Ferrell in Dynamical Aspects of Critical Phenomena, ed. by J.I. Budnick and M.P. Kawatra (Gordon and Breach, New York, 1972).

58. R. Perl and R.A. Ferrell, Phys. Rev. Lett. 29, 51 (1972); Phys. Rev. A 6, 2358 (1972).

59. R. Perl and R.A. Ferrell, to be published.

60. R.W. Gammon, H.L. Swinney, and H.Z. Cummins, Phys. Rev. Lett. 19, 1467 (1967).

61. N.C. Ford, Jr., K.H. Langley, and V.G. Puglielli, Phys. Rev. Lett. 21, 9 (1968).

62. A.H. Chynoweth and W.G. Schneider, J. Chem. Phys. 20, 1777 (1952).

63. M. Fixman, J. Chem. Phys. 33, 1363 (1960); ibid. 36, 1961(1962).

64. D.S. Cannell and G.B. Benedek, Phys. Rev. Lett. 25, 1157 (1970).

65. H.Z. Cummins and H.L. Swinney, Phys. Rev. Lett. 25, 1165 (1970).

66. C.W. Garland, D. Eden, and L. Mistura, Phys. Rev. Lett. 25, 1161 (1970).

67. K. Kawasaki, Phys. Rev. A 1, 1750 (1970).

68. R. Perl, Ph.D. Thesis, University of Maryland, 1972 (unpublished); Naval Ordnance Laboratory Technical Report 72-208 (Silver Spring, Maryland, 1972).

69. L. Mistura, Ref. 8, p. 563.

70. D. Eden, C.W. Garland, and J. Thoen, Phys. Rev. Lett. 28, 726 (1972).

71. K. Kawasaki, Int. J. Magnetism 1, 171 (1971).

72. L. Mistura, J. Chem. Phys. 57, 2312 (1972); G. D'Arrigo, L. Mistura, and P. Tartaglia, Phys. Rev. A 3, 1718 (1971); P. Tartaglia, G. D'Arrigo, L. Mistura, and D. Sette, Phys. Rev. A 6, 1627 (1972); L. Mistura, Nuovo Cimento 12B, 35 (1972); L. Guidoni et al., Phys. Rev. Lett. 31, 583 (1973).

73. J.M. Vaughan, this volume.

74. J. Sandercock, Light Scattering in Solids, ed. by B.M. Balkanski (Flammarion Sciences, Paris, 1971), p. 9.

75. D.S. Cannell, J.H. Lunacek, and S.B. Dubin, Rev. Sci. Instr. 44 (Nov. 1973).

76. J.B. Lastovka and G.B. Benedek, in Physics of Quantum Electronics, ed. by J. Singer (Columbia University Press, New York, 1961).

77. J.B. Lastovka, Ph.D. Thesis, M.I.T., 1967 (unpublished).

78. D. Eden and H.L. Swinney, to be published.

79. H.Z. Cummins and H.L. Swinney, Progress in Optics 8, 133 (1970).

80. H.Z. Cummins, unpublished.

81. I.W. Smith, M. Giglio, and G.B. Benedek, Phys. Rev. Lett. 27, 1556 (1971); I.W. Smith, Ph.D. Thesis, M.I.T., 1972 (unpublished).

82. P.C. Hohenberg and M. Barmatz, Phys. Rev. A 6, 289 (1972).

83. J. Thoen and C.W. Garland, private communication.

SEMINARS

MACROMOLECULAR DIFFUSION

P N Pusey

Royal Radar Establishment

Great Malvern, Worcs, England

1 INTRODUCTION

Cummins (this publication) has given a broad review of the field of macromolecular diffusion as studied by photon correlation spectroscopy. In this article I will cover in more detail three aspects of the field: (1) Diffusion of identical, non-interacting macromolecules, spherically symmetric and/or small compared to the wavelength of light; (2) The effects and characterization of macromolecular polydispersity (non-identical macromolecules); and (3) The effects of inter-macromolecular interactions. The first topic is probably the simplest that can be conceived of under the heading of diffusion. As Cummins (this publication) has shown, the theory is straightforward, and I hope to convince you that experiments to date show adequate agreement with the theoretical predictions. From measurements on such systems one can obtain directly the translational diffusion coefficient D of the macromolecules. From D one can obtain, via the Einstein equation, the macromolecular frictional coefficient. For spherical macromolecules this yields through the Stokes equation the macromolecular radius. Also, combining D in the Svedberg equation with independently determined values for the sedimentation rate and partial specific volume one obtains the macromolecular weight.

The theory of light scattering by polydisperse non-interacting systems is also straightforward. The scattered electric field correlation function $|g^{(1)}(\tau)|$ is, in this case, a sum of exponentials, rather than the single exponential found in the identical-particle case. The difficulty when studying polydisperse systems arises from the fact that the obvious approach of Laplace inversion of the experimental data to yield the distribution of diffusion coefficients is extremely sensitive to unavoidable statistical errors in the data. In fact it appears that this approach is valueless in a large majority of cases. One is forced therefore to adopt an approach which recognizes the insensitivity of photon correlation spectroscopy to macromolecular polydispersity, yet yields as much information as possible in a meaningful form. I will discuss one such approach, which yields a well-defined average diffusion coefficient and one or two higher moments of the distribution of diffusion coefficients.

Considering the third topic, if one "turns on" an interaction between the macromolecules in solution, one clearly goes from a simple single-particle Brownian-motion problem to a many-body problem, which in the general case can be expected to be complicated. There has been some success in explaining experimental results particularly for short-ranged interactions using a hydrodynamic/thermodynamic approach as outlined by Cummins (this publication). However, detailed applications of the newer kinetic theories (see, for example, Lallemand, Pike and Swinney, this publication) to this problem have not yet been made. With regard to the experimental situation, some experiments on interacting systems have recently been performed. However, unlike the independent-particle situation, diffusion in an interacting system will depend not only on individual particle sizes and shapes, but also on the nature of the interactions, which in turn can be influenced by a host of environmental factors. Thus there is virtually an infinity of potential experiments to be done in this area. Nevertheless, despite the paucity of experimental and theoretical information on this topic, I think there is still sufficient input from laboratory and computer experiments and simple theory to give a partially coherent picture.

For convenience, the experimental examples given in this article will largely be taken from work with which I have been associated over the last few years.

2 DIFFUSION OF NON-INTERACTING IDENTICAL MACROMOLECULES

For simplicity this discussion will be limited mainly to macromolecules which are spherically symmetrical and/or small compared to $1/K$ where $K \equiv (4\pi/\lambda)\, n \sin\theta/2$ is the magnitude of the scattering vector, λ is the wavelength of the incident light in vacuo, n is the refractive index of the solution and θ is the scattering angle. Cummins (this publication) has shown that the electric field scattered by a solution containing a large number of such macromolecules should have Gaussian statistics and a correlation function of the form

$$|g^{(1)}(\tau)| = \exp(-\Gamma\tau) \qquad (1)$$

where

$$\Gamma = DK^2 \ . \qquad (2)$$

Thus we have a Gaussian-Lorentzian light field and all the comments made by Jakeman and Oliver (this publication) concerning measurements on such fields apply here. We can obtain an experimental estimate of C $|g^{(1)}(\tau)|^2$, (where C is the parameter characterizing degree of spatial coherence, the effect of clipping, etc), either by full or single-clipped correlation.

As experimental examples I will discuss recent measurements on several spherical viruses, the bacteriophages R17, Qβ, PM2 and T7 and the plant virus, Tomato Bushy Stunt Virus, BSV, (Pusey et al, 1973a; Camerini-Otero, et al, 1973a).

Fig 1 shows a plot of ℓn C $|g^{(1)}(\tau)|^2$ for a dilute solution of R17 virus in 0.015M NaCl. This measurement was made using a 20-channel single-clipped correlator (clipping level 1) at a scattering angle of 90° and a temperature of 25°C. The sample time T was 11.25 μsec. The data can be fitted by a straight line indicating

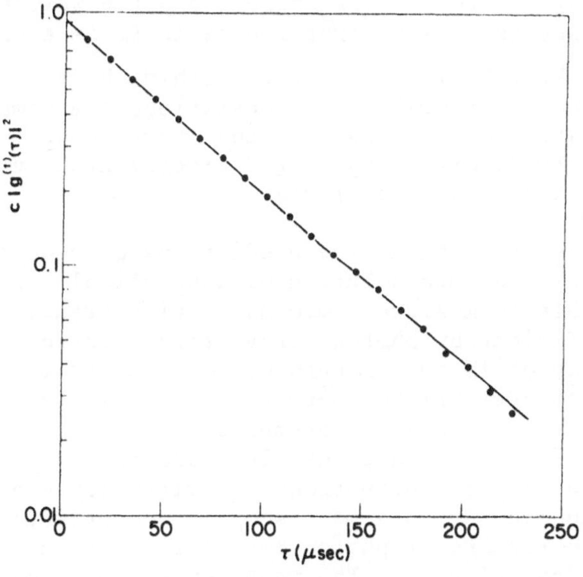

FIG 1. Semi-log plot of the measured correlation function for a dilute solution of R17 virus, (taken from Pusey et al, 1973a).

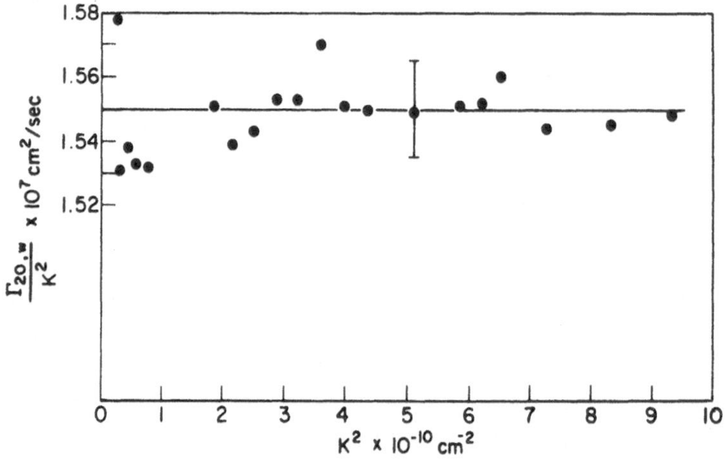

FIG 2. Angular dependence of the decay rate of the correlation function for a dilute solution of R17 virus, (taken from Pusey et al, 1973a).

that $|g^{(1)}(\tau)|$ is well described by a single exponential, as predicted by Eq 1. (More will be said in Sec III about testing the goodness of an exponential fit.) Fig 2 shows an experimental test of Eq 2, which predicts Γ/K^2 independent of K^2, again for a dilute solution of R17 virus. The scattering vector K was varied both by altering the scattering angle θ and the wavelength λ of the Krypton Ion laser used as a light source in these experiments. For most values of K, Γ/K^2 is constant to within about 1%, the expected statistical error. At small K the deviations are somewhat larger due to poor statistics arising from the longer coherence times which were not fully offset by longer experimental run-times. Run-times were typically 1-5 minutes.

Not surprisingly, then, the predictions of equations 1 and 2 are fulfilled by a dilute solution of a spherical virus. One can expect to be able to measure a macromolecular translational diffusion coefficient by photon correlation spectroscopy typically with an accuracy of 1% in a minute or so. The speed and accuracy of this method constitute an enormous advantage over conventional boundary spreading techniques for measuring D. An additional advantage is, of course, that the light scattering method, probing as it does spontaneous fluctuations, is virtually non-perturbative. There is no need to set up a macroscopic concentration gradient. Fig 3 shows the results of photon correlation measurements on a range of spherical viruses. The measurements were made at 25°C in various different buffer solutions. The values of D have been corrected to standard conditions, 20°C and water as solvent, making the usual assumption that D scales as the ratio of absolute

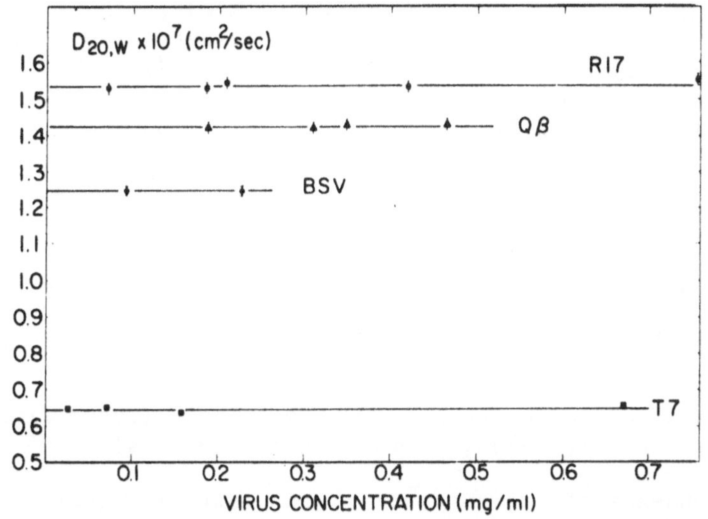

FIG 3. Diffusion constant of several spherical viruses as a function of virus concentration, (taken from Camerini-Otero et al, 1973a).

temperature to solvent viscosity (see Eqs 3 and 4, below). At these low concentrations, < 1 mg/ml, there is no evidence of concentration dependence of D, indicating that, on average, the macromolecules are far enough separated that the effect of intermacromolecular interactions is negligible.

Assuming now that D can be measured quickly and accurately, the next question is what information can be obtained from D. Two useful pieces of information are macromolecular size and weight. For non-interacting Brownian particles in a homogeneous solvent, D is given by the Einstein relation

$$D = kT/f \qquad (3)$$

where k is Boltzmann's constant, T is the absolute temperature and f is the frictional coefficient of the particle (see, for example, Tanford, 1961,p 349). For spherical particles, f is given by the Stokes relation

$$f = 6\pi \eta R_H \qquad (4)$$

where η is the solvent viscosity and R_H the hydrodynamic radius of the particle. In Table 1 we list R_H, calculated from Eqs 3 and 4, for the viruses shown in Fig 3. Also in Table 1 are given values for the particle radii determined by such techniques as electron microscopy, small-angle X-ray scattering from solutions and

Virus	$D \times 10^7$ (cm^2/sec)	R_H (Å)	Electron Microscope (Å)	X-ray, light scattering (Å)
R17	1.534 ± 0.015	140 ± 2	115, 125, 135	133
Qβ	1.423 ± 0.014	151 ± 2	125	
BSV	1.246 ± 0.013	172 ± 2		168 ± 7
PM2	0.650 ± 0.007	330 ± 3	319 ± 16	300
T7	0.644 ± 0.007	333 ± 3	330 ± 16	325

TABLE 1. Virus diffusion coefficients and radii (taken from Camerini-Otero et al, 1973a, qv for further details).

angular dependence of conventional light scattering (more detail concerning these latter measurements is given in Camerini-Otero et al, 1973a). When account is taken of the rather large error (say 5-10%) expected in the latter measurements, agreement between the values given in Table 1 is adequate. Nevertheless the values of R_H do appear to be generally a few per cent higher than the other radius values. Several tentative reasons have been given for this: (1) There is considerable evidence that many types of virus particle have open spaces, penetrable by solvent, near their surfaces, (see, for example, Camerini-Otero et al, 1973a). Thus the radii obtained by electron microscopy and X-ray scattering can be underestimates of the true outer radius, due to penetration into these spaces of the staining medium in electron microscopy and "spherical averaging" in the analysis of X-ray data. (2) There is the possibility of one or two monolayers of external hydration of the virus particle. Also, it should be remembered that the derivation of Eq 4 is based on a continuum picture of the solvent. Thus one might expect R_H only to have meaning to within the dimensions of one or two solvent molecules (3-6Å). It should be mentioned that recent investigations by photon correlation spectroscopy of an adenovirus (Oliver et al, 1973), Reovirus (Harvey, 1973) and ribosomes (Koppel, 1973) have found a somewhat larger discrepancy between R_H and the radius obtained by electron microscopy than that shown in Table 1. This is most likely due to distortion of the particles during sample preparation for the electron microscope.

D can be used to calculate macromolecular weights M by combining it with values of the sedimentation rate S and partial specific volume $\bar{v}$ in the Svedberg Equation:

$$M = \frac{N_A kTS}{D(1 - \bar{v}\rho)} \quad , \qquad (5)$$

where N_A is Avogadro's number (see, for example, Tanford, 1961, p 380). Table 2 shows the values of M so calculated for these viruses; also listed for comparison are values obtained by such techniques as sedimentation-equilibrium, conventional light scattering, turbidity, and direct calculation from the known composition. (The turbidity method involves measurement of the attenuation due to macromolecular scattering of a light beam on passage through a sample. This method was first investigated some thirty years ago (see, for example, Oster, 1946), and subsequently fell into disuse. However it has recently been resurrected and refined by Camerini-Otero et al, 1973b, with encouraging results.) Throughout this Table there is almost perfect agreement to within experimental error.

Thus, at least for spherical viruses, there is adequate agreement between values of radii and molecular weights calculated from D and those determined by other techniques. This fact, however, raises a question: If no information is being obtained from measurement of D which cannot be obtained in other ways, what, if any, are the advantages of photon correlation measurements? An attempt to answer this question is given in Tables 3 and 4, where the salient advantages and disadvantages of each technique are listed.

Virus	Sedimentation-Diffusion (a)	Sedimentation-Equilibrium	Turbidity (b)/ Light scattering	Composition (e)
R17	4.02 ± 0.17	3.65 ± 0.17 (b)	3.85 ± 0.25	3.82 ± 0.1
Qβ	4.55 ± 0.16			4.26 ± 0.2
BSV	8.81 ± 0.17			8.80
PM2	47.9 ± 1.7	44.0 ± 2.5 (c)	45.4 ± 2.0	
T7	50.9 ± 1.1	49.4 ± 1.5 (d)	52.2 ± 2.2	

TABLE 2. Virus molecular weights in millions (taken from Camerini-Otero et al, 1973a). (a) Camerini-Otero et al (1973a), (b) Camerini-Otero et al (1973b), (c) Camerini-Otero and Franklin (1973), (d) Bancroft and Freifelder (1970), (e) for further detail see Camerini-Otero et al (1973a).

METHOD	ADVANTAGES	DISADVANTAGES	TYPICAL ACCURACY
From D	- Fast - Accurate - Absolute - Non-perturbative	- No direct shape information	1-2%
Electron-microscopy	- Information on shape and surface structure	- Instrument must be calibrated - Possible distortion of sample during preparation	5% ?
Small-angle X-ray scattering in solution	- Non-perturbative - Potentially information on internal structure	- Time-consuming	5% ?
Angular dependence of conventional light scattering	- Non-perturbative	- No direct shape information - Limited range of particle size	5-10% ?

TABLE 3. Relative merits of methods of measuring macromolecular radii.

METHOD	ADVANTAGES	DISADVANTAGES	TYPICAL ACCURACY
Sedimentation-Diffusion (Svedberg equation)	- Fairly fast - Absolute	- Requires three separate measurements (D, S and $\bar{v}$)	< 5%
Sedimentation-Equilibrium	- Absolute - Requires only $\bar{v}$ and centrifuge measurement	- Time-consuming measurement and interpretation - Somewhat limited range of MW	5%
Conventional light-scattering/ Turbidity	- Fairly fast	- Usually not absolute (turbidity measurements are absolute) - Solution concentration must be accurately known - Must measure dn/dc	5%

TABLE 4. Relative merits of methods of measuring macromolecular weights.

With regard to measurement of radii the major advantage of the D measurement appears to be its speed and accuracy. It is worth noting that light scattering probes detail on a scale comparable to a wavelength of light (~ 5000Å), whereas electron microscopy and X-ray scattering probe detail on a scale several orders of magnitude smaller. Thus with the latter methods one can potentially obtain information, lacking with the former method, on particle shape and internal and surface structure. This is particularly important when studying biological macromolecules whose biological function is frequently determined by structure on the scale of a few Å. On the other hand, photon correlation spectroscopy seems ideally suited to the study of synthetic polymers, where as much interest is centred on the macroscopic conformation of, say, a random-coil, as on its microscopic structure.

With regard to molecular weights, the major disadvantage of the sedimentation-diffusion method seems to be that it requires independent measurements using two expensive pieces of apparatus, a centrifuge and photon correlation apparatus. However, given the requisite apparatus, the method is absolute, fairly quick and interpretation is straightforward.

For spherical particles the values of R_H, M and $\bar{v}$ can be combined to yield another valuable quantity, the degree of solvation of the particle, δ_1. The excess of the "hydrodynamic volume", $4\pi R_H^3/3$, over the "dry volume", $M\bar{v}/N_A$, is assumed to be due to hydrodynamically bound solvent (see, for example, Tanford, 1961, p 340). For the bacteriophages R17, Qβ, PM2 and T7 the degree of solvation calculated in this manner was about 1.10 ± 0.15 ml solvent/gm virus, (Camerini-Otero et al, 1973a). For the plant virus BSV δ_1 was significantly lower, 0.75 ± 0.04. A similar value has recently been found for another plant virus, Turnip Yellow Mosaic Virus, (Harvey, 1973). On the other hand recent studies of two animal viruses, adenovirus (Oliver et al, 1973) and Reovirus (Harvey, 1973), show significantly greater solvations; for adenovirus δ_1 is in excess of 2 ml/gm. It is thus tempting to suppose that viruses have a degree of solvation which reflects their origin: $\delta_1 \approx 0.7$ for plant viruses, $\delta_1 \approx 1.10$ for bacteriophages and $\delta_1 \approx 2$ or greater for animal viruses. Time will judge the correctness of this supposition. It should be noted that, taking into account the fact that the various radii in Table 1 are of similar magnitudes, the virus solvation must be largely internal. Viruses must be thought of as structures containing large internal spaces, frequently as much as 50% or more of the total volume, penetrable by the solvent (Camerini-Otero et al, 1973a).

3 EFFECTS AND CHARACTERIZATION OF MACROMOLECULAR POLYDISPERSITY

Cummins (this publication) has mentioned various approaches to the problem of polydispersity. In this section I will discuss one such approach in some detail (Koppel, 1972; Pusey et al, 1973a; Pusey, 1973) and illustrate it with an experimental example. For simplicity the discussion will again be limited mainly to macromolecules small compared to 1/K.

If several species of macromolecule are present in a solution, $|g^{(1)}(\tau)|$ will no longer be a single exponential. For non-interacting macromolecules it is straightforward to show that $|g^{(1)}(\tau)|$ is given by a sum or distribution of exponentials:

$$|g^{(1)}(\tau)| = \int G(\Gamma)\, e^{-\Gamma\tau}\, d\Gamma \tag{6}$$

where $G(\Gamma)$, the appropriate normalized distribution of decay rates, can be either continuous or discrete. Thus the obvious solution to the polydispersity problem is Laplace inversion of the experimental data to give $G(\Gamma)$. It turns out, however, and I hope this will become apparent shortly, that, in general, data of unattainably high precision are necessary to perform this inversion to a reasonable degree of accuracy. The method we adopt here can be obtained by expanding $\exp(-\Gamma\tau)$ about $\exp(-\bar{\Gamma}\tau)$ where $\bar{\Gamma}$ is the mean of $G(\Gamma)$,

$$\bar{\Gamma} = \int \Gamma\, G(\Gamma)\, d\Gamma \, . \tag{7}$$

From Eq 6,

$$|g^{(1)}(\tau)| = \exp(-\bar{\Gamma}\tau) \left[1 + \frac{\mu_2\tau^2}{2!} - \frac{\mu_3\tau^3}{3!} + \frac{\mu_4\tau^4}{4!} + \ldots\right] \tag{8}$$

where μ_2, μ_3 etc are the moments about the mean of $G(\Gamma)$,

$$\mu_2 = \int (\Gamma - \bar{\Gamma})^2\, G(\Gamma)\, d\Gamma \qquad \text{etc.} \tag{9}$$

Thus

$$\ell n[C^{\frac{1}{2}}\, |g^{(1)}(\tau)|] = \tfrac{1}{2}\,\ell n\, C - \bar{\Gamma}\tau + \frac{1}{2!}\frac{\mu_2}{\bar{\Gamma}^2}(\bar{\Gamma}\tau)^2$$

$$- \frac{1}{3!}\frac{\mu_3}{\bar{\Gamma}^3}(\bar{\Gamma}\tau)^3 + \frac{1}{4!}\left(\frac{\mu_4 - 3\mu_2^2}{\bar{\Gamma}^4}\right)(\bar{\Gamma}\tau)^4 + \ldots \, . \tag{10}$$

Eq 10 is exact if all the terms in the expansion are kept. For terms higher than the cubic term, it is the cumulants rather than the moments of $G(\Gamma)$ which enter Eq 10. In fact $\ell n\ |g^{(1)}(\tau)|$ is the cumulant generating function of $G(\Gamma)$, (Koppel, 1972).

For a polydisperse system, therefore, $\ell n\ C^{\frac{1}{2}}\ |g^{(1)}(\tau)|$ is a power series in τ rather than the linear function obtained in a monodisperse system. In a typical experiment, the longest delay time used will be given by $\bar{\Gamma}\tau_{max} \leqslant 4$. The usefulness of this data analysis method arises from the fact that, for many $G(\Gamma)$'s, the terms in Eq 10 fall off rapidly with increasing order over this range of τ. For such systems the observed correlation function can usually be described to within experimental error by the first three or four terms of Eq 10. A direct fit of the data to a polynomial will then yield values of $\bar{\Gamma}$ and one or two higher moments of $G(\Gamma)$.

The data analysis procedure is complicated by the fact that it is not a priori obvious what order of polynomial should be used to describe a given set of experimental data. This will depend on various factors, among them the degree of polydispersity, the statistical accuracy of the data and the value of $\bar{\Gamma}\tau_{max}$. If one uses a polynomial of too low order, one will obtain values of $\bar{\Gamma}$ and μ_2, etc, with relatively small statistical error but with a large systematic error, since here data is being force-fitted to a certain polynomial when it would be better described by one of higher order. On the other hand, if one uses a polynomial of too high order, one will obtain statistical errors on some of the parameters larger than the magnitude of the parameters themselves. There is therefore a trade-off between systematic and random errors. We have adopted an approach involving taking data over a range of $\bar{\Gamma}\tau_{max}$ and fitting each set of data to linear, quadratic and cubic functions of τ. The exact procedure will be discussed further below, in connection with the experimental example.

An advantage of this data-analysis method is that, in many cases, the information is obtained in immediately useful form. For non-interacting molecules small compared to $1/K$, the mean intensity of light scattered by N macromolecules of molecular weight M is proportional to NM^2 (see, for example, Tanford, 1961, p 278). Thus for a solution containing several species i of macromolecules,

$$G(\Gamma_i) = \frac{N_i M_i^{\ 2}}{\sum N_i M_i^{\ 2}} \quad . \tag{11}$$

Here we are using, for convenience, a discrete notation. We have assumed each species to have the same refractive index increment. Defining an average diffusion constant by $\bar{D} \equiv \bar{\Gamma}/K^2$ and taking $\Gamma_i = D_i K^2$, we have

$$\bar{D} = \frac{\sum N_i M_i^2 D_i}{\sum N_i M_i^2} \quad . \tag{12}$$

Thus from the mean decay rate $\bar{\Gamma}$ of the correlation function one obtains the average diffusion coefficient defined by Eq 12. It is a fortunate coincidence of nature that this average D, the so-called z-average, is a useful quantity. When combined in the Svedberg equation (Eq 5) with the easily measured weight-average sedimentation rate $\bar{S}_W$, it yields the useful weight-average molecular weight $\bar{M}_W$, assuming the partial specific volume $\bar{v}$ to be the same for all species (Koppel, 1972; Pusey et al, 1973a). This fact allows, for the first time, accurate use of the sedimentation-diffusion method of determining molecular weights of polydisperse solutes.

For systems described by Eq 11, $\mu_2/\bar{\Gamma}^2$ is the z-average normalized variance of the distribution of D's, and thus provides a measure of the degree of polydispersity of the sample. For relatively narrow $G(\Gamma)$'s $\mu_2/\bar{\Gamma}^2$ can be related to the more usual indexes of polydispersity, the ratios of the molecular weight averages, $\bar{M}_Z/\bar{M}_W$ and $\bar{M}_W/\bar{M}_N$ (Pusey, 1973). For a random-coil polymer sample with any $G(\Gamma)$, $(\mu_2/\bar{\Gamma}^2) + 1$ is roughly equal to the less common index of polydispersity, $\bar{M}_W/\bar{M}_V$ (Pusey, unpublished), where $\bar{M}_V$ is the viscosity-average molecular weight (see, for example, Flory, 1953, p 313).

Recently we have studied, by photon correlation spectroscopy, some polydisperse samples of random-coil polystyrene in cyclohexane, using the data-analysis approach outlined above (Brown et al, 1973; Pusey, 1973). We chose this system because synthetic polymers, unlike many homogeneous biological systems, are by nature polydisperse. Thus, if photon correlation spectroscopy is to be of any use in the study of such systems, it is essential to be able to detect and characterize sample polydispersity. The samples studied had been characterized fairly well by other techniques such as sedimentation-equilibrium and conventional light scattering (Dietz, 1973). The major aim of the work was to test the use of the sedimentation-diffusion method of molecular weight determination for polydisperse systems. The sample concentrations were typically 1 mg polystyrene/ml solution. Experiments were performed at scattering angles $\theta = 60^{\circ}$, 90° and 120° and at a temperature of $35^{\circ}C$, the theta temperature for this system, where the effect of intermolecular interactions is minimal. Nevertheless the effect of interactions in these expermments, though small, was not entirely negligible with respect to experimental error. This point has been discussed elsewhere (Brown et al, 1973) and will not be pursued further here. Suffice it to say that neglect of this effect should not seriously affect our conclusions.

The samples studied are listed in Table 5. Sample 1 was a sharp fraction of polystyrene of molecular weight $\bar{M}_W = 4.1 \pm 0.2 \times 10^5$. This virtually monodisperse sample was studied as a test of our experimental and data-analysis procedures. Fig 4 shows a semi-log plot of a typical measured correlation function. As expected, the points lie fairly well on a straight line. Fig 5 shows the first numerical derivative with respect to τ of the data of Fig 4. Here we have taken all possible sets of seven consecutive data points in Fig 4, performed fits of these sets to straight lines,

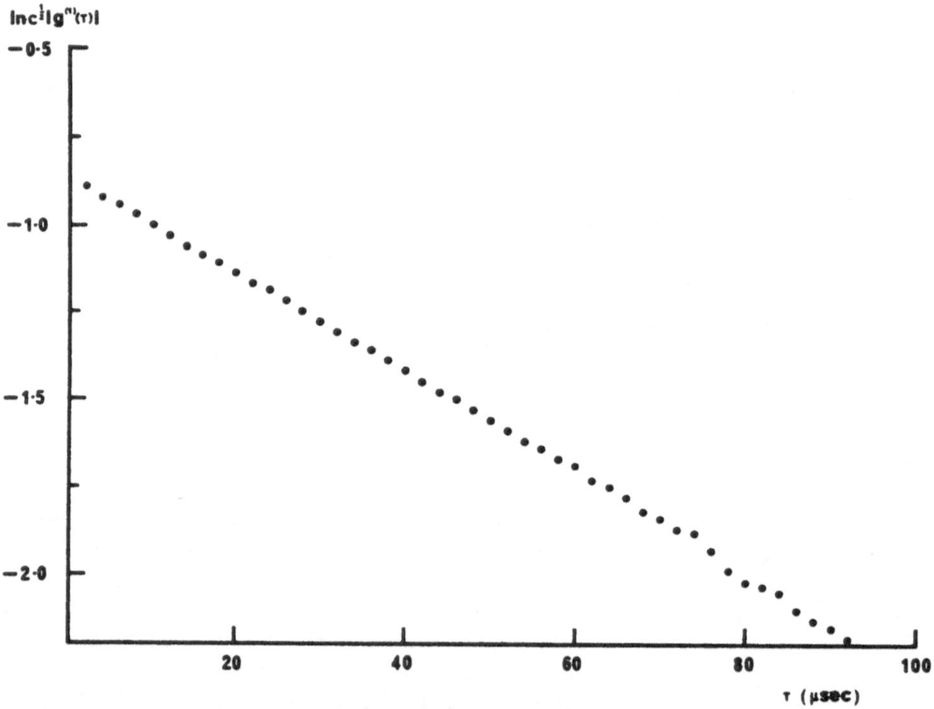

FIG 4. Semi-log plot of electric field correlation function for a sharp fraction of polystyrene ($\bar{M}_W = 4.1 \times 10^5$) in cyclohexane at $35^{\circ}C$, (taken from Brown et al, 1973).

SAMPLE	$\bar{M}_W \times 10^{-5}$ [a] (Sedimentation-equilibrium, light scattering etc)	$\bar{M}_Z/\bar{M}_W$ [a] (Sedimentation-equilibrium)	$\bar{D}_Z \times 10^7$ (Photon correlation spectroscopy)	$\bar{M}_W \times 10^{-5}$ (Sedimentation-Diffusion)	$\frac{\mu_2}{\bar{\Gamma}^2}$
1 "Monodisperse"	4.1 ± 0.2	< 1.1	2.07 ± 0.02	4.1 ± 0.2	0.026 ± 0.02
2 Mixture of 37.5% $\bar{M}_W = 4.1 \times 10^5$ 62.5% $\bar{M}_W = 0.51 \times 10^5$	1.86 ± 0.1	1.87 (calculated)	2.55 ± 0.05	1.98 ± 0.1	0.25 ± 0.05
3 "Polydisperse"	2.54 ± 0.13	1.24 ± 0.05	2.56 ± 0.03	2.76 ± 0.16	0.07 ± 0.02

TABLE 5. Molecular weights, diffusion coefficients etc for random-coil polystyrene in cyclohexane (taken from Brown et al, 1973; see also Pusey, 1973).
(a) Dietz, 1973.

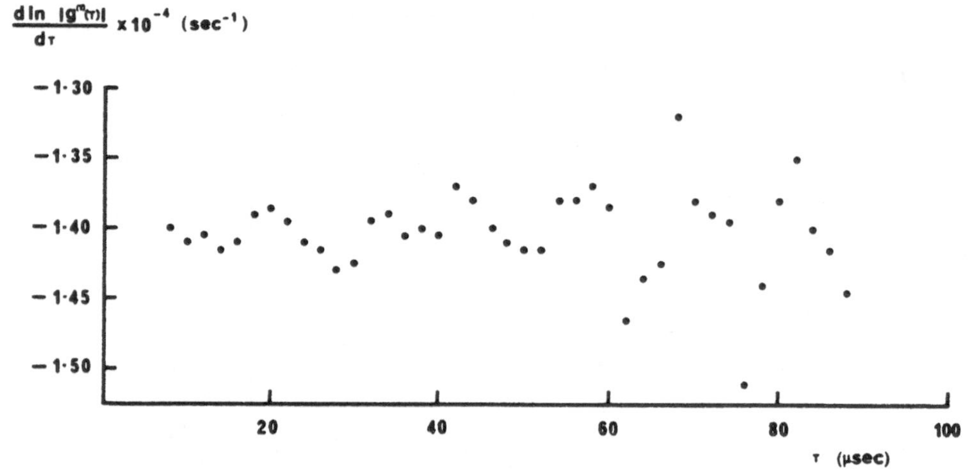

FIG 5. First numerical derivative of data of Fig 4, (taken from Brown et al, 1973).

and plotted the slopes so obtained against the average delay time of each seven-point segment. It can be seen from Eq 10 that for a monodisperse sample, μ_2, μ_3 etc = 0, this plot should be a horizontal line. Indeed the data in Fig 5 show no systematic trend away from the horizontal. The oscillations seen in this plot do not reflect a real effect, but are an artifact of the procedure of numerical differentiation. Despite this evidence of monodispersity we analyzed the data by fitting $\ell n\ C^{\frac{1}{2}}\ |g^{(1)}(\tau)|$ to a quadratic in τ. The values of $\bar{D}_Z$ and $\mu_2/\bar{\Gamma}^2$ given in Table 5 are the averages of several experimental runs. The result $\mu_2/\bar{\Gamma}^2$ = 0.026 ± 0.02 indicates a barely detectable degree of polydispersity. The value of $\bar{M}_W^{S-D}$ obtained from the Svedberg equation is in excellent agreement with that obtained by other techniques.

Samples 2 and 3 were true polydisperse samples. Sample 2 was a mixture of sample 1 and another narrow fraction of polystyrene in the actual weight proportions 40% of MW 4.1 x 10^5 and 60% MW 5.1 x 10^4. This sample was studied mainly at scattering angle $\theta = 90^o$. At this angle the random-coils are not entirely negligible in size with respect to 1/K. It has been shown elsewhere, however, that the particles in this sample can be regarded as point particles and the theory outlined above applied directly, if we take the weight proportions to be 37.5:62.5 (Pusey, 1973). This approach will be adopted here. This sample was chosen because the individual sharp fractions had been well characterized by other techniques, so that such quantities as $\bar{S}_W$ and $\bar{M}_W$ for the sample could be calculated directly from those of the constituents.

Fig 6 shows a typical correlation function for sample 2. This plot shows distinct upward curvature. Fig 7 is the numerical derivative of the data of Fig 6. In contrast to Fig 5, there is, in this plot, a distinct upward trend. Figs 6 and 7 therefore give definite evidence of polydispersity. It can be seen from Eq 10 that the initial slope of the plot of Fig 7 is a measure of $\mu_2/\bar{\Gamma}^2$.

The data for sample 2 were analyzed as follows: some 40 experimental runs were performed spanning a range of total delay time $0.4 \lesssim \bar{\Gamma}\tau_{max} \lesssim 4$. Routinely $\ell n\ C^{\frac{1}{2}}\ |g^{(1)}(\tau)|$ was least-squares fitted to linear, quadratic and cubic functions of τ. The values of the parameters $\bar{\Gamma}$, μ_2 etc obtained from these fits are then

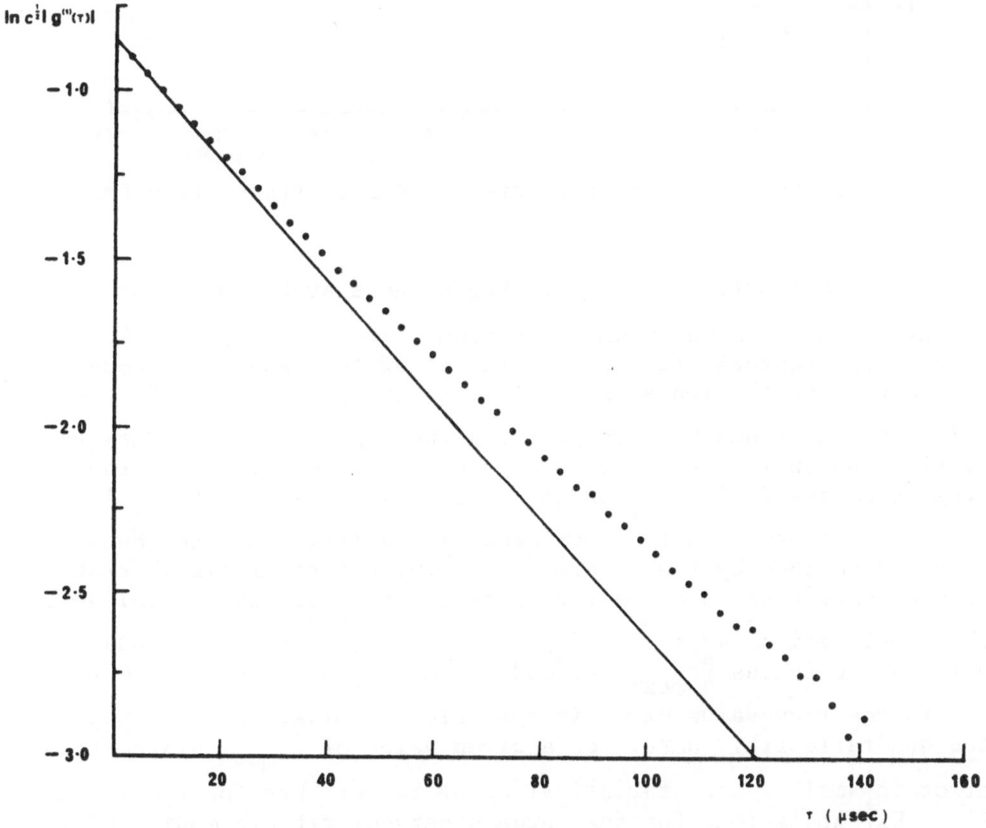

FIG 6. Semi-log plot of correlation function for a mixture of two sharp fractions of polystyrene in cyclohexane, (taken from Brown et al, 1973).

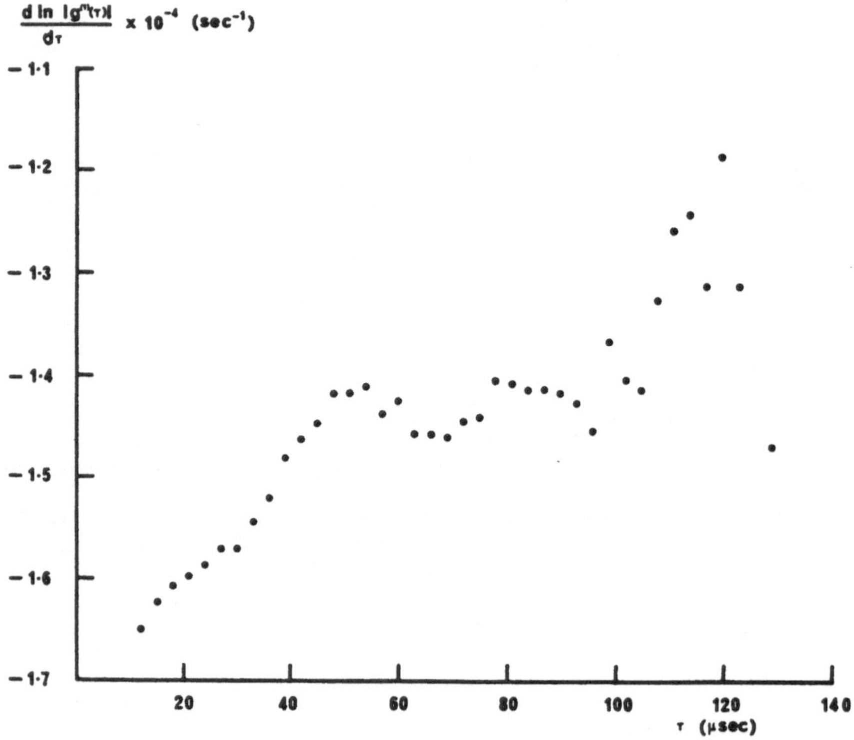

FIG 7. First numerical derivative of Fig 6, (taken from Brown et al, 1973).

plotted as a function of $\bar{\Gamma}\tau_{max}$. Fig 8 shows such a plot for the values of $\bar{\Gamma}/K^2$ obtained from the various fits. This plot illustrates many of the features (and difficulties) of this analysis procedure. Consider first the linear fit. At, for example, $\bar{\Gamma}\tau_{max} = 2$, the value of $\bar{\Gamma}$ obtained by force-fitting the non-exponential data to a single exponential has a systematic error of some 12%. On the other hand statistical error is relatively small. As $\bar{\Gamma}\tau_{max}$ is decreased the systematic error becomes smaller, since the data are better described by fewer terms in Eq 10, but statistical error becomes larger since a smaller range in $\bar{\Gamma}\tau_{max}$ is spanned (Oliver, this publication; Koppel, 1972). Nevertheless the plot tends towards a linear region as $\bar{\Gamma}\tau_{max} \to 0$, and extrapolation to $\bar{\Gamma}\tau_{max} = 0$ to obtain the true value of $\bar{\Gamma}$ is possible. Similar comments apply for the quadratic fit. Here, for a given value of $\bar{\Gamma}\tau_{max}$, systematic error is smaller and statistical error larger than for the linear fit. Extrapolations for the three orders of fit are indicated in Fig 8, leading to the result $\bar{D}_Z = 2.55 \pm 0.05 \times 10^{-7}$ cm^2/sec. The

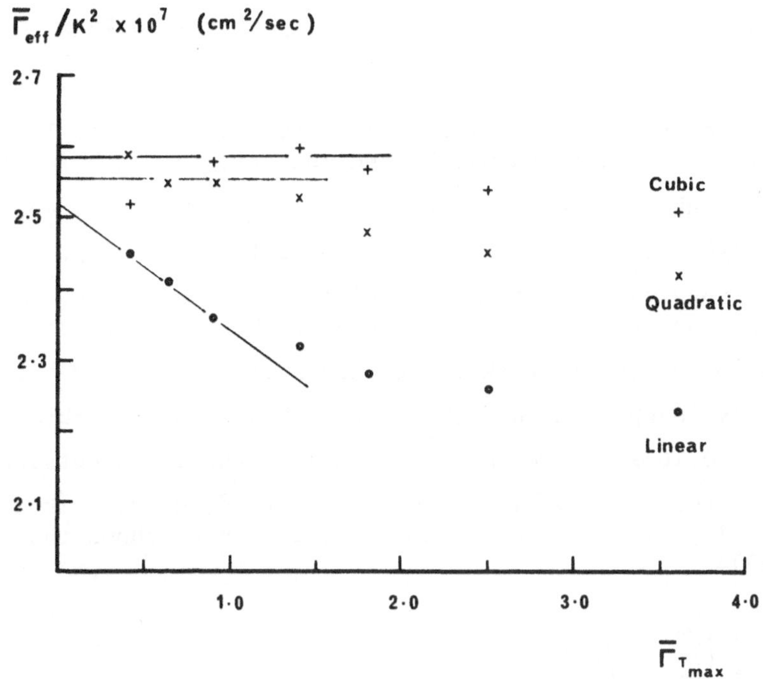

FIG 8. $\bar{\Gamma}_{eff}/K^2$ as a function of $\bar{\Gamma}\tau_{max}$ for sample 2 (see Table 5) for linear, quadratic and cubic fits of the data. Systematic error decreases and random error increases from bottom right to top left in this figure, (taken from Brown et al, 1973).

error quoted here, 2%, is larger than that for the monodisperse sample 1, 1%, since it takes into account uncertainty inherent in the extrapolation procedure.

A similar analysis can be followed to obtain $\mu_2/\bar{\Gamma}^2$. For this sample we found $\mu_2/\bar{\Gamma}^2 = 0.25 \pm 0.05$, indicating as expected significant polydispersity. Even with a sample as polydisperse as sample 2 statistical error in μ_3 was large. Our results indicated $\mu_3 > 0$ but little can be said with any confidence about its magnitude.

The value of the weight-average molecular weight calculated from the Svedberg equation is $1.98 \pm 0.1 \times 10^5$, which overlaps the value $1.86 \pm 0.1 \times 10^5$ calculated from the known proportions and molecular weights of the individual components. For comparison, the calculated values of $\bar{M}_N$ and $\bar{M}_Z$ were respectively 0.76 x 10^5 and 3.48 x 10^5.

Sample 3 contained a continuous distribution of molecular weights. Studies on this sample by sedimentation-equilibrium and conventional light scattering gave $\bar{M}_W = 2.54 \pm 0.13 \times 10^5$, and a degree of polydispersity $\bar{M}_Z/\bar{M}_W = 1.24 \pm 0.05$ (Dietz, 1973). Several experimental runs were performed on this sample over a range of $\bar{\Gamma}\tau_{max}$. In all cases the data could be described by a quadratic in τ, as evidenced by the fact that the values of $\bar{\Gamma}$ and μ_2 so obtained were virtually independent of $\bar{\Gamma}\tau_{max}$. For this sample we obtained $\bar{M}_W^{S-D} = 2.76 \pm 0.16 \times 10^5$ which again overlaps the value obtained by other means. For relatively narrow molecular weight distributions of random-coil polymers such as sample 3, it can be shown that (Pusey, 1973)

$$\frac{\mu_2}{\bar{\Gamma}^2} \approx \frac{1}{4}\left(\frac{\bar{M}_Z}{\bar{M}_W} - 1\right). \tag{13}$$

The values of $\mu_2/\bar{\Gamma}^2$ and $\bar{M}_Z/\bar{M}_W$ listed in Table 5 fulfil this prediction.

These results indicate, therefore, that, using the data analysis approach outlined above, the sedimentation-diffusion method of molecular weight determination can be used in polydisperse systems. Photon correlation spectroscopy apparently measures the z-average diffusion coefficient. However, the results also illustrate the insensitivity of photon correlation spectroscopy to sample polydispersity. Even for the quite polydisperse sample 2, we were able to obtain only two useful parameters, $\bar{\Gamma}$ and μ_2, with any degree of precision. Refinements of experimental technique can, no doubt, improve the situation somewhat. However, the knowledge of even three or four such parameters is far from providing a complete characterization of $G(\Gamma)$.

Throughout this section it has been assumed that the macromolecules under study are small compared to 1/K. In theory this condition can be fulfilled by using small scattering angles θ, although one can expect to encounter experimental problems due to scattering from dust, spurious flare, etc, at small angles. Alternatively the theory could be extended to include particle scattering factors in Eq 11, though this will probably make the method model-dependent to some degree.

Cummins (this publication) has mentioned other methods which have been suggested and used to study polydispersity by intensity fluctuation spectroscopy. I will not discuss these further except to say that it is certainly possible that these methods could have advantages over the moments-analysis method in certain specific situations, particularly if some a priori knowledge of the form of $G(\Gamma)$ is available. For instance if it is known that a sample contains, say, only two, individually monodisperse, components a direct fit of the data to the sum of two exponentials may prove viable (see, for example, Foord et al, 1970). The advantages of the moments-analysis approach appear to be: (1) It is not necessary to assume a priori a form for $G(\Gamma)$; (2) Well-defined quantities, $\bar{D}_Z$, $\mu_2/\bar{\Gamma}^2$, etc are obtained; and (3) Weight-average molecular weights can be determined using the Svedberg equation.

For completeness, it should be mentioned that the moments-analysis approach can be adapted to study other non-exponential correlation functions besides those arising from macromolecular polydispersity (Pusey et al, 1973a). Firstly, even if one has good reason to expect a single-exponential correlation function in a given experiment, it is good practice to test each set of data for single-exponential behaviour. A quadratic fit of $\ell n\ C^{\frac{1}{2}}\ |g^{(1)}(\tau)|$, provides such a test, in that, only if μ_2 is found to be zero to within experimental error, can the data be said to fit a single exponential. In this way such spurious effects as light scattered by dust and flare from the cell walls, acting as a small heterodyne signal, can be detected (see, also, Oliver, this publication). Secondly, the method can be adapted to the study of rotational and internal motion effects where multi-exponential correlation functions are expected. Indeed, the frequency-domain analogue of the moments-analysis method has been used in the light-scattering study of Tobacco Mosaic Virus (Schaefer et al, 1971).

To conclude this section one very recent approach to the polydispersity problem, namely a combination of photon correlation spectroscopy with band sedimentation (Koppel, 1973) will be mentioned. In biological systems, in particular, one frequently encounters polydisperse samples which consist of a few individually monodisperse components. Sedimentation of such a sample in a density gradient, usually a sucrose solution, will cause separation of the components into physically distinct bands. This technique is often used in sample preparation, where each band is collected separately. Koppel has taken "the next logical step: the light scattering analysis is carried out immediately after centrifugation directly on the different macromolecular bands in the sucrose gradient in the centrifuge tube". The intensity of the light scattered by each band allows determination of the macromolecular concentration, and the individual diffusion coefficients can be determined by photon correlation spectroscopy. Koppel has applied this method successfully to a study of E coli ribosomes.

Thus there appears to be a pattern emerging concerning the study of polydisperse systems by photon correlation spectroscopy. A direct measurement on the sample is relatively insensitive to the polydispersity yielding only a few moments of G(Γ). Nevertheless these latter are useful quantities which can be obtained quite quickly. However, for systems described by discrete G(Γ)'s, application of an external field, either electric (see, for example, Cummins, this publication) or gravitational, can provide considerably more information. The relative merits of the electrophoresis and band sedimentation methods remain to be fully evaluated. With the present state of the art electrophoresis on charged systems gives results quickly and simply, but band sedimentation experiments appear to be more amenable to useful quantitative analysis. It also remains to be established whether application of external fields has any advantages in the study of systems described by continuous G(Γ)'s.

4 INTERACTIONS IN MACROMOLECULAR SOLUTIONS

4.1 Introduction

Interactions in macromolecular solutions are frequently regarded simply as a nuisance whose effect can be removed or minimized by extrapolating measurements of, for instance, diffusion coefficients or sedimentation coefficients to zero solution concentration, thus giving single-particle values. Obviously, however, there is information to be obtained from the study of such solutions at finite concentration. To mention but two of a host of possible examples, intermacromolecular interactions play an important role in the stability of colloidal systems and the formation of liquid crystals in binary systems. Consider, then, an aqueous suspension of spherical colloidal particles of polystyrene of diameter a few thousand Å. At neutral and basic pH's such particles typically carry a charge of several thousand electronic charges. At low solution concentrations the particles are, on average, far enough separated that the effect of interactions is small, and each particle executes its individual Brownian motion. As the solution concentration is increased the particles must start to interact through long-range repulsive coulombic forces whose energy can exceed the thermal energy kT. Indeed at concentrations of 5-10% solids, where the mean distance between particles is only a factor of two or so greater than the particle diameter, the free energy of the solution is minimized by the formation of a liquid-crystal-like structure. Here each particle is constrained to a particular site in a close-packed array (see, for example, Hiltner and Krieger, 1969). Diffusion is thus completely hindered; the sole effect of the random thermal forces should be coupled displacements of the particles about their mean position. On the other hand, if an attempt is made at any solution concentration to suppress the coulombic interaction by the addition of electrolyte (eg 1M NaCl), then coagulation of the polystyrene particles can occur. Here the salt causes strong screening of the electrostatic forces, greatly reducing their range. Two particles can then approach each other close enough that the strong short-range Van der Waals attraction can take over, causing coagulation. This simple example should provide motivation for a more detailed look at interparticle interactions.

In this section I will give a far from complete discussion of intermacromolecular interactions ranging from simple theoretical ideas, using both macroscopic and microscopic approaches, to the results of recent laboratory and computer experiments.

4.2 Theoretical Considerations

For simplicity the discussion will be limited mainly to two types of interaction, direct hard-sphere interactions and repulsive coulombic interactions such as those mentioned above. It will be argued that coulombic interactions can frequently be described to a first approximation by an effective hard-sphere interaction. In what follows it will be instructive to keep in mind the relative magnitudes of three lengths relevant to the problem. These lengths are (Egelstaff, 1970; Pecora, 1970): (1) $1/K$, the reciprocal of the scattering vector. This quantity is a measure of the spatial scale being probed by a light-scattering experiment; (2) R, the range of interaction, which can loosely be defined as the spatial range over which the macromolecular pair correlation function is significantly different from one; and (3) $L \equiv (V/N)^{1/3}$, a measure of the interparticle spacing; here N is the number of particles contained in volume V.

We start by considering the situation $R \ll 2\pi/K$ or $RK \ll 2\pi$. (For the effect of interactions on the measured quantities to be appreciable, L must not be too much larger than R.) This regime appears to be directly analogous to the hydrodynamic regime discussed by Lallemand, Pike and Swinney (this publication) in connection with pure fluids and critical systems. We can therefore use a macroscopic thermodynamic/hydrodynamic approach to the problem. By analogy with a binary liquid mixture, we expect the mean scattered intensity to be given by

$$\langle I \rangle \quad \propto \quad \left(\frac{\partial\mu}{\partial c}\right)^{-1} \tag{14}$$

and the electric field correlation function to be given by Eqs 1 and 2 with diffusion coefficient given by

$$D \quad \propto \quad \left(\frac{\partial\mu}{\partial c}\right)/f \, . \tag{15}$$

Here μ is the chemical potential, c the concentration and $(\partial\mu/\partial c)$ is the "driving force" for diffusion. Eqs 14 and 15 are exactly those used to describe conventional light scattering and macroscopic diffusion in macromolecular systems (see, for example, Tanford, 1961, pp 285 and 348). Thus in the limit $KR \ll 2\pi$ we expect the electric field correlation function measured by photon correlation spectroscopy to reflect simple local diffusive behaviour (Eqs 1 and 2) and to give a value of D identical with that which would be measured by macroscopic boundary spreading techniques. This fact is usually assumed in photon correlation spectroscopy.

Eqs 14 and 15 are expected to apply for any type of interaction provided $KR \ll 2\pi$. Still following a thermodynamic approach let us consider as an example the first-order departure from zero-concentration behaviour in a solution of hard spherical macromolecules. The chemical potential of the solution can be calculated from the entropy of mixing, which, for this simple system, is just related to excluded volume due to the finite size of the particles (see, for example, Tanford, 1961, p192; also Cummins, this publication). Thus

$$\langle I \rangle \quad \propto \quad \langle I \rangle_o \; (1 + 8\Phi + \ldots\ldots)^{-1} \tag{16}$$

and

$$D \quad = \quad \frac{kT}{f} \; (1 + 8\Phi + \ldots\ldots). \tag{17}$$

Here Φ is the fraction of solution volume occupied by the macromolecules; it it clearly proportional to the macromolecular concentration. The subscript zero on a quantity indicates the value of that quantity in the absence of interactions. The factor 8 in these equations arises from the fact that a given particle in dilute solution excludes from occupation by the centre of mass of another particle a spherical volume of radius equal to the particle diameter.

To complete the calculation of the concentration dependence of D, the concentration dependence of f must be determined. It appears that this calculation must be approached from a microscopic (particle) viewpoint. There have been several treatments of this problem over the last thirty years which can be applied directly to a dilute solution of spherical particles. The result is

$$f = f_o(1 + K_f\Phi + ...) \tag{18}$$

where the predicted value of K_f ranges from 6.55 (Batchelor, 1972) through 6.86 (Burgers, 1941, 1942) to 7.2 (Pyun and Fixman, 1964). Combining Eqs 17 and 18

$$D = D_o[1 + (8 - K_f)\Phi + ...]. \tag{19}$$

Assuming $K_f \approx 7$, it can be seen that, for hard spherical molecules, the individual effects on the concentration dependence of D of $(\partial\mu/\partial c)$ and f almost cancel. Thus D in such systems can be expected to show a much smaller concentration dependence than, for instance, conventional light scattering or the sedimentation coefficient (which is proportional to 1/f).

It is of interest to attempt a derivation of Eqs 16 and 17 using a particle approach. This can be done, in part, following the approach of Komarov and Fisher (1963). Consider a large number N of particles in a scattering volume V. The electric field correlation function can be written

$$|\langle E^*(0)E(\tau)\rangle| = AP(K)\left\langle\sum_{i=1}^{N}\sum_{j=1}^{N} e^{i\underline{K}.[\underline{r}_i(0)-\underline{r}_j(\tau)]}\right\rangle$$

$$\equiv NAP(K)\int d\underline{r}\; e^{i\underline{K}.\underline{r}}\, G(\underline{r},\tau) \quad , \tag{20}$$

where A is a constant for a given experiment, P(K) is the single-particle scattering factor (see, for example, Cummins, this publication), $\underline{K}$ is the scattering vector and $\underline{r}_i(t)$ is the position of the centre of mass of particle i at time t. The van Hove space-time correlation function $G(\underline{r},t)$ is the probability that, given a particle at the origin at time zero, a particle (possibly the same one) will be found at point $\underline{r}$ at time t. As is commonly done in neutron scattering (see, for example, Berne, 1971), it is instructive to divide this expression into the self and distinct parts:

$$|\langle E^*(0)E(\tau)\rangle| = AP(K) \left\langle \sum_{i=1}^{N} e^{i\underline{K}.[\underline{r}_i(0)-\underline{r}_i(\tau)]} \right\rangle$$

$$+ AP(K) \left\langle \sum_{\substack{i=1 \\ i\neq j}}^{N} \sum_{j=1}^{N} e^{i\underline{K}.[\underline{r}_i(0)-\underline{r}_j(\tau)]} \right\rangle \qquad (21)$$

$$\equiv NAP(K) \int d\underline{r}\, e^{i\underline{K}.\underline{r}} [G_s(\underline{r},\tau) + G_d(\underline{r},\tau)] \ .$$

For non-interacting particles the distinct (second) term is zero, since different particle positions are uncorrelated at all times, and, as shown by Cummins (this publication), the first term gives, for Brownian particles,

$$|\langle E^*(0)E\ (\tau)\rangle| = NAP(K)\ \exp(-DK^2\tau) \ , \qquad (22)$$

since, for this system,

$$G_s(\underline{r},\tau) = (4\pi D\tau)^{-3/2} \exp(-r^2/4D\tau).$$

In the presence of interparticle interactions the distinct term is obviously not zero. Let us consider first the average scattered intensity, ie the value of Eq 21 at $\tau = 0$. It is straightforward to show (see, for example, Komarov and Fisher, 1963) that

$$\langle I\rangle = NAP(K) \left\{ 1 + \frac{4\pi N}{V} \int [\,g(r) - 1] \ \frac{\sin Kr}{Kr}\, r^2\, dr \right\} \qquad (23)$$

where g(r) is the macromolecular pair correlation function. This result was first derived by Zernicke and Prins (1927) in connection with X-ray scattering. It was first applied to macromolecular solutions by Doty and Steiner (1949) and Oster (1949); see, also, Doty and Edsall (1951) and Doty and Steiner (1952). The first term in parentheses in Eq 23 is the contribution of the self term, and has the same value as for non-interacting particles (see Eq 22). Thus the effect of interactions on the mean scattered intensity arises entirely from the non-zero distinct term. We will calculate this effect for the same example as above, namely a dilute solution of hard spheres. For such a system, to a first approximation,

$g(r) = 0$ for $r < R$, the hard sphere diameter, and $g(r) = 1$ for $r > R$. Then Eq 23 gives, in powers of KR,

$$\langle I \rangle = NAP(K) \left\{ 1 - \frac{4\pi}{3} R^3 \frac{N}{V} + \frac{4\pi}{3} R^3 \frac{N}{V} \frac{(KR)^2}{10} + \ldots \right\} . \qquad (24)$$

The first correction term in this expansion is just the excluded volume effect calculated by the thermodynamic approach used above, (Eq 16). Interestingly, however, this approach shows that, for KR not negligible with respect to unity, the thermodynamic approach breaks down as expected and the mean scattered intensity becomes dependent on K and hence scattering angle. Eq 24 shows that the first order correction due to repulsive inter-particle interactions gives an <u>increasing</u> intensity with <u>increasing</u> K. This can be contrasted with the effect of finite particle size, the single-particle scattering factor P(K), which invariably gives an effect in the opposite direction.

A calculation of the time dependence of the correlation function in the presence of interactions using a particle approach does not appear to have been performed. Nevertheless several relevant comments can be made. Firstly it is still instructive to consider separately the self and distinct terms in Eq 21. In the presence of short range, $KR \ll 2\pi$, (but not necessarily weak compared to kT) interactions a given particle is presumably still performing a random-walk diffusion of many small steps through the solvent. Thus one can still expect the time-dependence of the self term to be given by an equation of the same form as Eq 22 with, however, a value of D which depends on the interaction. Calculation of the time dependence of the distinct term in Eq 21 will in general be more complicated. One possible approach is to use the Vineyard (1958) approximation (see, also, Berne, 1971). Here the distinct part of the van Hove correlation function is expressed by the identity

$$G_d(\underline{r},\tau) = \int d\underline{r}' \; g(\underline{r}') \; H(\underline{r},\underline{r}',\tau) \qquad . \qquad (25)$$

The pair correlation function $g(\underline{r}')$ is the probability that, given a particle at the origin at time 0, another particle will be found at point $\underline{r}'$ at time 0. $H(\underline{r},\underline{r}',\tau)$ is the probability that this second particle has then moved to point $\underline{r}$ at time τ. The Vineyard approximation consists of setting

$$H(\underline{r},\underline{r}',\tau) = G_s(\underline{r}-\underline{r}',\tau) \qquad , \qquad (26)$$

thus ignoring the effect of the particle at the origin on the subsequent motion of the second particle. This approximation therefore relates the time-dependence of the distinct term to that of the self term. Eqs 21, 25 and 26 then lead, in a straightforward manner, to the result:

$$|\langle E^*(0)E(\tau)\rangle| \approx \langle I\rangle \exp(-D_s K^2 \tau) \quad . \tag{27}$$

Here <I> is the mean intensity given by Eq 23 and D_s is the self-diffusion coefficient in the presence of interactions. The Vineyard approximation fails for small $\underline{r}'$, but can be expected to be quite good for paths in which $\underline{r}'$ is large (see, for example, Berne, 1971). Therefore, for short-range interactions Eq 27 should hold. Thus the particle approach does predict a single diffusive mechanism for the decay of $|\langle E^*(0)E(\tau)\rangle|$ in the limit KR << 2π, in agreement with Eq 15. However a calculation of the actual value of D_s using a particle approach appears to be lacking.

4.3 Some Experimental Results

With this theoretical background let us turn to some experimental results. Fig 9 shows the results of photon correlation

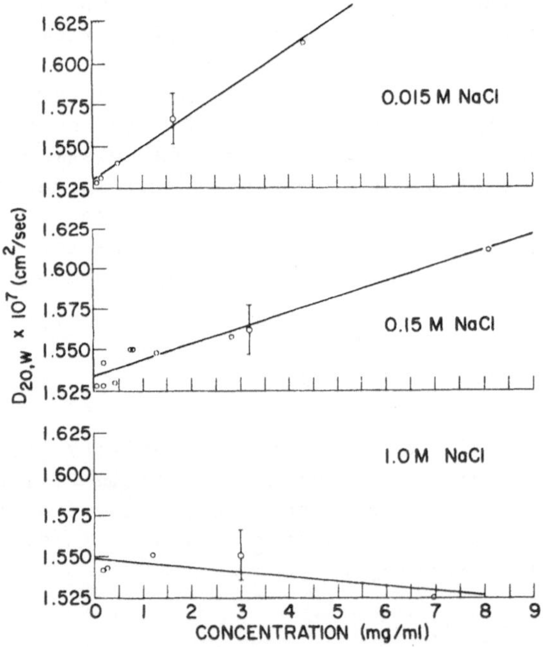

FIG 9. Diffusion coefficient of R17 virus as a function of virus concentration and solution ionic strength, (taken from Pusey et al, 1972).

spectroscopy measurements again on the spherical virus R17 (Pusey et al, 1972; Camerini-Otero et al, 1973a). Here the maximum virus concentration is some ten times greater than that shown in Fig 3. Experiments were performed at several different electrolyte concentrations. Away from the isoelectric point, the R17 particle can be expected to carry a charge of many electron units. This charge will be shielded by a diffuse atmosphere of small ions of opposite charge (see, for example, Verwey and Overbeek, 1948). The spatial extent of this atmosphere will depend on the solution ionic strength. Particles separated at any instant by more than their diffuse layer dimensions will not exert a significant mutual force. On the other hand, when the ionic atmospheres overlap a repulsive force can be expected; on sufficiently close approach the energy of interaction can greatly exceed the thermal energy kT. Such close approaches under the influence of Brownian motion are therefore extremely unlikely. Thus it seems reasonable that, to a first approximation, the coulombic interaction can be represented by an effective hard-sphere interaction, with effective hard sphere radius roughly equal to the mutual separation at which the interaction energy is equal to kT (see, also, Doty and Steiner, 1952; Burchard and Cowie, 1972). Returning to Fig 9, at 1 M NaCl the extent of the ionic atmosphere will be quite small compared to the virus radius ~ 140Å; the Debye-Hückel shielding length for 1 M NaCl is about 3Å. Thus, at this salt concentration, the virus solution should behave as a solution of "true" hard spheres and Eq 19 should apply. In fact, a least-squares fit to the data for 1 M NaCl gives a slight downward trend for D as a function of virus concentration. However when account is taken of experimental error the one per cent or so increase in D predicted by Eq 19 for the range of concentration spanned in Fig 9 is not in serious disagreement with the data.

At lower salt concentrations the ionic atmosphere will have a larger spatial extent. For .15 M NaCl, the Debye-Hückel length is about 8Å, whereas for 0.015 M NaCl it is about 25Å. Fig 9 shows that, at a given virus concentration, D increases with decreasing salt concentration, and therefore with increasing effective hard sphere size. From the arguments given above, it is clear that the thermodynamic term $(\partial\mu/\partial c)$ in Eq 15 should increase in this way. It is not immediately obvious that f should not show a similar increase leading to no change in D. It appears, however, that the effect of repulsive coulombic interactions on f is smaller than their effect on $(\partial\mu/\partial c)$. This is because, although two particles cannot approach so that their ionic atmospheres overlap significantly, the solvent within the atmospheres can still be expected to be quite mobile. For dilute solutions these effects can be incorporated simply into Pyun and Fixman's (1964) treatment of f (see, also, Goldstein and Zimm, 1971) with the result (Pusey, unpublished),

$$D = D_o [1 + h(x)\Phi + ...] \quad , \qquad (28)$$

where

$$h(x) \approx \frac{1}{2} + 2(1 + x)^2(1 + 4x) - \frac{15}{8}(1 + x)^{-1}$$

and

$$x = \frac{R_{eff}}{R_H} - 1 .$$

Here R_{eff} is the effective hard-sphere radius and R_H is the true hydrodynamic radius. Thus D is expected to increase with x and hence with decreasing salt concentration. Due to the various assumptions mentioned above, Eq 28 cannot be expected to apply exactly to the data of Fig 9. However there is qualitative agreement between Fig 9 and Eq 28. Fitting the 0.015 M NaCl data to Eq 28 we get an effective hard sphere radius of about 70Å greater than the true sphere radius. This difference is the same order of magnitude as the Debye-Hückel shielding length, ~ 25Å, for this solvent. Eq 28 represents an improvement over the theory given by Pusey et al (1972) where f was assumed to be independent of electrolyte concentration. It should also be mentioned that the observed electric field correlation functions for the data of Fig 9 could, as expected, all be well described by single exponentials with decay rates linear in K^2.

Since the results of Fig 9 indicated the existence of coulombic interparticle interactions we felt it might be worthwhile to look at a sample at very low salt concentration, essentially in pure water. Figs 10 and 11 show the results of such a measurement, where the average intensity and apparent diffusion coefficient Γ/K^2 are plotted against K^2. These plots show large deviations from normal behaviour. The mean intensity increases by a factor of about 2 with increasing scattering angle, over the range K^2 studied. This result indicates that inter-particle interactions are causing a considerable departure from a random distribution of virus particles in the solution. It should be pointed out that (1) The onset of such an increase in intensity is predicted by Eq 24. (2) In the absence of interactions the single-particle scattering factor P(K) would lead to a decrease in mean intensity of a few percent. The apparent diffusion coefficient (Fig 11) decreases also by a factor of about two. In addition the electric field correlation function could no longer be described by a single exponential, as evidenced by the fact that the initial decay rate was considerably greater than that obtained by force-fitting the data to a single exponential. These results indicate that one is clearly no longer in the hydrodynamic/thermodynamic regime $KR << 2\pi$. The

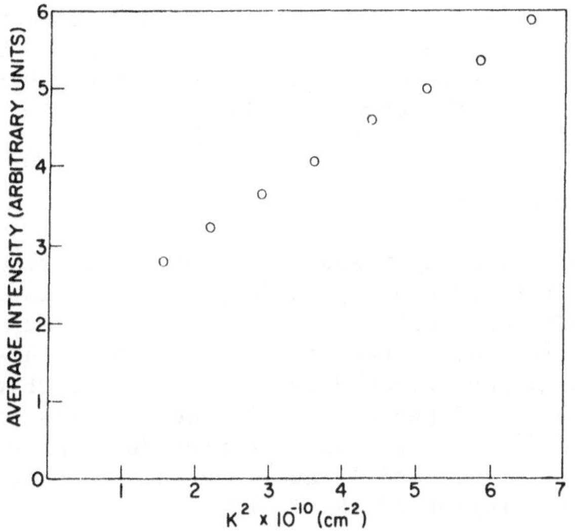

FIG 10. Average intensity of light scattered by R17 virus in pure water, as a function of scattering angle, (taken from Pusey et al, 1972).

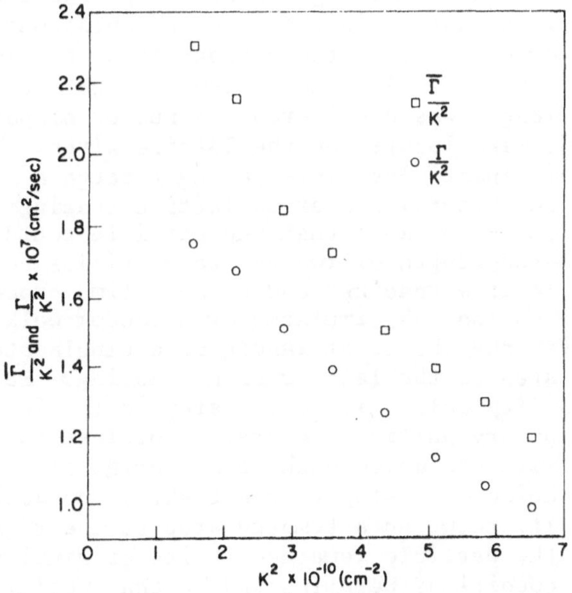

FIG 11. Decay rates of the electric field correlation function for R17 virus in water, as a function of scattering angle. □ initial or mean decay rate. ○ decay rate obtained by force-fitting data to a single exponential. (Taken from Pusey et al, 1972).

magnitudes of the relevant lengths for this measurement are 400Å $< 1/K <$ 800Å and L ≈ 2,200Å. Since large effects are being observed, it must be assumed that the extent of the virus ionic atmospheres (~R) is of the order of 1000Å.

4.4 Computer Experiments

There appears to be at present no theory with which Fig 11 can be compared. It thus seemed a good idea at this point to attempt a computer experiment on a well-defined system. The main aim of the computer calculation was to see if effects such as those shown in Figs 10 and 11 could be produced by the simple effective hard-sphere interaction described above. Indeed this was found to be the case. In addition the calculation provided confirmation of the predictions of Eqs 14 and 15 and gave some idea of what might be expected for the region KR, KL $\geq 2\pi$.

The problem considered was a very simple model of one-dimensional diffusion of hard rods (hard spheres in one dimension). It might be questioned whether a one-dimensional problem has any relevance to the true three-dimensional system. This question remains to be answered fully, though the results are encouraging. It can be noted that a normal light scattering experiment actually has a one-dimensional nature in that the instantaneous scattered electric fields depend only on the components of particle position parallel to the K vector. A chain of 100 lattice sites spaced equally by unit length was considered. A number of particles, usually 10 or 20, were located on the lattice sites. These were assumed to be point-particles, interacting through a hard-rod interaction of range an integral number of lattice spacings. Thus for a hard-rod length of one no more than one particle could occupy each site; for a hard-rod length of two no two particles could approach closer than two lattice spacings and so on. Time proceeded in unit steps. Brownian motion was simulated by a random walk. Each particle was given the choice at random of a single step to the right, a single step to the left or of not moving. Each possibility had probability 1/3. If, say, a step to the left led to a collision, a temporary particle separation of less than the hard-rod length, the particle was moved back to its original position and then given the choice of a step to the right or no motion, each with probability ½. If, then, an attempted step to the right also led to a collision, the particle remained in its original position. Cyclic boundary conditions were imposed so that position 101 was equivalent to position 1, etc. K vectors were chosen according to $K = 2\pi m/100$ with the integer m given by $4 \leq m \leq 20$. After each particle had attempted to move once, the instantaneous intensity I was calculated according to the equation

$$I = \left| \sum_i e^{iKr_i} \right|^2 \tag{29}$$

where r_i is the number of the lattice site occupied by particle i. This value of the intensity was stored. Each particle then attempted to move again, and the intensity was calculated and stored etc. From the stored values of I, estimates of the mean intensity and intensity correlation function for an "experimental run" were obtained.

Initially the particles were spaced at equal intervals along the lattice. 500 time units were allowed to ellapse before beginning computation of $\langle I \rangle$ and $\langle I(0)I(\tau) \rangle$ to allow complete randomization of particle positions. Experimental runs were typically 10^4 time units long. Observed correlation times ranged from about 4 to 50 time units. Thus statistical accuracy, though adequate for the purposes of the experiment, was not particularly high. Plots of the observed correlation functions were analyzed by eye, the initial slope $\bar{\Gamma}$ and an estimate of $\mu_2/\bar{\Gamma}^2$ being determined. It should be emphasized that "natural" units were used: the lattice sites were spaced by unit length, each particle attempted to move once in unit time and a free particle scattered, on average, unit intensity (Eq 29).

Before proceeding to results, several comments should be made concerning limitations and complications inherent in this model. Firstly it is evident that if a particle is not actually in contact with another particle it diffuses as a free particle. Thus the model assumes a constant frictional coefficient, independent of particle concentration and interactions. Secondly we consider a finite number of particles and a finite number of lattice sites. The former means that the scattered light is not truly Gaussian. The latter means that at small K the "scattering volume" is only a few times larger than 1/K and at large K space-quantization effects may become important.

Fig 12 shows some results for 20 particles (L = 5) with hard-rod length R' of 0, 1 and 2. (We call the hard-rod length R' to distinguish it from R, the definition of which is given at the beginning of this section. For R' << L, R and R' should be roughly equal, but for R' approaching L the formation of nearest-neighbour shells means that R can be greater than R'.) The maximum value of KR' for these data is 1.26 so that we should be in the hydrodynamic regime where Eqs 14 and 15 apply. It can be seen that, as expected, neither $\langle I \rangle$ nor D shows significant K-dependence. The time-

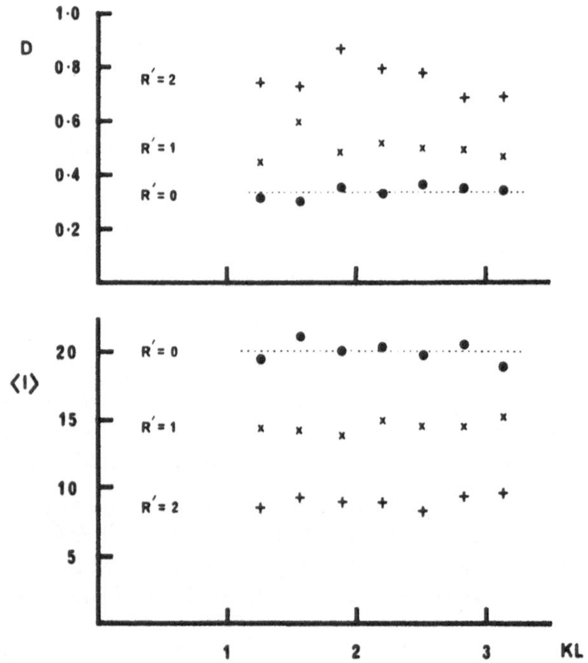

FIG 12. Diffusion coefficient D and mean scattered intensity <I> for one-dimensional Brownian motion of hard rods. Mean inter-particle spacing L = 5 and hard-rod length R' = 0, 1 and 2.

dependent parts of the correlation functions were found to be exponential to within the rather large experimental error. For free diffusion, R' = 0, $\langle I \rangle \approx 20$ as expected and $D \approx 1/3$, the theoretical value for this model. For R' = 1, $\langle I \rangle \approx 14.5$ and $D \approx 0.5$, ie, <I> is decreased and D is increased by roughly the same factor as predicted by Eqs 14 and 15, for constant f.

For R' = 2, similar but larger effects are observed. In Fig 13 we have plotted <I>D. According to Eqs 14 and 15 this should have the value 6.67 for this model. Again adequate agreement is seen. Thus in the region $KR' \ll 2\pi$, where theoretical predictions concerning interacting Brownian systems exist, this model fulfills the predictions. We can therefore have some confidence in extending the model to more complicated regimes.

Fig 14 shows results for 10 particles (L = 10) with hard-rod length 5. The maximum value of KR' for this data is 2π. We therefore expect to see distinct departures from "hydrodynamic" behaviour. At small KR', hydrodynamic behaviour is again observed, the apparent diffusion coefficient, $\bar{\Gamma}/K^2$, and <I> being roughly independent of K, but at higher KR' <I> starts to increase and $\bar{\Gamma}/K^2$ to decrease. In addition the time-dependence of the

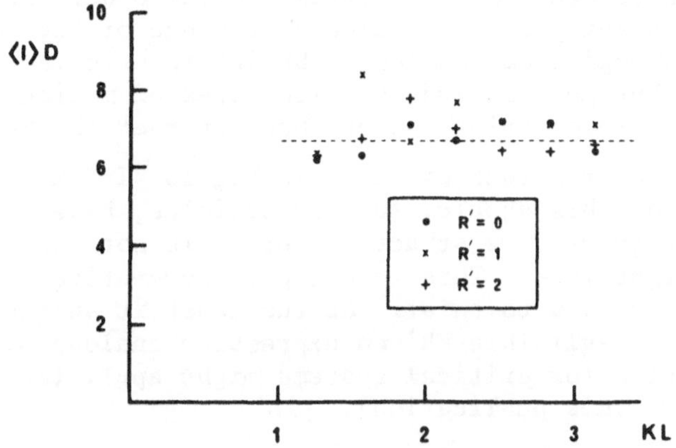

FIG 13. Product of mean intensity and diffusion constant for data of Fig 12.

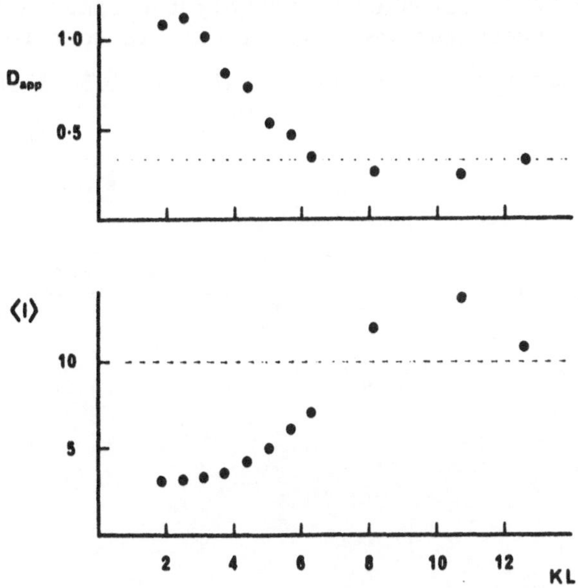

FIG 14. Apparent diffusion coefficient $\bar{\Gamma}/K^2$ and mean intensity <I> for L = 10, R' = 5.

correlation function in this region showed definite departures from single exponential behaviour. The increase of <I> with K can be interpreted as a tendency of the system towards order and the maximum as a broad Bragg peak. The time-dependence of the correlation function in this regime can no longer be interpreted in terms of local diffusion but must include coupled modes of motion of the particles. It is interesting to note however that the behaviour of $\bar{\Gamma}/K^2$ seems to mirror that of $\langle I \rangle$. In Fig 15 $\langle I \rangle \bar{\Gamma}/K^2$ is shown. For no interaction this should, for 10 particles, have the value 3.33. This plot shows some structure but it is not far from a horizontal straight line. This is a highly suggestive result for which no theory appears to exist. At the least it suggests that for small but non-negligible KR' an expression analogous to the Fixman modification for critical systems might apply (see, for example, Swinney, this publication).

One reason for choosing the particular parameters of Fig 14 was to attempt to produce data similar to that of Figs 10 and 11. For the R17 sample, L ≈ 2,200A. Thus the data of Figs 10 and 11 span roughly the range KL = 2.6 to KL = 5.6 in Fig 14. For this range of KL for the computer calculation, $\langle I \rangle$ increases by roughly a factor of two and $\bar{\Gamma}/K^2$ decreases by roughly the same factor. The major difference between the two sets of data is that for the virus data $\bar{\Gamma}/K^2$ dips below its free-particle value ($\approx 1.54 \times 10^{-7}$ cm^2/sec)

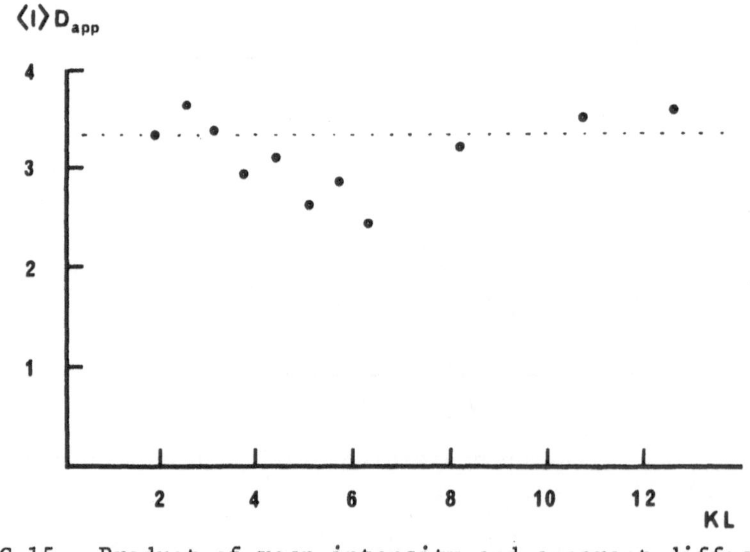

FIG 15. Product of mean intensity and apparent diffusion coefficient for data of Fig 14.

whereas for the computer calculation $\bar{\Gamma}/K^2$ remains above its free-particle value. Nevertheless it is this qualitative agreement which gives confidence in the explanation in terms of effective hard-sphere interactions advanced above for the virus data.

Fig 16 shows data for 4 particles, L = 25, with hard rod length R' = 20. Thus $(KR')_{max} \approx 15$. For this example the effects mentioned above due to a small number of particles and small K vectors should be considerable and conclusions from these results should be drawn with caution. Nevertheless <I> shows the type of behaviour one might expect, with "Bragg peaks" at KL ≈ 2π and 4π. This result is exactly analogous to that found in neutron or X-ray

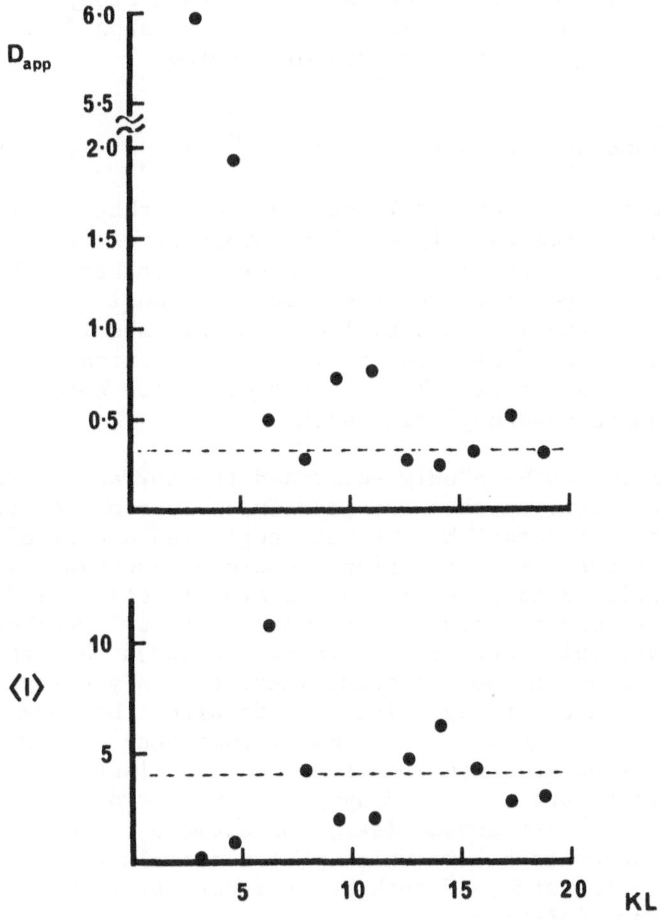

FIG 16. D_{app} and <I> for L = 25, R' = 20.

scattering from liquids. Again $\bar{\Gamma}/K^2$ seems to mirror the behaviour of $\langle I \rangle$, and again there appears to be no theoretical prediction with which the data can be compared. For KR', KL >> 2π, $\langle I \rangle$ and $\bar{\Gamma}/K^2$ appear to be tending towards their free-particle values, 4 and 1/3. This is to be expected, since at large K, a particle is unlikely to suffer a collision in the time it takes to diffuse a distance 1/K.

It is quite likely that photon-correlation experiments will, in the future, show the type of behaviour of Fig 16. It is an interesting regime for study in that one should be able to obtain the same kind of information concerning macromolecular interactions and motions as is obtained at present for atomic potentials and motions in liquids by neutron scattering. In retrospect a major omission in the computer experiment was a separate study of the self and distinct terms, which can be expected to show different types of behaviour in the non-hydrodynamic regime.

4.5 Discussion and Conclusions

I hope that it now seems at least reasonable that experimental data on Brownian systems experiencing coulombic interactions can be qualitatively explained in terms of an effective hard-sphere interaction. A more complete treatment, besides considering more realistic forms for the inter-particle potential, should also assess the variation of f with macromolecular concentration and degree of interaction; it is likely that f will also show a K-dependence in the non-hydrodynamic regime.

In this section we have only scratched the surface of the subject of interactions in that one particular type of interaction has been discussed in detail but we have neglected a host of other possible interactions. Many questions remain unanswered. Firstly it should be mentioned that Eqs 14 and 15 are strictly valid only for two-component (macromolecule + solvent) systems. When considering repulsive coulombic interactions we have assumed that the system is effectively a two-component system, where the only effect of the third component, the electrolyte ions, is to alter the form of the macromolecular interaction. In real multi-component systems one has not one, but several, mutual diffusion coefficients to consider. The thermodynamic theory of multi-component systems has been studied extensively over the last decade (see, for example, Eisenberg, this publication). The contribution which photon correlation spectroscopy can make to the study of such systems has yet to be determined (see, however, Camerini-Otero et al, 1973a).

Stephen (1971) has given a description of macromolecular diffusion in charged systems where, unlike the treatment given here, the predominant effect on diffusion of the macromolecules is assumed to be due to macromolecule-small ion interactions. The range of applicability of this theory appears limited in that the theory is restricted to relatively low macromolecular charges and the regime where the interparticle spacing is small compared to the Debye-Hückel shielding length. A recent attempt has been made to interpret data on charge effects in bovine serum albumin according to Stephen's theory (Doherty, 1972). Agreement was not particularly good, whereas the data can almost certainly be described in terms of effective hard-sphere interactions (see, for example, Doty and Steiner, 1952).

For completeness we conclude with two experimental examples of macromolecular interactions in uncharged systems. King et al (1973) have studied in the hydrodynamic regime random-coil polystyrene in the good solvent butan-2-one, for a range of polymer molecular weights and concentrations. The different effects of interactions on $(\partial\mu/\partial c)$ and f are evidenced by the fact that at small molecular weight, $\lesssim 2 \times 10^5$, D decreases with increasing polymer concentration, whereas at large molecular weight an increase in D is observed. Recent studies by Pusey et al (1973b) again on polystyrene ($\bar{M}_W \approx 200{,}000$) in a good solvent (toluene) have shown that $\bar{\Gamma}/K^2$ increases by almost a factor of three in the polymer concentration range 0-10% (Fig 17). At low concentrations this almost certainly reflects an excluded volume effect similar to that observed in charged systems. At higher concentrations, the distribution of decay rates $G(\Gamma)$ required to describe the data became increasingly broad, though $\bar{\Gamma}$ remained linear in K^2. This broadening might be due to polymer entanglements leading to increasing effective polydispersity; alternatively it might reflect the onset of viscoelastic effects such as those described by Ostrowsky (this publication). It is interesting and encouraging to note that the data of Fig 17 are in reasonable agreement with those of Rehage et al (1970) who studied diffusion in this system by conventional methods.

ACKNOWLEDGEMENTS

Much of the work presented above is the result of several enjoyable collaborations. The author is indebted to colleagues too numerous to mention by name. However special thanks are due to Dr R D Camerini-Otero and Dr D E Koppel who read a draft of the paper and made many valuable suggestions which have been incorporated in this final version. The computer calculations were performed when the author was with IBM.

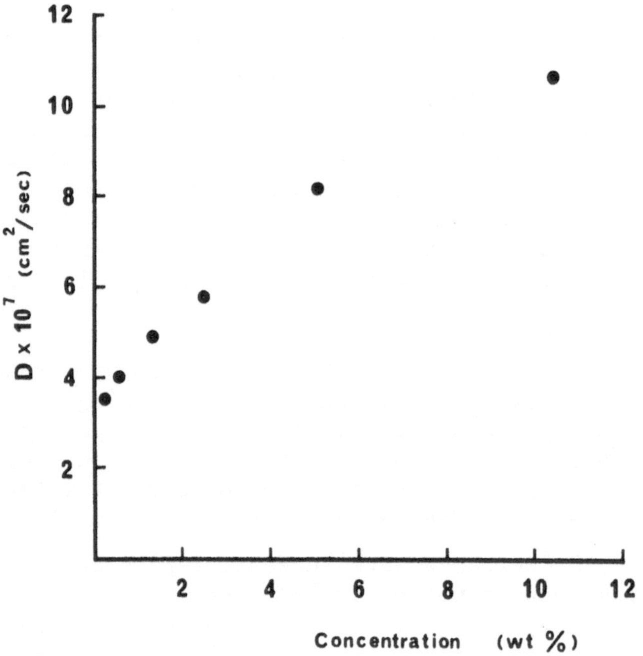

FIG 17. Diffusion coefficient $\overline{\Gamma}/K^2$ of polystyrene in toluene as a function of polymer concentration, (taken from Pusey et al, 1973b).

REFERENCES

G K Batchelor, 1972, J Fluid Mech 52, 243.

F C Bancroft and D Freifelder, 1970, J Mol Biol 54, 537.

B J Berne, 1971, in "Physical Chemistry: An Advanced Treatise", Vol VIIIB, Eds H Eyring, D Henderson and W Jost, Academic Press, New York, p 539.

J C Brown, R Dietz and P N Pusey, 1973, to be published.

W Burchard and J M G Cowie, 1972, in "Light Scattering from Polymer Solutions", Ed M B Huglin, Academic Press, London, p 725.

J M Burgers, 1941, Proc Acad Sci Amsterdam 44, 1045, 1177; also 45, 9, 126, (1942).

R D Camerini-Otero, P N Pusey, D E Koppel, D W Schaefer and R M Franklin, 1973a, Biochemistry, to be published.

R D Camerini-Otero, R M Franklin and L A Day, 1973b, to be published

R D Camerini-Otero and R M Franklin, 1973, to be published.

R Dietz, 1973, Private Communication.

P Doherty, 1972, Talk at MIT Conference, April 1972.

P Doty and J T Edsall, 1951, Advances in Protein Chemistry, Vol 6, p 35.

P Doty and R F Steiner, 1949, J Chem Phys 17, 743.

P Doty and R F Steiner, 1952, J Chem Phys 20, 85.

P A Egelstaff, 1970, Disc Farad Soc 49, 280.

P J Flory, 1953, "Principles of Polymer Chemistry", Cornell University Press.

R Foord, E Jakeman, C J Oliver, E R Pike, R J Blagrove, E Wood, A R Peacocke, 1970, Nature, 227, 242.

B Goldstein and B H Zimm, 1971, J Chem Phys 54, 4408.

J D Harvey, 1973, Virology, to be published.

P A Hiltner and I M Krieger, 1969, J Phys Chem 73, 2386.

T A King, A Knox and J D G McAdam, 1973, Polymer, 14, to be published (July).

L I Komarov and I Z Fisher, 1963, Soviet Physics JETP, 16, 1358.

D E Koppel, 1972, J Chem Phys, 57, 4814.

D E Koppel, 1973, Thesis, Columbia University.

C J Oliver, K F Shortridge and G Belyavin, 1973, to be published.

G Oster, 1946, Science 103, 306.

G Oster, 1949, Rec Trav Chim 68, 1123.

R Pecora, 1970, Disc Farad Soc, 49, 281.

P N Pusey, D W Schaefer, D E Koppel, R D Camerini-Otero and R M Franklin, 1972, J de Physique, 33, Colloque C1, c1-163.

P N Pusey, D E Koppel, D W Schaefer, R D Camerini-Otero and S H Koenig, 1973a, Biochemistry, to be published.

P N Pusey, J M Vaughan and G Williams, 1973b, to be published.

P N Pusey, 1973, in "Industrial Polymers: Characterization by Molecular Weight", Eds J H S Green and R Dietz, Transcripta Books, London.

C W Pyun and M Fixman, 1964, J Chem Phys 41, 937.

G Rehage, O Ernst and J Fuhrmann, 1970, Disc Farad Soc 49, 208.

D W Schaefer, G B Benedek, P Schofield and E Bradford, 1971, J Chem Phys 55, 3884.

M J Stephen, 1971, J Chem Phys 55, 3878.

C Tanford, 1961, "Physical Chemistry of Macromolecules", John Wiley and Sons, Inc, New York.

E J Verwey and J Th G Overbeek 1948, "Theory of the Stability of Lyophobic Colloids", Elsevier Publishing Co, Inc, Amsterdam.

G H Vineyard, 1958, Phys Rev 110, 999.

F Zernicke and J A Prins, 1927, Z Physik 41, 184.

CORRELATION COMPARED WITH INTERFEROMETRY FOR LASER SCATTERING SPECTROSCOPY

J M Vaughan

Royal Radar Establishment, Malvern, Worcs, UK

In the early 1960's high resolution optical spectroscopy was a venerable and well established subject. Topics of study included amongst others fine and hyperfine structures, isotope shifts, collision broadening, and line broadening in plasmas due to ions and electrons; the dispersing instruments most commonly used were the Fabry-Perot etalon and the diffraction grating. At that time a resolving power of $\sim 10^7$, corresponding to a resolving limit of $\sim 2 \times 10^{-3}$ cm^{-1} (60 MHz) would have been considered extremely high (Kuhn 1964). Instruments of higher resolution were in fact rarely required; the limits of resolution were almost always set by the Doppler widths of the spectral lines themselves, arising from the motion of the emitting or absorbing atoms and molecules. The Doppler width, for example, of Gaussian spectral form of the 546 nm line from a low pressure water cooled Hg198 isotope lamp is about 17×10^{-3} cm^{-1}; the widths observed for lighter atoms are of course considerably larger. It is worthy of note however that the Lamb shift in atomic hydrogen was observed in 1938 by Williams using a Fabry-Perot technique but received little attention (see eg Series 1957).

This picture of optical spectroscopy changed considerably with the great upsurge of interest in light scattering following the development of relatively inexpensive, stable lasers. Something of this new vast field of laser scattering spectroscopy is indicated in figure 1. Doppler broadening no longer provides a barrier and interferometric techniques using Fabry-Perot etalons have developed such that resolving limits of a few MHz are now attained in many laboratories. At the same time fast photo-

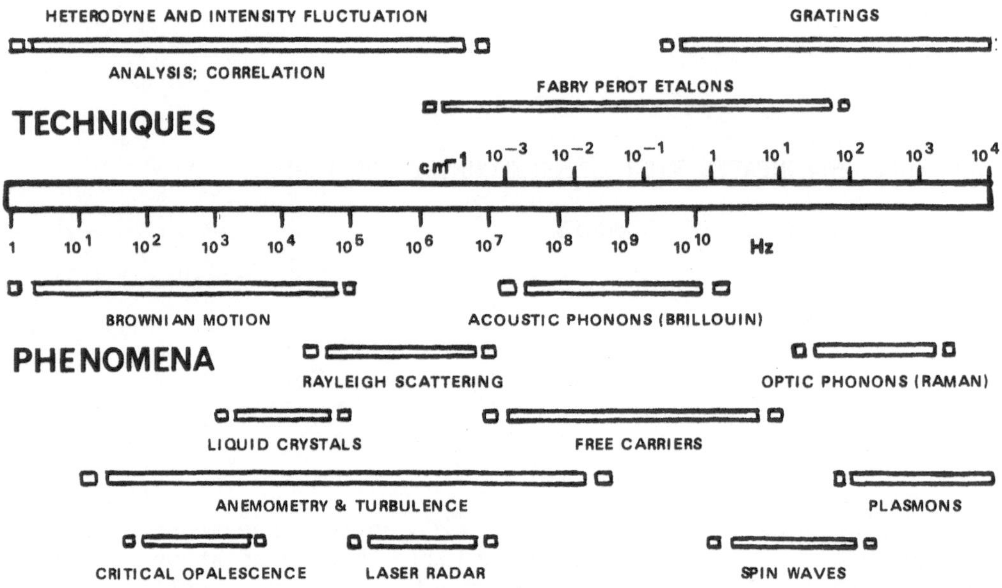

FIG 1 The field of laser scattering spectroscopy showing the range of phenomena and techniques.

multipliers have been perfected and techniques borrowed from the microwave region have been extensively applied. If we consider in particular intensity correlation with a sampling time of 10 n sec (corresponding to an upper frequency limit of ~ 100 MHz) we see that there exists a spectral region of ~2 to 100 MHz resolution where in principle at least experiments may be conducted either by interferometry or by correlation. It is thus important to examine the principles of the two techniques and to compare their relative merits in practice.

In this article the first section is devoted to a simple outline of Fabry-Perot interferometry with some account of the most recent developments of particular value for laser scattering spectroscopy. The second section provides a comparison of the principles of correlation and interferometry and the third discusses experiments conducted in both domains. The overall conclusion, unexceptionably, is that few hard and fast rules can be drawn. The application of either method to any particular experiment is best considered in individual detail; ideally one should call on both techniques.

FABRY PEROT INTERFEROMETRY

In ordinary spectroscopic use the significant questions one asks of an interferometer are the dependence on angle and wavelength of the light transmitted through the instrument. The distribution of field over the plates, requiring cavity mode analysis, is almost always ignored and has received sophisticated treatment only with the advent of the laser (Fox and Li, 1961, Boyd and Gordon 1961, Boyd and Kogelnik 1962, Bergstein and Schuchter 1964, for a review see Toraldo di Francia 1970). The treatment of the modes that can be established differs from that of a microwave cavity because the Fabry-Perot etalon is not completely enclosed and so the fields do not vanish upon some completely enclosed boundary. Use of Kirchoff diffraction formulae and Huyghen's principle investigates successive reflections at the plates. Iterative and self consistency procedures have been applied and for various geometries eg confocal paraboloidal mirrors, analytic solutions may be found for the various longitudinal, transverse and angular modes; other configurations including plane parallel plates require numerical solutions. There are further difficulties in the calculation of diffraction loss over finite plates.

For spectroscopic purposes the geometric treatment is probably more instructive (see, eg Kuhn 1951, Jacquinot 1960). Consider successive reflections between identical plane optical plates, Figure 2, with relative amplitudes r^2 and phase lag $\phi = 2\pi(2\mu d\cos\theta)/\lambda$, where r and t are the amplitude reflection and transmission factors of the surfaces, μ the refractive index of the medium between the plates, d the plate separation, and we ignore any phase changes at the surfaces. These beams are combined together in the focal plane of the fringe forming lens, which thus sums the complex amplitudes

$$t^2 \sum_{n=0}^{\infty} r^{2n} e^{in\phi}$$

The square of the modulus of this expression gives the intensity distribution.

$$I(\phi) = \frac{T^2}{(1-R)^2} \times \frac{I_o}{1 + [4R/(1-R)^2]\sin^2\phi/2}$$

where $R = r^2$ and $T = t^2$. This is the well known Airy pattern giving Haidinger rings, bright circular fringes for $\phi = 2\pi n$, that is for

$$2\mu d\cos\theta = n\lambda$$

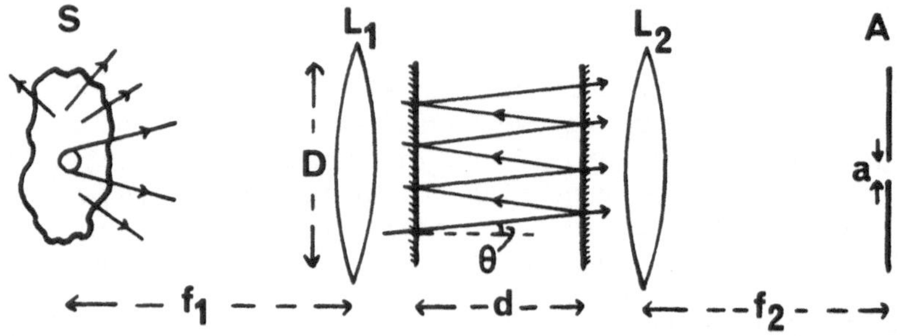

FIG 2: A plane Fabry-Perot etalon spectrometer. S is the luminous source, L_1 and L_2 the collimating and fringe forming lens of focal lengths f_1 and f_2 respectively. The highly reflecting etalon plates have a separation d and an effective diameter D. The scanning aperture in the image plane A is of diameter a.

The fringes may be recorded photographically (for a suitable area source) or photometrically. In the latter case the fringes are usually swept through a small central aperture, either by varying the plate separation with the use of piezo-electric scanning elements, or by changing the refractive index of the medium by pressure scanning (see eg Figure 3).

From these expressions one notes the following points:

(a) The mode separation, that is the smallest frequency separation such that the components exactly overlap, is given by $c/2\mu d$ Hz. This is known as the free spectral range (the FSR, or one order of interference).

(b) The fringes are narrow for high reflectivity coatings; the ratio of the FSR to the fringe width at half height is called the finesse. The theoretical reflectivity finesse is given by $\pi\sqrt{R}/(1-R)$, but this is reduced in practice by many factors including plate imperfections, lack of parallelism, and the effect of the aperture function.

(c) The fringe contrast or extinction, that is the ratio of peak to trough, is quite low and given by $[1 + 4R(1-R)^2] : 1$. In practice this is additionally reduced by scattering in the surface and is usually $\sim 10^3$; the study of weak satellites thus requires multiple or multipass instruments.

(d) The transmission of the instrument is high; for $T = 1-R$, that is for no absorption or scatter, there is no loss of intensity in the fringes.

(e) The scanning aperture is an important experimental parameter defined in angular terms by $\theta_a = a/f_2$. This corresponds to a rectangular band pass of width ν_a equal to $(\theta_a/\theta_1)^2$ times the FSR where θ_1, the angular diameter of the first bright fringe from a bright centre, is given by $2\sqrt{\lambda/\mu d}$. As θ_a is made larger the total light flux increases but the resolving power decreases. The final choice of aperture must depend on the constraints of the individual experiment; in general if ν_a is made less than one fifth of the width of the spectral feature under study the line distortion and effective loss of resolution are small. When, as is now common, the instrument is used in the scanning mode with photon counting and multi-channel recording (see eg Jackson and Pike 1968, Vaughan, Vinen and Palin 1972) the channel width may conveniently be made equal to an integral fraction of ν_a.

(f) The light gathering power is characterised by the etendue- the product of the luminous area observed and the collection solid angle. From figure 2 this is given by

$$\frac{\pi}{4} \cdot (a\, f_1/f_2)^2 \cdot \frac{\pi}{4}\ (D/f_1)^2$$

This establishes the well known result that for a given band pass (a/f_2 constant) the light flux collected is independent of the lens geometry, and can only be increased by the use of larger diameter etalon plates or a brighter source. In this connection it should be noted that the techniques of using cylindrical or conical optics in the light collecting arrangement prior to the interferometer can only give very limited increases in flux gathering while vastly increasing in many cases the prospects of collecting spurious flare. (However these techniques may in certain applications have some advantage in reducing the effective spread of K vectors involved in the scattering process, thus reducing the band width of the light collected).

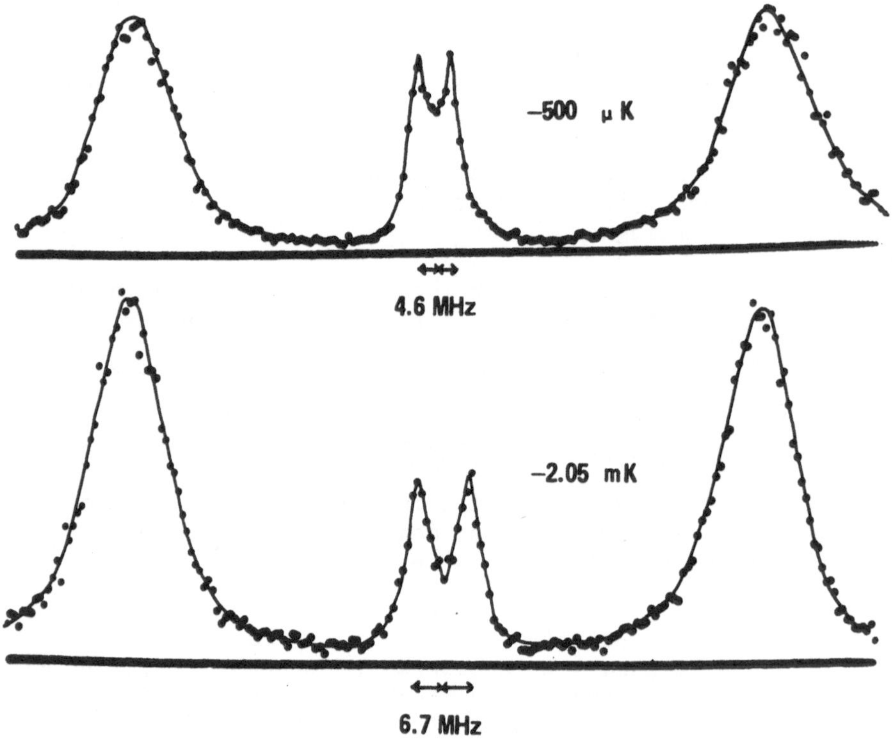

FIG 3: Spectra of the light scattered by superfluid helium close to the λ point (-500 μK, -2.05 mK; pressure 20 bars). The laser frequency is at the centre of the recording, the outer peaks are due to Brillouin scattering from first sound and overlap ~5½ orders of interference. The central doublet is due to scattering from second sound (a propagating entropy fluctuation). Each channel is 1.28 MHz wide, the instrumental line width is ~5.6 MHz and there are about 500 counts per channel at the peaks of the lines. These spectra were obtained using a highly stabilised argon ion laser at 488 nm and ~70 mw power, and a plane etalon of spacing 833 mm employing fast piezo-electric scanning. The recording time was ~30 minutes (Vaughan, Vinen and Palin 1972). Similar spectra have been obtained by Winterling, Holmes and Greytak (1972, 73) using a helium-neon laser at 633 nm and a confocal etalon of 500 mm spacing with slow pressure scanning.

One thus sees that the plane Fabry-Perot interferometer is a versatile, high resolution, high efficiency, pre-detection, angular filter. In normal scanning operation it is a single channel instrument; use of a multi-annular diaphragm in the fringe plane could in principle provide multi-channel operation, the instrument then serving as a parallel bank of filters. This is possible however only for suitable area sources.

In recent years the confocal spherical mirror interferometer first described by Connes (Connes 1956, 58, Fork et al 1964, for a useful theoretical and practical guide see Hercher 1968) has been extensively applied to light scattering spectroscopy. In the confocal mode of operation, after 4 successive reflections rays are travelling approximately on their original paths independently of the angle of incidence. The instrument thus serves as a true angle-independent filter; the fixed FSR is given by $c/4\mu r$, where r is the plate separation and common radius of curvature, and the finesse is $\pi\sqrt{R}/2(1-R)$. The increase in light gathering can be very large; it may be shown that the etendue is proportional to the resolving power rather than inversely proportional as in the case of the plane etalon (Connes 1964). In many scattering experiments however the luminous region is a streak or filament source and thus the full theoretical gain in etendue of the spherical etalon cannot be realised. The light gathering may be further reduced in practice by the limitations on solid angle imposed by spherical aberration. Nevertheless there is usually some gain at the highest resolution; however the particular advantage of the instrument is its relative insensitivity to tilting of the plates. The strict maintenance of parallelism, vital for successful operation of the plane interferometer, is not essential.

In summary the analysis of a narrow spectral feature $B(\nu)$ using an interferometer presents a number of technical difficulties. One has (a) the overlapping of different orders of interference and of neighbouring components (b) the maintenance of a linear frequency scan (c) frequency stabilisation of the laser source (d) distortion due to the collection of parasitic light and photodetector dark current (e) problems of instrumental line shape - including plate imperfections and departures from plate parallelism etc. The deconvolution of all these effects to obtain $B(\nu)$ from the observed spectra can be extremely complicated. In general no analytic solutions are possible and extensive computer simulation and fitting is required. In certain cases one may approximate plate imperfections by small Gaussian functions, lack of parallelism - a circular function - as a near Gaussian, the Fabry-Perot Airy pattern as a succession of overlapped Lorentzians and the rectangular aperture function as a small correction. Observed profiles may then be analysed in terms of Voigt functions

(Davies and Vaughan 1963) and precision parameters obtained (eg Kuhn and Vaughan 1964, Vaughan 1968).

Possibly because of the large systematic errors that are suggested by these problems remarkably little attention has been devoted in the literature to the fundamental limitations to the accuracy. As pointed out by Connes (1964) this limit must be due to the noise in the signal itself due to its statistical nature.

A simple approach to this problem may be made as follows. We consider the product of an experiment as a histogram of photodetections distributed among a number of recording channels each of width $\delta\nu$. Suppose we have a symmetric spectral feature, of form $S(\nu)$ and half-value width ν_s (whole width at half height) for which the instrument accumulates a total number of counts N_c in the experimental time $\mathfrak{J}$. The experimental recording is thus a histogram approximating the function $S(\nu)$; clearly successive experiments each accumulating N_c counts will give slightly different distributions. In the low count rate limit the variance on the number of counts N_n in the nth channel is given by

$$[\mathrm{Var}\ N_n] = N_n$$

From a single experiment a measurement of the central frequency may be made by determining the frequency at which there are equal numbers of counts on either side $(N_c/2)$. Similarly the half-value width may be found by establishing the symmetrically placed frequency interval containing the appropriate fraction of the counts, which for a Lorentzian feature is also $N_c/2$. From the variance of $N_c/2$ one can immediately establish the variance on the frequency interval and position for the width and centre respectively. For the line centre this is given by

$$\frac{[\mathrm{Var}\ N_c/2]^{\frac{1}{2}}}{N_c/2} = \frac{[\mathrm{Var}\ \nu]^{\frac{1}{2}} \cdot S(\nu_o)}{\frac{1}{2}\int_{-\infty}^{+\infty} S(\nu)\, d\nu}$$

where ν_o is the centre-peak frequency of the symmetric $S(\nu)$; $S(\nu_o)$ is thus the peak height and the integral is of course the area under the curve. Expanding we have

$$\frac{[\mathrm{Var}\ N_c/2]^{\frac{1}{2}}}{N_c/2} = \frac{1}{(N_c/2)^{\frac{1}{2}}} = \frac{[\mathrm{Var}\ \nu]^{\frac{1}{2}} \cdot S(\nu_o)}{\frac{1}{2} S(\nu_o) \cdot (\nu_s/2) . \pi}$$

and rearranging

$$\frac{[\mathrm{Var}\ \nu]^{\frac{1}{2}}}{\nu_s} = \frac{\pi}{2\sqrt{2}} \cdot N_c^{-\frac{1}{2}}$$

A similar derivation for the half-value width gives a standard deviation fractional error of

$$\frac{\Delta\nu}{\nu_s} = \frac{\pi}{\sqrt{2}} \cdot N_c^{-\frac{1}{2}}$$

Comparable results are readily obtained for other spectral forms. We note that the error is independent of the experimental duration,which is clearly true for an ideal stable instrument. In addition, while more sophisticated treatments may give a somewhat different arithmetic factor they are unlikely to alter the $N_c^{-\frac{1}{2}}$ dependence. The results are much what one might expect; they indicate that for a Lorentzian spectrum containing 10^4 counts the sd error on the half-value width is

$$\frac{\Delta\nu}{\nu_s} = \frac{\pi}{\sqrt{2}} \approx 2\%$$

and the line centre should be determined to ~1% of the half width of the line.

In practice, as we have observed, the situation is more complicated. The observed spectrum $S(\nu)$ is the convolution of $B(\nu)$ with the instrumental function $I(\nu)$; in principle at least the latter may be obtained with infinite precision. An additional consideration will be the number of recording channels or resolution elements one arranges relative to the width of the spectral feature.

As examples of the applications of interferometry to laser scattering in addition to the experiments outlined in figure 3, one might mention the work on second order Raman scattering in liquid helium by Greytak and his colleagues (1969, 1970, 72) using a plane etalon of small spacing and FSR greater than

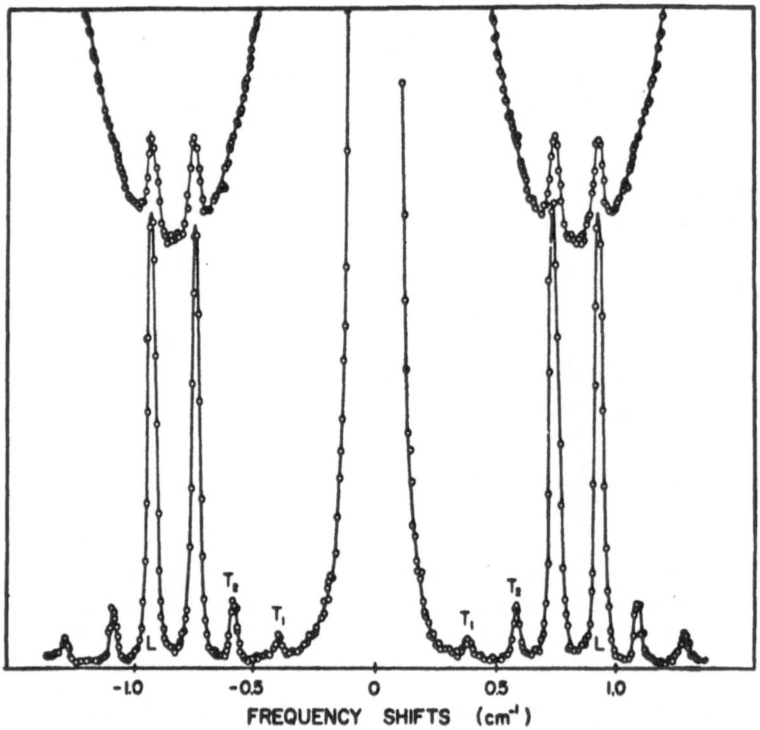

FIG 4 Brillouin back scattering showing the gain in contrast with a multipass interferometer (Sandercock 1970). The upper recording shows the spectrum with a single pass. The lower recording (on a common base line) shows the same spectrum obtained using a double pass on the same interferometer. The vertical scale is linear and the longitudinal peaks overlap from neighbouring orders. Instruments of 5 or more passes have now been developed.

40 cm^{-1}; the early experiments of Pike, Vaughan and Vinen on liquid helium where parasitic flare light was eliminated (1969, 70, 71); the Rayleigh and Brillouin scattering from gases due to Greytak and Benedek (1966, 67); the studies of light scattering from organic liquids of Jackson and his colleagues (1970); Brillouin scattering from liquids in the vicinity of the critical point (Gammon, Swinney and Cummins 1967, Mohr, Langley and Ford, 1969, Cannell and Benedek 1970) the work on light scattering in liquids of Gornall et al 1966, Volterra 1969); and many other studies. (see eg 1972 J de Phys Supp, C1, T33).

In concluding this review of Fabry-Perot interferometry one should mention the recent developments of multi-pass instruments by Sandercock (1970) in the study of weak satellites in the presence of strong elastic scattering. Separate double and triple etalons of

STATISTICAL MECHANICS OF CHAIN MOLECULES — Paul J. Flory

ERRATA

p. 5, line 2: Replace "l" by "$\{l\}$"

p. 46: Remove "R. Chiang, J. Phys. Chem., **70**, 2348 (1966)" from footnote a, and append to footnote b.

p. 85, line 10: Replace "chain" with "chains"

p. 122: The first two lines should read:
"illustration in its application in Chapter VI to stereoirregular vinyl polymers, which may be treated as copolymers of units that are mirror images"

p. 157, line 3: Replace "kcal mole^{-1}" by "cal mole^{-1}"

p. 162: The first two lines should read:
"represent them. For the bond pair i, i + i centered about O in Fig. 15, and for bond pair i - 1, i centered about CH_2, we have respectively"

line 4: Replace "former" with "latter"

line 7: Replace "ω_a" with "ω_b"

line 12: Replace "ω_b" with "ω_a"

p. 193 third line following Eq. (34.1) replace "group 1" with "group 0"

p. 255, line 7: Replace "(Chap. I, p. 3)" with "(Chap. I, p. 13)"

p. 321 last line of Eq. (70) should read:

$$+(33/175)(r/nl)^4 + \ldots]\} \quad (70)$$

p. 381: Replace "$\mathbf{E}_3$" with "$\mathbf{E}_9$" at each of the two places where $\mathbf{E}_3$ occurs in Eq. (128)

integral spacing have been described in the past, however they do not have the essential simplicity of operation of the multipass single etalon. In the instrument described by Sandercock (1971) a contrast in excess of 10^9 has been observed with 5 passes. This type of instrument has been used in the study of thin films and solids (Sandercock 1972, et al 1972) and is likely to be increasingly applied. Something of the possibilities are indicated in figure 4.

CORRELATION AND INTERFEROMETRY

As we have seen an interferometer operates directly on the light field itself; a scanning Fabry-Perot etalon acts as a pre-detection filter and typically produces a spectrum directly - a plot of intensity versus frequency. On the other hand correlation provides a system of post-detection processing in which the int
in suit
just the ... ation
of the

Form ... of
photodet
It thus ... T.

$G^{(2)}$

where τ
cathode.

We r
function

$G^{(1)}$

which is
are rela
Siegert

$g^{(2)}$

For ligh
this bec

$g^2(\tau$

Bücher im Werte von unter ord. 20,— DM zu remittieren, weil sie wegen eines Herstellungsfehlers unverkäuflich sind, ist unwirtschaftlich,und zwar sowohl für mich wie auch für Sie.Bitte erkennen Sie daher diese Blind-Remittendenfaktur an; das Buch halte ich einen Monat hier zu Ihrer Verfügung. Ich bitte um Nachricht, wenn Sie - abweichend von meinem Vorschlag - auf Rücksendung bestehen sollten.Ein Ersatzexemplar (Umtausch) brauche ich nicht. Ich habe das Buch inzwischen der Eile wegen vom Barsortiment/Verlag nachbezogen.Das defekte Exemplar vernichte ich einen Monat nach Ausstellung dieser Remittendenrechnung.

Stern-Verlag Janssen & Co.,4 Düsseldorf,Friedrichstr.26

where C is a correction factor depending on detector area, dead time effects and 'clipping' and Γ is the half width at half peak intensity in angular frequency units (equal to $\pi\nu_s$ in the preceeding section).

We thus have a very simple conceptual problem - the error arising in Γ due to the experimental parameters of count rate, experiment time etc. This problem has been treated in great detail by Jakeman, Pike and Swain (1971, see also the articles in this volume by Oliver and Jakeman).

They find most generally that

$$\mathrm{Var}\,[g^{(2)}(0,\tau)] = \frac{1}{N\gamma}\left[A + \frac{B}{r} + \frac{C}{r^2\gamma}\right]$$

where A, B, C are constants and N is the total number of samples, γ the ratio of sample time to coherence time (T.Γ), and r is the mean number of counts per coherence time. The coherence time is of course just the reciprocal of the half width in angular frequency units (Γ^{-1}). For $r<1$ and hence $r^2\gamma \ll 1$ (which is of course just the low count regime met at frequency widths > 1MHz) and for an infinite number of channels, they find that

$$\frac{[\mathrm{Var}\ \Gamma]^{\frac{1}{2}}}{\Gamma} = \frac{4.6}{r(N\gamma)^{\frac{1}{2}}}$$

which may be expressed alternatively as

$$\frac{\Delta\nu}{\nu_s} = \frac{4.6}{N_c}\,.\,\mathcal{J}^{\frac{1}{2}}\,.\,\Gamma^{\frac{1}{2}}$$

or

$$\frac{\Delta\nu}{\nu_s} = 4.6\,.\,N_c^{-\frac{1}{2}}\,.\left[\frac{N_c}{N_{coh}}\right]^{-\frac{1}{2}}$$

where $\mathcal{J}$ is the total duration of the experiment (NT), and N_{coh} is the total number of coherence times in the experiment ($\mathcal{J}\,\Gamma$). This result applies to the correlation of an intensity fluctuation or self beating signal. The error is reduced by a numerical factor of ~4 in the case of a homodyne experiment employing a coherent reference beam.

That these results have the correct form may be seen from the first expression as follows: if the number of counts N_c

for a constant experimental time is increased by a factor F then the number of correlation events entering corresponding channels of the correlator is increased by F^2. The error on each channel contents will be multiplied by F^{-1} and thus the overall fitting error multiplied by a factor F^{-1}. If on the other hand the experimental time is increased by a factor F while accumulating the same total number of counts N_c the rate of accumulation of correlation events is divided by a factor F^2. However since the experimental time is now F$\mathcal{T}$ the contents of each channel are only divided by a factor F and the line width error is multiplied by $F^{\frac{1}{2}}$. Similar arguments can be used to establish the square root error dependence for Γ.

This of course gives us some physical insight into the behaviour of the correlator - one count is of itself not an 'event' as it is for the interferometer. Only a pair of counts gives an 'event' and indeed one could imagine for a very low count rate and a limited number of channels that no events would be registered at all. For the correlator a single photodetection requires a subsequent detection within a few coherence times, or within a few cycles to provide spectral information. Comparing the expressions for the line-width error in interferometer and correlator we see that the latter contains the additional factor $[N_c/N_{coh}]^{-\frac{1}{2}}$; this term is just the number of photodetections per coherence time.

The first comparisons of conventional interferometric spectroscopy with optical mixing and heterodyne spectroscopy using wave analysers appear to have been made by Connes (1964) and Lastovka (1967). Both authors emphasize that the significant parameter is the number of photons detected per second within a single coherence area (an etendue of λ^2) per unit band pass; that is the number of detections from one cell of phase space, equivalent to the number of counts per coherence time. For classical sources and for laser scattering sources in the frequency range ($\Delta\nu >$ 1 MHz) with which we are concerned this number is always small. Connes notes that with classical methods the optical filter acts directly on the light field and one measures the direct current component of the signal collected in a beam of etendue U. Comparison of the two methods for instruments of equivalent band pass gives a factor of

$$\frac{(\text{Signal/Noise})_{CLASS}}{(\text{Signal/Noise})_{Het}} = \sqrt{\frac{U}{\lambda^2}} = \sqrt{m}$$

where m can be made $\gg 1$ for classical instruments. He concludes that the gain thus lies in the information content of the light beams. In the classical methods the light beam having m degrees of freedom can transmit m distinct pieces of information. In heterodyne spectroscopy the detector and analyser can only operate on the information content of a single optical mode.

Lastovka's treatment also deals with the amount of information that any device may extract from the electro-magnetic field. However he additionally emphasizes the conjugate nature of the quantities measured by a detector - the intensity and knowledge of the phase. A single measurement from which both the phase α and the number of quanta n in the beam can be deduced implies an uncertainty in these quantities which must satisfy the inequality.

$$\Delta n \ \Delta\alpha \geq \tfrac{1}{2}$$

He shows that an ideal optical mixing spectrometer operating at its limit of sensitivity and necessarily requiring phase information is extracting from the incident field the maximum amount of information allowed by the uncertainty principle. This is to be compared with the classical interferometer where the information required and provided is contained in the dc component of the detector output; the uncertainty in α is infinite. Thus in principle the presence of a signal can be deduced from a single photodetection counted in an arbitrarily long time. Lastovka is thus led to the conclusion that the inherent sensitivity of a conventional optical spectrometer is infinitely greater than that of an optical mixing spectrometer and that the origin of this advantage lies in the amount of phase information which is obtained by each method.

The application of these ideas to time domain correlation spectroscopy does not require much elaboration since a real time correlator is in essence equivalent to a parallel bank of filters - an ideal multi-channel wave analyser. As an illustration, consider a low-level light field composed of two frequencies ν_1 and ν_2 beating together at the surface of a photodetector. From two adjacent photon counts at a well defined time interval Δt a difference frequency $\nu_1 - \nu_2$ can only be deduced provided one has knowledge of the relative phase. Clearly large phase uncertainty leads to a large frequency uncertainty. The correlator output will contain an event in the channel corresponding to the delay Δt, but this single event can in no way define the spectrum. On the other hand consider an ideal multi-channel classical spectrometer of equivalent etendue and noiseless detection (zero dark current). Two detections in different channels at arbitrary times measure the frequency difference to within the width of the channels themselves.

TABLE 1 Experimental Parameters in the Comparison of Correlation and Interferometric Spectroscopy

	PARAMETER	FABRY-PEROT	CORRELATOR
1	Laser stability	Requires a single mode frequency stability much higher than the required resolution over times equal to a scanning period.	Frequency stability relativity unimportant. Requires good amplitude stability over periods of the length of the correlator store.
2	Detector and recording electronics:	Requires reasonable linearity.	Must be completely free of internal correlation. (see Oliver this volume).
3	Light grasp	Limited only by acceptable spread of K vector and required resolution.	Strictly limited by one coherence area condition; etendue λ^2
4	Operating mode:	Single channel (when operating as a scanning filter).	Multichannel for real time processing.
5	Systematic errors:	Need to unfold instrumental width, problems of overlap of orders and scanning linearity etc.	Usually only found in the first channel of the correlator due to dead time effects.
6	Parasitic light (dust, flare etc)	Introduces gross distortion of the spectra	Can be added to a homodyne reference beam and discounted in the first approximation.
7	Interpretation of complicated spectra; (multi component of compound spectral form):	Relatively straightforward within the limitations of instrumental width, and items listed in 5.	Interpretation requires correlation data of extremely high precision.

We conclude this section with an example and a list of practical comparisons. From the earlier expressions one finds that the correlator processing 10^5 counts per second from a Lorentzian line of half-value width 16 MHz would give a line width error of ~4.6% in an experiment time of 100 s. On the other hand an interferometer would give ~ 2% error from 10^4 counts accumulated over an indefinitely long period. However this advantage of signal utilisation is only one aspect; a full comparison in practice shows several corresponding advantages for the correlation technique as indicated in table 1. Some of these points are elaborated in the following section.

EXAMPLES OF INTERFEROMETRY AND CORRELATION

1 Rayleigh Scattering

The unshifted Rayleigh component of the light scattered from pure liquids is most simply considered as arising from non-propagating isobaric entropy fluctuations. The half-width at half-height of the Rayleigh line (of Lorentzian spectral form) is given by

$$\Gamma = \frac{\Lambda}{C_p^*} \left(2n\, k_o \sin \frac{\theta}{2}\right)^2 \quad \text{rad s}^{-1}$$

where k_o is the magnitude of the incident wave vector, θ the scattering angle, n the refractive index, C_p^* the specific heat per

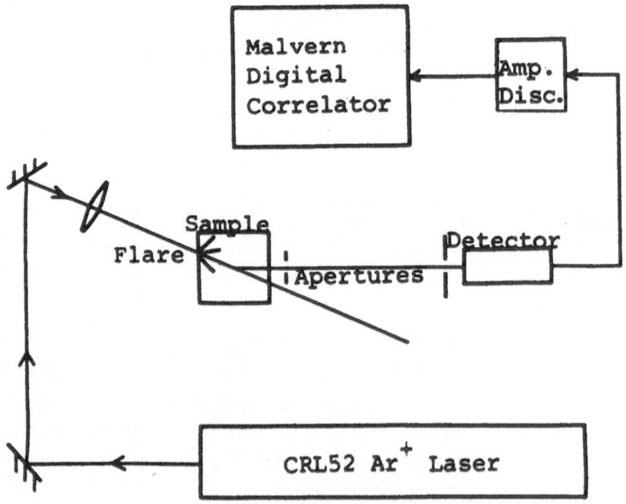

FIG 5 Experimental arrangement for measuring the homodyne photon-correlation spectrum for non-critical Rayleigh scattering of laser light by pure liquids.

unit volume and Λ the thermal conductivity. The thermal diffusivity depends on the specific heat and the thermal conductivity and is given by Λ/C_p^* (but see also the article by Pike, this volume). Measurement of the infinite wavelength thermodynamic variable for pure liquids are subject to large discrepancies as can be seen from tables of constants. Measurement of the Rayleigh line-width provides potentially a very accurate determination of the thermal diffusivity. Replacing typical values of Λ/C_p^* ($\sim 10^{-3}$ cm^2 s^{-1}) in the equation, the Rayleigh full width at half-height can be seen to be about 15 to 30 MHz at a scattering angle of 90° for visible light.

Several measurements have in fact been made for various pure organic liquids by Greytak (1968), Oliver and Pike (1970) and Searby (1971) using Fabry-Perot interferometers. This work was done at large scattering angles to obtain the greatest possible ratio of Rayleigh width to instrumental width.

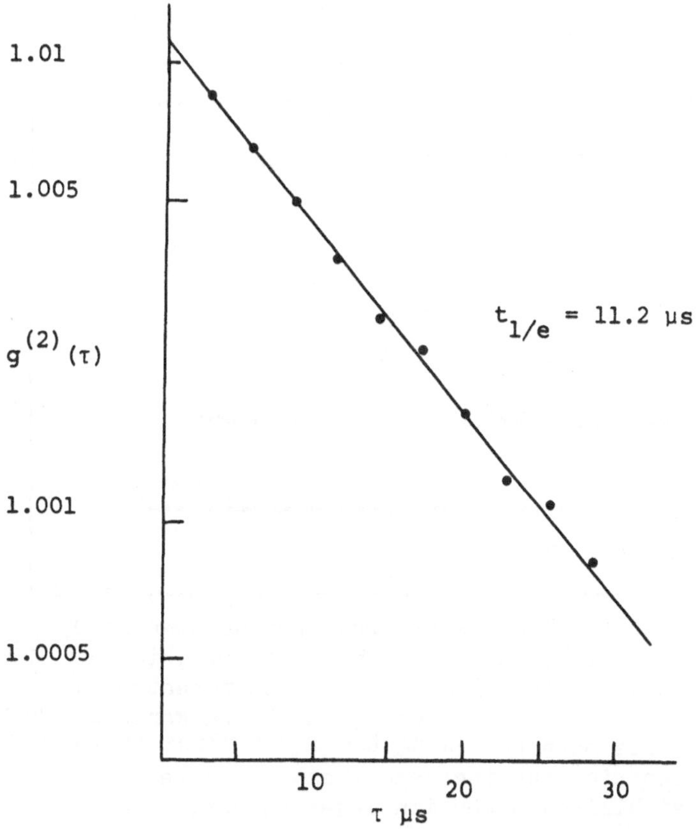

FIG 6 A typical autocorrelation function for scattering off carbon tetrachloride. The scattering angle was 3° 23' and the temperature 20°C. (Oliver et al 1972).

The scattered light is quite weak and clearly larger signals per unit bandwidth (counts per coherence time) are attained at smaller scattering angle. This has been used in the narrow-angle measurements with homodyne techniques and wave analysers carried out by Lastovka and Benedek (1966) and Berge and Dubois (1969). Correlation, providing as it does an efficient use of signal in this regime, is thus particularly suitable and a preliminary account of measurements on carbon tetrachloride has been published (Oliver et al 1972). The experimental arrangement is shown in figure 5; unshifted flare light at the laser frequency provides a coherent reference beam. A typical auto-correlation function is shown in figure 6. The intercept on the y axis is small because of the large homodyne reference beam, and also the large incoherent component due to Brillouin scattering. Clearly line width measurements of high precision can be made from such data. From a series of these recordings at different scattering angles the thermal diffusivity was calculated to be $7.6 + 0.3 \times 10^{-4}\ cm^2\ s^{-1}$.

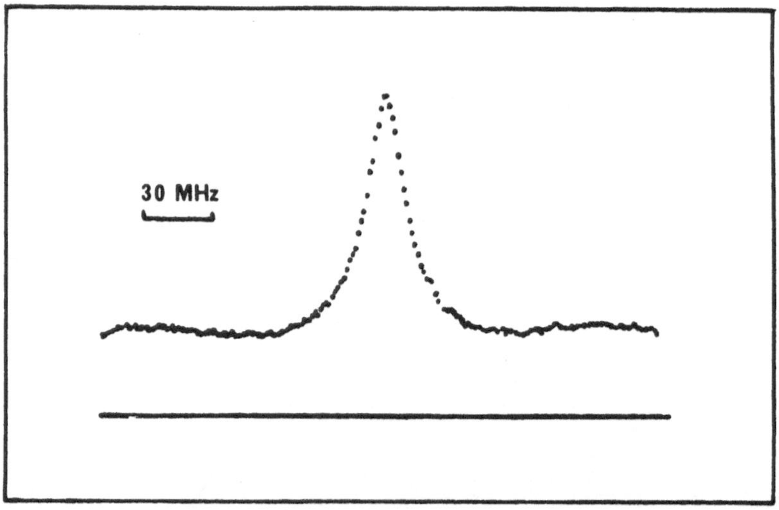

FIG 7 Spectrum of the light scattered from benzene at 90°. The central Rayleigh line is ~25 MHz wide of which 5.6 MHz is instrumental width. The Brillouin components and depolarised scattering overlap in the background. This spectrum was obtained using the large plane etalon with piezo-electric scanning described in figure 3, and a highly stabilised argon ion laser source.

A corresponding spectrum of the Rayleigh line in benzene is shown in figure 7. Some of the problems of precision analysis are immediately apparent. One has first to subtract the large background due to depolarised scattering and the Brillouin lines (whose broad shape, overlapping many order of intereference is discernible in the background). One must then de-convolve the instrumental function and any contribution due to spread of k-vector. Remembering that there is the possibility of large systematic error due to the presence of even a very small amount of parasitic light, it is difficult to reduce the ultimate error below ~ 10%. Clearly in the case of the analysis of a line of simple profile the correlation technique has some considerable advantages.

2 Scattering From Binary Mixtures

In mixtures light scattering can also arise from non-propagating fluctuations in relative concentration. This additional spectral component has a line width given by:

$$\Gamma = D\,(2nk_o \sin\frac{\theta}{2})^2 \text{ rad s}^{-1}$$

where D the mass diffusion coefficient is typically two orders of magnitude smaller than thermal diffusivities. Measurements have been made of this parameter using homodyne techniques and wave analysers; [(Berge et al (1970), Dubois and Berge (1971), Arefeev et al (1967), Jamieson and Walton (1973)]. In the system 10% by volume acetone in carbon disulphide results differing by an order of magnitude have been reported. The ratio of the light scattered by the concentration fluctuation to that scattered by the entropy fluctuation is also of considerable interest. This latter parameter can be deduced with some uncertainty from the wave analyser measurements. Intensity correlation measurements have recently been carried out on this binary mixture (Vaughan 1973) and the results of a series of experiments are shown in figure 8. In this case at the relatively large scattering angles examined (~ 10° - 30°), the experiments were conducted in the intensity fluctuation or self beating mode without local oscillator The linear dependence of line width on K^2 is established, and the value of the mass diffusivity of 2.44×10^{-5} cm^2 s^{-1} is close to that found by Berge et al. However while the intensity ratio could be deduced from the correlator data, it seems likely that a much

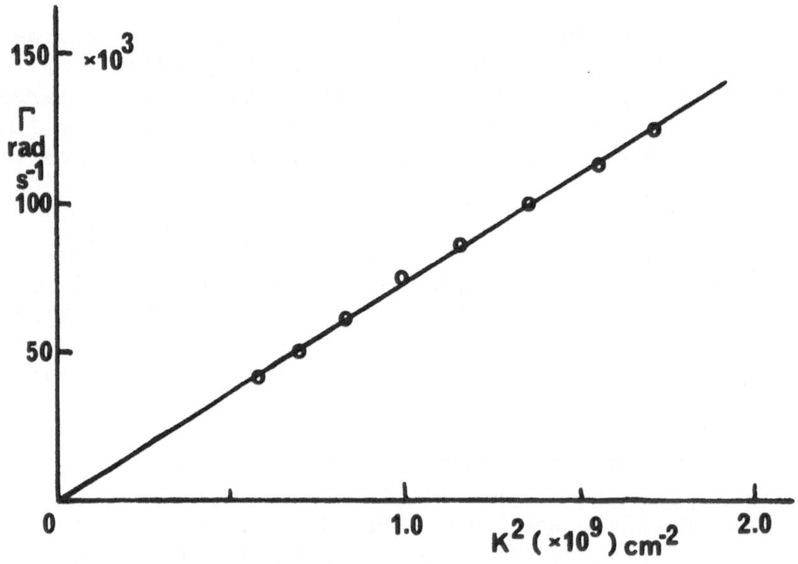

FIG 8 Correlation analysis of light scattered from a 10% V/V acetone/carbon disulphide mixture at 22°C.

more reliable value would be obtained by spectral analysis using an interferometer. Rudimentary analyses of spectral profiles at different concentrations would be required, followed by calculation of the integrated intensities within an appropriate frequency range

3 Chemical Reactions

Some initial experiments on the light scattered by various salt solutions have been reported (Yeh and Keeler 1969). The component of the light scattered by fluctuations due to a chemical reaction would be expected to be of width related to a reaction relaxation time τ by

$$\Gamma = \frac{1}{\tau} \quad \text{rad s}^{-1}$$

This expression is independent of the angle of scattering; strong central components of width 10 to 30 MHz (corresponding to $\tau \sim 10^{-8}$ sec) were found using a Fabry-Perot spectrometer and He-Ne laser. Experiments in this laboratory have been unable to duplicate these results (Clarke, Oliver, Vaughan, 1973). Figure 9 shows recordings of the instrumental function, and of scattering from 1.0M $MnSO_4$ with good signal to noise using a very high resolution Fabry-Perot etalon. No difference in width is apparent. Detailed investigations were made for different concentrations, for various states of polarisation of the incident and scattered light, and also in the wings of the line, but no appreciable additional component was observed. The scattered light was also analysed with a correlator and the results are shown in figure 10. A scattered component of quite high frequency is observed (< 200 kHz), however due to its diffusion-like angular dependence it cannot be attributed to scattering from a chemical reaction but must almost certainly be due to scattering from the moving ions themselves and their surrounding hydration sheath.

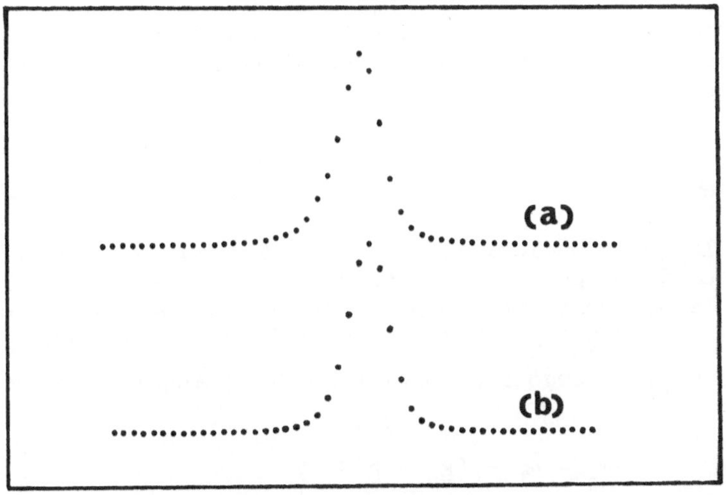

FIG 9 Spectra of (a) light scattered from 1.0M $MnSO_4$ at 90° and (b) the instrumental function of light scattered from black card. Each channel is of width 1.15 MHz and the recordings were made using the instrument described in figure 3. No differences could be discerned in the two recordings.

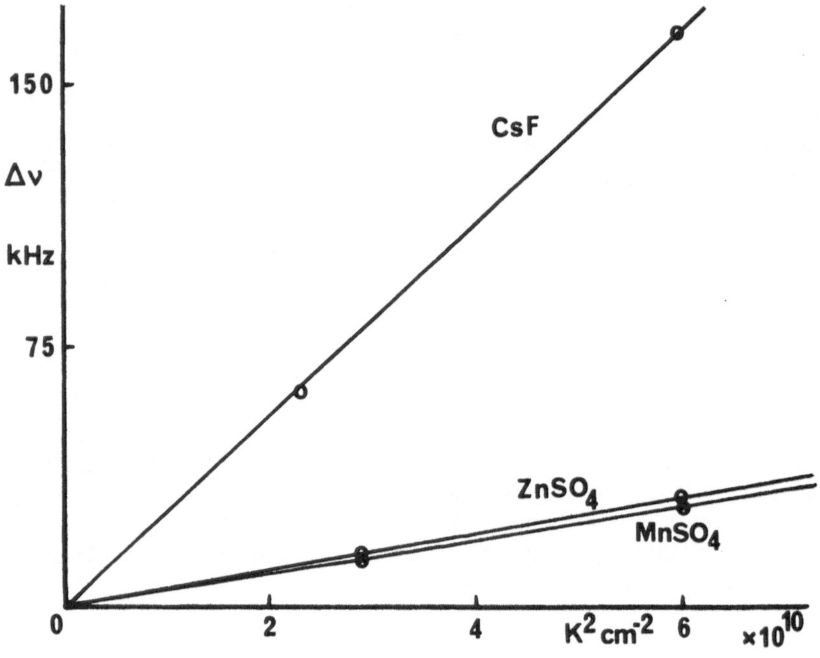

FIG 10 Correlation analysis of light scattered by various chemical solutions showing the diffusion like angular dependence.

4 Laser Anemometry

In this large technological field a wide variety of methods have now been developed. The underlying principle of laser anemometry is of course the Doppler effect : light scattered into a wave vector $\underset{\sim}{k}_s$ by a particle of velocity $\underset{\sim}{V}$ from an incident beam $\underset{\sim}{k}_i$ undergoes a Doppler shift of angular frequency given by

$$\Delta_{si} = \omega_s - \omega_i = (\underset{\sim}{k}_s - \underset{\sim}{k}_i) . \underset{\sim}{V}$$

In the visible region the Doppler shifts are quite large - characteristically ~4 MHz in back scatter for a line-of-sight velocity component of ~ 1 ms^{-1} at a wavelength of 0.5μm. The shifts can of course be measured by heterodyne mixing techniques or by direct observation with an interferometer. The widely used 'differential Doppler' or real fringe technique is in effect a device for reducing the magnitude of the shift. In this method two incident beams overlap at the point of investigation. Light, scattered into a given vector, from each beam has a slightly different Doppler shift; the two components beat together to give a differential Doppler signal. The difference frequency is independent of the scattering vector and thus scattered light can be collected over a large solid angle. A less rigorous way of considering this method is to think of the overlapping beams as forming a system of real interference fringes. Light scattered by a particle passing through the fringes is consequently modulated at the difference frequency.

The technique is particularly suitable for correlation spectroscopy (Pike 1972) and has now been applied to measurements in a high speed wind tunnel (Abbiss et al 1972) and to the measurement of simple turbulent flows (Birch et al 1973). Figure 11 shows the arrangement as used in the latter experiment. A typical correlation function is shown in figure 12; the amplitude decay arises from the finite number of of fringes and any spread of velocities (turbulence) which of course reduces correlation at longer delay time. For particles of a simple velocity distribution (in this case assumed Gaussian), the mean flow and turbulence parameters were obtained. In the experiments of Birch et al good agreement was found between correlation measurements in a free jet and hot wire results. Similarly in the wind tunnel trials good agreement was obtained

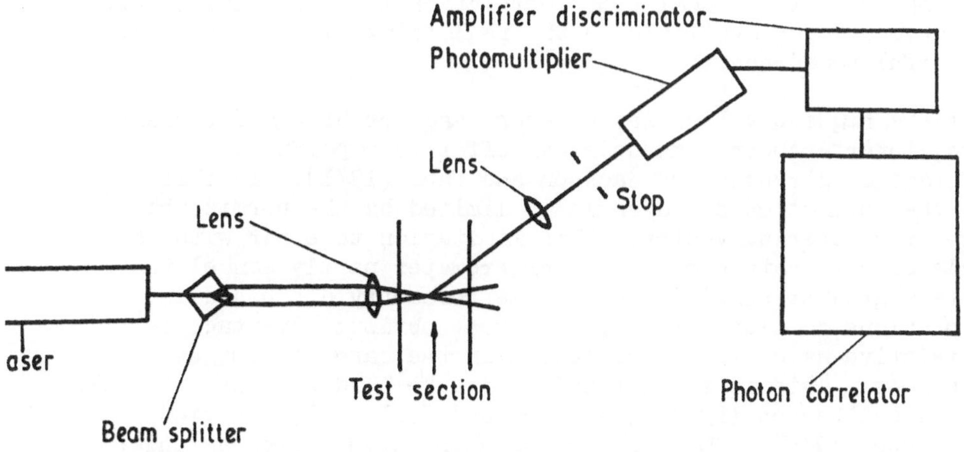

FIG 11 A differential Doppler or fringe mode laser anemometer. The incident beams overlap in the test section

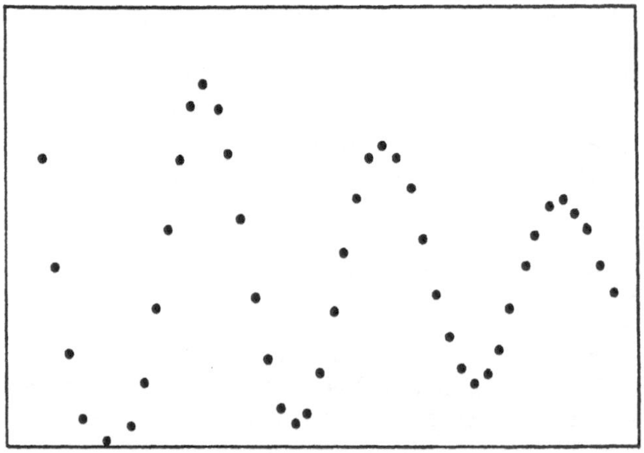

FIG 12 A typical photon correlation function obtained in a free jet for a fringe mode anemometer. The channel time was 50 ns

with an aerodynamic calibration; velocities up to Mach number 2.5 were examined. It should be remarked that in these types of measurements one requires the highest efficiency in the use of available signal from naturally occurring particles. Artifical seeding for instance is often undesirable or impractable in many cases. Correlation experiments have been successfully carried out with average signals as low as one detection per 500 Doppler cycles. Real time correlation methods combined with the large etendue available in the real fringe mode thus provide a powerful technique.

At the highest velocities however, and for highly turbulent flows, interferometric methods can offer some points of advantage as discussed by Jackson and Paul (1971). In this case the collection solid angle is limited by the permissible range of scattering vector. This limitation together with the single-channel nature of the interferometer partly cancel the inherent greater sensitivity of interferometry discussed in the previous section. Perhaps the most obvious advantage is the relative ease of interpretation in the case of complex flow situations. This is illustrated in figure 13 due to Dr D A Jackson and his colleagues (Eggins, Jackson and Paul, 1973; Eggins and Jackson 1973). The diagram shows the signal from an under expanded free jet with the measuring volume straddling the normal shock wave. In terms of practical analysis correlation methods

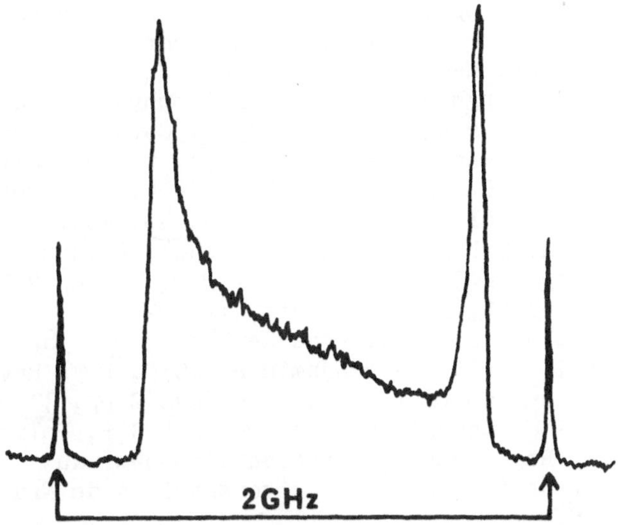

FIG 13 Spectrum of light scattered from an under expanded free jet using a confocal Fabry-Perot spectrometer of FSR 2GHz. The measuring volume straddled the normal shock wave and the spatial resolution was 200 μm. The incident light was from a single-mode argon ion laser at 514 nm and a power level of ~200 mW. The scattering angle was 90° and a shift of 1 order corresponded to 730 ms^{-1}. The scan rate was 1 minute per order and no artifical seeding of the flow was used. (Eggins and Jackson 1973).

applied to such a situation would provide a correlation function of considerable complexity.

REFERENCES

Abbiss J B, Chubb T W, Mundell A R G, Sharpe P R, Oliver C J, Pike E R, 1972, J Phys D:Appl Phys, 5, L100.

Arefèev I M, Kopylovskii B D, Mask D Sh, and Fabelinskii I L, 1967, JETP Letters, 5, 355.

Berge P and Dubois M, 1971, Phys Rev Lett, 26, 121; and Calmettes P, and Laj C, 1970, Phys Rev Lett, 24, 89.

Berge P and Dubois M, 1969, C R Acad Sci (Paris) 269, 842.

Bergstein L and Schuchter H, 1964, J Opt Soc Am, 54, 887.

Birch A D, Brown D R, Thomas J R and Pike E R, 1973, J Phys D : Appl Phys, 6, L71.

Boyd G D and Gordon J P, 1961, Bell Syst Tech J, 40, 489.

Boyd G D and Kogelnik H, 1962, Bell Syst Tech J, 41, 1.

Cannell D S and Benedek G B, 1970, Phys Rev Lett, 25, 1157.

Clarke J H R, Oliver C J, Vaughan J M, 1973, to be published.

Connes P, 1956, Rev Opt, 35, 37; 1958, J Phys Radium, 19, 262; 1964 Quantum Electronics and Coherent Light, p207 Academic Press, New York and London.
Davies Tudor and Vaughan J M, 1963, Astrophys Journal, 137, 1302.
Eggins P L, Jackson D A, Paul D M, 1973, Opto Electronics, 5, 91; Eggins P L and Jackson D A 1973, to be published.
Fork R L, Herriott D R and Kogelnik H, 1964, Appl Opt, 3, 1471.
Fox A G and Li T 1961, Bell Syst Tech J, 40, 453.
Gammon R W, Swinney H L, Cummins H Z, 1967, Phys Rev Lett, 19, 1467.
Gornall W S Stegeman G I A, Stoicheff B P, Stolen R H and Volterra V, 1966, Phys Rev Lett, 17, 297.
Greytak T J and Yan J, 1969, Phys Rev Lett, 22, 987; with Woerner R L, Yan J and Benjamin R, 1970, Phys Rev Lett 25, 1547; and Woerner 1972, J de Phys Supp, 33, C1.
Greytak T J, 1968, unpublished see Benedek G B, Polisation Matière et Rayonnement, Livre en l'honneur du Professeur Kastler Presses Universitaires de France, Paris 1968.
Greytak T J and Benedek G B 1966, Phys Rev Lett, 17, 179; Greytak T J, 1967, Ph D Thesis MIT.
Hercher M, 1968, Applied Optics, 7, 951.
Jackson D A and Paul D M, 1971, J Phys E, 4, 173.
Jackson D A and Pike E R, 1968, J Phys E: J Sci Inst, 1, 394.
Jackson D A and Simic-Glavaski B, 1970, Molec Phys, 18, 393; Lucas H C (nee Craddock), Jackson D A, Powles J G and Simic-Glavaski 1970, Molec Phys, 18, 505.
Jacquinot P, 1960, Rep Prog Phys, 23, 267.
Jakeman E, Pike E R and Swain S, 1971, J Phys A General Physics 4, 517.
Jamieson A M and Walton A G, 1973, J Chem Phys, 58, 1054.
Kuhn H G, 1951, Rep Prog Phys, 14, 64; 1964, Acta Physica Polonica, 26, 315.
Kuhn H G and Vaughan J M, 1964, Proc Roy Soc, 277A, 297.
Lastovka J and Benedek G B, 1966, Phys Rev Letts, 17, 1039.
Lastovka J, 1967, Ph D Thesis MIT.
Mohr R, Langley K H and Ford N C Jr, 1969, J Acoust Soc Am, 49, 1030.
Oliver C J, Pike E R and Vaughan J M, 1972, Coherence and Quantum Optics, Plenum, New York, p457.
Oliver C J and Pike E R, 1970, Phys Lett, 31A, 90.
Pike E R, 1972, J Phys D:Appl Phys, 5, L23.
Pike E R, Vaughan J M and Vinen W F, 1969, Phys letters 30A, 373, 1970, J Phys C Solid State Physics, 3 37, with Palin C J, 1971, J Phys C Solid State Physics, Solid State Phys, 4, L225.
Sandercock J R, 1970, Opt Comm, 2, 73; 1971, Light Scattering in Solids, Flammarion Paris, 1972 Phys Rev Letts, 29, 1735, with Palmer S B, Elliott R J, Hayes W, Smith S R P, and Young A P, 1972, J Phys C : Solid State Phys, 5, 3126.

Searby G M, 1971, D Phil Thesis, Oxford University.
Series G W 1957, The Spectrum of Atomic Hydrogen, Oxford University Press, Oxford.
Toraldo di Francia G, 1970, Quantum Optics, Academic Press, London New York.
Vaughan J M, 1968, Phys Rev, 166, 13, 1973, to be published.
Vaughan J M, Vinen W F and Palin C J, 1972, Proc LT13, in the press.
Volterra V, 1969, Phys Rev. 180, 156.
Winterling G, Holmes F S, Greytak T J, 1972, Proc LT13 in the press; 1973 Phys Rev Lett 30, 427.
Yeh Y and Keeler R N, 1969, J Chem Phys, 51, 1120.

INVESTIGATION OF LAMINAR AND TURBULENT FLOWS BY MEANS OF LIGHT SCATTERING

Bruno Crosignani and Paolo Di Porto

Fondazione Ugo Bordoni, Istituto Superiore Poste e Telecomunicazioni, Roma, Italy

1. Introduction

The advent of laser sources and of sophisticated frequency shift measurements as the ones provided by optical heterodyne and self-beating (homodyne) techniques (see, for example, Cummins and Swinney 1970, Benedek 1969) has opened the way to a broad class of applications in the frame of "quasi-elastic" light scattering. This term refers in particular to the small frequency Doppler-shift associated with the motion of scatterers, whose typical dimension is large compared with that of an ordinary molecule.

We shall deal in the following with a situation in which the scatterers are embedded in an otherwise homogeneous fluid, whose dielectric constant fluctuation gives rise to a diffused field negligible with respect to that due to the seeding particles, so that the analysis of the scattered radiation furnishes a direct information on their motion. Two situations are encountered in practice, according whether the host fluid possesses or not a macroscopic motion. In the second case, the scatterers undergo only a Brownian diffusion because of their interaction with the surrounding molecules and the scattered field is connected with the size and shape of the scatterers and with the transport properties of the fluid (see Cummins, this Course). In the first case, each particle follows the motion of the fluid element in which it is embedded, an hypothesis verified in practice whenever it is very small compared with the smallest scale of the macroscopic velocity variation (see Hinze, 1959, pg. 352). Thus, if the associated frequency shifts are larger than the broadening pertaining to the Brownian motion, one can observe the part of the electromagnetic field connected with the dynamics of the medium.

The usual scattering experiments deal with the measurement of the optical spectrum, since one is led to take for granted that the statistics of the diffused radiation is a Gaussian one. All information available from the electromagnetic field $\underline{E}(t)$ is then contained in the ensemble average $\langle \underline{E}(t) \cdot \underline{E}^*(0) \rangle$ (see Jakeman, this Course). The simplest higher order quantity is the intensity correlation $\langle |\underline{E}(t)|^2 |\underline{E}(0)|^2 \rangle$ and does not contain further information, although its measurement may be sometimes preferable for experimental reasons. This is for example the case of the Gaussian field diffused by a suspension of Brownian particles (see Cummins, this Course), where the homodyne technique is usually employed.

We treat here the situation of a seeded fluid undergoing a macroscopic motion. First we consider a laminar regime and evaluate the diffused field, which turns out to possess a Gaussian statistics. Then we consider a regime of hydrodynamical turbulence, so that the particles are subjected to a chaotic wandering in the fluid. As we shall see, a comparison between the correlation functions $\langle \underline{E}(t) \cdot \underline{E}^*(0) \rangle$ and $\langle |\underline{E}(t)|^2 |\underline{E}(0)|^2 \rangle$ shows a remarkable departure of the statistical behavior of the scattered electromagnetic field from the Gaussian one. This allows to perform experiments apt to investigate the spatial structure of the turbulent field, which cannot be determined by examining the diffused optical spectrum alone. It is finally hardly necessary to underline the advantages of this method over the traditional hot-wire anemometer technique, such as the absence of perturbation introduced in the system and the possibility of remote sensing.

2. The quasi-elastically scattered field

The problem of determining the electromagnetic field diffused by an ensemble of moving particles has been first investigated by Pecora (1964). If a linearly polarized, monochromatic plane wave

$$\underline{E}_0(\underline{r}, t) = \underline{E}_0 \exp\left[i(\underline{k}_0 \cdot \underline{r} - \omega_0 t)\right] \tag{1}$$

of angular frequency ω_0 and wavenumber $k_0 = \omega_0/c$ impinges on the scatterer, one can write in full generality the scattered field $\underline{E}_i(t)$ observed at a point $\underline{R}$ in the wave zone as (see Fig. 1)

$$\underline{E}_i(t) = \underline{A}_i(t) \exp\left\{i\left[\underline{k}_1 \cdot \underline{r}_i(t) - \omega_0 t\right]\right\} \quad . \tag{2}$$

Here $\underline{k}_1 = \underline{k}_0 - k_0 \underline{R}/R$, $\underline{r}_i(t)$ represents the trajectory of a convenient point of the i-th scatterer and $\underline{A}_i(t)$ is in general a time-dependent vector, whose magnitude and direction are related to its size, shape and rotational motion around $\underline{r}_i(t)$.

The expression of $\underline{A}_i(t)$ greatly simplifies if one considers a spherical particle or one obeying the conditions (see Van de Hulst, 1957)

$$d \ll 2\pi/k_0 \tag{3}$$

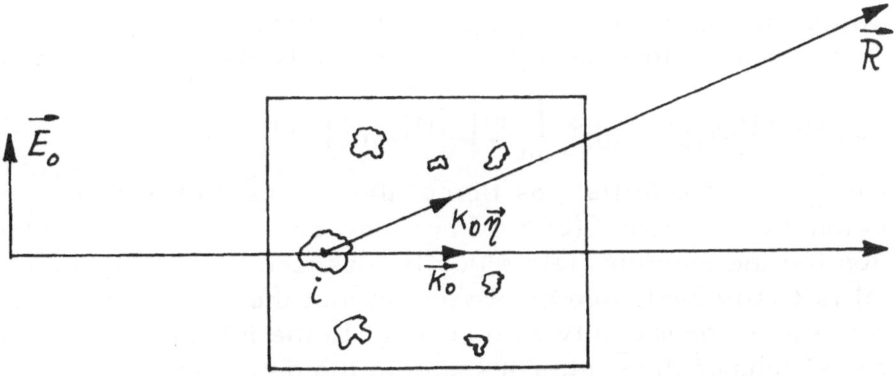

Fig. 1 Geometry of the scattering experiment.

and

$$|n - n_0| << 1 \quad , \tag{4}$$

where d is a typical dimension of the scatterer, n its index of refraction and n_0 that of the host fluid. In both cases, $\underline{A}_i(t)$ turns out to be independent from time, irrespectively of the rotational motion. Its expression can be obtained by means of Mie's theory for a spherical particle (Born and Wolf, 1970), while in the other case it reads

$$\underline{A}_i = - k_0 \underline{\eta} \times (\underline{\eta} \times \underline{E}_0)\, 2n_0(n - n_0)V_i/(4\pi R)\, \exp(in_0 \omega_0 R/c), \tag{5}$$

where V_i is the volume of the scatterer and $\underline{\eta} = \underline{R}/R$. For our diagnostic purpose, we can choose this kind of seeding particles, so that the field diffused by N identical scatterers reads

$$\underline{E}(t) = \underline{A} \sum_{i=1}^{N} \exp\left\{ i \left[\underline{k}_1 \cdot \underline{r}_i(t) - \omega_0 t \right] \right\} \quad , \tag{6}$$

where $\underline{r}_i(t)$ is the position of the centre for a spherical particle. As usual, the linear dimension of the total scattering volume V_S, that is of the zone where the particles are confined, is assumed to be small with respect to the observation distance R.

3. Scattering by particles suspended in a laminar flow

In order to apply Eq. (6) to a specific situation, it is necessary to write the expression of $\underline{r}_i(t)$. Let us consider the case of a stationary laminar flow described by a velocity field $\underline{U}(\underline{r})$. One has

$$\underline{r}_i(t) \equiv \underline{r}(t, \underline{r}_{i0}) = \underline{r}_{i0} + \int_0^t \underline{U}\left[\underline{r}(t', \underline{r}_{i0})\right] dt' \quad , \tag{7}$$

where $\underline{r}_{i0}$ is the initial position of the i-th particle. The correlation function $< E(t)E^*(0) >$ (from now on, we omit the vector notation for the electric field since it possesses a well defined direction) is easily evaluated by observing that the ensemble average has to be performed only with respect to the initial distribution of the positions of the scatterers, due to the deterministic character of $\underline{U}(\underline{r})$. One can immediately write

$$\langle E(t)E^*(0)\rangle = |\underline{A}|^2 \sum_{i,j}^{1,N} \Big\langle \exp\Big\{ i\underline{k}_1\cdot(\underline{r}_{i0} - \underline{r}_{j0}) + i\underline{k}_1\cdot\int_0^t \underline{u}\big[\underline{r}(t',\underline{r}_{i0})\big]\,dt'\Big\}\Big\rangle \exp(-i\,\omega_0 t)$$

$$= |\underline{A}|^2 N \Big\langle \exp\Big\{ i\underline{k}_1\cdot\int_0^t \underline{u}\big[\underline{r}(t',\underline{r}_{i0})\big]\,dt'\Big\}\Big\rangle \times \exp(-i\,\omega_0 t) \quad , \qquad (8)$$

having assumed the distribution $P(\underline{r}_{i0},\underline{r}_{j0})$ of the initial particle positions to be uniform. Strictly speaking, a $P(\underline{r}_{i0},\underline{r}_{j0})$ independent from $\underline{r}_{i0}$ and $\underline{r}_{j0}$ does not always describe in a correct way our stationary situation, but in practice its spatial variation takes place on a scale much larger than $1/k_1$, thus justifying the neglection of the terms with $i \neq j$ in Eq. (8).

Let us consider for example the practical situation of a parabolic profile of the velocity field (see Fig. 2)

$$u_x = u_y = 0 \quad , \qquad u_z(\varrho) = B\,(R^2 - \varrho^2) \quad , \qquad (9)$$

where B is a suitable constant, as the one present in a tube of con-

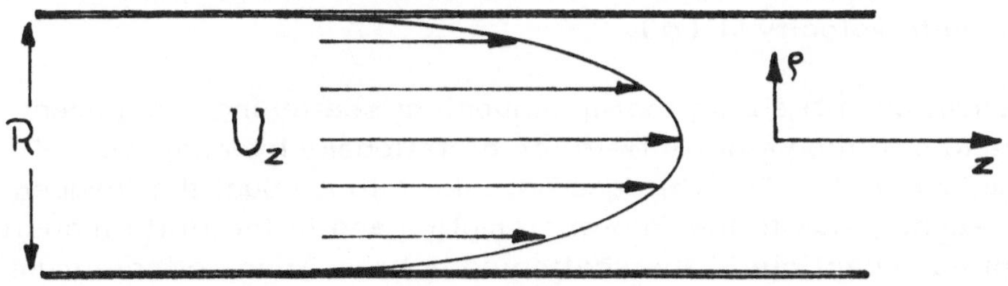

Fig. 2 Parabolic laminar velocity profile.

stant section. By introducing Eq. (9) into Eq. (8), we obtain

$$\langle E(t)E^*(0)\rangle = |\underline{A}|^2 N \exp(-i\,\omega_0 t)\langle \exp[ik_{1z}u_z(\varrho)t]\rangle$$

$$= |\underline{A}|^2 N \exp(-i\,\omega_0 t)\,(2\pi/V_s)$$

$$\times \int_{V_s} \exp\left[ik_{1z}u_z(\varrho)t\right] dV_s \quad . \qquad (10)$$

If the transverse dimension of V_s is small with respect to the tube diameter, Eq. (10) reduces to

$$\langle E(t)E^*(0)\rangle = |\underline{A}|^2 N \exp\left\{i\left[k_{1z}u_z(\bar{\varrho}) - \omega_0\right]t\right\} \quad , \qquad (11)$$

where $\bar{\varrho}$ is an average radial coordinate of the scattering region.

The analysis of the lowest significant order statistical properties of a stationary radiation is usually made by investigating its optical spectrum $S(\omega)$, which is related to $\langle E(t)E^*(0)\rangle$ by means of the Wiener-Kintchine theorem through the equation

$$S(\omega) = \left[c/(16\pi^2)\right] \int_{-\infty}^{+\infty} \exp(i\omega t)\langle E(t)E^*(0)\rangle\, dt \quad . \qquad (12)$$

By inserting Eq. (11) into Eq. (12) we deduce

$$S(\omega) = \left[c/(8\pi)\right]|\underline{A}|^2 N\delta\left\{\omega - \left[\omega_0 - k_{1z}u_z(\bar{\varrho})\right]\right\} \quad , \qquad (13)$$

which describes the Doppler shift associated with the particles moving with velocity $u_z(\bar{\varrho})$.

Equation (13) shows that a convenient scattering experiment allows to map the velocity field for a stationary hydrodynamical regime. Actually, Eq. (13) does not take into account the frequency broadening due to the Brownian motion and to the finite transit time of each particle in the scattering volume. These effects must be considered case by case, since they can limit the sensitivity of the method.

The first experimental investigation has been performed in a liquid by Yeh and Cummins (1964) by employing the heterodyne technique and, successively, repeated in air by Foreman et al. (1965, 1966), (see Figs. 3 and 4). Since then, this kind of measurements has been widely applied and has given rise to a now-a-days standard technique known as Laser Anemometry (see, for example, for a good bibliography on recent applications, F. Durst et al. (1972)).

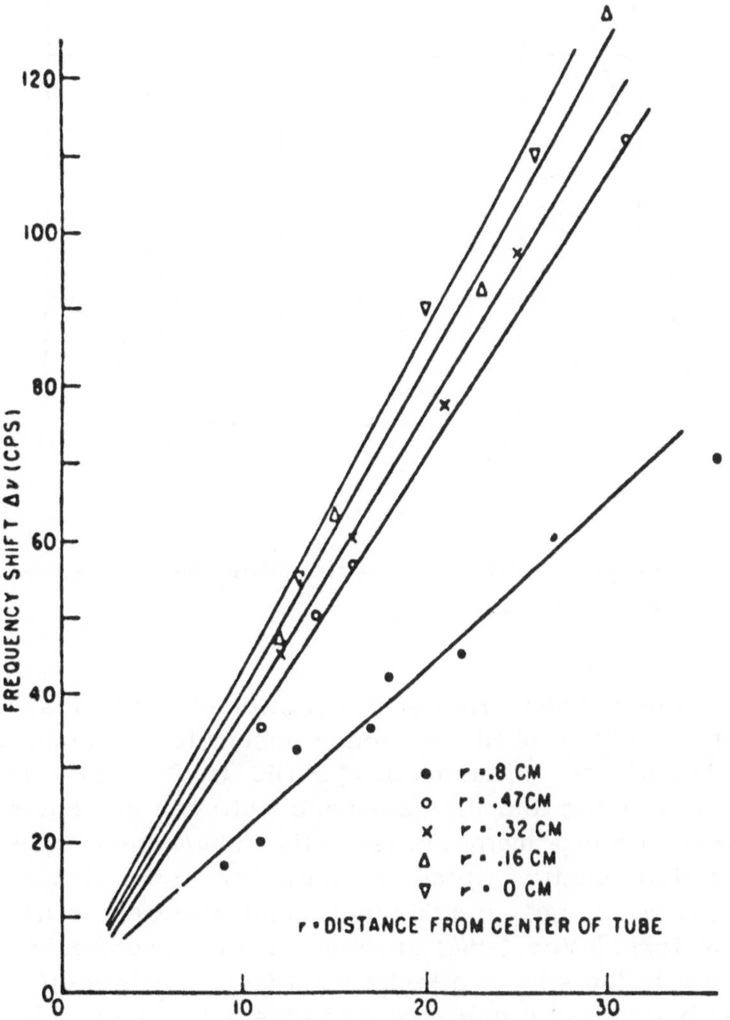

Fig. 3 Experimental frequency shifts at several radial positions as a function of flow rate (after Yeh and Cummins, 1964).

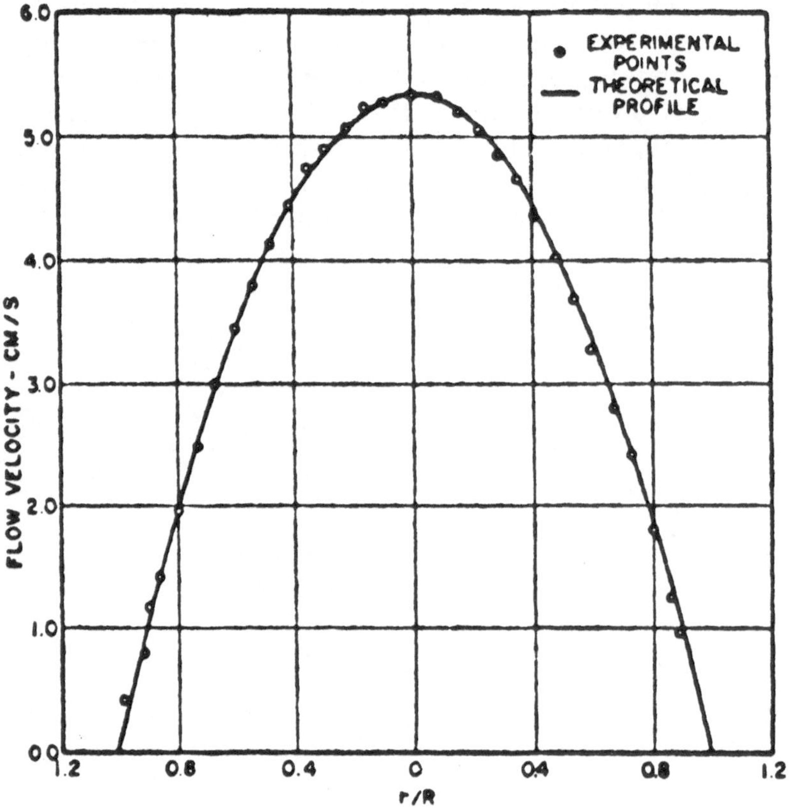

Fig. 4 Velocity profile of laminar water flow through a circular pipe (after Foreman et al., 1966).

The statistical properties of the scattered field, that is essentially the behavior of higher order ensemble averages as $\langle |E(t)|^2 |E(0)|^2 \rangle$ in terms of $\langle E(t)E^*(0) \rangle$, are simply obtained by observing that the scattered field can be considered as the superposition of many statistically independent and equivalent contributions each of which originates from the single scatterer. This allows to apply the Central Limit Theorem in its general vectorial form (Rice, 1964) stating that the M components of a vector, which is the sum of a large number of statistically independent and equivalent contributions, possess a joint-Gaussian (normal) distribution. Thus, the diffused field has a Gaussian statistics and higher order measurements do not contain a new information. In particular, the characteristic factorization of the normal distribution furnishes

$$\langle |E(t)|^2 |E(0)|^2 \rangle = \langle |E(t)|^2 \rangle^2 + |\langle E(t)E^*(0) \rangle|^2 . \quad (14)$$

Then, in the present case, the measurement of the intensity correlation is not useful, since this quantity does not depend on the velocity field as it is seen by inserting Eq.(11) into Eq.(14).

4. Scattering by particles suspended in a turbulent fluid

The introduction of higher order measurements, as that of the intensity correlation associated with homodyne technique, is essentially a matter of experimental convenience, as far as the Gaussian statistical behavior of the radiation is assumed to be known a priori. The situation which is described in this section constitutes the first example in which a significant departure of the statistics from a normal one allows us to obtain a more complete description of the medium by means of a homodyne measurement than that obtainable by employing the heterodyne technique alone (Di Porto et al., 1969).

Let us consider a turbulent flow described by a stochastic velocity field $\underline{U}(\underline{r}, t)$ which, under the hypotheses of homogeneity and stationariety, can be written as

$$\underline{U}(\underline{r}, t) = \underline{U}_0 + \underline{U}'(\underline{r}, t) \quad , \quad (15)$$

where $\underline{U}_0$ is the average velocity independent from space and time. The evaluation of $\langle E(t)E^*(0) \rangle$ and $\langle |E(t)|^2 |E(0)|^2 \rangle$ follows the same line as in the laminar case, Eq.(7) being now substituted by

$$\underline{r}_i(t) = \underline{r}_{i0} + \underline{U}_0 t + \int_0^t \underline{U}'\left[\underline{r}_i(t'), t'\right] dt' \quad . \quad (16)$$

Furthermore, one has to take into account the fact that each determination of the statistical variable $E(t)$ depends on the initial positions $\underline{r}_{i0}$ of the scattering particles, assumed to be uniformly distributed, and on the particular realization of the velocity field $\underline{U}'(\underline{r}, t)$. Thus, the ensemble average is performed by means of two independent averaging operations, one over the initial positions and one over the turbulent field.

The evaluation of the lowest order correlation function furnishes (Di Porto et al., 1969)

$$\langle E(t)E^*(0)\rangle = |\underline{A}|^2 N \exp\left[i(\underline{k}_1 \cdot \underline{U}_0 - \omega_0)t\right]$$
$$\times \exp\left[-(1/2)k_1^2\langle U_1^2\rangle \int_0^t dt' \int_0^t dt'' R_L(t''-t')\right] , \quad (17)$$

where the relation

$$\langle U_1^2\rangle R_L(t''-t') = \langle U_1[\underline{r}_i(t'), t'] \, U_1[\underline{r}_i(t''), t'']\rangle_T \quad (18)$$

defines the Lagrangian correlation function R_L of the velocity (Hinze, 1959, pg.42). Here the symbol $\langle \ldots\ldots \rangle_T$ indicates the average over turbulence, $U_1 = \underline{k}_1 \cdot \underline{U}'/k_1$ is the component of $\underline{U}'$ along the $\underline{k}_1$-direction, $\langle U_1^2\rangle$ being its mean square value, and use of the joint-Gaussian distribution hypothesis for the velocity field has been made. This amounts to say that the displacement $\underline{r}_i(t) - \underline{r}_{i0}$ of the generical particle is a Gaussian variable (for a discussion of this hypothesis see Hinze, 1959, pg.324).

The correlation time t^*, that is the time for which $R_L(t^*) \simeq 0$, is in practical situations such that $t^* \gg (k_1^2\langle U_1^2\rangle)^{-1/2}$, thus allowing us to write $R_L(t''-t') \simeq R_L(0) = 1$ for all significant times concerning Eq. (17). Thus, Eqs. (12) and (17) furnish

$$S(\omega) = \left[c\,|\underline{A}|^2 N/(16\,\pi^2)\right]\left[2\pi/(k_1^2\langle U_1^2\rangle)\right]^{1/2}$$
$$\times \exp\left\{-\left[\omega - (\omega_0 - \Omega_0)\right]^2/(2k_1^2\langle U_1^2\rangle)\right\} , \quad (19)$$

where $\Omega_0 = \underline{k}_1 \circ \underline{U}_0$ is the Doppler shift associated with the average fluid velocity. Equation (19) shows that, through the measurement of the diffused optical spectrum, one is able to obtain the intensity $\langle U_1^2\rangle$ of the turbulence but no information on its spatial correlation.

The first experimental results on the spectrum of light scattered by particles suspended in a turbulent medium were obtained by Goldstein and Hagen (1967) and by Pike et al. (1968) in liquids and by Lewis et al. (1968) in air by using the heterodyne technique. Their data were consistent with a Gaussian shape of the spectrum (see Fig. 5).

To proceed further, we evaluate the intensity correlation function which turns out to be (Di Porto et al., 1969)

$$\langle |E(t)|^2 |E(0)|^2 \rangle = |\underline{A}|^4 N^2 + |\underline{A}|^4 N(N-1)/(V_S^2) \times \int_{V_S} d\underline{r}' \int_{V_S} d\underline{r}'' \exp\left\{ -k_1^2 \langle U_1^2 \rangle \left[1 - f(\underline{r}'-\underline{r}'') \right] t^2 \right\} \quad (20)$$

where

$$\langle U_1^2 \rangle f(\underline{r}'-\underline{r}'') = \langle U_1(\underline{r}', 0)\, U_1(\underline{r}'', 0) \rangle_T \quad , \quad (21)$$

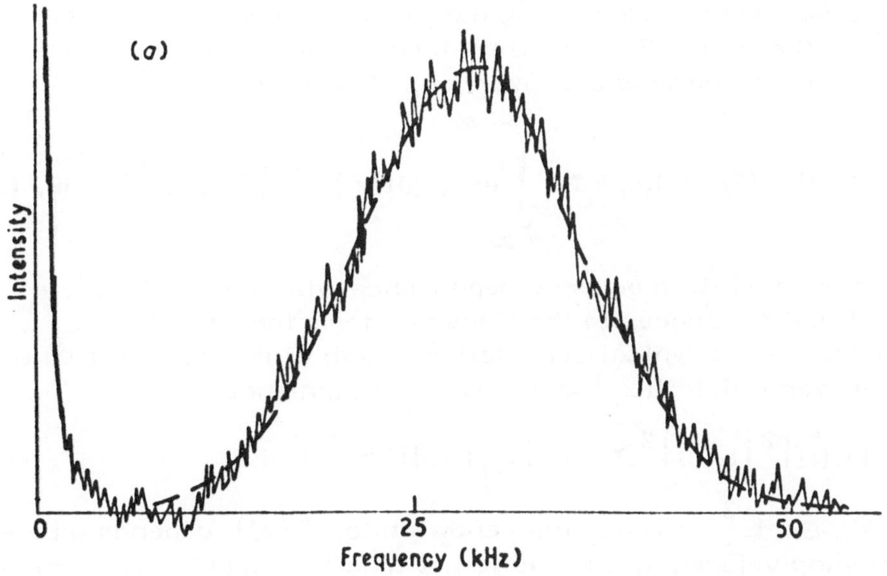

Fig. 5 Typical Gaussian optical spectrum $S(\omega)$ in a turbulent case (after Bourke et al., 1969).

and use has been made, as before, of the relation $t^* \gg (k_1^2 \langle u_1^2 \rangle)^{-1/2}$, which allows us to consider only equal time velocity correlation. Furthermore, it has been assumed that the distribution of the displacements of two generical particles is a joint-Gaussian one. In the limit $N \gg 1$ (this assumption has been already tacitly made in evaluating the optical spectrum, since the fluctuation of the number of particles in the scattering volume has been neglected; the influence of the scatterer number fluctuation on the statistics of the diffused field has been first put into evidence by Schaefer and Berne (1972) in the case of Brownian diffusion and, more in general, it is important whenever the average number of scatterers is comparable or smaller than unity) Eq. (20) can be rewritten, with the help of Eq. (17), in the form

$$\langle |E(t)|^2 |E(0)|^2 \rangle = \langle |E(t)|^2 \rangle^2 + |\langle E(t)E^*(0) \rangle|^2 \times (1/V_S^2) \int_{V_S} d\underline{r}' \int_{V_S} d\underline{r}'' \exp\left[k_1^2 \langle u_1^2 \rangle f(\underline{r}'-\underline{r}'')t^2\right]. \quad (22)$$

Equation (22) clearly shows a departure from the Gaussian behavior described by Eq. (14) and this circumstance allows us to obtain information not contained in the optical spectrum. This can be accomplished by measuring the intensity fluctuation spectrum $P(\omega)$, that is the Fourier transform of the intensity correlation function (see Cummins and Swinney, 1970)

$$P(\omega) = (1/2\pi)(c/8\pi)^2 \int_{-\infty}^{+\infty} \exp(i\omega t) \langle |E(t)|^2 |E(0)|^2 \rangle \, dt, \quad (23)$$

and comparing its frequency dependence with that of $S(\omega)$. In fact, Eq. (22) reduces to the Gaussian form for $V_S \gg L_c^3$, where L_c is the typical correlation length of the turbulent velocity field for which $f(L_c) \simeq 0$, while it furnishes

$$\langle |E(t)|^2 |E(0)|^2 \rangle = 2\langle |E(0)|^2 \rangle^2 \quad (24)$$

for $V_S \ll L_c^3$. Thus, the bandwidth of $P(\omega)$ depends on the scattering volume, while this is not true for $S(\omega)$, and decreases from $\sqrt{2}$ times the bandwidth of $S(\omega)$ (Gaussian limit) to zero.

This property has been experimentally verified by Bourke et al. (1968, 1969), who were able to give an estimate of the vortex dimension associated with the turbulent flow of water in a pipe (see Figs. 6 and 7). This is the first experiment of light scattering in which the non-Gaussian statistical behavior of the diffused field has been used for diagnostic purposes.

While Eq. (22) depends in general on the overall structure of the velocity correlation, it can be used for investigating in detail f($\underline{\varrho}$) by suitably choosing the scattering volume V_S. In particular, one can employ the geometry sketched in Fig. 8, that is one can collect the light scattered by the two volumes V_1 and V_2 much smaller than L_c^3 and separated by a distance ϱ, and measure the intensity fluctuation spectrum of the field superposition of the diffused fields. In fact, by specifying Eq. (22) to this situation one obtains (Bertolotti et al., 1971)

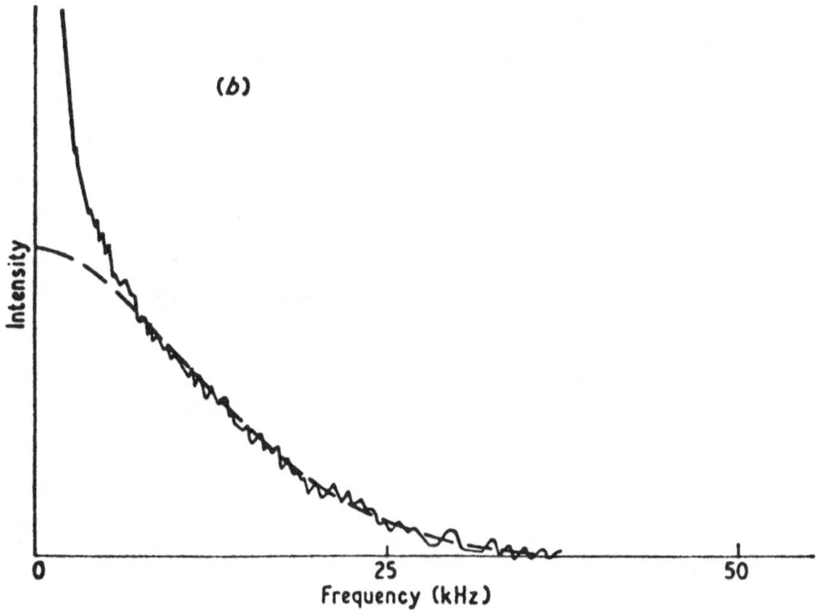

Fig. 6 Typical Gaussian intensity fluctuation spectrum $P(\omega)$ for large scattering volumes. The bandwidth is $\sqrt{2}$ times that of $S(\omega)$ given in Fig. 5 (after Bourke et al., 1969).

$$\langle |E(t)|^2 |E(0)|^2\rangle = |\underline{A}|^4 N^2 + |\underline{A}|^4 N^2 \Big\{ V_1^2 + V_2^2 + 2V_1 V_2 \times \exp\Big[-k_1^2 \langle u_1^2\rangle (1 - f(\underline{\varrho}))t^2\Big]\Big\} / (V_1 + V_2)^2, \quad (25)$$

so that, taking for simplicity $V_1 = V_2$, we obtain

$$P(\omega) = |\underline{A}|^4 N^2 (1/4\pi)(c/8\pi)^2 \Big\{\pi / \big[k_1^2 \langle u_1^2\rangle (1 - f(\underline{\varrho}))\big]\Big\}^{1/2} \times \exp\Big\{-\omega^2 / \big[4k_1^2 \langle u_1^2\rangle (1 - f(\underline{\varrho}))\big]\Big\}, \quad (26)$$

having neglected the δ-function contribution at $\omega = 0$. Thus, by varying the distance ϱ between V_1 and V_2, one can obtain the behavior of $f(\underline{\varrho})$ in terms of the bandwidth $\Delta(\underline{\varrho})$ of $P(\omega)$ according to the relation

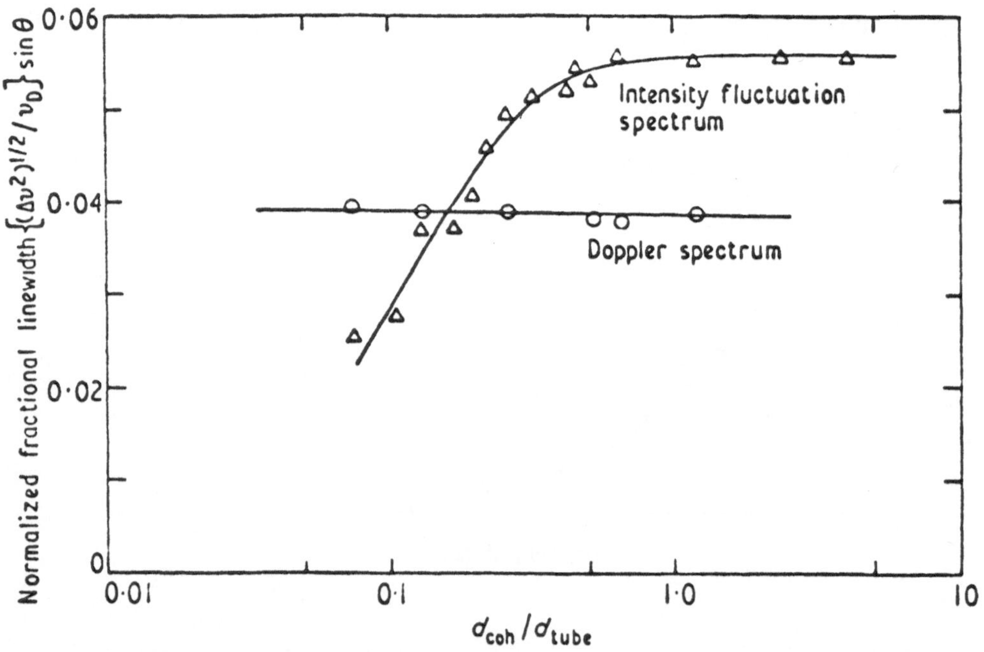

Fig. 7 Variation of spectral width with visible length for optical and intensity fluctuation spectra (after Bourke et al. 1969).

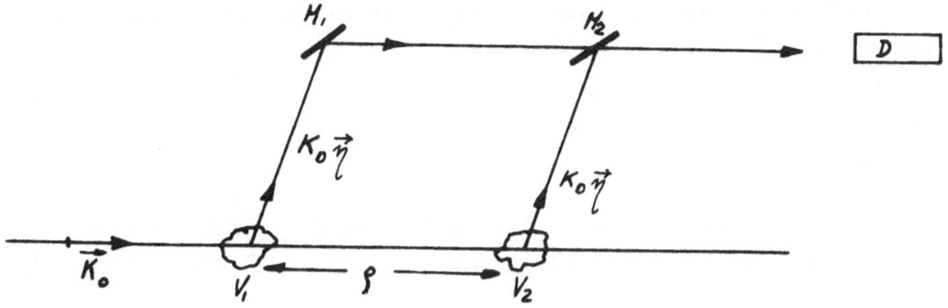

Fig. 8 Scattering geometry for measuring the spatial velocity correlation function.

$$f(\underline{\varrho}) = 1 - \Delta^2(\underline{\varrho}) / \Delta^2(\infty) \ . \qquad (27)$$

5. Conclusions

We have given here a brief outline of the analytical basis for the application of Laser Anemometry to the study of the dynamical properties of laminar and turbulent flows. As far as the measurement of the laminar velocity and of the turbulent strength $\langle U_1^2 \rangle$ is concerned , this technique has undergone a rapid growth and various optical arrangements are now used which have not been treated in this Lecture. The reader is referred for an up-to-date review and a wide bibliography to the paper of Durst et al. (1972) and to that of Whitelaw (1973). Recent relevant contributions have been given by Abbiss et al. (1972), Meneely et al. (1972) and Birch et al. (1973).

We wish finally to note that one could in principle investigate turbulence by measuring the spectrum of light scattered by the hydrodynamical density fluctuations associated with the turbulent motion itself, that is by considering an unseeded fluid. This possibility has been examined by Frisch (1967), who has shown that the ratio of the total cross-section due to turbulence alone to that due to molecular thermal fluctuations can be made greater than one. Since the fluids one deals with in practice are often naturally seeded, this kind of measurement can find a suitable field of application in the investigation of the vortices present in the upper atmosphere (Villars and Weisskopf, 1954).

References

1. Abbiss J.B., Chubb T.W., Mundell A.R.G., Sharpe P.R., Oliver C.J., and Pike E.R., 1972, J. Phys. D (Appl. Phys.), 5, L100.
2. Benedek G., 1969, (A. Kastler LX Birthday Volume) " Polarization Matter and Radiation ",(Presse Universitaire de France, Paris).
3. Bertolotti M., Crosignani B., Daino B., and Di Porto P., 1971, J. Phys A (Gen. Phys.), 4, L47.
4. Birch A.D., Brown D.R., Thomas J.R., and Pike E.R., 1973, J. Phys D (Appl. Phys.), 6, L71.
5. Born M., and Wolf E., 1970, " Principles of Optics ", (Pergamon Press, London).
6. Bourke P.J., et al., 1969, Phys. Letters 28 A, 692; 1970, J. Phys. A (Gen. Phys.), 3, 216.
7. Cummins H.Z., and Swinney H.L., 1970, " Progress in Optics ", edited by E. Wolf, vol. VIII, (North Holland Publishing Co., Amsterdam).
8. Di Porto P., Crosignani B., and Bertolotti M., 1969, J. Appl. Phys., 40, 5083.
9. Durst F., Melling A., and Whitelaw J.H., 1972, J. Fluid Mech., 56, 143.
10. Foreman J.W., Jr., George E.W., Jetton J.L., Lewis R.D., Thorton J.R., and Watson H.J., 1966, IEEE Journal of Quantum Electronics QE-2, 260.
11. Foreman J.W., Jr., George E.W., and Lewis R.D., 1965, Appl. Phys. Letters, 4, 77.
12. Frisch H.L., 1967, Phys. Rev. Letters, 19, 1278.
13. Goldstein R.J., and Hagen W.F., 1967, Phys. Fluids, 10, 1349.
14. Hinze O., 1959, " Turbulence ", (Mc Graw Hill Book Co., New York).
15. van de Hulst H.C., 1957, " Light Scattering by Small Particles ", (John Wiley and Sons, New York).
16. Lewis R.D., Foreman J.W., Jr., Watson H.J., and Thorton H.J., 1968, Phys. Fluids, 11, 433.

17. Meneely C.T., She C.Y., and Edwards D.F., 1972, Opt. Communications, 6, 380.
18. Pecora R., 1964, J. Chem. Phys., 40, 1604.
19. Pike E.R., Jackson D.A., Bourke P.J., and Page D.I., 1968, J. Phys. E (J. Sci. Instrum.), 1, 727.
20. Rice O., 1954, " Selected Papers on Noise and Stochastic Processes ", edited by N. Wax, (Dover Publication Inc., New York).
21. Schaefer D.W., and Berne B.J., 1972, Phys. Rev. Letters, 28, 475.
22. Villars F., and Weisskopf V.F., 1954, Phys. Rev., 94 , 232.
23. Whitelaw J.H., April 1973, " Turbulence Models and Their Experimental Verification ", Course given at the Department of Mechanical Engineering, Imperial College, London.
24. Yeh Y., and Cummins H.Z., 1964, Appl. Phys. Letters, 4, 176.

LIQUID CRYSTALS

J. D. Litster

Center for Materials Science and Engineering, Department of Physics, Massachusetts Institute of Technology, Cambridge, Massachusetts 02139.

Introduction

Liquid crystals represent a rather new area of study to me, but these rather interesting compounds have been known for a long time. In fact it was nearly a century ago that the first man-made liquid crystal, cholesteryl benzoate, was synthesized by F. Reinitzer, an Austrian botanist[1]. However, it also appears that liquid crystals have been known to nature before they were synthesized by man. It might amuse you to learn that the shells of certain Scarab beetles show both the irridescent reflection of colors and the spiral layered structure of cholesteric liquid crystals[2]. These beetle shells contain a large amount of uric acid, and therefore it is perhaps not too surprising to learn that the excrement from a common household pet (the budgerigar or parakeet) also forms a liquid crystal phase[3]. I do not intend this information to be a comment on my opinions of the biological importance of liquid crystals; there seem to be more significant examples. Solutions of various plant viruses, such as tobacco mosaic virus[4], form liquid crystalline phases. It is not clear if this is related to their biological function in any way, but perhaps more important is evidence that the "sickling" of red blood cells in those afflicted with sickle cell anemia results when the reduced mutant hemoglobin forms a liquid crystalline state[5]. I shall not say any more about naturally occurring liquid crystals and their biological significance; it appears there may be some, but it likely occurs with lyotropic materials (those which show liquid crystalline order when in solution).

Let me turn now to discuss thermotropic liquid crystals. These are usually pure organic substances consisting of anisotropic molecules that pass through one or more distinct thermodynamic states intermediate between solid and isotropic liquid states. These mesomorphic states are characterized by long-range orientational order (and sometimes some translational order as well) of the anisotropic molecules. Mechanically, they are very similar to liquids, but the anisotropic polarizability of the molecules results in anisotropic optical properties of the ordered phase, such as one would expect to see in crystalline solids. This is the origin of the name "liquid crystal".

Thermotropic liquid crystals are classified according to the structure of the ordered state (Fig. 1). The simplest type is the nematic, characterized by long-range orientational order of the molecules, but no ordering occurs in the positions of the center of mass of the molecules. Optically, they exhibit a uniaxial birefringence. Most of the molecules that form nematics possesses permanent electric dipole moments, but there appears to be no ordering of these. No ferroelectric nematics have been observed, and it is not possible to generate the second harmonic of intense laser light — indicating that at least on a scale down to the wavelength of light, the nematic state does have a center of inversion symmetry.

A special case is the class known as cholesteric liquid crystals (most are esters of cholesterol). Thermodynamically similar to nematics, they are formed from chiral molecules and order like nematics but with a spiral superstructure. Within a given plane (containing the optic axis) the molecules align parallel, but they make a very slight angle with their neighbors out of this plane. As a result there is a helical twist imparted to the local optic axis of the ordered state. This twist is slight on a molecular scale (several thousand intermolecular distances being required for one turn), but there is a strong periodic variation in the refractive index with the pitch of the helix; this is often about the wavelength of visible light and leads to Bragg reflections of the light and beautiful irridescent colors such as are observed on the shells of the Scarab beetles.

The smectic liquid crystals are a class that possesses some long-range translational ordering and are therefore closer to solids. The simplest type are the smectics A, in which the molecules tend to form layers. Their centers of mass tend to be located on planes separated by about 20 Å perpendicular to the alignment direction of the molecular axes. Optically uniaxial, these show a solid-like order in one dimension but within the planes of the layers are two-dimensional liquids. A slightly more complicated ordering occurs in smectic C materials. The molecules show a translational order in layers similar to the A type, but the molecular axes are at an

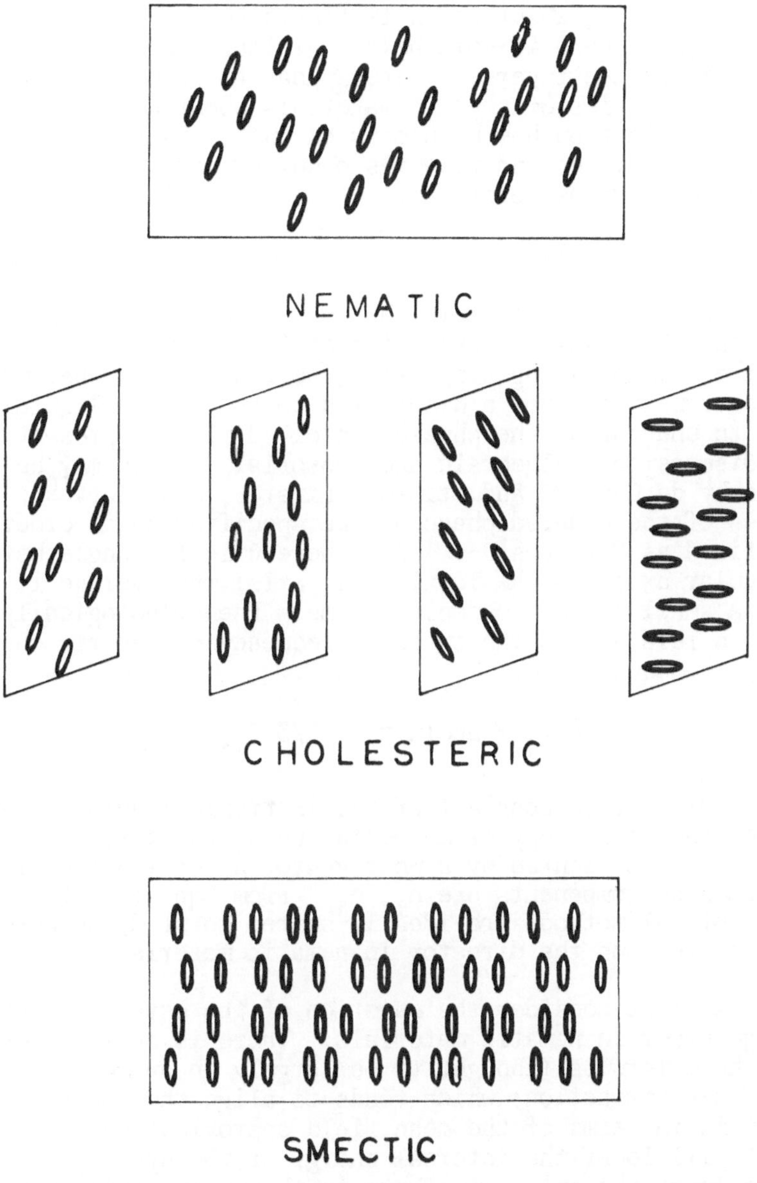

Fig. 1: Schematic of the arrangement of molecules in nematic, cholesteric, and smectic A liquid crystals.

angle to the plane of the layers. Because of this tilt angle, smectics C are usually biaxial optically[6]. There also exist smectic B materials, about which little is known, in which x-ray studies[7] suggest there is some hexagonal long-range order of the molecules within the layers. There appear to be perhaps half a dozen other varieties of smectic materials about which little is known. Classification has been carried out by Sackmann and Demus[8] based on miscibilities and textures observed under the polarizing microscope for various materials.

Nematic Liquid Crystals

Nematics are the materials that have been most intensively studied, and about which we know the most. I would like to discuss some of their interesting properties, beginning with what we know about the transition from a disordered to an ordered liquid phase. In order to understand the physics involved, I will present a simplified discussion. Algebraic and tensorial details may be found in papers by de Gennes[9] and Stinson, Litster, and Clark[10]. The properties of the ordered phase may be specified by an order parameter defined as $S = \langle \cos^2\theta - 1/3 \rangle$, where θ is the angle between the molecular axis and the local optic axis; the average is carried out over a small but microscopic volume. Phenomenologically it is possible to relate S to the optical frequency dielectric constant tensor by the equation

$$\varepsilon_{ij} = \bar{\varepsilon}\delta_{ij} + \Delta\varepsilon S[n_i n_j - 1/3\ \delta_{ij}] \tag{1}$$

with $\bar{\varepsilon}$ the dielectric constant of the isotropic liquid and $\Delta\varepsilon$ a measure of the anisotropy of molecular polarizability. The local optic axis is represented by a unit vector $\vec{n}$ (called the director) whose Cartesian components are n_i, n_j. From Eqn. (1) it is apparent that optical methods are ideally suited to studying both the order parameter and the director in nematic materials.

Let us first consider the question of the establishment of long-range order in nematic materials. There is an interaction between the molecules (thought to be largely an anisotropic van der Waals attraction) which tends to align them parallel. We discuss this in terms of the mean field approximation. Such an alignment will lower the internal energy of the system, but at the same time lower the entropy. There is thus a competition between these two effects that can be expressed by writing the free energy per unit volume as[9]

$$\Phi = \Phi_0 + \frac{A}{2}S^2 - \frac{B}{3}S^3 + \frac{C}{4}S^4 + \frac{D}{2}|\vec{\nabla}S|^2 - \frac{\chi_a}{3}SH^2 + \ldots \tag{2}$$

This expression includes a gradient term since the lowest energy state will be one of spatially uniform order parameter, and a term which reflects the anisotropic diamagnetic susceptibility of the molecules; the volume susceptibility along the long axis of the molecules is greater than that orthogonal to it by an amount $\chi_a \sim 10^{-7}$ in cgs units. As usual, in this Landau type[12] expansion of the free energy, A = $a(T - T_c^*)$ and thus changes sign at temperature T_c^*. All other coefficients are assumed to be weak functions of temperature.

Readers familiar with the Landau model of second order phase transitions, as for example applied to ferromagnets, may be surprised by the presence of the cubic term in Eqn. (2). This arises because of a fundamental symmetry difference between ferromagnets and nematic liquid crystals. If Eqn. (2) were to be written for a magnet, the order parameter would be the magnetization $M \sim \langle \cos \theta \rangle$ where θ is the angle between a spin and the local direction of magnetization. In this case, changing the sign of M corresponds to reversing the direction of all the spins and should have no effect on the free energy (aside from a Zeemann term); thus only even powered terms enter the expression. In a nematic liquid, with its quadrupolar symmetry, positive S corresponds to positive birefringence (molecules tend to be parallel to the optic axis and to each other) while negative S corresponds to negative birefringence (molecules randomly oriented in a plane normal to the optic axis) and a quite different interaction energy. Thus odd powered terms in S enter the free energy; the presence of these terms has a pronounced effect on the phase transition. From Eqn. (2) it can be seen that the S term will introduce a second minimum in the free energy that will occur at a finite value of S. As the coefficient of the quadratic term becomes smaller ($T \to T_c^*$ from above), this second minimum occurs at a lower value of Φ and eventually will correspond to a lower free energy than at the minimum for S = 0. At this point, T somewhat larger than T_c^*, a first order (discontinuous) transition to an ordered state occurs. It is interesting that this argument was first advanced by Landau in 1937[12]; at that time there was some speculation that He II was a liquid crystal and Landau used this argument to explain that this could not be correct because the phase transition was not first order.

The nematic isotropic transition is in fact only weakly first order, and I would like to discuss some experiments that show this as well as verify the model in Eqn. (2). As we can see from Eqn. (1), optical methods should be very useful for studies of nematic liquids. They can be simple classical ones, such as refractive index measurements to measure the static properties of the order parameter. An example is the magnetic birefringence induced in the isotropic phase. By minimizing Eqn. (2) in a magnetic field, one may calculate a birefringence (for H along z)

$$\Delta n = \varepsilon_{zz}^{\frac{1}{2}} - \varepsilon_{xx}^{\frac{1}{2}} \simeq \frac{\Delta\varepsilon\chi_a H^2}{6\bar{\varepsilon}^{\frac{1}{2}} a(T - T_c^*)} \qquad (3)$$

This predicts, so long as the quadratic term dominates in the free energy, a diverging susceptibility and magnetic birefringence. Experimental results are shown in Fig. 2 for two samples of p-methoxy benzylidene p-n-butylanilene (MBBA) that have slightly differing T_c^* values. They confirm the prediction of Eqn. (3); a first order transition to the nematic phase occurs 1.0°C above T_c^* (which is the temperature where the solid lines extrapolate to zero) in each sample.

Another feature predicted by the model is a critical divergence of the fluctuations in order parameter. From Eqn. (2) one may calculate the mean squared fluctuations in the Fourier component of S with wave vector $\vec{q}$ to be (in a sample of volume V)

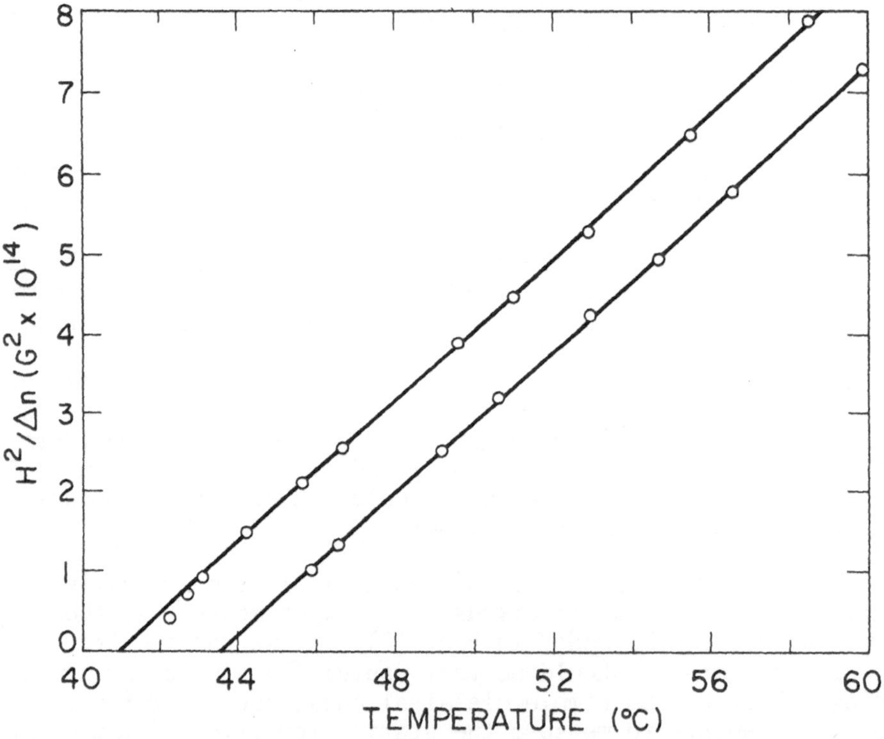

Fig. 2: The diverging magnetic birefringence in the isotropic phase of two MBBA samples with different transition temperatures.

$$\langle S^2(\vec{q})\rangle = \frac{kT}{Va\left(T - T_c^*\right)\left(1 + \xi^2 q^2\right)} \qquad (4)$$

where $\xi = (D/A)^{\frac{1}{2}}$ is the correlation range for short-range order in the isotropic phase. The intensity of the scattered light is proportional to Eqn. (4), and measurements of the intensity are given in Fig. 3. They agree with the predictions of the model for $\xi q \ll 1$. In fact, ξq is very small for visible light, but very precise measurements of the angular dependence of the scattered light have made it possible to determine ξ and the results are in agreement with $\xi^2 = D/A$ as predicted by de Gennes' model[9].

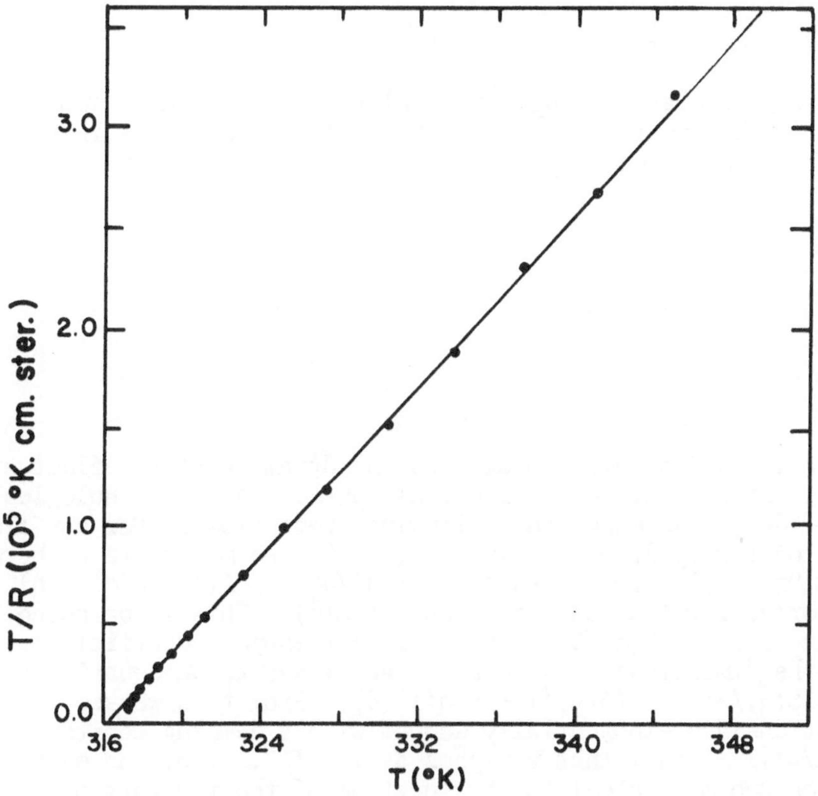

Fig. 3: Reciprocal of the intensity of light scattered by order parameter fluctuations in the isotropic phase of MBBA.

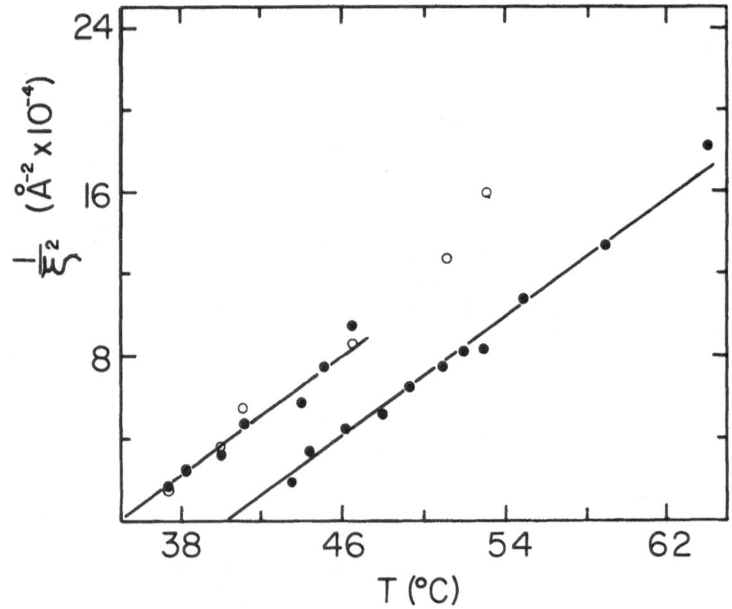

Fig. 4: Correlation length for order parameter fluctuations in the isotropic phase of two MBBA samples with differing transition temperatures.

Let me turn briefly to discuss the dynamics of the fluctuations of short-range order in the isotropic phase. A phenomenological model of de Gennes makes the following predictions. For any fluctuation of S away from equilibrium (S = 0), we may write a thermodynamic "restoring force" equal to $-\partial\Phi/\partial S = -A(1 + \xi^2 q^2)S(\vec{q})$ to lowest order for the Fourier component $S(\vec{q})$. This is balanced by a damping force $\nu\partial S(\vec{q})/\partial t$, where ν is a transport coefficient having the dimensions of viscosity, and leads to an equation of motion $\partial S(\vec{q})/\partial t = -(A/\nu)(1 + \xi^2 q^2)S(\vec{q})$. From this we see the fluctuations are exponentially damped with a damping constant $\Gamma = (A/\nu)(1 + \xi^2 q^2)$ that vanishes as $T \rightarrow T_c^*$. From the equation of motion one may calculate the spectrum of the fluctuations to be

$$\langle S^2(\vec{q},\omega)\rangle = \frac{2\Gamma}{\Gamma^2 + \omega^2} \langle S^2(\vec{q})\rangle \qquad (5)$$

One can directly measure Γ either by measuring the linewidth of the scattered light, or using a correlator, the decay time of the fluctuations; for MBBA results are shown in Fig. 5. In this substance Γ is too large to be measurable with an autocorrelator, and it was necessary to use a high resolution spherical Fabry-Perot interferometer. The upper curve shows the raw data (including the finite resolution of the interferometer); if one assumes that the transport coefficient ν has the same temperature dependence as the viscosity, the results can be corrected to give the lower curve — in agreement with the model. In some materials the relaxation time is sufficiently long that an autocorrelator can be used to measure the decay time, thereby giving much more precise results than can be achieved

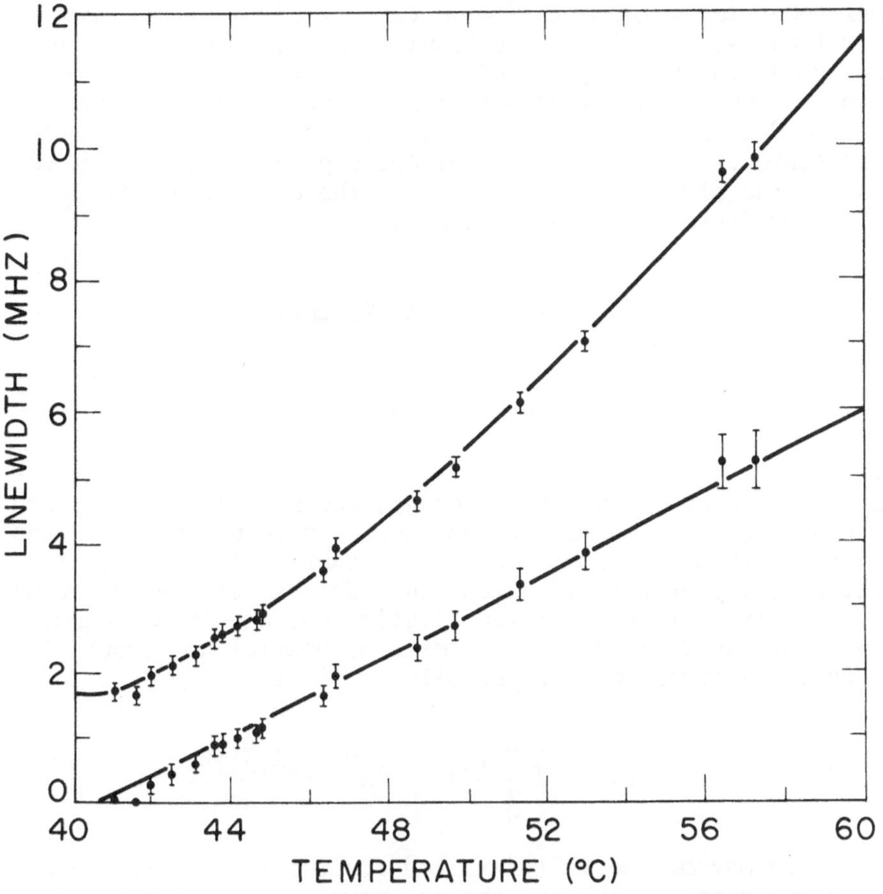

Fig. 5: Linewidth (Γ) of light scattered by order parameter fluctuations in the isotropic phase of MBBA. The upper curve is uncorrected data and includes the instrumental resolution. The lower curve is corrected for temperature variation of the transport coefficient ν.

with an interferometer. Such an experiment was carried out by Yang[13] in the isotropic phase of cholesteryl-ethoxy-ethoxy-ethyl-carbonate. He was able to make sufficiently precise measurements to determine the correlation length from the angular dependence of the linewidth.

Let me now turn briefly to a discussion of the ordered phase of nematic materials. Because of the long-range order there exists a new set of normal modes of the substance, and whose time dependence falls within the range to be studied by photon correlation spectroscopy. An excellent phenomenological discussion of the normal modes of nematics is to be found in a 1969 paper by the Orsay group[14]. More recently a general discussion of the types of normal modes to be found in a wide variety of materials from isotropic liquids, all classes of liquid crystals, through solids is given by Martin, Parodi, and Pershan[15]. A very simplified discussion of the normal modes of nematics can be given based upon the elastic model of Frank[16]. In the ordered phase ($S \neq 0$) the lowest energy state of the liquid crystal corresponds to a uniform vector field for the director $\vec{n}$ of Eqn. (1). The extra free energy required by distortions in $\vec{n}$ can be written as

$$\Phi_{el} = \frac{1}{2}\Big\{K_{11}(\mathrm{div}\,\vec{n})^2 + K_{22}(\vec{n}\cdot\mathrm{curl}\,\vec{n})^2 + $$

$$+ K_{33}(\vec{n}\times\mathrm{curl}\,\vec{n})^2 - \chi_a S(\vec{n}\cdot\vec{H})^2\Big\} \qquad (6)$$

Here div $\vec{n}$ (splay), $\vec{n}\times$ curl $\vec{n}$ (bend), and $\vec{n}\cdot$curl $\vec{n}$ (twist) are the only possible types of distortions that can occur in $\vec{n}$, and the K_{ii} are three phenomenological elastic constants. The stabilizing effect of a magnetic field, from Eqn. (2), has also been included in Eqn. (6). If we take all three elastic constants to be equal and express the distortions in $\vec{n}$ in terms of Fourier components $\delta n(\vec{q})$, then Eqn. (6) takes on an especially simple form

$$\Phi_{el} = \frac{1}{2}\sum_{\vec{q}}\left(Kq^2 + \chi_a H^2\right)|\delta\vec{n}(\vec{q})|^2 \qquad (7)$$

Thus, there may be thermally excited fluctuations in the director in the ordered phase, which are the new normal modes of the ordered phase. Their mean squared amplitude is

$$\langle\delta\vec{n}^2(\vec{q})\rangle = \frac{kT}{Kq^2 + \chi_a H^2} \qquad (8)$$

These correspond to rotations in the local optic axis and give rise to intense scattering of light; to first order the selection rules for the scattering require either the incident light to be polarized along $\vec{n}$, with the scattered light polarized normal to $\vec{n}$, or vice versa. Measuring the intensity of scattered light provides a means to determine the elastic constants, although measurement of the distortion produced by a magnetic field is usually more accurate. These director fluctuations are very highly damped by the rather large viscosity coefficient of the liquid crystal, and the mass of the molecules plays no significant effect in the normal mode dynamics. The normal modes (director fluctuations) of nematics are overdamped, decaying exponentially with a decay constant of the form $\Gamma = (Kq^2 + \chi_a H^2)/\eta$ where η is a viscosity coefficient. The relaxation times are usually in the millisecond to microsecond region, ideally suited to study by autocorrelation techniques. In a real situation things can become complicated as there are three independent elastic constants and five viscosity coefficients in nematics; the combination determining the relaxation times depends on the relative orientations of $\vec{q}$ and $\vec{n}$, as well as the polarization of the scattered light. These have been studied by the Orsay group[17] and Haller and Litster[18].

Of great practical importance for the design of visual displays are the electrohydrodynamic effects produced when electric fields are applied to samples in the nematic phase. If a nematic, whose molecules have permanent dipole moments directed perpendicular to the long axis, is made into a thin film (several tens of microns thick) with the molecules aligned parallel in the plane of the film, then rather striking optical effects can be observed by applying a field of a few volts across the film. What was originally a transparent film that reflected light only in a specular direction becomes very turbid and scatters a great deal of light. This is the effect used to make display devices using liquid crystals. An explanation of the physical processes involved was provided by the Orsay group and may be found in a paper by Dubois-Violette, de Gennes, and Parodi[19]. Let me try to summarize what takes place. These electrohydrodynamic (EHD) effects are quite complicated.

In d.c. or very low frequency (10 Hz) a.c. fields, the dominant effect seems to be the result of charge injection at the electrodes. The charge carriers flowing through the liquid crystal across the film and perpendicular to the molecular axes tend to upset the alignment. Above a threshold this effect is enough to overcome the viscous and elastic forces that oppose it and produce fluctuations in the optic axis orientation and scatter light. This effect, explained by Felici[20], can also be observed in the isotropic phase where it is less spectacular. It is neither reproducible nor of much practical importance for displays, except that for d.c. excitation it can mask the effects I shall now discuss. With a.c.

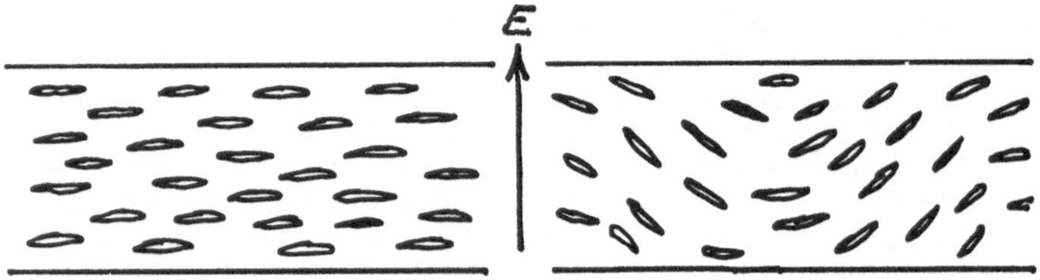

excitation there are two interesting regimes depending upon the frequency of excitation; the effects do not depend on charge injection for they are seen even with electrodes insulated from the liquid crystal. For low frequencies one observes a pattern of striations of the liquid crystal whose period is about the same as the sample thickness; these were first observed by Williams[21]. Consider an unperturbed sample as shown in the left-hand sketch above. The molecules prefer to lie perpendicular to the field and should be stable in moderate applied fields. However, if a bend distortion is produced as in the right sketch, the anisotropy of conductivity of the liquid crystal will lead to build up of space charges and the molecules try to lie perpendicular to the local field resulting from these space charges. This has a destabilizing effect as it increases the amount of bend distortion. The resulting instability is an oscillation of space charge density that follows the applied field. Since the space charge changes sign with the field, the force has always the same sign and produces a static director bend distortion to scatter light. The period is that requiring the lowest elastic energy (longest wave length) and corresponds to the thickness of the sample. The threshold voltage increases with frequency as the build-up of space charge is not able to follow the field so well. At a high enough frequency (perhaps 50 to 100 Hz), the space charge is not able to follow the field, and a second type of instability called the fast turn-off mode by the discoverers[22] occurs. Here, the charge density is static and, therefore, a sinusoidal force produces an oscillating bend distortion in the director. The spatial period is much smaller than for the Williams striations and is determined by the relationship between the relaxation time for the distortions (Kq^2/η) and the frequency of the field necessary to maintain the oscillation about $\pi/4$ out of phase with the driving field. (Thus making about equal forces available to overcome viscous and elastic forces opposing the director rotations.) The wavelength of the distortion produced decreases with increasing excitation frequency until the point where diffusion of the space charge becomes important and a distortion is no longer observed.

EHD changes in the optical properties of nematics can be produced with other geometrical arrangements, and the essential physics of what happens appears to be understood. There is still a considerable amount of more detailed work to be done in order to be able to predict what happens in a given situation, and just what will be the properties of the scattered light. The latter are important for the design of optical display devices and Jakeman and Pusey report some recent progress at this meeting.

Smectic Liquid Crystals

I have very briefly reviewed some of the more interesting features of nematic materials and would like to conclude with a discussion of the various types of smectic liquid crystals. It seems that nematics will perhaps be more useful for device applications, but the smectics are a fascinating class of materials and some very interesting physics can be done with them.

The simplest type is smectic A illustrated in Fig. 1. A very coherent discussion of the elastic properties of these materials can be found in an article by de Gennes[23]. The essential point is that a very large amount of energy is required to compress the layers, and so to the same approximation as we used for nematics in Eqn. (6) one must traverse an equal number of layers going from one point to another along any path in the sample. If we choose the director $\vec{n}$ perpendicular to the layers, then the constant layer thickness is expressed by

$$\oint \vec{n}\cdot d\vec{\ell} = 0 \quad \text{or} \quad \text{curl}\ \vec{n} = 0 \qquad (9)$$

Thus a smectic liquid crystal will not support twist or bend of the director. The only type of distortion in a smectic A that will scatter light intensely is a splay, and this can occur only with a wave vector $\vec{q}$ parallel to the layers. Other types of distortion involve compression of the layers and give rise to an extra propagating wave with a velocity comparable to a normal sound wave. Because of the higher energy involved these excitations are much weaker when thermally excited than are the director fluctuations in a nematic. They have been observed by the Harvard group[2] using a double-pass Fabry-Perot interferometer. Their results appear to confirm de Gennes' prediction of a propagating longitudinal and transverse excitation for most directions of $\vec{q}$. The change of the transverse mode from a propagating one to an intense overdamped one as $\vec{q}$ becomes in the plane of the layers is predicted by the theory[23] and suggested by the experiments[24], but has not yet been directly observed due to the presence of large static distortions in the liquid crystal samples[25,26].

It is also interesting to study the onset of smectic order in smectic A liquid crystals. Simple models that make helpful predictions have been proposed by Kobayashi[27] and McMillan[28]. In these models one may write the smectic order parameter as the amplitude of a plane wave in density with the period of the layer spacing. Thus if the smectic planes are perpendicular to the z direction (which is determined by the optic axis of the nematic phase, hence one cannot have smectic order without the simultaneous presence of nematic orientational order), one may write the density as the real part of

$$\rho = \rho_o[1 + \psi(r)e^{iq_s z}] \tag{10}$$

where d is the smectic layer spacing and $q_s = 2\pi/d$. The smectic order parameter is

$$\psi(\vec{r}) = |\psi|e^{i\phi} \tag{11}$$

and contains a phase factor ϕ that locates the position of the layers. McMillan constructs a mean field model analogous to the one we have discussed for the nematic-isotropic transition, but symmetry dictates that only even powers of $|\psi|$ appear in the free energy. Thus, we may write the smectic free energy terms as

$$\Phi_s = \frac{\alpha_o}{2}|\psi|^2 + \frac{\beta_o}{4}|\psi|^4 + \frac{\delta_V}{2}\left|\frac{\partial\psi}{\partial z}\right|^2 + \frac{\delta_T}{2}\left(\left|\frac{\partial\psi}{\partial x}\right|^2 + \left|\frac{\partial\psi}{\partial y}\right|^2\right) + \ldots \tag{12}$$

where α_o will go to zero at some temperature. If β_o is positive, there will be a second order nematic-smectic phase transition at this point. If β_o is negative, more terms must be added to Eqn.(12) for stability, and the transition will be first order. This is similar to the Bean and Rodbell[29] model for magnetic phase transitions, or Devonshire's[30] model of ferroelectricity.

We can see from Eqn. (12) that there will be two correlation lengths $\xi_V = (\delta_V/\alpha_o)^{\frac{1}{2}}$ and $\xi_T = (\delta_T/\alpha_o)^{\frac{1}{2}}$ for short range smectic order, depending upon direction. One prediction of the model is a divergence in the fluctuations of density at wave vector q_s as the phase transition is approached; this has been observed in x-ray scattering by McMillan[31]. In addition, the existence of short range smectic order can alter the properties of the nematic phase. To see this, let us proceed as follows[32]. The terms in $\vec{\nabla}\psi$ can be separated into two parts according to Eqn. (11). For example,

$$\left|\frac{\partial\psi}{\partial x}\right|^2 = \left(\frac{\partial|\psi|}{\partial x}\right)^2 + |\psi|^2\left(\frac{\partial\phi}{\partial x}\right)^2 \tag{13}$$

The role of terms, such as $\partial|\psi|/\partial x$, is fairly obvious; they represent variations in the magnitude of the smectic order. The term $\partial\phi/\partial z$ represents a compression of the layers which costs a lot in free energy; so we expect δ_V is large compared to δ_T. The term $\partial\phi/\partial y$ represents a displacement of the layers u such that $\partial u/\partial y = (1/q_s)(\partial\phi/\partial y)$, and thus a constant value of $(\partial\phi/\partial y)$ represents a rotation of the layers by an angle $(1/q_s)(\partial\phi/\partial y)$ about the x axis relative to the long axes of the molecules. Thus, the terms $(\delta_V|\psi|^2/2)[(\partial\phi/\partial x)^2 + (\partial\phi/\partial y)^2]$ represent the cost in free energy to rotate the molecules relative to the layers. If we also permit director fluctuations in the nematic (these are local rotations of the optic axis or alignment direction of the molecules), then the important quantity to use in evaluating the gradient terms of Eqn. (11) is the net rotation of molecules relative to the layers. A change δn_y in n represents a molecular rotation of $-\delta n_y$ about the x axis; thus, the net rotation relative to the layers is $(1/q_s)(\partial\phi/\partial y) + \delta n_y$. Therefore, if nematic director fluctuations are to be included, Eqn. (12) may be rewritten as

$$\Phi_s = \frac{\alpha_o}{2}|\psi|^2 + \frac{\beta_o}{4}|\psi|^4 + \frac{\delta_V}{2}\left|\frac{\partial\psi}{\partial z}\right|^2 + \frac{\delta_T}{2}|\nabla_T\psi + iq_s\delta\vec{n}\psi|^2 \qquad (14)$$

(where $\nabla_T = \partial/\partial x + \partial/\partial y$ and $\delta n_z = 0$ to first order since $\vec{n}$ lies along z).

de Gennes[32] has pointed out the striking similarity between Eqn. (14) and the Landau-Ginsberg equation for the mean field model of superconductivity, viz

$$\Phi_{super} = \frac{\alpha}{2}|\psi|^2 + \frac{\beta}{4}|\psi|^4 + \frac{h^2}{2m}\left|\nabla\psi + \frac{ie^*\vec{A}}{c}\psi\right|^2 \qquad (15)$$

This suggests an analogous role of $\vec{n}$ and the vector potential $\vec{A}$ of a superconductor. In fact, curl $\vec{n}$ is analogous to curl $\vec{A}$, and the smectic A excludes bend and twist the same way as the superconductor excludes a magnetic field. There are a number of interesting ways to pursue this analogy, but perhaps it is wise to ask how much of it has been confirmed. Not too much as yet, but if we examine Eqn. (14), we may see a prediction for the effects of short range smectic order on the nematic elastic constants. The value of ψ remains essentially constant over a coherence distance ξ, and the value of δn over this distance is $\delta n \sim \xi$ curl n. Thus, the director distortions contribute an amount of free energy, from Eqn. (14) of

$$\delta\Phi \sim \delta_T q_s^2 \xi^2 (\text{curl } \vec{n})^2 \langle|\psi|^2\rangle \qquad (16)$$

and within the correlation volume of dimension ξ one estimates from Eqn. (14) that

$$\langle|\psi|^2\rangle \sim \frac{kT}{\alpha_o \xi^3} \tag{17}$$

Combining Eqn. (16) and Eqn. (17), and adding the result to the nematic expression of Eqn. (6), we find a bend or splay represents a free energy change

$$\delta\Phi_{el} = K(\text{curl } \vec{n})^2 + kTq_s{}^2\xi(\text{curl } \vec{n})^2 \tag{18}$$

And so the presence of short range smectic order causes an increase in the nematic bend or twist elastic constants that is proportional to the coherence length, $\xi = \sqrt{\delta_T/\alpha_o}$, for this order. According to the mean field approximation this should increase as $(T - T_c)^{-1/2}$. Such a divergence has recently been observed by three groups[33-35]; all report the interesting fact that the divergence varies as $(T - T_c)^{-2/3}$. This indicates a breakdown of the mean field approximation, the first observed in a liquid crystal system, and behavior like that commonly observed near critical points.

There are many other facets of liquid crystals that could be discussed, such as smectic C or B materials, or two-dimensional liquid crystal phases. I have attempted to provide you with an introduction to the field and references to the most exciting literature. The models proposed for smectic materials are at present quite crude and will undoubtedly require some refinement. Nonetheless, they suggest some interesting physical behavior which you might wish to pursue.

References

1. Reinitzer, F., Monatsch. Chem. 9, 421 (1888).
2. Caveney, S., Proc. Roy. Soc. London Ser. B 178, 205 (1971).
3. Lonsdale, K., and Sutor, D. J. Science 172, 958 (1971).
4. Bernal, J. D., and Fankuchen, I., J. Gen. Physiology 25, 111 (1941).
5. Perutz, M. F., Liquori, A. M., and Eirich, F., Nature 167, 929 (1951).
6. Taylor, T. R., Fergason, J. L., and Arora, S. I., Phys. Rev. Lett. 24, 359 (1970).
7. Levulet, A. M., and Lambert, M., Compt. Rend. Acad. Sci. Paris 272B, 1018 (1971).
8. Sackmann, H. and Demus, D., Molecular Crystals 2, 81 (1966).
9. de Gennes, P. G., Mol. Cryst. Liq. Cryst. 12, 193 (1971).
10. Stinson, T. W., Litster, J. D., and Clark, N. A., J. de Physique 33, C1-69 (1972) (Colloque C1, supplement to Tome 33); and Stinson, T. W., and Litster, J. D., Phys. Rev. Lett. 30, 688 (1973).

11. Maier, W. and Saupe, A., Z. Naturforsch A 15, 287 (1960).
12. L. D. Landau, in Collected Papers of L. D. Landau, edited by D. Ter Haar (Gordon and Breach, New York, 1965), p. 193.
13. Yang, C. C., Phys. Rev. Lett. 28, 955 (1972).
14. Group d'Etudes des Cristaux Liquides, J. Chem. Phys. 51, 816 (1969).
15. Martin, P. C., Parodi, O., and Pershan, P. S., Phys. Rev. A 6, 2401 (1972).
16. Frank, F. C., Disc. Faraday Soc. 25, 19 (1958).
17. Orsay Liquid Crystal Group, Phys. Rev. Lett. 22, 1361 (1969).
18. Haller, Ivan, and Litster, J. D., Phys. Rev. Lett. 25, 1550 (1970).
19. Dubois-Violette, E., de Gennes, P. G., and Parodi, O., J. de Physique 32, 305 (1971).
20. Felici, N., Revue Générale d'Electricité 78, 717 (1969).
21. Williams, J., J. Chem. Phys. 39, 384 (1969).
22. Heilmeier, G. H., and Helfrich, W., Appl. Phys. Lett. 16, 1955 (1970).
23. de Gennes, P. G., J. de Physique 30, C4-65 (1969).
24. Liao, York, Clark, Noel A., and Pershan, P. S., Phys. Rev. Letters 30, 639 (1973).
25. Ribotta, R., Durand, G., and Litster, J. D., Sol. St. Comm. 12, 27 (1973).
26. Clark, N. A., and Pershan, P. S., Phys. Rev. Letters 30, 3 (1973).
27. Kobayashi, K., J. Phys. Soc. Japan 29, 101 (1970).
28. McMillan, W. L., Phys. Rev. 4, 1238 (1971).
29. Bean, C. P., and Rodbell, D. S., Phys. Rev. 126, 104 (1962).
30. Devonshire, A. F., Phil. Mag. Suppl. 3, 85 (1954).
31. McMillan, W. L., Phys. Rev. A 6, 936 (1972).
32. de Gennes, P. G., Sol. St. Comm. 10, 753 (1972).
33. Cheung, L., Meyer, R. B., and Gruler, Hans, Phys. Rev. Letters 31, 349 (1973).
34. Delaye, M., Ribotta, R., and Durand, G., Phys. Rev. Letters 31, 443 (1973).
35. Léger, L., and Martinet, A., to be published.

THE PHASE TRANSITION APPROACH TO LASER STATISTICS

H.Haken

Institut für theoretische Physik

Universität Stuttgart, Germany

§ 1 INTRODUCTION

At this summer school various applications of laser light statistics to problems in rather different disciplines are given, so that this school has - at least to some extent - the character of an interdisciplinary meeting. For this reason I would like to present here the theory of laser statistics from a viewpoint which goes considerably beyond laser statistics alone. As we will see below, at the laser threshold a remarkable transition from disorder to order occurs. Order-disorder transitions occur in many disciplines. In physics we mention the class of phase transitions of different kinds, e.g. in ferromagnets, in superconductors or in liquids. In recent years it has become clear that these phase transitions are governed by quite similar rules (e.g. scaling laws or Wilson's work on phase transitions). While in the just mentioned systems phase transitions occur at thermal equilibrium and the ordered state is reached by lowering the temperature, ordered states in the biological domain are maintained by a flux of energy and matter. It is in this respect that the laser serves also as an excellent example how an energy flux may establish an ordered state in a system far from thermal equilibrium.

In our lecture we want to develop the basic ideas rather than the mathematical apparatus which has been represented elsewhere [1]. For illustration we treat three examples: the single mode laser, the continuous mode laser and laser light emission from interacting chaotic fields, which seems to establish a new type of parametric processes.

§ 2 INTERPRETATION OF SINGLE-MODE LASER ACTION BY CONCEPTS OF PHASE TRANSITION THEORY AND SOME GENERALIZATIONS

We consider a solid state laser as an example. It consists of a set of laser active atoms embedded in a solid state matrix. The end-mirrors serve for two purposes. They select modes mainly in axial direction and with discrete cavity frequencies. In the following we assume that only one mode may be ultimately supported by the laser process. Laser action now proceeds in principle as follows (Figs. 1 and 2).

For simplicity we consider two-level atoms. In thermal equilibrium the levels are occupied according to the Boltzmann distribution function. If we excite now the atoms by an external pump, e.g. incoherent light, we obtain an inverted population which may be described by a negative temperature. The atom now starts to emit light, which is ultimately absorbed by the surroundings, whose temperature is much smaller than $\frac{\hbar\omega}{k}$ (where ω is

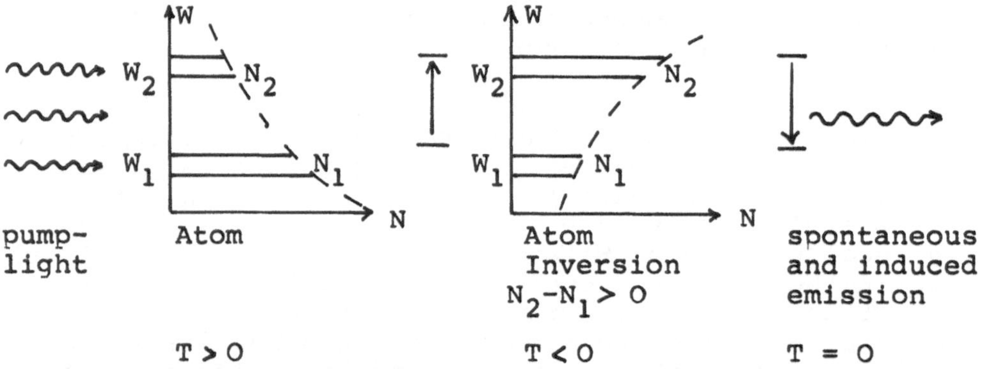

Fig. 1 Atomic occupation numbers at room temperature and in an inverted state

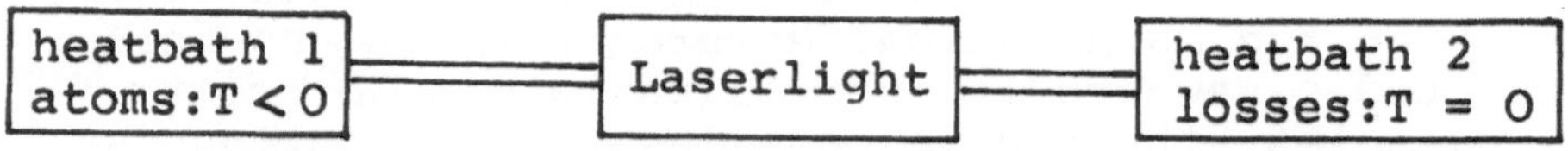

Fig. 2 Laser as an example for a system far from thermal equilibrium

the light frequency of the atomic transition and k is Botzmann's constant). Thus from the thermodynamic point of view the laser is described as a system composed of atoms interacting with the field, which is coupled to two reservoirs at different temperatures. Thus the laser is a system far from thermal equilibrium. Nevertheless we want to demonstrate that it shows all features of a second order phase transition (Fig. 3)

thermodyn. phase	ferromagnet	laser
unordered		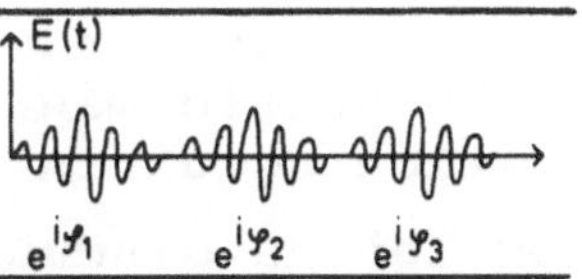 E(t) $e^{i\varphi_1}$ $e^{i\varphi_2}$ $e^{i\varphi_3}$
ordered		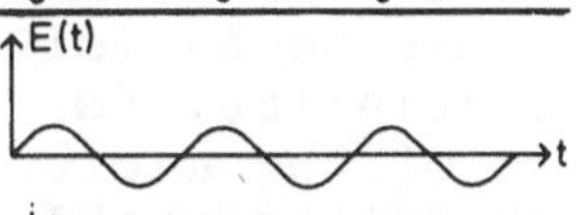 E(t) t $e^{i\varphi_0}$, φ: optical phase
broken symmetry	preferrence of one direction	preferrence of an optical phase
	spin waves 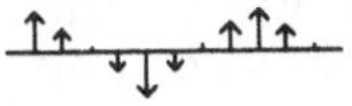	phase diffusion ϕ (t)

Fig. 3 Comparison between the ferromagnet and the laser

To do this let us compare it with the ferromagnet, In the ferromagnet the spins are completely disordered above the Curie temperature pointing in all directions. If we decrease the temperature below the Curie temperature, the spins get aligned pointing all in one direction so that we obtain a spatially ordered state with a magnetization of macroscopic size. Now consider the electric field as a function of time. If the laser is operated as a lamp in the usual sense the atoms emit independently of each other wave tracks with random phases. The coherence time of about 10^{-11} sec is evidently on a "microscopic" scale. Now consider the laser above threshold. Then its linewidth is extremely sharp of the order of 1 Hz so that the phase of the field remains unchanged on the macroscopic scale of 1 sec. The laser is thus in a macroscopic state in the time domain. In both the ferromagnet and the laser we observe the phenomenon of broken symmetry:While in the disordered states all spin directions or all light phases are acquired, in the ordered state just one spin direction or just one phase is preferred. In both cases this broken symmetry can be restored in the ferromagnet by spin waves and in the laser by phase diffusion. So far we established an analogy between spins and wave tracks of the field. The analogy between the ferromagnet and laser can be viewed at from a still different point of view if we set in analogy the magnetic moment of each spin with the electric polarization of each laser atom. In the disordered phase, the dipole moments oscillate with random phases while in the laser they oscillate coherently. Thus, similar to the static magnetization,the oscillating electric polarization acquires a macroscopic state.

Now let us discuss if this more or less superficial analogy may be exploited to get a deeper insight into laser statistics. For this end we explain first some basic features of the microscopic theory which then will allow us to study the problem again from a phenomenological point of view. In the microscopic theory one starts with equations which describe the interaction of a set of atoms with a single-mode field. The dynamical variables are the mode amplitude, the complex polarization p_μ , the inversion $(N_2-N_1)_\mu = d_\mu$ of each atom with index μ. One then readily establishes that the field amplitude interacts with the atoms only via the total polarization P of all atoms with the same wave vector k of the field:

$$P = \sum_\mu e^{ikx} p_\mu \tag{2.1}$$

Thus the total atomic polarization appears as a suitable macroscopic variable, a procedure which is quite typical for the treatment of cooperative systems: One has first to look for the correct macroscopic variables out of which one will undergo the phase transition. A further macroscopic variable is the total inversion of all atoms. We have now to discuss the interplay between these three variables which, in the quantum mechanical domain,are represented by operators:

mode amplitude:	b^+, b	with $[b,b^+] = 1$
atomic polarization:	S^+, S^-	with $[S^+,S^-] = 2S_z$
atomic inversion:	$N_2-N_1 = 2S_z$	(2.2)

We use a notation which clearly exhibits the analogy to spin-systems, where the S's have the usual meaning. Note that the electric polarization is obtained from S^+, S^- by multiplication with a factor, which contains essentially the atomic dipole matrix element. The interaction Hamiltonian reads (we assume exact resonance and work in the interaction representation)

$$H = \hbar g \ (b^+S^- + bS^+) \tag{2.3}$$

g is a coupling constant, which is proportional to the atomic dipole moment matrix element and proportional to $V^{-1/2}$, where V is the volume of the cavity. In the following we assume that the interaction between atoms and field is dominant over their individual interaction with the reservoirs. In a first step we consider only (2.3). It allows for several constants of motion h_j:

the interaction energy	$h_o = H$	
the total number of photons and inverted atoms	$h_1 = b^+b + S_z$	(2.4)
operator identity	$h_2 = S_z^2+\frac{1}{2}S^+S^-+\frac{1}{2}S^-S^+$	

The third quantity follows from the fact that S^+, S^-, S_z behave like spin operators. We now use the method of quantum-classical correspondence, which we will explain later in more detail. It allows us to establish a one-to-one correspondence between operators and classical quantities: like field amplitudes, classical polarization and the occupation number of atoms.

field amplitude	$b, b^+ \longleftrightarrow u, u^*$	
polarization	$S^-, S^+ \longleftrightarrow v, v^*$	(2.5)
inversion	$2S_z \longleftrightarrow D$	

The integrals of motion may be expressed by the classical quantities as may be shown by detailed calculations

$$\begin{aligned} h_o &= u^*v + uv^* \\ h_1 &= u^*u + D/2 \\ h_2 &= D^2/4 + v^*v \end{aligned} \tag{2.6}$$

(Note that we have dropped the factor $\hbar g$ in h_o, because it is not needed in our following considerations).

Now let us look at the stationary distribution function f(u,v,D) which gives the probability of finding a configuration u,v,D. With no coupling to the external world all configurations may be realized which are compatible with the conservation laws. Thus the probability of finding a certain configuration of u,v,D will depend on these coordinates via the conservation laws:

$$f = f(h_o(u,v,D),\ h_1(u,v,D),\ldots) \tag{2.7}$$

where f may be an arbitrary function. In this sense the system is highly degenerate.

Now consider the coupling of the proper laser system of atoms and field to its surroundings ("reservoirs"). Then depending on the temperatures of the reservoirs e.g. certain energies are favoured compared to others. In thermal equilibrium, f would thus acquire the Boltzmann distribution function $\sim\exp(-H/kT)$. Similarly we expect that in our more general case f is no more a completely arbitrary function but will acquire a specific form. If we admit that the system finds a stationary state we expect that the values of h_o, h_1, h_2 are centered around certain values giving a maximum probability around which the distribution function will fall off in a Gaussian way. An example for f would be a product of functions of the form

$$\exp\left(-\alpha_j(\beta_j - h_j)^2\right) \tag{2.8}$$

where α_j, β_j are constants.

Let us assume now further that the phase of the polarization is heavily damped by its coupling to external heatbaths like lattice vibrations etc. We expect that the constant of motion h_o is also heavily damped out, because it depends on the phases of v and we are left only with the distribution function of the two other constants of motion. We further use the fact that the action of the heatbaths on the polarization stems from many statistically independent events, and that the effect of this reservoir is dominant over the other ones. Under these conditions $f\sim\exp(-C|v|^2)$, i.e. a Gaussian distribution in v (C is a parameter). Because, however, v must occur via the constant of motion

$$h_2 = |v|^2 + D^2/4 \tag{2.9}$$

we expect the distribution function in the form

$$f = N\exp\left(-C(|v|^2 + D^2/4)\right) = N\exp(-C\,h_2) \tag{2.1o}$$

The factor N may still depend on h_1. Assuming for N the form (2.8), as discussed above, we obtain finally

$$f_o = N_o' \exp\left(-\alpha_1(\beta_1 - h_1)^2\right) \exp(-C\, h_2) \qquad (2.11)$$

or, after a slight rearrangement+)

$$f_o = N_o \exp\left(A\, h_1 - B\, h_1^2 - C\, h_2\right) \qquad (2.12)$$

which is indeed found by a detailed calculation from first principles. (2.11), (2.12) exhibits a strong correlation between photon number and inversion. When we integrate over the inversion we obtain a distribution function of the form

$$\tilde{f}(n) = \overline{N_o} \exp\left(an - bn^2\right) \qquad (2.13)$$

($n = |u|^2$) which evidently is a consequence of the conservation laws. In our derivation we did not make assumptions about the decay rates of u and D, which may be even equal, so that our analysis applies also to mode-mode coupling theory. (Note that hitherto (2.13) could be derived only using an adiabatic elimination of the inversion D).

As we will discuss below in detail, $\tilde{f}(n)$ may be interpreted in the framework of the Landau theory of phase transitions. Here we anticipate the following: The "constants" a and b will depend in general on the temperature of the reservoirs. Now assume $a \sim T - T_c$, where T_c is a critical temperature. Because a changes its sign at $T = T_c$, the behaviour of $\tilde{f}(n)$ as a function changes even qualitatively completely (compare fig. 9) We shall call a the pump parameter, a = 0 the laser threshold, $a < 0$ the region below threshold, and $a > 0$ the region above threshold.

The distribution function (2.12) whose parameters are determined by a microscopic theory may now serve as a first step of an expansion of the type

$$f = f_o + \lambda\, f_1 + \lambda^2\, f_2 + \ldots \qquad (2.14)$$

+) Our reasoning, which allowed us to derive (2.12), can be easily generalized to arbitrary systems far from thermal equilibrium. For a microscopic theory see Haken (1973a), for a phenomenological theory Haken (1973e)

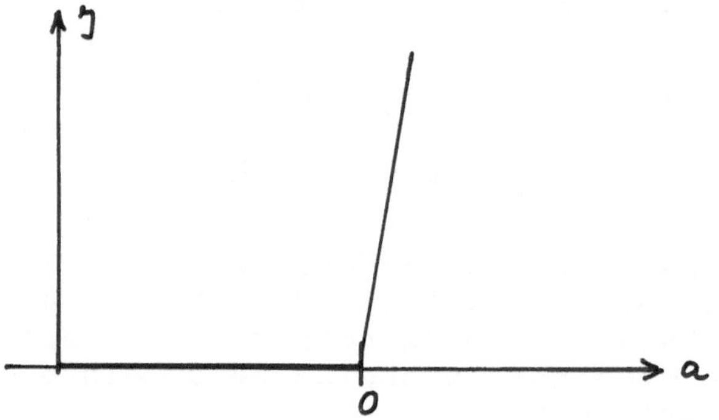

Fig.4 light emission versus pump (<u>without fluctuation</u>) "mean field theory"

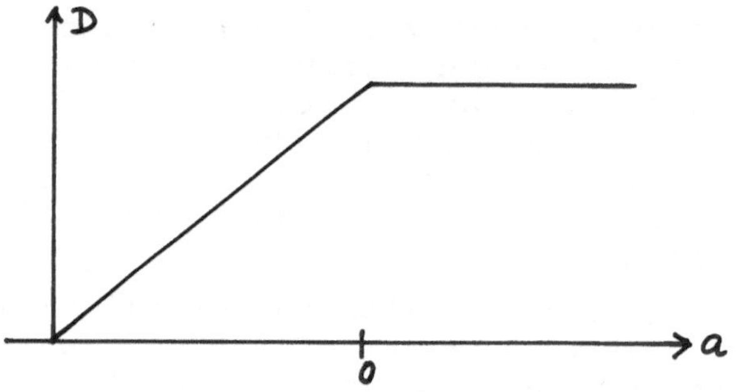

Fig.5 atomic inversion versus pump (without fluctuation) "mean field theory"

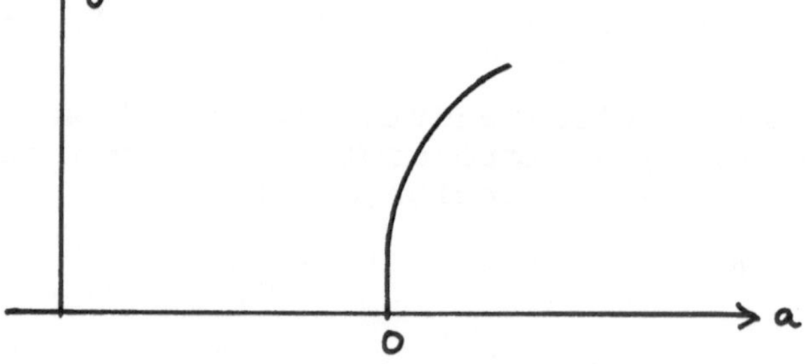

Fig.6 atomic polarization versus pump (without fluctuation) "mean field theory"

where the expansion parameter λ is essentially a fraction of the coupling constants $\kappa, \gamma, \ldots$ of the external reservoirs and of the internal coupling constant g of the laser, e.g. $\lambda \sim \gamma/g$. Instead of determining f_1, f_2 etc. directly, it is sufficient in important cases to determine the moments containing the higher order corrections. Thus one obtains for the correlation between field u and polarization v

$$\langle u\, v^* \rangle = \frac{\kappa}{ig} \left(1 - \frac{n_{th}}{\langle |u|^2 \rangle}\right) \langle |u|^2 \rangle \qquad (2.15)$$

where $\langle \cdots \rangle$ denotes the average over the distribution function (2.11). Because well below threshold $\langle |u|^2 \rangle = n_{th}$ (= thermal number of photons), u and v are uncorrelated below threshold. On the other hand, u,v are completely correlated, as may be checked by a direct solution of the Langevin equations (see appendix) without fluctuations ("mean field equations"). In Figs. 4 - 6 we exhibit the qualitative dependence of the mean photon number $\langle |u|^2 \rangle$ or, equivalently, of the laser output power J, the inversion D and the atomic polarization $|v|$ on the pump parameter a. (In the framework of the present considerations, a depends on the temperatures of the reservoirs and also on certain decay constants. One may show, that by means of higher orders of λ, a becomes "renormalized": $a \rightarrow a - \delta |\lambda|^2$ where $\delta \sim N^{-1}$, N number of laser atoms). Looking at the behaviour of these variables the analogy between the laser threshold and a phase transition in the framework of the molecular field theory becomes evident. At this stage we may easily invoke the Landau theory of phase transitions where we may construct the analogy according to table I. Apparently the specific heat finds its analogue in the quantum efficiency of the laser. The electric field strength u plays the role of the order parameter as does the magnetization in the magnetic field case. We may also easily reinterprete the pump parameter a: Because the temperature, at which the surrounding, by which the photons are absorbed, is practically zero ($\hbar\omega \gg kT$!), the only relevant temperature T must be that of the atomic system, where $T < 0$ in the inverted state, where laser action can take place. Expressing $T - T_c$ by the inversion $D_0 - D_c$, we find the analogy of table I.

Table I

Landau : $e^{-F/hT}$		laser : $e^{-B/Q}$
free energy		
$F(T,\psi) = F_o(T) + \alpha(T)\psi^2 + \beta(T)\psi^4$		$B(d,u) = \underbrace{B_o(d)}_{\circ\circ} + \tilde{\alpha}(d)u^2 + \tilde{\beta}u^4$
$\alpha(T) = a(T - T_c)$		$\tilde{\alpha}(d) = \tilde{a}(d_c - d)$
temperature	$T \leftrightarrow -d$ $T_c \leftrightarrow -d_c$	pump parameter
	order parameter $\psi \leftrightarrow u$	
entropy $S = -\dfrac{\partial F}{\partial T}$		$\tilde{S} = +\dfrac{\partial B}{\partial d}$
$S = S_o$	disordered	$\tilde{S} = \tilde{S}_o$
$= S + \dfrac{a}{2\beta}(T - T_c)$	ordered	$= \tilde{S}_o + \dfrac{\tilde{a}^2}{2\tilde{\beta}}(d_c - d)$
	continuous	
specific heat $C = T(\partial S/\partial T) = \begin{cases} T\partial S_o/\partial T \\ T\partial S_o/\partial T + \dfrac{a T_c}{2\beta} \end{cases}$	disordered ordered	$\tilde{C} = d\dfrac{\partial \tilde{S}}{\partial d} = \begin{cases} d\dfrac{\partial \tilde{S}_o}{\partial d} \\ d\dfrac{\partial \tilde{S}_o}{\partial d} - \dfrac{\tilde{a}^2}{2\tilde{\beta}} d_c \end{cases}$
	discontinuous	

As may be shown by the microscopic theory, D_o is the "unsaturated" inversion, D_c the critical inversion. Note that the pump parameter $a \propto D_o - D_c$. The second order phase transition is characterized by a discontinuity of the specific heat. This discontinuity may be found for systems at thermal equilibrium in the so-called thermodynamic limit where the number of particles N and the volume tend to infinity letting the ratio N/V finite. In a fruitful analysis V.Dohm has shown that such a limit exists indeed for the single mode laser thus proving, that there exists a phase transition of the laser in the exact sense. In practical cases, the laser behaves like a system of restricted geometry, however, exhibiting a smooth transition (compare Fig. 7).

To determine an equation for the mean value of the order parameter we seek the maximum value of the distribution function (2.13), i.e. the maximum value of the exponent. We readily obtain the so-called mean field equation

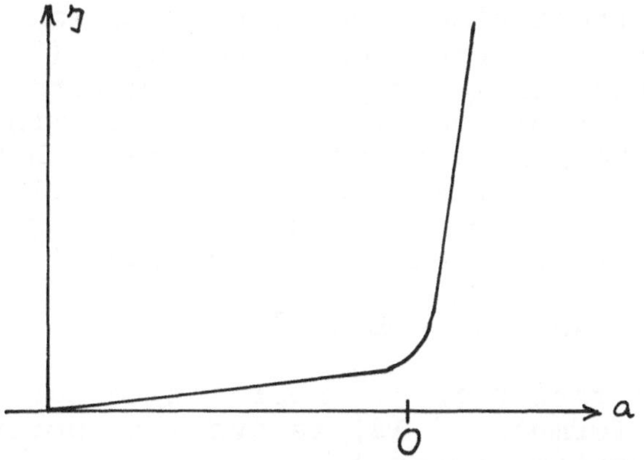

Fig. 7 light emission versus pump (with fluctuation) ("Landau theory", "restricted geometry")

$$u(a - 2b|u|^2) = 0 \tag{2.16}$$

The solution reads

$$\begin{aligned} u &= 0 \qquad \text{for } a < 0 \\ |u|^2 &= \frac{a}{2b} \qquad \text{for } a > 0 \end{aligned} \tag{2.17}$$

i.e. we find that the order parameter u is equal 0 in the disordered state and increases $\propto$ a in the ordered state. From (2.16) we may proceed to a time dependent equation by putting $\frac{du}{dt}$ = const.x (left hand side of equ. (2.16)), i.e.

$$\frac{du}{dt} = ua' - 2b'|u|^2 u, \text{ where } \begin{aligned} a' &= \text{const.}a \\ b' &= \text{const.}b \end{aligned} \tag{2.18}$$

which is quite often done in a phase transition theory.

In order to treat time dependent phenomena it is comforting to observe that an equation of the type (2.18) may be obtained for the single mode laser from first principles provided the atomic variables may be eliminated by the assumption that their decay rates are much bigger than the decay constants of the laser mode. Eq.(2.18) has been interpreted by the present author as an equation of a particle moving overdamped in a potential of the form

$$V(u) = -(a'|u|^2 - b'|u|^4) \tag{2.19}$$

(For generalization it is important to note that the exponent in formula (2.13) is just the potential (2.19), besides a constant factor).

By means of this potential we may easily make contact with stability theory and with the nomenclature often used in phase transition theory. Before doing so we interpret equation (2.18) in the following manner, which sheds light on the physical meaning of the constants a' and b'. We consider the temporal change of the mode amplitude u (note that throughout this paper we are working in the interaction representation. Consequently, we decompose the electric field strength of the mode as

$$E = u\, e^{-i\omega t} + u^{*} e^{i\omega t} \tag{2.2o}$$

where u is now a slowly varying amplitude and ω the atomic resonance frequency). u in the cavity decreases due to its transmission through the endmirrors (and other losses), the mirrors with the decay constant κ giving rise to the terms $-\kappa u$. On the other hand u is supported by spontaneous and stimulated emission. The stimulated emission is proportional to $G \cdot D \cdot u$, where G is a gain constant which depends essentially on the square of the optical dipole matrix element, and D is the atomic inversion, $N_2 - N_1$. Now we have to take into account that the inversion D is decreased by laser action. We assume that this process takes place instantaneously. Because this depletion is the bigger, the higher the field intensity, we expand D as a power series of the field intensity $|u|^2$. Confining ourselves to lowest order, we obtain

$$D = D_o - s|u|^2 \tag{2.21}$$

where D_o is the "unsaturated inversion". Inserting this into $G \cdot D \cdot u$ and adding $-\kappa u$, we obtain the equation

$$\frac{du}{dt} = (G\,D_o - \kappa)u - Gs\,|u|^2\,u \tag{2.22}$$

Comparing this with (2.18), we find $a' = GD_o - \kappa$,
$b' = GS$

Finally we take into account the spontaneous emission processes which give contributions at random times t_j with random phases ϕ_j to the field u by a term of the form

$$F(t) = \sum_j c_j\, e^{i\phi_j}\, \delta\,(t-t_j) \tag{2.23}$$

Our laser equation thus reads

$$\frac{du}{dt} = a'u - b'\,|u|^2\,u + F(t) \tag{2.24}$$

Now let us discuss this equation using the potential (2.19) and interpreting u as coordinate of a fictitious particle. Consider Fig. 8. If the laser is operated below threshold then the fictitious particle falls back after each push to the equilibrium position u = O in accordance with (2.17). If we are, however, above threshold, the solid lines apply. The particle falls down, after each push by F(t), to an equilibrium value $|u_o| \neq 0$. At threshold, the system becomes unstable, and acquires a new stable position $|u| = |u_o|$. This local instability is connected with symmetry breaking. In principle the particle has now several equivalent equilibrium positions. If we take into account that the potential possesses rotational symmetry around the V-axis, the field amplitude u may still have a fixed but arbitrary phase, as long as we neglect the random pushes by F(t). These pushes restore the full symmetry

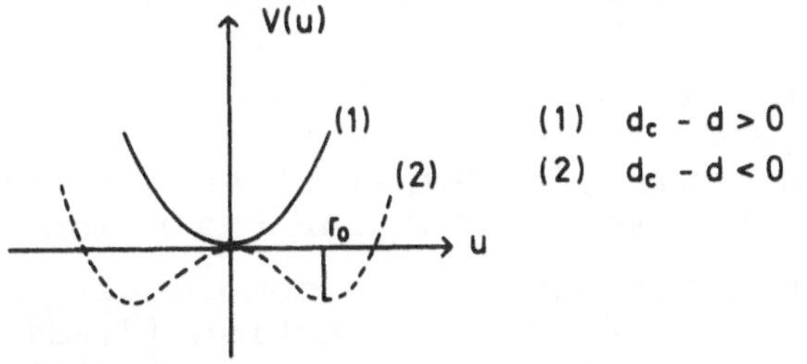

Fig. 8 Potential of fictitious particle below and above threshold

by letting the phase diffuse (i.e. the "particle" diffuses in the potential valley). If we proceed from the region below threshold to the region above threshold the potential curve becomes flatter and flatter. Thus if the fictitious particle is pushed by random pushes by F, it comes down the potential curve more and more slowly. This effect is well known as "critical slowing down", The mode behaving in this way is the so-called "soft mode". At the same time, because the restoring force becomes smaller and smaller, the net effect of the fluctuating force F(t) on it becomes bigger and bigger. The mode amplitude u shows critical fluctuations. The symmetry of the potential allows to apply in a simple manner the basic symmetry principle of the Landau theory of phase transitions. Let us assume that there exists one nondegenerate distribution function. Then it must possess the same symmetry of the problem, i.e. as the potential (2.19). According to the Landau theory one puts the distribution function f = N exp U, expands the exponent into a power series in the field amplitude u which shows the same symmetry as the problem. This means that the leading terms must be

$$U = a|u|^2 - b|u|^4 \tag{2.25}$$

in accordance with (2.13).

We now exhibit some further analogies between laser physics and approaches presented in the contributions to this book by Cummins, Lallemand, Pike and others. As had been made quite clear by these authors the physical properties of a system are exhibited in modern approaches by correlation functions. So in light scattering theory correlation functions of the type

$$\langle E^{(-)}(t)\; E^{(+)}(0)\rangle \tag{2.26}$$

appear, where $E^{(-)}$ and $E^{(+)}$ are the negative and positive frequency parts of the electric field, respectively. Similarly in fluid dynamics e.g. correlation functions of the density

$$\langle \rho(t)\; \rho(0)\rangle \tag{2.27}$$

play an important role. For equal times the correlation functions reduce to expressions for the moments.

The situation in laser physics is completely analogous. So the stationary distribution function (2.13) (compare Fig. 9) becomes manifest by its moments for the photon number

$$\langle n^\nu\rangle \quad , \nu = 1, \ldots \tag{2.28}$$

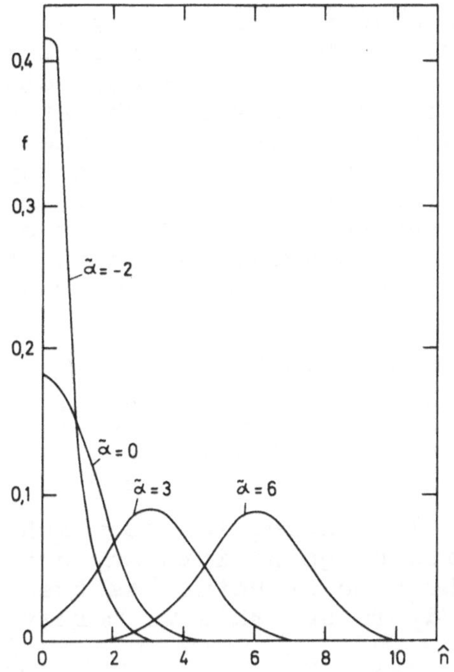

Fig. 9 Photon distribution function f(n) for different values of the pump parameter below, at and above threshold

The correlation function (2.26) plays a fundamental role for the determination of the coherence properties with respect to phase fluctuations. Indeed (2.26) can be shown to be proportional to

$$\propto \exp \quad (-\hat{\gamma} t) \tag{2.29}$$

where $\hat{\gamma}$ is the optical linewidth.

Beside phase fluctuations, the amplitude fluctuations or, in other words, the intensity or photon fluctuations play a decisive role, when one wants to distinguish light produced by the laser from light produced by thermal sources. Here e.g. correlation functions for the photon number of the form

$$\langle n(t)\; n(o) \rangle \tag{2.3o}$$

appear which have their formal analogue in equation (2.27). We now proceed to explain how correlation functions of the form just discussed may be calculated in laser theory.

Again there exists a very nice analogy e.g. to the theory of liquids as presented by the paper of Lallemand. For this end we establish a formal analogy between the field strength u, the polarization v and the atomic inversion D on the one hand, with the density ρ, the pressure p, and the velocity component v of a liquid on the other hand. The fully nonlinear equations referring to u,v,D are exhibited in § 5 as the so-called Langevin equations. If we linearize these Langevin equations below threshold around their equilibrium values u = O, v = O, D = D_o, we obtain the equation

$$\begin{aligned} \dot{u} &= -\kappa u - ig\, v + \Gamma_u(t) \\ \dot{v} &= -\gamma_\perp v + ig\, D_o\, u + \Gamma_v(t) \\ \dot{D} &= \gamma_\parallel (D_o - D) + \Gamma_D(t) \end{aligned} \tag{2.31}$$

where the Γ's are fluctuating forces which are δ-correlated ("Markoffian"). Without going into any details we observe that (2.31) has the same structure as the equations discussed by Lallemand. To see this we merely introduce

$$A = \begin{pmatrix} u \\ v \\ D \end{pmatrix} \tag{2.32}$$

and

$$F(t) = \begin{pmatrix} \Gamma_u(t) \\ \Gamma_v(t) \\ \Gamma_D(t) \end{pmatrix} \tag{2.33}$$

so that (2.31) acquires the form

$$\dot{A} = -MA + F \tag{2.34}$$

This analogy makes it superfluous for us now to discuss the method of solution and of calculating the correlation functions for the type e.g. (2.26), which is done in Lallemands paper. However, one important difference should be noted here. While in liquids one assumes thermal equilibrium and thus may determine the correlation functions

$$\langle F(t)\, F(o) \rangle \tag{2.35}$$

e.g. by the usual Einstein relations, the laser represents a system far from thermal equilibrium and the correlation function (2.35) must be now determined in a more general manner (see e.g.Haken 197o). Above threshold the method of linearization breaks down (which had been overlooked by quite a number of authors) and the method of quasi-linearization must be used. It consists in looking for stationary solutions of the Langevin equations (compare) § 5) without fluctuations and then assuming that the fluctuating forces F give rise to only small fluctuations of amplitude and phase. Thus in the region above threshold one makes the hypothesis

$$\begin{aligned} u &= (u_o + \delta u)\, e^{i\Psi} \\ v &= i(v_o + \delta v)\, e^{i\Phi} \\ D &= D_o + \delta D \end{aligned} \tag{2.36}$$

and establishes linear equations for

$$A = \begin{pmatrix} \delta u \\ \delta v \\ \Psi \\ \Phi \\ \delta D \end{pmatrix} \tag{2.37}$$

From now on essentially the same analysis as described e.g. by Lallemand might be applied. A historical remark may be in order. The above ahalogy uses the fact that the quantities u,v,D of the laser could be treated as classical variables which has become possible by the method of quantum classical correspondence. Before this method was known, u,v, and D were indeed operators obeying nonlinear Heisenberg equations of motion. It is interesting to note, however, that even such operator equations,where also the fluctuating forces were operators, could be solved by methods of linearization and quasi-linearization.

Both in the classical and in the quantum mechanical domain these methods are applicable to a wide laser region, where one has to exclude the threshold region. In particular for this region the application of the Fokker-Planck equation (see § 5) proved to be superior over the other methods. While the stationary solution has been found analytically, as was shown in our preceding derivation, the time dependent Fokker-Planck equation could be solved even for the single mode case only by computer calculation. This may have also some impact on the general theory of phase transitions showing that at the critical point new methods still have to be developed when dealing with the multi-mode case. Perhaps a combination of the exact numerical results for the soft mode combined with approximate methods for the other modes might offer a reasonable approach.

§ 3 THE CONTINUOUS MODE LASER, ANALOGON TO THE GINZBURG-LANDAU THEORY TO SUPERCONDUCTIVITY

Because I conjectured since a long time that a close analogy between the laser threshold and phase transitions exists and because a more sophisticated theory of phase transitions in the familiar systems like superconductors etc. must take care of the local field dependence (instead of a constant field) we investigated the case of an infinitely one-dimensional laser. Without going into the details I just represent two typical results. First I exhibit the stationary distribution function found by Graham and myself which we compare with the distribution function of superconductivity (Table II). Evidently the

Table II

The distribution function of the laser

$$f = N \exp\left(-\frac{B}{Q}\right)$$

$$B = \int \left\{\tilde{\alpha}\,|E(x)|^2 + \tilde{\beta}\,|E(x)|^4 + \tilde{\gamma}\,\left|\left(\frac{d}{dx} - i\frac{\omega_0}{c}\right)E(x)\right|^2\right\}dx$$

E : electric field strength

$\tilde{\alpha} = \tilde{a}\,(d_c - d)\,;\ \tilde{a}, \tilde{\beta}\ \tilde{\gamma} > 0$ are laser constants

c : velocity of light

ω_0: atomic line frequency

N : normalization factor

d : atomic inversion

d_c : critical atomic inversion

B : is a generalized thermodynamic potential

Q : is the thrength of the fluctuations

The distribution function of the Ginzburg-Landau theory of superconductivity

$$f = N \exp\left(-\frac{F}{kT}\right)$$

$$F = \int \left\{\alpha\,|\psi(\underset{\sim}{x})|^2 + \beta\,|\psi(\underset{\sim}{x})|^4 + \frac{1}{2m}\left|\left(\underset{\sim}{\nabla} - \frac{2ei}{c}\underset{\sim}{A}\right)\psi(\underset{\sim}{x})\right|^2\right\}d^3x$$

ψ : pair wave function

$\alpha = \alpha'(T - T_c)$

α', b': superconductor constants

k : Boltzmann's constant

T : absolute Temperature

T_c : critical Temperature

F : the free energy

m, e: mass and charge of electrons respectively

Comparison of the statistical distribution function of the laser light amplitude with that of the Ginzburg-Landau theory of superconductivity.

analogy is complete. Of course, we may go back to the single mode case by projecting out a single mode. We have also derived from first order principles the time dependent laser equations which again show a close resemblance to the time dependent Ginzburg-Landau equations. It is amusing to note that the methods of solution of these equations in the quasilinear range were developed simultaneously and independently for the laser and for the superconductor by making for the laser - field amplitude and for the pair-wave function of the superconductor, respectively, the hypothesis

$$u(x,t) = (r_o + \rho(x,t))\ e^{i\phi(x,t)}$$

which shows especially the phase diffusion. Because the continuous mode laser corresponds to an one-dimensional superconductor no second order phase transition may be found here in the strict thermodynamic sense. The situation here corresponds to phase transitions in restricted geometries. Here the slope of the laser output, i.e. the quantum efficiency, changes continuously.

§ 4 LASER LIGHT EMISSION FROM INTERACTING CHAOTIC FIELDS - A NEW TYPE OF PARAMETRIC PROCESSES

In this paragraph we indicate by means of an example how the concepts developed in § 2 may be applied to interacting Bose fields. We chose a parametric process in which a pump quantum with frequency ω_p decays into a signal quantum with frequency ω_s and an idler quantum with frequency ω_i. It seems to be generally adopted to assume that the signal shows laser light only if the pump shows laser action, i.e. amplitude stabilized oscillation with a fixed phase. In our model we assume for pump, signal and idler spatially fixed modes which are coupled to external reservoirs, which would render each of these fields to a Gaussian distribution typical for the emission of lamps, if the modes were not coupled among each other. Now assume that there is an interaction between these fields given by a Hamiltonian by the form

$$H = \hbar g\ (b_p^+ b_s b_i + b_p b_s^+ b_i^+)$$

where b_p^+, b_p, b_s^+, b_s, b_i^+, b_i are the creation and annihilation operators of pump, signal and idler, respectivly. As in § 2, we use the principle of quantum classical correspondence and put

$$b_p,\ b_p^+ \longleftrightarrow v,\ v^*$$
$$b_i,\ b_i^+ \longleftrightarrow u_1,\ u_1^*$$
$$b_s,\ b_s^+ \longleftrightarrow u_2.\ u_2^*$$

We assume that the effective coupling between the fields is bigger than between the fields and their reservoirs. Then we may find the distribution function of the total system looking first for the constants of motion. If we first assume that one of the field phases is heavily damped, the "surviving" constants of motion are

$$h_o = |v|^2 + |u_1|^2$$
$$h_1 = |u_2|^2 - |u_1|^2$$

If we further assume that one of the fields, for instance the pump field, is heavily damped, so that it obeys a Gaussian distribution, the total distribution function must be of the form

$$f = N \exp\ (U(h_1))\exp(-Ch_o),$$

or, if we expand the exponent around the most probable value:

$$f = N' \exp\ (Ah_1 - Bh_1^2 - Ch_o)$$

The evaluation by means of a microscopic theory, in which the decay constants and equilibrium values of the photon numbers due to the heatbaths enter, demonstrates that the constants a and b are such that in fact finally after integration over unwanted variables the photon number distribution of the signal arises

$$\tilde{f}(n_2) = N' \exp\ (a\ n_2 - b\ n_2^2)$$

where $a > 0$, if the losses of the signal are smaller than those of the idler, and if the mean photon numbers of signal and idler are sufficiently small. This proves that laser action is possible.

§ 5 SOME MATHEMATICAL TOOLS

We quite briefly sketch some of the equations and methods used implicitly in the above discussions.

a) quantum-classical correspondence

This method allows us to calculate quantum mechanical expectation values and correlation functions by means of classical expectation values and correlation functions by establishing the following correspondence:

quantum mechanical	classical
operators	variables
$\Omega_1, \ldots, \Omega_n$	$v_1, \ldots, v_n$
density matrix ρ	quasi-distribution function f
density matrix equation	generalized or usual Fokker-Planck equation
$\dot{\rho} = L\rho$	$\dot{f} = Lf$
(L:Liouville operator)	(L:differential operator)
expectation values	expectation values
$\overline{\Omega} = \mathrm{tr}\ (\Omega\rho\ )$	$\overline{\Omega} = \overline{v} = \int vf\ dV$
(tr = trace)	(dV = volume of v-space)
correlation functions	correlation functions
$\mathrm{tr}(\Omega(t_1)\Omega(t_2)\rho)$	$\iint v\ dv\ v'\ dv'\ f(v,v';t_1,t_2)$

In principle the transformation establishing this correspondence reads

$$f(v_1,..,v_n) = \int \mathrm{tr}\ (\prod_{\lambda=1}^{n} \exp\ (\xi_\lambda(\Omega_\lambda - v_\lambda))\rho)\ d^n\xi$$

Starting from a given density matrix equation the Fokker-Planck equation has been derived for the quasi-distribution function f in the following cases:

α) Bose fields
β) two-level atoms (or spin 1/2)
γ) arbitrary quantum systems characterized by projection operators

b) The Fokker-Planck equation of the single mode laser with atoms, field and atomic inversion

It reads in the interaction representation at exact resonance:

$$\dot{f} = (L_o + L_1)f$$

where

$$L_o = -ig \left\{ \frac{\partial}{\partial u^*} v^* - \frac{\partial}{\partial u} v + \left(\frac{\partial}{\partial v} u - \frac{\partial}{\partial v^*} u^* \right) D - 2 \frac{\partial}{\partial D} (uv^* - u^* v) \right\}$$

$$L_1 = \kappa \left(\frac{\partial}{\partial u} u + \frac{\partial}{\partial u^*} u^* \right) + 2\kappa\, n_{th} \frac{\partial^2}{\partial u \partial u^*}$$

$$+ \gamma_\perp \left(\frac{\partial}{\partial v} v + \frac{\partial}{\partial v^*} v^* \right) + \left(N\, w_{12} + \frac{N}{2} \eta + D \frac{\eta}{2} \right) \frac{\partial^2}{\partial v \partial v^*}$$

$$+ \gamma_\parallel \frac{\partial}{\partial D} (D - D_o) + \gamma_\parallel N \frac{\partial^2}{\partial D^2}$$

u: field amplitude, v: complex atomic polarization,
D: inversion

$D_o = N \dfrac{w_{12} - w_{21}}{w_{12} + w_{21}}$; N: total number of laser atoms,

w_{ij} : incoherent transition rate from level i to level j

κ : cavity decay constant, $\gamma_\parallel, \gamma_\perp$ longitudinal and transverse atomic decay constant, respectively

η : decay constant of atomic phases

In lowest order, i.e. $(\sim \frac{i}{g})^o$ the stationary distribution function near threshold is given by

$$f = N_o \exp \left\{ - \frac{4\kappa}{\gamma_\parallel N} \left(n + \frac{D}{2}\right)^2 + \frac{2D_o}{N} \left(n + \frac{D}{2}\right) - \frac{2}{N} \left(|v|^2 + \frac{D^2}{4} \right) \right\}$$

(N_o: normalization constant)

The moments may be expanded into a power series of $(\frac{1}{g})$. We exhibit two examples of equations for the moments[9]

$$\overline{u\, v^*} = \frac{\kappa}{ig} \left(\overline{|u|^2} - n_{th} \right)^{(o)}$$

holds in first order and

$$2\, \overline{|v|^2} + \overline{2D\, |u|^2} = \frac{2\kappa(\kappa + \gamma_\perp)}{g^2} \left(\overline{|u|^2} - n_{th} \right)^{(o)}$$

results in second order. We also find e.g. the following exact relations

$$\overline{u\,v^*} + \overline{u^*v} = 0$$

$$4\kappa\,(n_{th} - \overline{|u|^2}) + \gamma_{\parallel}(D_0 - \bar{D}) = 0$$

c) Langevin equations connected with the Fokker-Planck equation

$$\dot{u} = -\kappa u - igv + \Gamma_u$$

$$\dot{v} = -\gamma_{\perp} v + iguD + \Gamma_v$$

$$\dot{D} = \gamma_{\parallel}(D_0 - D) - 2\,ig(uv^* - u^*v) + \Gamma_D$$

$\Gamma_u, \Gamma_v, \Gamma_D$ are the fluctuating (Langevin) forces. If we drop them, we obtain the "fully coherent" laser equations. In this case we find as stationary solutions

$$v = -\frac{\kappa}{ig}u \qquad \text{or} \qquad uv^* = \frac{\kappa}{ig}|u|^2 \quad \text{(compare § 2)}$$

$$D = \frac{\gamma_{\perp}\kappa}{g^2}$$

$$|v|^2 = \frac{\gamma_{\parallel}}{4g^2}\left(D_0 - \frac{\gamma_{\perp}\kappa}{g^2}\right)$$

EXAMPLES FOR THREE-MODE PROCESS (§ 4)

pump v	signal u_1	idler u_2	effect
light	light	light	optical parametric process
light	light	phonon	Raman + Brillouin }
light	phonon	light	coherent phonon generation
excitons	light	phonons	laser action of excitons

PHASE TRANSITION ANALOGY
Single mode laser

<table>
<tr><th>Laser theory</th><th>phase transition theory</th></tr>
<tr><td colspan="2">fully nonlinear, statistical theory
seek macroscopic variables</td></tr>
<tr><td>distribution function
(2.12)
for field, polarization,
inversion</td><td>+)</td></tr>
<tr><td colspan="2">seek slowly varying macroscopic
variables (order parameter)</td></tr>
<tr><td>reduced distribution
function (2.13)</td><td>Landau theory</td></tr>
<tr><td>Langevin equations of
the laser (§ 5c) or,
reduced (2.24)</td><td>mean field theory
with fluctuations</td></tr>
<tr><td colspan="2">linear or quasilinear solution
with respect to fluctuations</td></tr>
<tr><td>semiclassical theory
(2.18) or § 5c without
the Γ's</td><td>mean field theory</td></tr>
</table>

+) To our knowledge, a procedure analogous to that of eqs. (2.6) - (2.13) is not known in phase transition theory

References

General laser theory:

H.Haken: Laser Theory, Encyclopedia of Physics, Vol. XXV/2c, Springer, New York 197o

H.Haken: in "Quantum Optics" ed.S.M.McKay and A. Maitland, Academic Press, New York, 197o

H.Haken and W.Weidlich: in "Quantum Optics", ed. R.J. Glauber, Academic Press, New York, 1969

M. Lax in "Statistical Physics, Phase Transitions and Superfluidity", eds. M.Chrêtien, E.P. Gross and S.Deser, Gordon and Breach, New York, 1968

E.R.Pike, in "Quantum Optics", ed. S.M. McKay and A.Maitland, Academic Press, New York, 197o

H. Risken, Progress in Optics, Vol. 8, ed.E.Wolf, North Holland, Amsterdam, 197o

M.O.Scully in "Quantum Optics", ed.R.J.Glauber, Academic Press, New York, 1969

Phase transition analogy:

R.Graham and H.Haken, Z.Physik 213, 42o (1968); 237, 31 (197o)

V.DeGiorgio and M.O.Scully, Phys.Rev. A2, 117o(197o)
S. Großmann and R.Richter, Z.Physik 242, 458 (1971); 249, 43 (1971)

S. Großmann in "Synergetics" ed. H.Haken, Teubner, Stuttgart, 1973

R. Graham, in Springer Tracs of Modern Physics, 66, 1 1973 and to be published

R. Graham in "Synergetics", ed. H.Haken, Teubner Stuttgart 1973

H.Haken in Festkörperprobleme X (197o) ed.O.Madelung Pergamon-Vieweg

H.Haken in "Synergetics, ed. H.Haken, Teubner Stuttgart 1973

K.Hepp and E.H.Lieb, Ann.Phys. N.Y. 76, 36o and several preprints

V. Dohm, Solid State Commun. 11, 1273 (1972)

Unpublished material:

a) H.Haken: Distribution functions of classical and quantum systems far from thermal equilibrium, Z.Physik, in press

b) H.Haken and H.G.Wöhrstein: Atom-field correlation, conservation laws and the phase transition of the laser, Optics Comm. in press

c) H.Haken: Laser light emission from interacting chaotic field, in press

d) H.Haken: Correlations in classical and quantum systems far from thermal equilibrium, in press

e) H.Haken: A phenomenological approach to systems far from thermal equilibirum

INTENSITY FLUCTUATION AUTOCORRELATION STUDIES OF THE DYNAMICS OF MUSCULAR CONTRACTION

Francis D. Carlson and Allan B. Fraser

The Thomas C. Jenkins Department of Biophysics

The Johns Hopkins University, Baltimore, Maryland U.S.A. 21218

I. Introduction

The past thirty to forty years of physiological, biochemical, and electron microscopic and x-ray diffraction structural studies on striated muscle have provided biologists with a molecular model of muscle. This model suggests a dynamic mechanism for the process which transforms chemical energy to mechanical energy in muscle. Recently, many of the molecular components characteristic of muscle have been found associated with other important physiological processes in other organs. For example, actin, a contractile protein, has been found in large amounts in nerve tissue, in blood platelets, and in amoeboid cells. The other most important contractile protein, myosin, or primitive forms of it, has also been found in these cells. The suspicion is that the streaming of cytoplasm and the release of various transmitter substances in nerve cells, amoeboid movement, and retraction in blood clots are all driven by some modified form of the actin-myosin contractile system that is found in its most highly organized form in muscle.

II. The sliding filament-cross bridge model of muscle

As background for the intensity fluctuation studies on muscle and contractile proteins we introduce the current structural and dynamic model of muscle.

<u>The sliding filament structure of muscle</u>

Striated muscle cells from all species are tubular structures ranging in length from a few millimeters to many centimeters and in diameter from 20 to 100 μm. Each cell is densely packed with long thin fibrils, called myofibrils, which extend axially within the cell and have irregularly shaped cross-sections with mean widths of about 1 μm. The myofibrils show a highly regular repeating axial structure of sarcomeres.

Sarcomeres in various unstretched muscles vary in length from 2 to 10 μm. These features of muscle and those described below are shown in Fig. 1.

The sarcomere itself consists of two sets of interdigitating filaments. In the central region of the sarcomere, the A-Band, there is a close-packed hexagonal array of thick filaments. The thick filaments contain about 95% myosin and 5% C-protein. Myosin is an enzyme that catalyzes the energy yielding hydrolytic reaction for contraction:

$$ATP \xrightarrow{myosin} ADP + Pi$$

where ATP is adenosine triphosphate, ADP is adenosine diphosphate and Pi is inorganic phosphate. The function of C-protein is not yet certain.

The sarcomere is terminated at each end by a dense protein structure, the Z-line. Projecting from each side of the Z-line toward the center of the adjacent sarcomeres are the thin filaments. They run toward the center of the sarcomere and pass into the hexagonal array of thick filaments at the trigonal points of the array. The thin filaments terminate within the array of thick filaments, thus forming a region in which the thick and thin filaments overlap. The region containing only thin filaments is the I-Band. The thin filaments contain F-actin, and two regulatory proteins.

In the muscle cell the myofibrils are aligned by the Z-discs arranging themselves into planes that run the whole way across the cell. Thus, the muscle cell has a striated appearance that arises from the transverse alignment of the optically different A- and I-Bands. This registration of the myofibrillar sarcomeres in a single muscle cell makes it possible to observe microscopically that when the muscle shortens or is extended there is no change in length of either the thick or thin filaments. The filaments slide past one another producing concomittant changes in the I-Band length, and the A-I overlap region.

Myosin and its organization into thick filaments

Myosin is a rod-like molecule with one globular end (see Fig. 2a). It can be dissociated into six subunits by various mild chemical treatments that do not involve breaking any covalent bonds. There are two heavy chains and four light chains. The heavy chains are identical and make up most of the weight of myosin. The two heavy

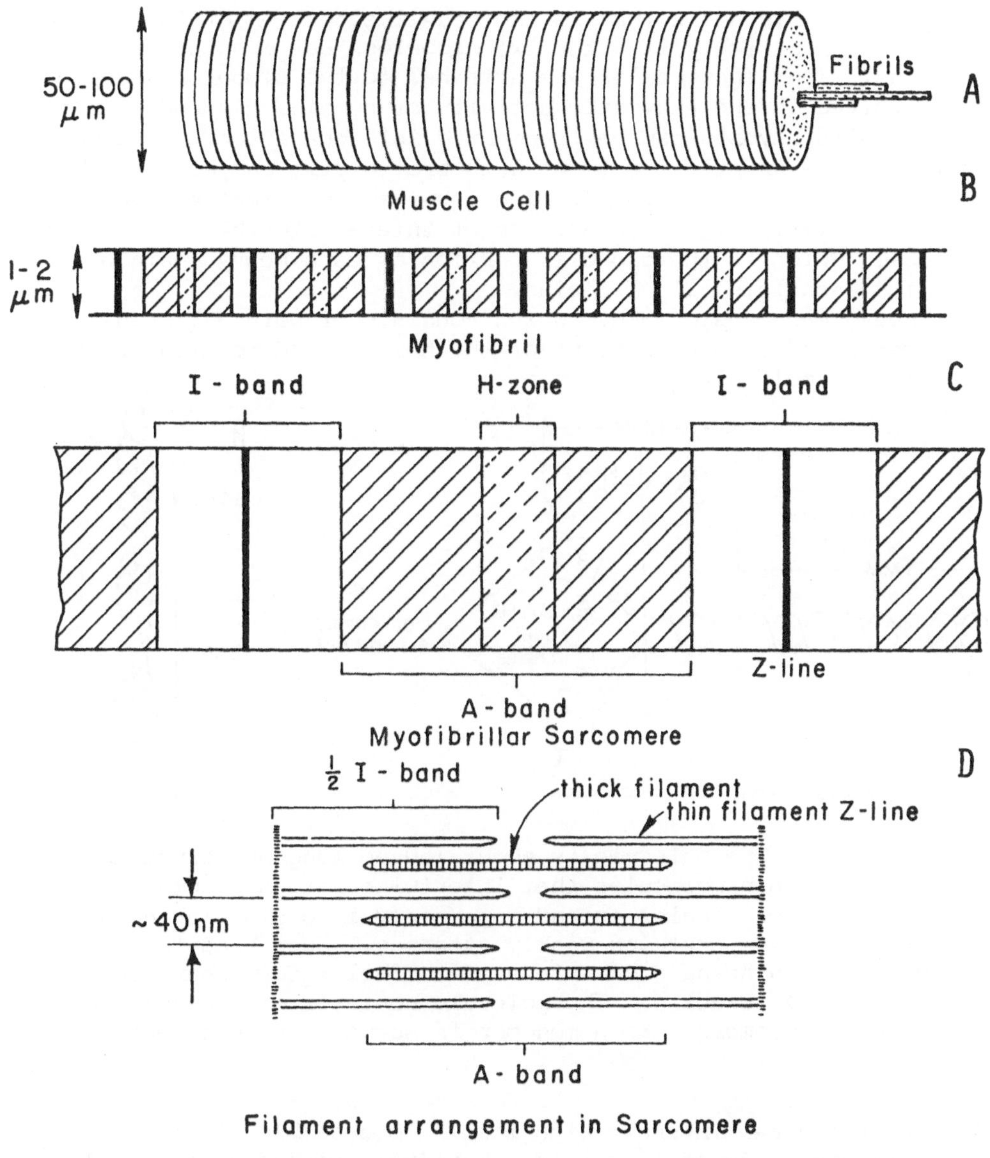

Fig. 1. Schematic of skeletal muscle. a) Muscles consist of very long cells that show striations due to sarcomeres and **are divided** longitudinally into myofibrils. b) Each myofibril is 1-2 μm in diameter and shows several bands in each **2.3μm** long myofibrillar sarcomere. c) The myofibrillar sarcomeres show optically different A- and I-bands. Shortening occurs by reduction in length of the I-band and H zone. d) This comes about by the thin filaments from the I-band sliding along and having increased overlap with the thick filaments in the A-band. Structure to level c is visible by light microscopy; level d requires electron microscopy for visualization.

chains form a double helix that is the rod-like portion of the molecule and extend separately into the two globules that constitute the globular end. The light chains are attached to the globular heads and are involved in the enzymatic activity of myosin. In addition to forming subunits, myosin can be fragmented into characteristic sub-fragments by more drastic treatment with proteolytic enzymes like pepsin and trypsin (see Fig. 2a). The sub-fragments that can be produced by proteolytic digestion and are of interest in this study are:

1) S-1, which is the globular portion of each heavy chain and its associated light chains.
2) Heavy meromyosin (HMM), which consists of both S-1 fragments and a short segment of the helical rod portion of the myosin molecule.

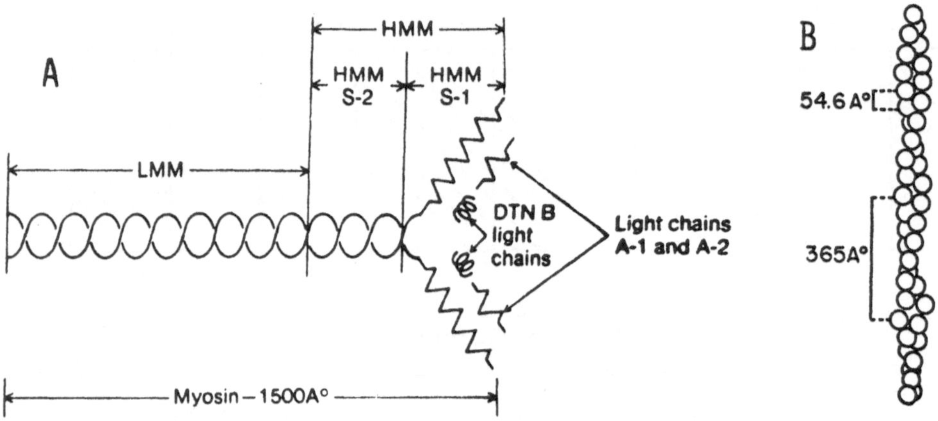

Fig. 2a) Myosin. The molecule is about 150 nm long and consists mostly of two protein molecules that intertwine into a double helix. The same protein chains extend separately into globular heads. Each head has smaller proteins weakly attached. Each head is an S-1 sub. unit. Both heads and a piece of the helix form heavy meromyosin (HMM). 2b) F-actin. The molecule consists of a double helix of globular monomers. Each monomer is about 5 nm in diameter.

Myosin apparently can interact with itself to assemble itself into the thick filaments found in muscle. Myosin is packed at the center of the thick filament in an anti-parallel orientation with the rod portions of the molecules abutting. Thus, the center of the thick filament has a smooth appearance. Away from the center of the thick filament, the myosin molecules are in a parallel configuration and the HMM fragments of the molecules project from the surface on an approximate 6/2 helical lattice. The helical rod portion of the myosin is organized to form the core of the thick filament. Pairs of HMM cross bridges project from the thick filament at intervals of 14.3 nm with each pair rotated 120° with respect to its nearest neighbor.

Actin and the thin filament structure

Thin filaments consist primarily of a polymer called F-actin. Its appearance is that of a double twisted string of pearls, each globule of which is a protein monomer (see Fig. 2b). Each filament is about 1μ m long with one end attached to the Z-line and the other projecting into the thick filament array. Thin filaments also contain two kinds of regulatory proteins that are arranged in the groove between the two actin monomer strands. They regulate the interaction of myosin and actin. They will not concern us extensively; their scattering is a part of that of the thin filaments to which they are attached.

Filaments of F-actin can be made in vitro. These filaments are presumably identical to thin filaments from the muscle from which the regulating proteins have been stripped. Artificial F-actin, and native filaments with the regulating proteins removed have all of the actin in its fully active myosin-combining configuration.

The cross bridge contraction theory

A projection down the axis of a sarcomere shows the HMM cross bridges extending from the thick filaments toward the thin filaments in resting muscle. In contraction, cross bridges move out and combine with thin actin filaments and ATP hydrolysis occurs in a manner that chemical energy is transduced into mechanical energy. Myosin alone, under the conditions existing inside the muscle, hydrolyzes ATP only insignificantly and without transduction.

The details of the cycling of a cross bridge as it moves along the thin filament generating force and causing shortening is presently under active investigation. General aspects of the scheme are widely accepted while exact chemical and physical changes in cross bridges, and thick and thin filaments are being studied. The general features of the cross bridge cycle are as follows:

1.) The cross bridges in the activated muscle leave their positions in the 6/2 helix about the thick filament and move about 10nm to come into contact with the thin filaments, concommitant with this there is a change from the 6/2 helix to another helical form, which may be 16/3.
2.) Each S-1 portion of the cross bridge interacts with an individual actin monomer in some way that involves transduction of energy from ATP. Precisely at what point in the cycle the ATP is hydrolyzed is not certain, although the best chemical evidence indicates that it is already in a hydrolyzed form when the actin and S-1 combine
3.) During the interaction of the S-1 and the actin force is generated by some sort of conformational change in one or both of the proteins.

4.) The S-1 dissociates from the F-actin filament as a result of interaction and combination with free ATP that is available in the surrounds, and begins another contraction cycle.

III. Possibilities for intensity fluctuation autocorrelation studies in muscle

The aspects of the cross bridge cycle that make it potentially susceptible to investigation by intensity fluctuation spectral studies are three fold. Firstly, the diffusive movement of the HMM cross bridge toward the thin filament might produce intensity autocorrelations of appreciable magnitude and with coherence times dependent on the properties of the two S-1 sub-fragments and the tightness of their coupling to the thick filament core. If it is assumed that the cross bridge movement is essentially free diffusion one would expect exponential intensity autocorrelations, $(g)^{(2)}(\tau)$, with coherence times linear in q^2 and with a constant amplitude that would depend on the heterodyning with the "stationary" core of the thick filaments. A better model for the cross bridge movement may be one that limits the extent of diffusive motion of the S-1 doublet. A model for bound diffusion for which intensity autocorrelations are easily calculated is the harmonically bound Brownian particle. The over-damped harmonically bound Brownian particle should have, as was mentioned by Dr. Cummins, the following intensity autocorrelation:

$$g^{(2)}(\tau) = 1 + \exp\left[(D_t q^2/\gamma)(1-\exp(2\gamma\tau))\right] - \exp D_t q^2/\gamma$$

Where D_t is the coefficient of diffusion of the S-1 doublet, γ is the relaxation rate of the overdamped system given by the ratio of the stiffness of the elastic element to the viscous damping coefficient, and q is the scattering vector, and τ is the delay time. This autocorrelation is not exponential, and has an amplitude that approaches zero as q approaches zero, a q dependent coherence time, and a non-zero negative slope for $\tau = 0$. Other models for "bound" diffusion systems also yield intensity autocorrelations which have coherence times and amplitudes that are q dependent.

The second feature of the cross bridge cycle that might be expected to produce intensity fluctuations during contractions is the possibility that the conformational change in myosin that occurs during the cycle could involve a change in the polarizability of the myosin molecule and thus give rise to intensity fluctuations with coherence times determined by the relaxation time of the chemical reaction sequence in which the myosin -ATP complex goes from one conformation to another. Such a purely chemical process would not be q dependent in amplitude or coherence time.

A third way by which the contractile cycle might produce intensity fluctuation spectra is through the structural rearrangements of the thick filaments that are initiated by activation. If individual thick filaments change from one form to another e.g., a (6/2 helix to 16/3 helix) during the contractile cycle there could be fluctuations in the A-Band dimensions or polarizability, resulting in intensity fluctuations in the scattered light.

IV. Studies of Muscle Proteins

Myosin

Monomer-n-mer equilibrium reactions of myosin are known to occur and structures very similar to intact native thick filaments have been produced from myosin solutions. In our laboratory, Dr. Thomas J. Herbert studied the apparent molecular weight and apparent diffusion constant of myosin at various concentrations of myosin. He measured apparent diffusion constants by fitting single lorentzians to scattered laser light spectra obtained by light-beating. He determined the dependence of the band widths on the square of the scattering vector, q, and from this data obtained the diffusion constant in the zero limit of q. Also, he measured apparent molecular weight by classical light scattering techniques. The apparent diffusion coefficient decreased and the apparent molecular weight increased with increasing myosin concentration. His data were well fitted with a with a monomer-dimer equilibrium scheme, and the radius of gyration of the resultant dimer was consistent with the idea that the dimers were forming with parallel orientation. This kind of dimer is like the side by side arrangement of myosin molecules along the cross bridge containing portion of thick filaments.

Actin

When artificial F-actin is made, a collection of fibrous molecules having a length from far less than 1 μm to several μ m results. Solutions of artificial F-actin show flow birefringence and are optically inhomogeneous. In spite of this inhomogeneity, biophysical light-scatterers study F-actin because of interest in its flexibility and to investigate F-actin that is interacting with the various fragments of myosin. The possibility that F-actin is flexible arises from electron microscope studies and has obvious implications in the dynamics of muscular contraction.

An argon ion laser and digital autocorrelator were used to measure intensity fluctuation autocorrelations from F-actin preparations that were provided by Dr. Evan Eisenberg of the National Institutes of Health, (U.S.A.). The same apparatus was used in the actin and myosin fragment experiments discussed below.

The intensity fluctuation autocorrelation functions that were measured in this and following actin and actin-myosin studies were fitted with a best single band width by a second degree fit to the log of $[g^{(2)}(\tau)-1]$ using the method of Koppel (1972). Figure 3 shows the band width of the light scattered from F-actin solutions as a function of q^2. This curve is presumably influenced by translational and rotational diffusion, polydispersity in length, and aggregation. It may depend on the flexibility of the longer molecules. The plot passes through zero band width at $q^2=0$ and is concave upwards. These measurements were made under conditions nearly identical to those of Fujime, et al (1971). Fujime and co-workers fit single lorentzians to data collected with a light beating spectrometer and they found that F-actin produced spectra having band widths linear in q^2 and having relatively large positive band widths in the limit $q^2=0$. Our results always extrapolated to zero bandwidth, were concave upwards, and behaved this way at temperatures between 0 and 23°C in various buffers. Fujime has also studied the flexibility of F-actin in combination with HMM, S-1, and tropomyosin. We will comment below on our measurements of F-actin combined with myosin fragments.

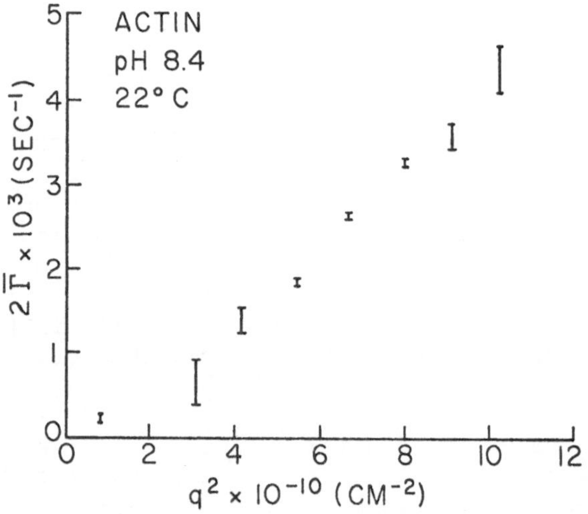

Fig. 3 Bandwidth of light scattered from actin. Each error bracket shows mean position and standard error of mean for 3 samples taken from one actin preparation. The temperature and ionic conditions were as reported by Fujime, et al.

According to Fujime's theory, flexible rod-like macromolecules should produce intensity fluctuation spectra whose single lorentzian bandwidths are given by:

$$\Gamma = 2D_t q^2 + 1/\tau \quad (1)$$

where τ is the average period of the lowest vibrational modes of the flexible macromolecules. Thus, at an extrapolated $q^2 = 0$ there would still be a positive bandwidth in contrast to the zero extrapolated bandwidth for rigid diffusing particles. From our measurements, we suppose that the magnitude of the coefficient (which was left unevaluated) of the $2D_t q^2 + 1/\tau$ term in Fujime's analysis may be small compared to a dominant but neglected term that limits to zero bandwidth with zero q. Our measurements are consistent with the notion that dominant phase fluctuations are governed by $\Sigma q \cdot r_i(\tau)$, where $r_i(\tau)$ is the displacement of the ith scattering segment in time τ. This form obviously converges to zero with q.

Myosin Fragments

Enzymatically active and highly pure preparations of heavy meromyosin (HMM) and S-1 fragments were also supplied by Dr. Eisenberg. Approximate diffusion constants were determined from bandwidth measurements in spite of non-idealities of the protein solutions. Bandwidths were fitted by the method described by Koppel (1972) to autocorrelation data at 63.8°, 90° and 116.2°. Heavy meromyosin gave an apparent diffusion constant of approximately 1×10^{-7} cm^2/sec and S-1 gave an apparent diffusion constant of approximately 6×10^{-7} cm^2/sec. These values are close to those expected for particles with molecular weights of approximately 350,000 and 120,000 daltons respectively. It should be noted that both of these diffusion constants are much larger than the approximate 4×10^{-8} cm^2/sec diffusion coefficient obtained for F-actin under the same conditions.

F-actin - subfragment interactions in the presence of ATP

The availability of soluble forms of F-actin and myosin subfragment makes it possible to study the interaction of these components of the contractile system under conditions of maximal ATP hydrolysis. At very low ionic strength the F-actin-sub-fragment system splits ATP at maximal rates. Solutions of F-actin and HMM or F-actin and S-1 in 0.1M KCl do not hydrolyze ATP nor do they form gels because ATP dissociates the acto-sub-fragment complex as fast as it is formed. The non-interacting KCl and ATP containing system is a model of relaxed muscle, while the interacting, ATP-splitting system is a model of contracting muscle. In the absence of ATP gels form that are biochemical models of rigor mortis.

We have investigated these model systems with the view of answering such questions as: What changes occur in the diffusion of the various components under conditions of maximal ATP hydrolysis?, What is the character of gelled acto-myosin systems in the absence of ATP and how does it relate to the maximally interacting system?

We mixed actin and HMM, and actin and S-1 under non-interacting conditions and we were able to detect the presence of the rapidly diffusing myosin fragment in the presence of F-actin. Fig. 4a shows the measured band widths plotted against the concentration of the myosin fragment at a fixed F-actin concentration for 3 scattering angles. The free small molecules significantly increased the bandwidth of the scattered light. Field autocorrelation functions, $(g^{(2)}(\tau) - 1)^{\frac{1}{2}}$, were computed for actin, S-1, and heavy meromyosin. F-actin and HMM field correlation functions were used to calculate the intensity autocorrelation of the non-interacting F-actin-HMM system and gave results that agreed well with the observed autocorrelations. They were added according to relative scattering power and weight fraction to predict the autocorrelation functions of the non-interacting mixtures. Success was had with HMM, but calculated bandwidths were too large for S-1 containing mixtures.

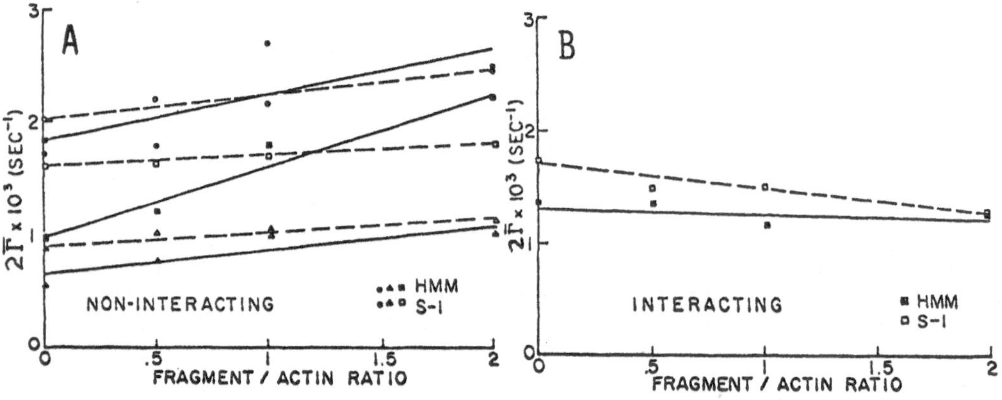

Fig. 4a Non-interacting acto-HMM and acto-S-1 mixtures. The myosin fragments increase the bandwidth of the mixtures. S-1 has less effect than HMM. Bottom curves 63.8°, middle curves 90°, and top curves 116.2° scattering angles.
Fig. 4b Interacting acto-HMM and acto-S-1 systems, Slight decreases in the bandwidth of the scattered light were caused by increasing quantities of myosin fragments.

The interacting system uses ATP so rapidly that there was only enough time to measure autocorrelations at one angle; 90° was chosen. Fig. 4b shows the band width of the scattered light plotted against the concentration of the myosin fragment. Heavy meromyosin decreased the band width slightly as its concentration increased. S-1 caused a somewhat greater decrease in bandwidth with increasing concentration. Thus, a small effect in the opposite direction to that found in the non-interacting system occured. Examination of the autocorrelation functions of the interacting mixtures showed that the part of the autocorrelation caused by the F-actin showed a decrease in bandwidth. That is, at delay times far greater than twice the longest times that the myosin fragments showed appreciable amplitudes in their autocorrelation functions the interacting mixtures still showed larger amplitudes and slower decay rates than did F-actin alone. Chemical turnover rates are only 1 sec^{-1} in these systems, and should not influence the spectra significantly. Parallel viscosity measurements done by Dr. Eisenberg showed that in the interacting HMM system a concentration of 2 gm/l of each protein produced about a doubling in viscosity compared to F-actin alone or F-actin not interacting with HMM. In the case of the S-1 interacting system the viscosity was increased only slightly. Interpretation of these finding appears below.

Bound Systems

Actin-myosin fragment systems were allowed to gel by hydrolyzing all of the initial ATP. The acto-HMM and acto S-1 were stiff enough to remain in an inverted cuvette.

Fig. 5 shows autocorrelation functions obtained from these gels. HMM and actin in equal weight fractions formed a gel that had an intensity fluctuation autocorrelation function with a very small amplitude. This implies that most of the scatterers in the system were immobilized, that is, there was virtually no movement over distances of the order of the wavelength of the incident light. Thus, binding of HMM immobilized actin as well as HMM. Less HMM with the same actin produced a gel having almost full mobility but a narrow bandwidth. S-1 and actin produced an intensity fluctuation autocorrelation that generally had almost as great an amplitude as did F-actin alone, and the bandwidth was similar to but smaller than that of F-actin alone. Apparently, the non-pouring acto-S-1 gel exhibited molecular movement over distances comparable to the wavelength of the incident light.

Fujime and co-workers (1972) made band width measurements on actin-myosin fragment systems in the absence of ATP. They also determined the various diffusion coefficients and interpreted their results as evidence for flexibility. The basis of this interpretation was their finding that these various systems gave non-zero bandwidths at q=0. Our measurements do not support the finding of non-zero intercepts. Further, they show a dramatic diminution of magnitude of the fluctuations of the light scattered from gel systems. Perhaps this decrease in amplitude was not noticed by Fujime et al because they used an a.c. coupled spectrum analyzer.

Interpretation of actin-myosin fragment spectra

The interpretation of these results on soluble model systems of the contractile process is difficult but some statements can be made:

1) Under conditions of nearly maximal ATP hydrolysis where there is more than enough F-actin to bind all the myosin subfragment present, very little if any of the "gel" structure found in the bound system exists.

2) The wide bandwidth of the actin-HMM interacting system, as wide as actin alone, and the fact that the viscosity doubled argue that there is some rapidly diffusing species present.

3) Wide varieties of gels can be made that show changes in amplitude as well as bandwidths of the modulation of the scattered light.

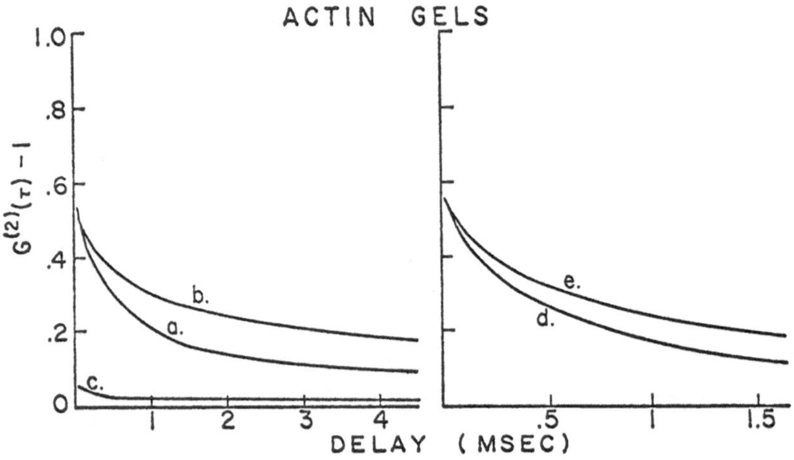

Fig. 5 Acto-HMM and Acto-S-1 gels. Curves a and d show autocorrelations of actin at 1gm/l, b shows that actin with 10% as much HMM formed a gel whose intensity fluctuation autocorrelation had almost full amplitude but a much longer coherence time. c shows that actin with half as much HMM makes a quite rigid gel. e shows that S-1 causes a much less tightly bound gel with actin, even though it remained in an inverted cuvette.

V. Measurements of intensity fluctuations from muscle.

Some time ago we began a study of the intensity fluctuation correlation of contracting muscle with the view of testing the highly simplified diffusion model of cross bridge motion which is described in the introduction. We sought an experimental answer to the question of whether a simple model of cross bridge diffusion could account for the observed light scattering data. Such a model assumes that the cross bridge motion is either simple free of bound diffusion and the coupling to the optical properties of the muscle are solely those due to the local refractive index fluctuations caused by diffusing cross bridges. The studies reported in brief here were conducted by Mr. Robert Bonner.

Experimental Methods

A sartorius muscle from a central North American bull frog was mounted with its long axis vertical and illuminated over a small area with a vertically polarized low intensity 465 nm beam from an argon laser. Light scattered at various angles in the horizontal or equatorial plane of the muscle diffraction pattern was coherently detected by a photo-multiplier. Discriminated output pulses from the photo-multiplier were fed into an on-line digital autocorrelator.

It is instructive to consider first the time dependence of the intensity fluctuations in the light scattered by resting, contracted, and relaxing muscle. Strip chart records of the intensity fluctuations for these three conditions appear in Fig. 6. During the onset of contraction and throughout the contraction there were large amplitude, rapid fluctuations in the intensity of the scattered light. These fluctuations persisted through the relaxation phase and on into the post relaxation period. Also shown in Fig. 6 are records of the tension developed by the muscle during the initiation, plateau, and relaxation phases when contraction occurs with both ends of the muscle clamped to prevent their shortening. Such a contraction is called an isometric tetanus and all the studies reported here were obtained during or near to the plateau of tension in contractions of this type.

Coherence times measured during the plateau phase of an isometric tetanus reached constant values of about 0.5 to 3.0 msec. after about 1 sec of tetanic contraction. To a fair approximation, therefore after 1 sec of plateau of a tetanus a quasi-stationary process is established. Coherence times measured during the rising and falling phases of tension ranged from 10 μsec. to 200 μsec. Coherence times measured during the post relaxation phase were in the 10 to 30msec range. During all phases of contraction, autocorrelation functions gave amplitudes equal to or slightly less than the maximum value imposed by coherence area limitations assuming a guassian field. We take this to mean that the bulk of the scattering material was involved in producing the intensity fluctuations observed, and that little or no heterodyning with stationary structures in the muscle occured.

Fig. 7 illustrates a typical plot of raw autocorrelation data taken from about 0.2 to 0.7 sec. in a tetanus. The data definitely show a zero initial slope and thus can not be properly fitted with an exponential. To obtain a measure of the coherence time and the amplitude of these autocorrelation functions a least squared error fit was made to a smoothing function of the form

$$G^{(2)}(\tau) = R(e^{-K\tau} - Me^{-K\tau/M}) + B, \qquad (2)$$

where all the arbitrary constants are taken as positive. The solid curve in Fig. 7 represents a fit of this function to the data points shown. The background, B, obtained from such fits agreed within experimental error with the background computed from the total counts, number of samples, and prescale level. A coherence time for this autocorrelation was defined as the value of τ at which the decaying component of the amplitude reached (1-M)R/2. That is: for $\tau_{\frac{1}{2}}$ $G^{(2)}(\tau)=(1-M)R/2 + B$, where $\tau_{\frac{1}{2}}$ =the coherence time. The angular dependence of the ratio, A/B, and $\tau_{\frac{1}{2}}$ was examined for muscles with their fiber axes oriented normal to the scattering plane, i.e., mounted vertically since the scattering plane was always horizontal, and with the long axes in the scattering plane, i.e., mounted horizontally.

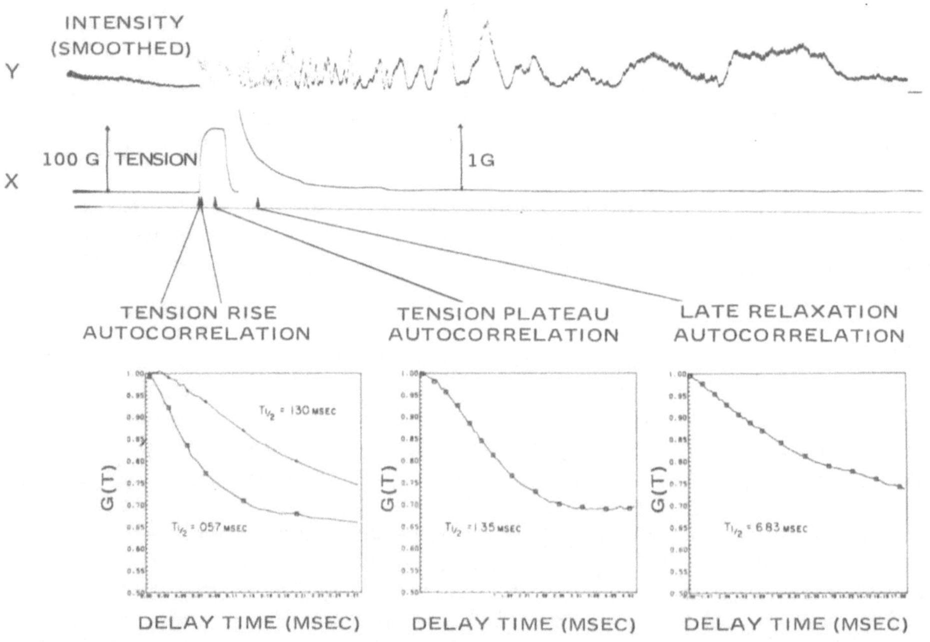

Fig. 6 Intensity fluctuations caused by active muscle. Muscle tension, x, and intensity fluctuations, y, recorded by strip chart recorder. Rapid fluctuations occur during contractions. After tension is maintained, quasi-stationary intensity fluctuations occur. Relaxing muscle produces fluctuations, and resting muscle produces almost no fluctuations. Sample normalized autocorrelations functions appear at bottom.

Control Experiments

To eliminate the possibility that the intensity fluctuations were due to unwanted disturbances such as room vibrations, bulk movement of the muscle, etc. various control experiments were conducted with the following results:

1) Dead muscle prepared by glycerol extraction, which preserved myofibrillar structure, gave absolutely flat intensity fluctuation correlation.
2) Relaxed living muscles gave variable, very low amplitude autocorrelations with coherence times in the 10msec. range.
3) A sixteen-fold change in scattering volume produced no appreciable change in coherence time and only the expected coherence area dependent variations in maximum amplitude of measured autocorrelations.
4) Recent, less extensive experiments on small bundles of muscle cells containing 1 to 10 single cells gave large amplitude autocorrelations with coherence times nearly the same as those of whole muscles. The scattering volume in the single muscle cell experiments was an extremely small fraction of the scattering volumes used in the whole muscle experiments.
5) In whole muscle, less than 10% of the scattered light was depolarized, indicating that multiple scattering was slight. In small bundles of cells multiple scattering was negligible.

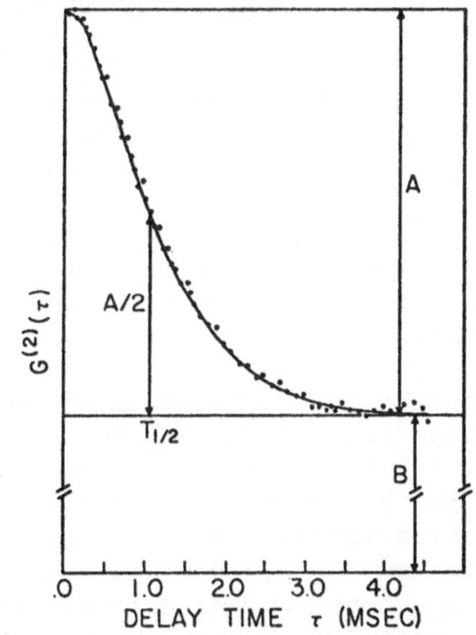

Fig. 7 Sample unnormalized autocorrelation from muscle. Coherence time was characterized as the time required for half of the initial amplitude to be reached $\tau_{\frac{1}{2}}$. Depth of modulation was about as large as for guassian light.

Collectively these control experiments establish that the intensity fluctuations observed in light scattered from contracting muscle were not the result of heterogeneous regions of the muscle moving through the scattering volume under the action of internal or external forces. They support the view that the intensity fluctuations are due to fluctuating changes in the optical properties of many independent scattering elements that are localized within volumes as small as that of a single muscle fiber.

Experimental Findings

The q dependence of the autocorrelation bandwidth, $1/\tau_{\frac{1}{2}}$, for scattering angles from 10° to 50° is shown for vertically mounted muscle in Fig. 8a and for horizontally mounted muscle in Fig. 8b.

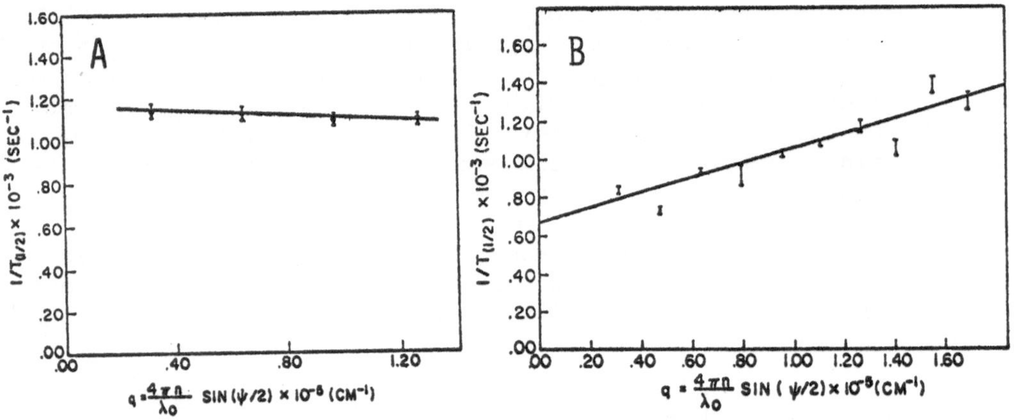

Fig. 8a Bandwidth vs. q for tetanized muscle in the vertical orientation with the fiber axis normal to the scattering plane and parallel to the polarization. The bandwidth is independent of the scattering angle below 50°.
Fig. 8b Bandwidth vs. q for tetanized muscle in the horizontal orientation. With the fiber axis in the scattering plane and perpendicular to the polarization vector the bandwidth decreases somewhat with q, but does not extrapolate to zero with q.

For the vertically oriented muscles slight axial motions of the muscle elements due to fluctuations in the tension developed by the muscle would not be detected in the q dependence of the bandwidth. Such motions are normal to the scattering plane and the q vector. For the horizontally mounted muscles, however, axial movement of scattering elements in the muscle would be in the scattering plane and hence could produce a q dependent bandwidth. The significant feature of both of these plots is that the bandwidth does not approach zero in the limit of zero scattering angle as theory predicts for both free and bound diffusion models of the scatterers. The fact that the bandwidths at q=0 for the vertical and horizontal orientations differ slightly is due to sample variability.

Figures 9a and 9b show the q dependence of the normalized amplitude of the autocorrelation, A/B, for vertically and horizontally mounted muscles, respectively. Here again the amplitude in the limit q=0 is large for both mounting configurations. The dependence of A/B on q for the vertically mounted muscle seems to be a genuine result. We believe the changes are due to a small amount of accidental heterodyning.

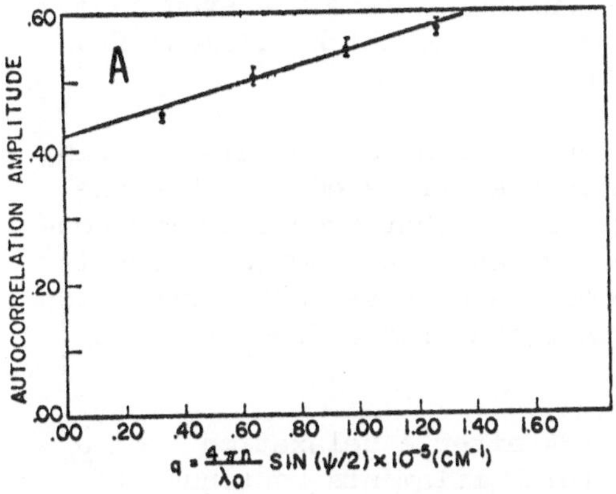

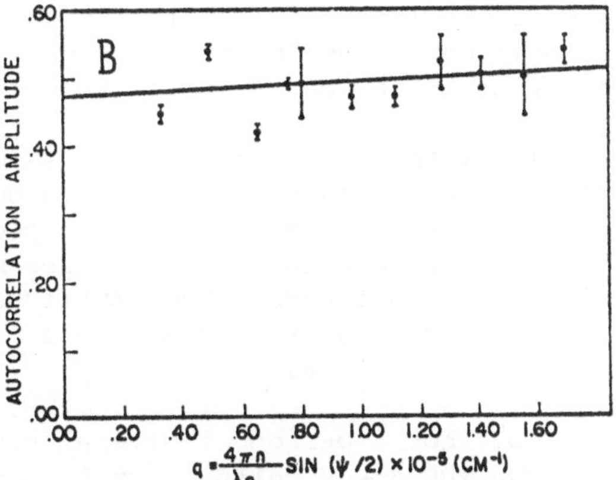

Fig 9a) Amplitude vs. q for tetanized muscle in the vertical direction. Autocorrelation amplitudes decreased somewhat with q, but not in direct proportion. The cause is thought to be heterodyning.
9b) Amplitude vs. q for tetanized muscle in the horizontal direction. Amplitude was independent of q.

Two other important features of the intensity autocorrelations obtained from muscle appeared. Firstly, bandwidth decreases with increasing extension of the muscle, there being about a 10-fold decrease in bandwidth with an extension of 33 percent.

The second additional fact is that large amplitude autocorrelations are obtained from activated muscle that had been stretched to the point of no overlap and hence to the point at which little or no additional tension is developed since in this state few if any active cross links form. These autocorrelations had the same amplitude as those obtained at full overlap and maximum tension, but bandwidths were smaller.

Correspondence between intensity fluctuation autocorrelation studies and structural and mechanical studies on muscle

There is a correspondence between the intensity fluctuations and the changes observed in the X-ray diffraction patterns obtained from muscle during a tetanic contraction and in the subsequent relaxation period. These are:

1) Resting muscle and muscle in rigor give sharp small angle X-ray diffraction patterns. Intensity fluctuations autocorrelations

from resting and rigor muscle are essentially flat. However, we can not assert with certainty that in resting living muscle there is no thermally excited bending of the thin filaments that would give rise to small amplitude autocorrelations.

2) During the steady state plateau phase of a tetanus the X-ray data indicate that the cross bridges move toward the thin filaments, and lose their helical order. Further, there may even be a transition of the organization of myosin in the thick filament from one kind of helical symmetry to another.

Prominent intensity fluctuation autocorrelations are obtained during this phase of contraction. These autocorrelations do not show the characteristics that one would expect if all that occurs during contraction is that the cross bridges behave as simple bound or freely diffusing scatterers. The amplitude of the observed autocorrelation is too large and its q dependence is not that for a free or bound diffusing particle.

3) For a period of several seconds after a relaxation from a tetanus the X-ray pattern due to the thick filaments continues to show signs of some of the changes associated with contraction. These changes gradually diminish and the X-ray pattern returns to that of the fully relaxed resting state. Similarly, intensity fluctuation autocorrelations in the post-relaxation phase persist and return slowly to that of resting muscle.

4) When a muscle is stretched to the point of no overlap between thick and thin filaments and is then stimulated to contract it develops little or no tension, but the X-ray patterns due to the thick filaments show changes nearly identical to those that occur at full overlap and maximum tension development. Correspondingly, intensity autocorrelations obtained from muscles stretched to the no overlap point are similar to those of the unstretched muscle.

5) If a muscle in the steady state of a tetanus is allowed to shorten a small amount it will show a transient phase of tension redevelopment which has half-time of 1 msec. or so. This time constant is believed to be associated with the rate constants of the cross bridge cycle. The values of $\tau_{1/2}$ obtained for the intensity fluctuation autocorrelations were in the 1 msec. range which makes it plausible to associate them with the relaxation time of the mechanical transients which have been attributed to events of the cross bridge cycle.

Because the intensity fluctuation autocorrelations show this close parallellism with the structural changes of the thick filament and the kinetics of tension development, we believe that they originate in the processes that are closely coupled to the elementary contractile mechanism. It is clear, however, that the elementary molecular events of the cross bridge cycle couple to the optical properties of muscle in a more complex fashion than directly through the refractive index fluctuations that arise from either the simple free bound diffusion model.

Conceivably fluctuations in the number of cross bridges active in a myofibrillar sarcomere occur and give rise to small fluctuations in the sarcomere's position about its mean value. Indeed, the slight q dependence of the bandwidth for horizontally mounted muscles suggests that such displacements of the sarcomere do occur. But if this were all that occurred then the system might be described by the bound harmonic oscillator model (see equation 2) and the amplitude of the autocorrelation should approach zero as the scattering angle goes to zero. This is not the case.

Our present working hypothesis is that the fluctuating elementary events of the cross bridge cycle produce fluctuations in the optical polarizability of the A-band and possibly the I-band. Such polarizability fluctuations could be due to refractive index changes, volume changes, or shape changes of the A-band, or any combination of these processes. A knowledge of the dimensions of the scatterers which we are "seeing" in our studies would be most useful and we are initiating efforts to obtain this information from single fiber studies. Other promising approaches to the question of the origin of the scattered light intensity fluctuations associated with contraction are to extend the intensity autocorrelation studies to muscles with altered contractile properties, to different muscle types, and to preparations of isolated myofibrils or thick filaments.

The results which we have reported hardly scratch the surface of the question of the molecular dynamics of muscular contraction. They are to us at least tantalizing.

We thank Drs. H. Z. Cummins and H. Swinney for many useful discussions. This research was supported by U.S.P.H.S. Grant RO1 AM12803.

BIBLIOGRAPHY

Carlson, F. D., B. Bonner, and A. Fraser (1972), Intensity Fluctuation Autocorrelation Studies of Resting and Contracting Frog Sartorius Muscle, Cold Spring Harbor Symp. Quant. Bio. 37, 389

Eisenberg, E. and W. W. Keilley (1972), Evidence for a Refractory State of Heavy Meromyosin and Subfragment-1 Unable to Bind to Actin in the Presence of ATP, Cold Spring Harbor Symp, on Quant. Bio. 37, 145.

Fujime, S. and S. Ishiwata (1971), Dynamic Study of F-actin by Quasielastic Scattering of Laser Light, J. Mol. Bio., 62, 251.

Koppel, D. E. (1972) Analysis of Macromolecular Polydispersity in Intensity Correlation Spectroscopy: The Method of Cumulants. J. Chem. Phys. 57, 4814.

Lymn, R. W. and E. W. Taylor (1971) Mechanism of Adenosine Triphosphate Hydrolysis by Actomyosin, Biochemistry 10, 4617

STRUCTURAL RELAXATION IN VISCOUS LIQUIDS

N. OSTROWSKY

Laboratoire de Spectroscopie Hertzienne de l'Ecole

Normale Supérieure - 24 Rue Lhomond 75005 Paris France

The usual techniques for studying structural relaxation phenomena involve ultrasonic and/or dielectric measurements. In highly viscous liquids such as supercooled glycerol, the structural relaxation time is so long that these conventional techniques are no longer applicable. We have therefore used a photon correlation method to study this problem.

We will begin (in § I) by giving an intuitive picture and a phenomenological treatment of these structural relaxation phenomena. We shall then (in § II) describe our light scattering study of these phenomena.

I ♦ STRUCTURAL RELAXATION PHENOMENA

a Simple picture

A liquid can be thought of as being formed of a large number of small ordered regions each containing a few (5~50) molecules. These regions are continually breaking up and reforming, and their average life time is the so-called structural relaxation time.This time is roughly proportional to the viscosity of the liquid, and in a highly viscous liquid such as supercooled glycerol, this time becomes increasingly large as the temperature goes down. When it reaches 30 minutes, the medium is considered to be a glass. One of the major practical reasons for studying this problem is the following : when the medium reaches its glassy state, the local inhomogeneities are "frozen in " and are responsible for the intrinsic loss in the optical transmission of glasses.

The questions we would like to answer are : What are the effects of this microscopic phenomena on the macroscopic properties of the liquid, and how can these phenomena be studied ?

b Response of a relaxing liquid to an external perturbation

Assume that at a given instant we brusquely impose a constant additional strain on the medium (see fig 1 a) ; the molecules will instantaneously get closer to each other, which

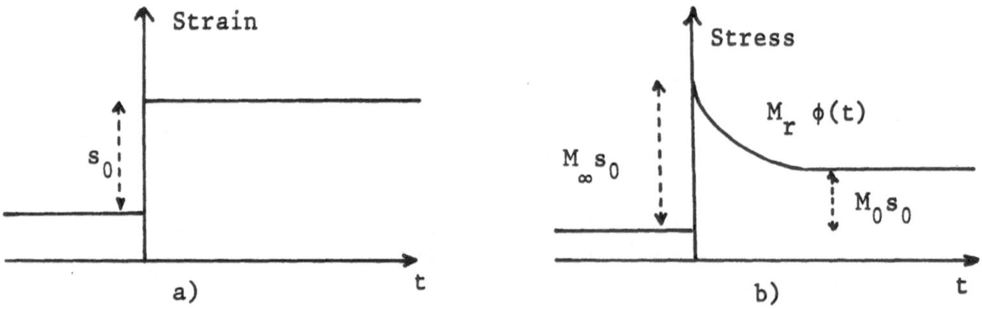

Figure 1 : Stress at time t (fig 1b) due to a sudden change of the strain at time t = 0 (fig 1a).

instantaneously results in a high stress situation ; the molecules will then move around and change the local structure of the liquid to reduce the stress to its equilibrium value (fig 1b). This last process takes a certain time, which is the structural relaxation time, τ.

If the externally imposed perturbation is oscillating at a frequency ω, the response of the liquid will depend on the ratio $\omega/(1/\tau)$. If $\omega \gg 1/\tau$, structural changes are too slow to follow the perturbation and cannot therefore contribute to the response of the system, which behaves like a "glass" with all structural rearrangements "frozen in ". The thermodynamic parameters of the liquid take on their "infinite frequency value" denoted by c_p^∞, κ^∞ for example for the specific heat at constant pressure, and the compressibility. On the other hand, for $\omega \ll 1/\tau$, structural rearrangements are fast enough to give a contribution to the response of the system, and the thermodynamic parameters will take on their "zero frequency values" denoted by c_p^0 and κ^0.

We see therefore that by studying the response of the liquid to an external oscillating perturbation with a variable frequency ω, one can determine the structural relaxation time τ. The most common perturbations applied to the system are :

- an oscillating electric field, in which case one measures the dielectric constant of the liquid as a function of the frequency of the applied field.

• an ultrasonic wave, in which case the measured quantities are the speed v and the absorption coefficient α of the wave. Since the starting point of our work was the ultrasonic studies of glycerol by Litovitz group [1], we shall discuss this method in the next paragraph. This will enable us to introduce, via a phenomenological treatment, the stress relaxation function which is the quantity we measure in our light scattering experiment.

c Ultrasonic measurements - Phenomenological treatment

The response of a viscoelastic fluid to a stress may be expressed by the following relation :

[stress] = [modulus] × [strain]

A fluid can be submitted to two kinds of stresses : a compressional (or bulk) stress and a shear stress. This results in a compressional and a shear strain, determined respectively by the bulk and the shear modulus of the liquid. The propagation of a sound wave involves a linear combination of these two kinds of stresses, and one talks about a "longitudinal" stress. We shall use a simplistic approach and assume that both shear and bulk parameters relax in the same way ; in what follows, we shall only worry about "longitudinal" parameters.

The amplitude A of an accoustic wave propagating along the x axis must obey the following equation :

$$\rho \frac{\partial^2 A}{\partial t^2} = M \frac{\partial^2 A}{\partial x^2} \qquad (1)$$

where ρ is the liquid density and M the longitudinal modulus which relates the longitudinal strain to the longitudinal stress.

To express the delay between the applied longitudinal strain s(t) and the resulting longitudinal stress $\sigma(t)$, one is led to assume that M is time dependent and one has :

$$\sigma(t) = M_0 s(t) + M_r \int_{-\infty}^{t} dt' \; \phi\,(t-t')\; \dot{s}(t') \qquad (2)$$

with : $M_r = M_\infty - M_0$

where M_0 and M_∞ are the zero and infinite frequency limit of the longitudinal modulus, M_r its relaxational part, and $\phi(t)$ is the stress relaxation function. This function gives the stress at time t due to a sudden change of the strain at time t=0 (see fig 1a and 1b).

By taking the Fourier transform of equation (2) one gets the longitudinal stress-strain relation :

$$\tilde{\sigma}(\omega) = \tilde{M}(\omega)\ \tilde{s}(\omega) \qquad (3)$$

with

$$\tilde{M}(\omega) = M_0 - i\omega M_r \int_0^\infty dt\ e^{i\omega t}\ \phi(t) \qquad (4)$$

Note that one very often defines :

$$\eta(\omega) = M_r \int_0^\infty dt\ e^{i\omega t}\ \phi(t) \qquad (5)$$

where $\eta(\omega)$ is the complex frequency dependent part of the longitudinal viscosity

We can now go back to the equation (1) for the propagation of the sound wave. Assuming an amplitude A of the form :

$$A(x,t) = A_0\ e^{-\alpha x}\ e^{i\omega(t-x/v)} \qquad (6)$$

equation (1) may be written as :

$$-\ \omega^2\rho = \tilde{M}(\omega)\ (\alpha + i\omega/v)^2 \qquad (7)$$

From this equation, we can relate the real (M') and imaginary (M'') part of the modulus to the measured quantities α and v. When $(\alpha\ \ v/\omega)^2 << 1$ one can write the approximate relations : [2]

$$M' = \rho\ v^2$$
$$M'' = \frac{2\rho v^3\alpha}{\omega} \qquad (8)$$

As both M' and M'' are frequency dependent,equations 8 explain the dispersion and absorption of an ultrasonic wave in a relaxing medium.

Ultrasonic measurements allow studies of these phenomena in the frequency domain 0.5 Mhz $<\omega<$ 1500 MHz. As we shall see in the next section, light scattering measurements allow a considerable extension of this domain, as both ends.

II ♦ LIGHT SCATTERING STUDIES

a Characteristic times of the problem

The spectrum of the light scattered from a fluid reflects the characteristic times of the density fluctuations:$(\omega_B)^{-1}$which is the period of the spontaneous sound waves, and$[(\lambda/\rho c_p)k^2]^{-1}$which is the characteristic time of the thermal diffusion. As we have just shown in the phenomenological treatment (see eq. 5), structural relaxation phenomena may be described by a frequency dependent viscosity. Consequently, as LALLEMAND has shown in his lecture [3], there is an additional relaxation mode for the density fluctuation, with a characteristic time τ. As τ can have almost any value, depending on the temperature, we will examine different cases, starting with high temperatures.

Case I :

$$1/\tau \gg \omega_B \gg \left(\frac{\lambda}{\rho c_p}\right)k^2$$

The relaxation time τ is so short compared with $(\omega_B)^{-1}$ that the structural relaxation contributes only to the Brillouin lines in the spectrum.

Case II :

$$\omega_B \gg 1/\tau \gg \left(\frac{\lambda}{\rho c_p}\right)k^2$$

At this lower temperature the structural relaxation is too slow to affect the Brillouin lines, but too rapid to be coupled to the thermal diffusion mode : Consequently it appears in the spectrum as an additional broad line, of width $1/\tau_{PS}$ centered at the laser frequency ; it is called the "Mountain line". The indices P and S of τ_{PS} express the fact that structural rearrangements must occur a constant pressure (since $\tau \gg (\omega_B)^{-1}$) and at constant entropy (since $\tau \ll (\frac{\lambda}{\rho c_p} k^2)^{-1}$).

Both of these cases have been studied extensively on various viscous liquids [4] but no measurements had been made on the remaining situation corresponding to temperatures close to the glass transition temperature :

Case III :

$$\omega_B \gg \left(\frac{\lambda}{\rho c_p}\right) k^2 \gg 1/\tau$$

The center line again consist of two components, but now it is the Mountain component that forms the narrow line while the normal Lorentzian component associated with heat transport appears as a relatively broad background. Moreover, because the viscoelastic relaxation process occurs quite slowly compared with the thermal diffusion rate, the relaxational behaviour responsible for the Mountain line shape is <u>isothermal</u> rather than adiabatic, and we denote the corresponding relaxation time by τ_{PT} .

Following a suggestion by T.A. LITOVITZ, we have studied [5] supercooled glycerol in a temperature region corresponding to this last case. On the time scale of our experiment $\left(t \gg \left(\frac{\lambda}{\rho c_p} k^2\right)^{-1}\right)$, as all the other relaxational modes have already died out, we expect the correlation function of the density fluctuations to be determined by the isothermal structural relaxation function of the liquid. This function has been previously introduced (see eq . 4). Assuming $\phi(t)$ normalized so that $\phi(0) = 1$, the average relaxation time is given by :

$$\tau_{PT} = \int_0^\infty dt \, \phi(t) \qquad (9)$$

The signal measured by the correlator is proportional to the second order correlation function of the density fluctuations, and may therefore be expressed as a function of the square of the first order correlation function

$$\langle n_k(0) \, n(t) \rangle = a + b \, (\phi(t))^2 \qquad (10)$$

where a and b are constants determined by the experimental conditions.

b Experimental procedure

A shematic diagram of the experimental set-up is shown on figure 2. Approximatively 800 mw of optical power from a coherent Radiation Laboratory Ar^+ laser (488.0 nm) was focused into a high purity, water-free sample of liquid glycerol from which dust

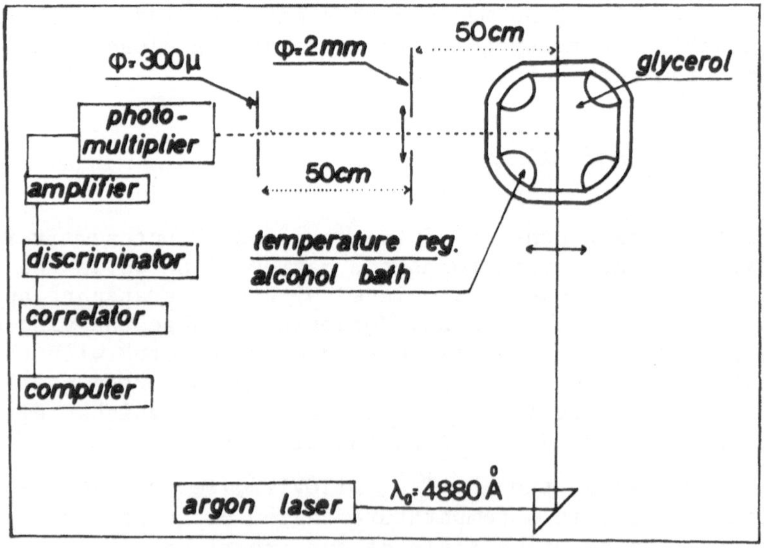

Figure 2 : Schematic diagram of the experiment.

particles had been removed by filtration. The vessel containing the sample was built in the form of an optical cross fused into a dewar and has been described by Pinnow et al [4]. The temperature was stable within ±0.1°. The light scattered through 90° was spatially filtered, focused on a 300μm diameter pinhole and then detected by a photomultiplier tube (Radiotechnique 56 DVP 03). The pulses coming from the photo-tube are amplified, discriminated and then fed to the 100-channel home-made digital correlator.

For each experimental condition (fixed values of the signal intensity and the sampling time) it was important to adjust the clipping level k of formula [10] to its optimum value, which is of the order of $\bar{n}$. When measuring very long time constants, we had $\bar{n} \gg 10$. As the maximum value of k given by our apparatus was 6, we used a method inspired by the scaling technique. We place before the "k gate" a scaler which divides the number of pulses entering the gate by a preset number s. On can show [6] that when $s \gg k$, this method amounts to measuring the signal $\langle n_\ell (0)\, n(t) \rangle$ where ℓ varies randomly from (k-1)s to ks and we have thus externally increased the capacity of the "k gate".

c Data analysis

The data given by the correlator (see figure 3) are analysed on a computer by a least square fit according to the equation [10].

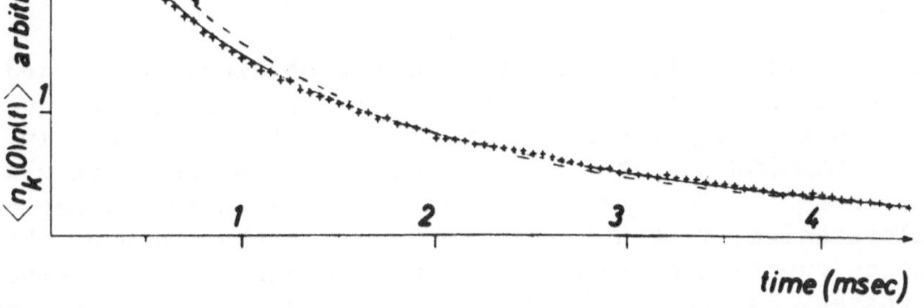

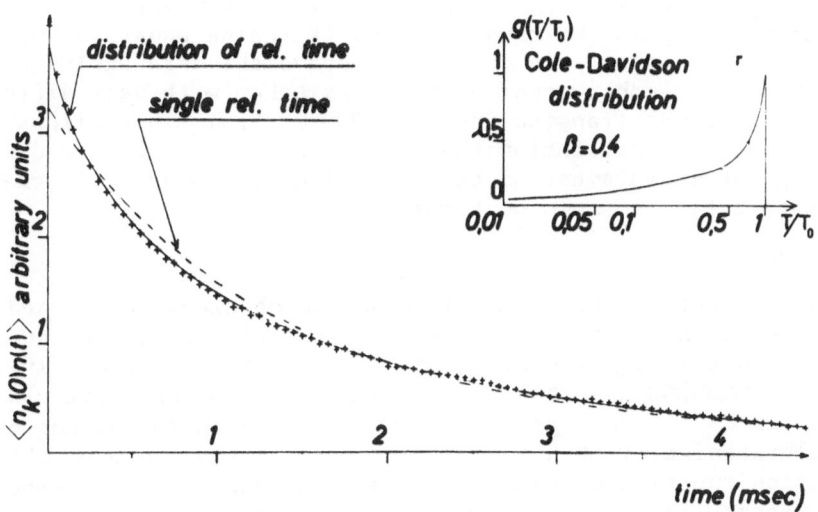

Figure 3 : Typical recording of the data given by the correlator.

If one assumes an exponential form for $\phi(t)$, the fit is rather bad (dashed curve on figure 3). Moreover, the best value for τ found by the computer decreases if the interval between each experimental point decreases, i.e. if the experimentally explored time domain decreases. This clearly shows that structural relaxation phenomena may not be described with a single relaxation time, but involve a distribution of relaxation times with a non-negligeable contribution of times much shorter than the average relaxation time τ_{PT}. This is well represented by a COLE-DAVIDSON type of distribution function [7]. These authors have used (to interpret their results on dielectric relaxation) a two parameter distribution function : τ_0 ,the values of the largest relaxation time present in the distribution and β,parameter varying between 0 and 1, and describing the width of the distribution (β =1 is equivalent to a single relaxation). In the interval dlog(τ/τ_0) this distribution function is given by :

$$g(\tau/\tau_0) = \frac{\sin \beta\pi}{\pi} \left(\frac{\tau/\tau_0}{1 - \tau/\tau_0} \right)^{\beta} \quad \text{if } 0 \leq \frac{\tau}{\tau_0} < 1$$

$$= 0 \qquad\qquad \text{if } \frac{\tau}{\tau_0} \geq 1$$

We have therefore assumed that the stress relaxation function was of the form :

$$\phi(t) = \int_0^\infty d \log (\tau/\tau_0) \quad g(\tau/\tau_0)\, e^{-t/\tau} \qquad (11)$$

We have computed this function numerically for different values of the parameter β , and we have compared it (using equation 10) to the experimental points. In order to improve the data analysis, we have fitted together experimental data taken with very different sampling times (ranging from $\tau/5000$ to $\tau/5$). This allows us to explore the correlation function in a very large time domain, and is a great improvement on the precision of the data. An example of such a fit is shown on figure 4

d) Results

The isothermal relaxation function was obtained at a dozen temperatures in the range(-80°C, -48°C),over which range the average relaxation time τ_{PT} was found to vary by a factor of nearly 10^5. For all the temperatures studied, we found that the best fit was obtained for β=0.4. This result is consistent with the values obtained ultrasonically at higher temperatures : the width of the distribution function of relaxation times increases as the temperature decreases, and tends toward a limiting value as the temperature approaches the glass transition temperature.

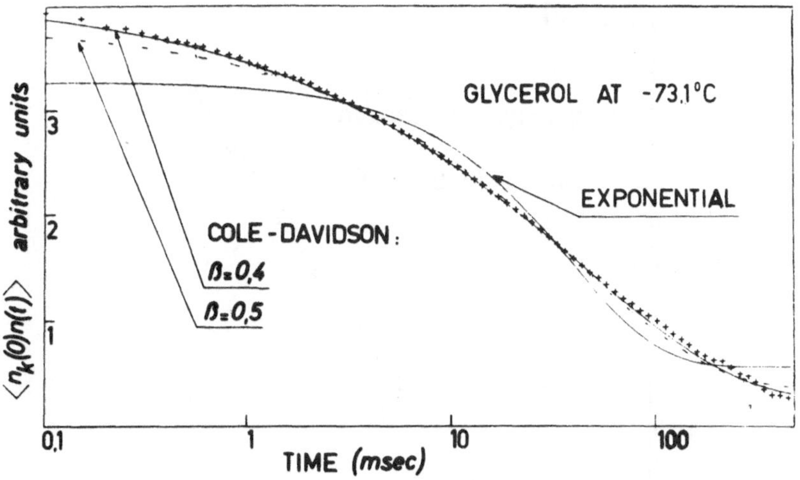

Figure 4 : Experimental data obtained by fitting together 7 experimental curves.

This result may be understood using a simple molecular picture [8] . We assume that the movements of the molecules follow the jump diffusion model : A molecule sits at a point (rattling and vibra - ting around its equilibrium position) and then jumps to a new site where it waits a while before jumping again. The important parameter is the ration of the average jump lenth, ℓ , and the average intermolecular distance, σ . At high temperature, $\ell >> \sigma$ and one jump is enough for a molecule to destruct the local structure ; we have then a single relaxation phenomena. At low temperature, $\ell < \sigma$ and a lot of "jumps" are then needed for a molecule to break up the local structure. This leads to a broad distribution of relaxation times.

From the best value of the parameter β , one deduces the average relaxation time :

$$\tau_{PT} = \beta \tau_0$$

The results obtained are shown on figure 5.

If we assume a temperature dependence of the Arrhenius form

$$\tau_{PT} \propto \exp (E/RT) \qquad \text{we get } E \approx 32 \text{ kcal/mole}$$

Figure 5 also shows the values obtained ultrasonically at higher temperature [1] and they join smoothly with those measured at lower temperature with the digital correlator.

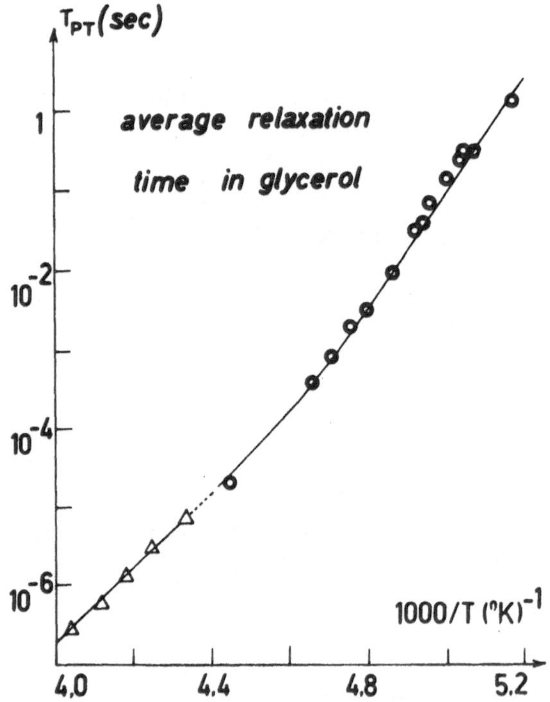

Figure 5 : Temperature dependence of the average relaxation time; the circles correspond to our light scattering study and the triangles represent the values deduced from the ultrasonic measurements of Piccirelli et al. (1).

It is worth noting that by slightly modifying the experimental set-up (to study the light scattered at very small angles), this same technique could be used to study the coupling between the structural relaxation and the thermal relaxation, i.e. the intermediate case :

$$\tau \sim \left(\frac{\lambda k^2}{\rho C_p} \right)^{-1}$$

This study has begun in our laboratory and as a preliminary experiment, we have been studying the decay rate of an externally imposed temperature fluctuation. The liquid is heated by a strong laser beam, and the return to equilibrium is studied by an optical method, using a Jamin interferometer. This preliminary study has already shown that, when the structural relaxation time is much faster (high temperature) than the expected thermal relaxation time, the observed decay rate varies as ($\lambda k^2/\rho C_p$)

On the contrary, when the temperature is low enough so that τ is the largest time of the problem, the temperature fluctuation decay follows the structural relaxation time.

In conclusion, we hope to have shown that digital correlation spectroscopy is a very useful tool in studying viscoelastic properties of viscous liquids ; it provides very fast and precise informations, in a time domain inacessible to the more conventional methods.

REFERENCES

[1] R. PICCIRELLI and T.A. LITOVITZ, J. Acoust.Soc. Am. 29,1009 (1957)

[2] See for ex. T.A. LITOVITZ and C.M. DAVIES in Physical Acoustic W.P. MASON Ed. (Academic Press, Inc., New York 1965) Vol 2, Part A Chap. 5

[3] P. LALLEMAND , this book.

[4] See for example D.A. PINNOW, S.J. CANDAU, J.T. LAMACCHIA and T.A. LITOVITZ, J. Acoust. Soc.Am. 43 , 131 (1968)

[5] C. DEMOULIN, C.J. MONTROSE and N. OSTROWSKY to be published

[6] C. DEMOULIN, Thèse de 3ème cycle, Université de Paris 1973

[7] D.W. DAVIDSON and R.H. COLE, J. Chem. Phys. 19, 1484 (1951)

[8] C.J. MONTROSE and T.A. LITOVITZ, J. Acoust. Soc. Am. 47 , 1250 , (1970)

[9] P. LALLEMAND and N. OSTROWSKY, to be published .

LIGHT SCATTERING INTENSITY STUDIES IN MULTICOMPONENT SOLUTIONS OF BIOLOGICAL MACROMOLECULES

Henryk Eisenberg

Fogarty International Center, National Institutes of Health, Bethesda, Maryland *and* Polymer Department,*
Weizmann Institute of Science, Rehovot, Israel

INTRODUCTION

A considerable fraction of the living cell (such as nucleic acids and proteins for instance) are macromolecules, carry electrical charges and are suspended in aqueous media containing low molecular weight ionic and non-ionic species. In the physico-chemical characterization and the analysis of the behavior of these large molecules many problems arise which may be solved by consideration of their special polyelectrolyte character. Polyelectrolytes (Armstrong and Strauss, 1969) are long chain molecules carrying a large number of ionizable sites. They may exist in solution in either linear open conformations, as do many nucleic acids for instance, or they may be folded into tertiary globular structures, such as most proteins are. In all cases, we should remember that solutions of either nucleic acids or proteins are electroneutral in the thermodynamic sense, and deviations from electro-neutrality are extremely rare. For every charged group fixed to the macromolecule there must therefore exist, somewhere in the vicinity of the macromolecule, a small mobile counterion of opposite charge. In the case of an ampholytic protein (a protein which carries both types of positive and negative charges), the mobile ions balance the net charge of positive and negative ions fixed to the macromolecular matrix. Often we shall be interested in the way in which the electrical charges influence the physicochemical behavior of these solutions. In many other instances though we shall be satisfied to conclude that if large enough test volumes in our solutions are considered, these volumes will contain both many macromolecules and counterions. They will therefore - for many purposes - be considered electro-neutral phases. It will be important to distinguish when this assumption

* Address for correspondence

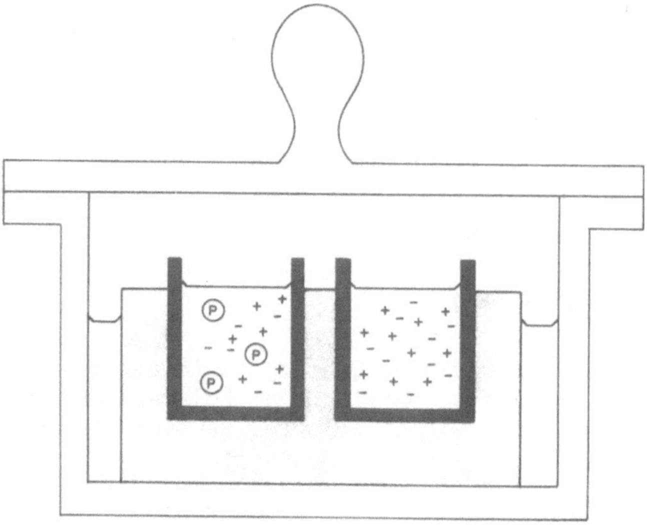

Figure 1. Schematic arrangement in isopiestic experiment. Protein solution containing low molecular weight salt is equilibrating in thermostated copper block against solution of low molecular weight salt;solvent molecules which equilibrate across water phase are not indicated.

no longer holds. In such cases we are speaking of electrochemical systems, a discussion of which lies outside of the scope of the present analysis.

In this seminar we will mostly discuss the thermodynamic basis of absolute molecular weight determinations in relation to light scattering intensity studies and in particular how these are affected by the multicomponent nature of the polyelectrolyte solutions under investigation.

OSMOTIC PRESSURE AND VIRIAL EXPANSIONS

Osmotic pressure is one of the basic phenomena of nature and in the laboratory. It involves solutions separated by membranes, which are "semipermeable", *i.e.* permeable to the solvent, and to some of the solutes only. In the laboratory the usual osmotic pressure experiment consists of a thin collodion membrane separating two liquid phases permeable to all solutes except the macromolecular component. An equally satisfactory semipermeable membrane, however, is afforded by the air space (or vacuum) in a closed desiccator in an isopiestic experiment (Figure 1). This membrane is permeable only to the principal solvent (usually water), which is allowed to equilibrate between a container holding a protein or nucleic acid solution, and another container enclosing a suitable reference solution containing

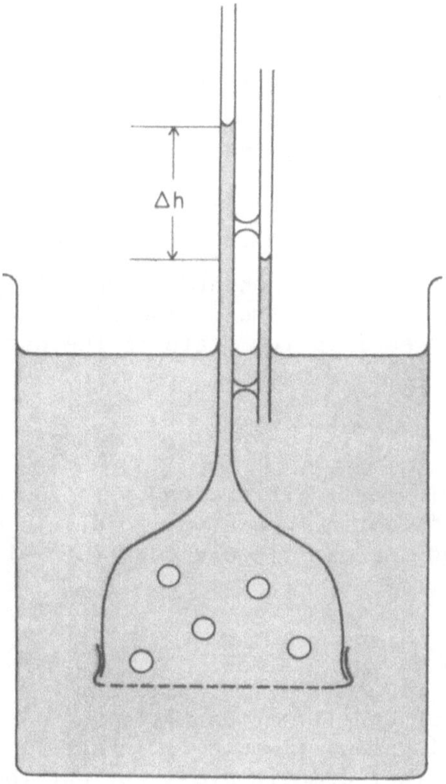

Figure 2. Schematic representation of simple osmometer.

a non-volatile solute. Distillation of water proceeds until the vapor pressure of both solutions are identical (distillation would proceed *ad infinitum* - at constant pressure - if the reference solution would be pure water). A more general statement for the equilibrium condition requires that the chemical potentials, μ, of all components (taken in neutral combinations of the ionic species) permeable through the membrane, be identical in contiguous phases. A simple osmometer is shown in Figure 2. In the simplest case of a neutral polymeric solute, component 2, which cannot pass through the pores of the semipermeable membrane, there is only solvent, component 1, on the outside of the compartment enclosed by the membrane, but both components 1 and 2 are present inside this compartment; component 1 is free to move across the membrane. What happens when the osmometer compartment is introduced into the solvent beaker is easily predictable on thermodynamic grounds. The chemical potential μ_1, of the solvent is lowered in the inner compartment because of the presence of the dissolved component 2. Therefore component 1 tends to flow from the outer into the inner compartment, from a higher to a lower potential.

This process would go on indefinitely were it not for the fact that the influx of component 1 raises the liquid level in the measuring capillary and therefore raises the pressure P in the inner compartment (a reference capillary identical to the measuring capillary dips in the solvent to allow for corrections due to capillary rise). Increase in pressure raises μ_1, and solvent influx ceases as soon as

$$\mu_1 = \mu_1' \tag{1}$$

Outer polymer-free solution components in the dialysis equilibrium experiment are designated by primes. The difference in pressure corresponding to this equilibrium state is the osmotic pressure Π and is equal to

$$\Pi = \Delta h \; \rho \; g \quad \text{dyne/cm}^3 \tag{2}$$

where Δh is the liquid level difference, ρ is the density and g is the gravitational constant. In analogy to dilute gaseous systems it is possible to expand the osmotic pressure in a virial series in powers of the concentration

$$\Pi/\rho_n kT = 1 + B_2\rho_n + B_3\rho_n^2 + \cdots \tag{3}$$

where ρ_n is the number density of particles and B_2 and B_3 are virial coefficients. At this stage I would like to call attention to an important point. Neither osmotic pressure, nor the other two major techniques which we intend to relate to it presently — equilibrium sedimentation and light scattering - measure molecular weights in a direct way. All of these methods *count* particles and determine weighted averages of the particle count. How therefore is the molecular weight derived in these procedures? Primarily because we do not know ρ_n in equation (3) and express particle density in terms of some experimentally measurable concentration unit related to particle mass. We thus use

$$c = \rho_n M/N_A \tag{4}$$

where c is the concentration of the macromolecular component in g/mℓ, M is the molecular weight in g/mole and N_A is Avogadro's number. Substitution of equation (4) into equation (3) yields the familiar form

$$\Pi/cRT = M^{-1} + A_2\, c + \cdots \tag{5}$$

where $A_2 = B_2\, N_A/M^2$ is the well known second virial coefficient in the units presently used. The intercept in the plot Π/cRT against c now yields the reciprocal of M and the slope of the plot is equal to A_2 and expresses particle-particle interaction (Tanford, 1961).

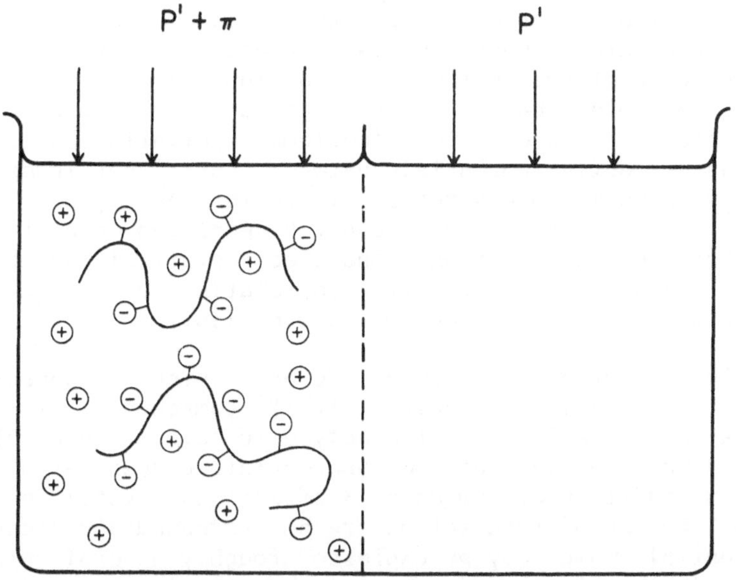

Figure 3. Schematic representation of osmotic membrane experiment; charged polyelectrolyte molecules are restricted to one side of semipermeable membrane and no low molecular weight salt is present.

The osmotic pressure Π itself (when $c \to o$) is proportional to c/M, which has units of mole/mℓ and is therefore not concerned with particle weight (this point will assume increased significance when dealing with charged polyelectrolyte particles). We also note that, for a given weight of dissolved particles (g/mℓ), Π decreases as M increases (for a given value of c the number of particles in solution decreases with increase of M) and the utility of osmotic pressure determinations is restricted to the low molecular weight range. We shall see that on the other hand the usefulness of both equilibrium sedimentation and light scattering increase as the molecular weight increases.

CHARGED POLYMERS AND DONNAN DISTRIBUTION

Our next problem refers to an *ionized* polymer, PX_z, containing Z fixed charges (taken as negative for the sake of the argument) and a complement of Z positive mobile counterions X (necessary to ensure electroneutrality). Schematically this situation is represented in Figure 3. The ionized component 2 is restricted to one side of the membrane and the counterions cannot escape because the electroneutrality condition cannot be easily relaxed. Polyelectrolyte theory (Katchalsky, 1971; Manning, 1972) is concerned with the detailed polyion-counterion interaction, but this does not concern us in our more phenomenological quest. Of major concern though is the fact that in the case of the system shown in Figure 3 the molecular weight may not be determined under these circumstances: because of long range electrostatic interactions (coulombic potentials decrease with r^{-1} whereas attractive potentials between neutral particles decay with r^{-6}) between macroions and small counterions, as well as between macroions, the osmotic pressure cannot be expanded in a virial series in powers of the concentration. The electrostatic interactions do not decay fast enough upon dilution in relation to the increase in linear distance between charged macromolecules.

How is this problem overcome? We add another component to the solution, usually a low molecular salt XY, component 3; if component 2, for instance, is a Na salt of a polymeric acid, then a suitable component 3 would be $NaCl$ at some appropriate concentration ($Na^+=X$; $Cl^-=Y$). The addition of component 3 affects the system in a profound way. Forces of interaction are now screened and decay with a Debye potential, which may be expressed roughly as $\exp(-\kappa r)/r$ (Harned and Owen, 1958); this is a short-range potential (the range of electrostatic forces now falls off faster than the distance between particles upon dilution) and it becomes possible to use a virial expansion. This is justifiable on rigorous statistical mechanical grounds (Hill, 1960).

The schematic osmotic pressure experiment now corresponds to the situation depicted in Figure 4. The low molecular weight component 3 dissociates into positive and negative ions; it is permeable through the membrane and it can be shown that its distribution is unequal on both sides of the membrane. The positive counterions to the macromolecule can of course not be distinguished from the positive ions of component 3. In addition to equation (1) another equilibrium condition applies now

$$\mu_3 = \mu_3' \qquad (6)$$

arising from the equilibration of the component 3. The equilibrium distribution of component 3 is known as the Donnan distribution and may be characterized by a distribution parameter

$$\Gamma \equiv (C_3' - C_3)/ZC_2 \qquad (7)$$

where the capital C's stand for concentrations in moles/ℓ and Z is the number of charges per macromolecule. Sometimes $Z\Gamma$ is given in the form of a differential coefficient $(\partial C_3/\partial C_2)_\mu$ where subscript μ signifies constancy of μ_1 and μ_3 (the components diffusible through the semipermeable membrane). An idealized calculation shows* that when $C_2 \to o$, Γ equals 0.5 and therefore $C_3' > C_3$: component 3 is excluded, or rejected, from the macromolecular solution. In practice, because of ion-binding and hydration Γ may assume different

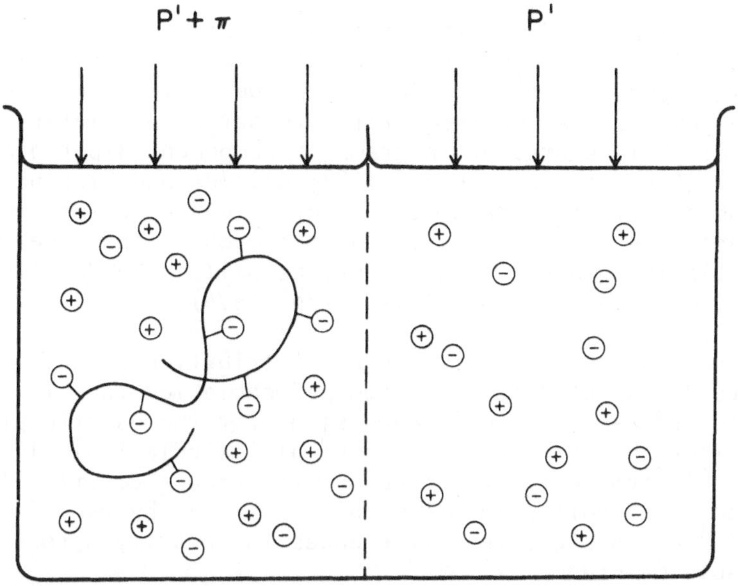

Figure 4. Same as Figure 3, but low molecular weight uni-univalent salt has now been added to system.

---

* For the purpose of this calculation we neglect all activity coefficients, express μ_3 and μ_3' (equation (6)) as the sum of the potentials $\mu_i = \mu_i^o + RT \ln C_i$ of the ionic species (the μ_i^o are standard chemical potentials independent of the phase) to yield

$$C_X C_Y = C_X' C_Y'$$

which transforms to

$$(C_Y + ZC_2)\, C_Y = C_Y'^2$$

because electroneutrality requires

$$C_X = C_Y + ZC_2 \quad \text{and} \quad C_Y' = C_X' \equiv C_3' \; .$$

Solution of the above quadratic equation for C_Y yields

$$\Gamma = 0.5 - (ZC_2/8C_3') + \cdots \tag{8}$$

values and be expressed in different forms (Casassa and Eisenberg, 1964). Particular care should be exercised when transforming from molar (weight per volume) to molal(weight per weight) concentration units. We have used a "preferential"interaction parameter ξ_3 in which "binding" is expressed in grams of component 3 per gram of component 2 (Cohen and Eisenberg, 1968; Reisler and Eisenberg, 1969). The distribution parameters Γ or ξ_3 may be determined by chemical analysis of the composition of the outer and inner phases in an equilibrium dialysis experiment, as well as by density and refractive index determinations (Eisenberg, 1974).

From considerations on the ideal distribution of a uni-univalent electrolyte component 3 across a semipermeable membrane it is also possible to arrive at a simple expression for the osmotic pressure. We assume (for the purposes of this simple calculation) all deviations from electroneutrality to be due to unequal distributions of ions. From Van't Hoff's law $\Pi = \Delta\rho_n kT$ we find by use of $\Delta C = \Delta\rho_n 10^3/N_A$ (Δ signifies differences in molarity across the semipermeable membrane)

$$10^3 \, \Pi/RT = \Delta C = (Z+1)C_2 + 2C_3 - 2C_3'$$

Substitution of Γ from equation (8) and transformation into g/mℓ concentration units for component 2, yields for the experiment described in Figure 4, an idealized expression

$$\Pi/c_2 RT = M_2^{-1} + (10^3/C_3' 4M_u^2)\, c_2 + \cdots \qquad (9)$$

where M_u is the molecular weight per macromolecular charge (M_2/Z). Equation (8) shows that the intercept in the plot of $\Pi/c_2 RT$ against c_2 correctly yields M^{-1}, which was indeed what we had planned to achieve. The value of the second virial coefficient due to unequal salt distribution is not entirely correct in the above idealized calculation (as a matter of fact it may be off by a large factor due to non-ideality corrections) but the calculation correctly indicates that this coefficient decreases with increasing C_3.*

A point which has sometimes baffled many investigators, and still baffles some, refers to the question whether the weight of the mobile counterions is included in the molecular weight M_2 or not. Alternatively, if we dissolve *NaDNA* in excess *CsCl*, do we determine the molecular weight of *Na* or of *Cs DNA* or some hybrid quantity? I hope that the attentive listener is by now convinced that this question is quite irrelevant as in this case, as before, we *count* the number of moles of component 2, irrespective of their ionic state, interactions with small particles, and so forth. If we express concentration in grams of *NaDNA* weighed into the solution, we shall obtain *M* of *NaDNA*, irrespective of the low molecular weight ionic solvent used. We may, for instance, express concentrations in terms of nitrogen (or phosphorus) per unit volume and obtain molecular weight in units of moles nitrogen (or phosphorus) per mole of macromolecules. Very often ultraviolet absorption coefficients are used and this may be related to any concentration units one may require, by a proper conversion factor.

You may wonder why this introduction involving the osmotic pressure and related quantities was emphasized in the above discussion. This was mainly done for two reasons. Firstly it helped to introduce the topic of charge-carrying biological and synthetic macromolecules and some of the problems associated with systems containing these. Secondly we shall show that equilibrium sedimentation (which we shall not discuss in detail) and light scattering are intimately related with the derivative of osmotic pressure discussed above with respect to concentration of the macromolecules.

EQUILIBRIUM SEDIMENTATION

In equilibrium sedimentation a macromolecular solution is spun at high speeds in the ultracentrifuge such that in the (initially uniform) solution a stable concentration gradient of component 2 (and sometimes of component 3) is set up. At sedimentation equilibrium the total chemical potentials (including the ultracentrifugal potentials) are uniform across the centrifugal cell. To produce the concentration gradients mentioned osmotic work (concentration work)

---

* The ionic contribution to the second virial coefficient can be exactly calculated for certain models (in particular for cylindrical symmetry) on the basis of solutions of the Poisson-Boltzmann equation (Katchalsky, Alexandrowicz and Kedem, 1966; Gross and Strauss, 1966) and the "ion condensation" concept (Manning, 1972) in highly charged systems. Other contributions to A_2 are a positive term due to physical volume exclusion and negative contributions due to attractive forces at short distances. The second virial coefficient (and therefore A_2) may be made to vanish by a suitable choice of concentration of mobile ions and temperature (Eisenberg and Woodside, 1962), following the concept of the "ideal" theta temperature in non-ionic polymers introduced by Flory (Flory, 1953). This is entirely analogous to the concept of the Boyle temperature in imperfect gases, and has led to useful applications in solutions of biological and synthetic macromolecules (Eisenberg and Felsenfeld, 1967; Raziel and Eisenberg, 1973). The aim of these excluded volume studies on selected polyelectrolyte systems has been to relate the phenomena observed to molecular parameters and to compare these with the conformational properties, for instance, of their non-ionic analogs.

under specified conditions has to be performed. This is how the osmotic pressure is involved. A calculation (Eisenberg, 1962) shows that, for any component J in a multicomponent solution, and at all distances r from the center of rotation of the ultracentrifuge rotor, the following differential equation holds

$$d\ln c_J/dr = \omega^2 r \; (\partial\rho/\partial c_J)_\mu \; / \; (\partial\Pi/\partial c_J)_\mu \tag{10}$$

Here ω is the angular velocity, $(\partial\rho/\partial c_J)_\mu$ is a density increment under conditions of dialysis equilibrium and $(\partial\Pi/\partial c_J)_\mu$ is the derivative of the osmotic pressure with respect to the concentration of component J. If we specialize now to the distribution of component 2, expand the osmotic pressure derivative in a virial series, restrict ourselves to the case $c_2 \to o$ and write dr^2 for 2 rdr, we find

$$d\ln c_2/dr^2 = (\omega^2/2RT)(\partial\rho/\partial c_2)_\mu \; M_2 \tag{11}$$

Thus, in this case, the slope of a plot of the logarithm of any quantity (ultraviolet absorption, number of interference fringes) proportional to c_2, versus r^2, is proportional to M_2 which may be derived if the accessory quantities are known. The slope increases with increase in M_2, improving the sensitivity of the method as molecular weight increases (Creeth and Pain, 1967). Note that the product $M_2(\partial\rho/\partial c_2)_\mu$ can be written $[\partial\rho/\partial(c_2/M_2)]_\mu$, which indicates that the density increment is naturally expressed in terms of an increment due to the number of particles, and not their weight. Considerations presented earlier with respect to molecular weight determinations by osmotic pressure apply here as well. The intuitive notion that the centrifugal field *weighs* the particles, is not correct.

For readers familiar with the expression of equilibrium sedimentation in terms of a classical buoyancy term we note that the density increment at dialysis equilibrium can be written as

$$(\partial\rho/\partial c_2)_\mu = (1 - \bar{v}_2\rho^o) + \xi_3 \; (1 - \bar{v}_3\rho^o) \tag{12}$$

where $\bar{v}_2$ and $\bar{v}_3$ are partial specific volumes of component 2 and 3 respectively and ρ^o is the density of the solvent. The second term on the right-hand side of equation (12) indicates the correction to be applied to the buoyancy term in a three component system, as a result of preferential interactions with respect to the third component (Casassa and Eisenberg, 1964). Use of equation (12) enables evaluation of the distribution parameter ξ_3 by way of density determinations (Cohen and Eisenberg, 1968; Reisler and Eisenberg, 1969).

LIGHT SCATTERING

Light scattering is an optical method which is also related to the osmotic pressure but which, in addition to yielding the molecular weight and the macromolecular solute-solute interactions, provides an entirely new piece of information. A homogenous material, such as a perfect crystal for instance, scatters almost no light, by virtue of the fact that light scattered from any atomic center undergoes destructive interference with scattered light coming from another scattering center in the regular crystalline lattice. For destructive interference to be incomplete and therefore for scattering to occur refractive index inhomogeneties are required in the material. Many years ago Einstein deduced that the relatively low scattering from pure liquids may be ascribed to refractive index inhomogeneities which derive from density fluctuations in the compressible liquids. In the case of solutions, Debye has shown that here inhomogeneities result from concentration fluctuations in small volumes of the solutions. The ease with which concentration fluctuations are produced depends on the osmotic work required, which establishes the sought-for connection. Light is only scattered if the local change in concentration is accompanied by a change in refractive index (similarly changes in concentration in the ultracentrifuge only occur if accompanied by changes in the density of the solution, *i.e.* $(\partial\rho/\partial c_2)_\mu \neq 0$). The analysis of small angle X-ray scattering is essentially equivalent to that of light scattering and depends upon changes in electron density subsequent to fluctuations in concentration (Eisenberg, 1971).

The additional information available from measurements of the angular dependence of scattering relates to parameters characteristic of the shape and size of the particles. Here small angle X-ray and light scattering are beautifully complementary methods and together cover a wide range of interesting information. We shall not discuss the angular dependence of scattering intensity at all and refer to two recent reviews (Eisenberg, 1971; Timasheff and Townend, 1970). In the concluding part of this work we will discuss a simplified version of the thermodynamic aspects only and refer to the above quoted discussions for specific examples in the fields of nucleic acids and proteins, respectively. We mention here the exciting new dimension added to the investigation of solution of biological macromolecules with the present-day use of coherent laser radiation, the latter topic being the main subject of this Study Institute.

Rayleigh's equation for the scattering intensity from N independent scatterers per unit volume, in a plane defined by the direction of propagation of the plane-polarized incident light and

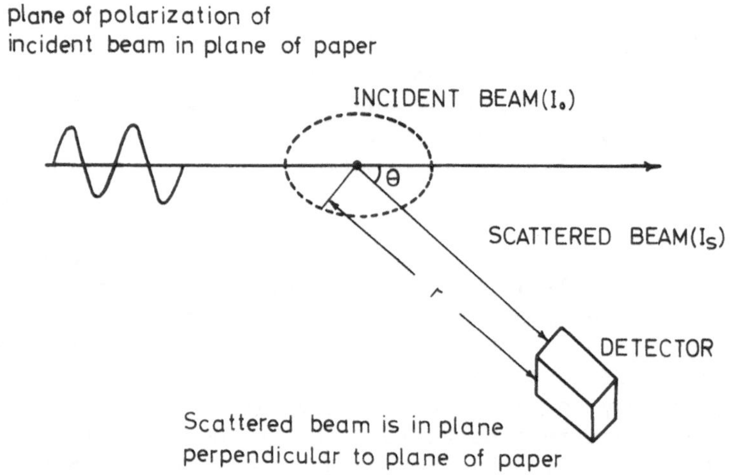

Figure 5. Geometry of light scattering experiment.

perpendicular to the direction of polarization (Figure 5), is given by

$$R = 16\pi^4\alpha^2 N/\lambda^4 \tag{13}$$

where the Rayleigh factor R equals $r^2 I_s/I_o$, r is the distance from the scattering center to the detector, I_s/I_o is the ratio of scattered to incident light intensity, α is the polarizability of the individual scatterers and λ is the wavelength. The value of R is independent of the scattering angle θ for scatterers much smaller than λ. For pure liquids and for macromolecular solutions, Einstein and Debye respectively approached the problem by dividing the solution into a large number of volume elements V, containing many molecules each, yet with linear dimensions small with respect to λ. The polarizability of each volume element fluctuates at each moment around its equilibrium value and is given at each instant by $\alpha+\delta\alpha$. The equilibrium value α is equal for all scattering volumes and the time averaged fluctuation $\langle\delta\alpha\rangle$ is equal to zero; only the mean square average fluctuation $\langle(\delta\alpha)^2\rangle$ contributes to the scattering. One finds, again for vertically polarized incident radiation, that equation (13) has to be substituted by

$$R = (16\pi^4/\lambda^4 V)\langle(\delta\alpha)^2\rangle \tag{14}$$

where V^{-1} is the number of volume elements per mℓ and the remaining problem becomes the evaluation of $<(\delta\alpha)^2>$. Changes in polarizability in the volume element V exposed to an external radiation field are related to changes in refractive index n (n_o is the refractive index of the solvent) by

$$n^2 - n_o^2 = 4\pi\alpha/V \tag{15}$$

from which we derive, in conjunction with equation (14)

$$R = (4\pi^2 n^2/\lambda^4)V<(\delta n)^2> \tag{16}$$

where $<(\delta n)^2>$ is the mean-square refractive index fluctuation.

In view of the set of thermodynamic variables to be used ultimately in multicomponent systems Stockmayer (1950) considered the fluctuations of index in a portion of solution containing a constant mass of one of the components. The choice of this component is arbitrary, but for dilute solutions it is convenient to choose the solvent, component 1, and to fix its quantity as one kilogram; the numbers of moles of component 2 and 3 are then expressed as weight molalities m, moles per kg of component 1 (Casassa and Eisenberg, 1964). We disregard here the variation of refractive index with pressure and temperature and show the variation with the molalities of the various components only

$$dn = \psi_2\, dm_2 + \psi_3\, dm_3 \tag{17}$$

where

$$\psi_i \equiv (\partial n/\partial m_i)_{T,P,m}$$

The aim of the calculation is to derive fluctuations of the m_i for a system in statistical equilibrium with surroundings in which constant values of T, P, and chemical potentials μ_2 and μ_3 are maintained; the quantity m_1 is fixed. Stockmayer used a modified grand canonical partition function $Z(\beta,P,\mu_2,\mu_3)$, $\beta = 1/kT$,

$$Z = \exp(-\beta\mu_1 m_1) = \sum_{E,V,m_2,m_3} \exp[-\beta(E + PV - \mu_2 m_2 - \mu_3 m_3)] \tag{18}$$

from which the desired fluctuations may be obtained by appropriate differentiation (Hill, 1960). One finds, for the mean-square refractive index fluctuation

$$\beta<(\delta n)^2> = \sum_i \sum_j \psi_i \psi_j (\partial m_i/\partial \mu_j)_{T,P,\mu} \tag{19}$$

where the summation, in our simple system, is over components 2 and 3 and subscript μ indicates constancy of the potential not appearing in the partial differentiation. The scattering equation, derived by use of equation (19) for the fluctuations in equation (16), is

$$\Delta R = KV_m \; RT \sum_i \sum_j \psi_i \psi_j / (\partial \mu_i / \partial m_j)_{P,T,\mu} \tag{20}$$

where K denotes $4\pi^2 n^2 / N_A \lambda^4$ and ΔR represents a quantity corrected for solvent scattering.

Thermodynamic considerations show that the rather unfamiliar derivative of the chemical potential with respect to molality is equal to within a good approximation to the osmotic pressure derivative

$$(\partial \mu_i / \partial m_j)_{P,T,\mu} \approx (V_m / m_j)(\partial \Pi / \partial m_j)_{T,\mu} \tag{21}$$

where V_m is the volume of solution, in milliliters, containing 1 kg of component 1.

For the three component system of interest to us some manipulations of the preceding equations, substitutions and transformation to the more customary g/mℓ concentration scale leads to the following expression

$$RT(\partial n / \partial c_2)^2_{P,T,\mu} \; Kc_2 / \Delta R = (\partial \Pi / \partial c_2)_\mu \tag{22}$$

which shows the direct relationship between the reciprocal scattering intensity and the osmotic work required to produce a change in concentration (it is well known that when, near the critical point, the osmotic pressure derivative vanishes, the scattering intensity diverges).

The osmotic pressure may be expanded in a virial series (*cf.* equation (5)) to give

$$(\partial n / \partial c_2)^2_{P,T,\mu_3} \; Kc_2 / \Delta R = M_2^{-1} + 2A_2 c_2 + \cdots \tag{23}$$

which shows that a plot of the quantity on the left-hand-side of equation (23) against c_2 yields an intercept M_2^{-1} and a slope of $2A_2$ (twice the osmotic virial coefficient). Note that (in distinction to osmotic pressure) scattering intensity increases with increasing molecular weight.

The major significance of equation (23) in terms of our analysis of multicomponent systems lies in the restrictions specifying the refractive index increment $(\partial n / \partial c_2)_{P,T,\mu_3}$. Thus, up to minor correction terms not dealt with here, equations for multicomponent systems can formally be represented by classical equations for two component systems, with the *proviso* that refractive increments be measured at constant chemical potentials of added electrolyte. This is easily achieved experimentally by an accessory experiment involving equilibrium dialysis and measurement of the refractive index difference (divided by concentration) corresponding to an equilibrium dialysate and complex solvent equilibrium solution. Large errors may occur in molecular weight determinations if this procedure is not properly taken into account.

Finally we would like to show that by light scattering, just as well as by equilibrium sedimentation, both methods which we have shown relate to the derivative of the osmotic pressure with respect to concentration, no information on ion binding, dissociation of small ions or physical interaction with other components present in the solution can be derived. Suppose we consider equation (23) in the limit $c \to o$, and define the molecular weight and the concentration of component 2 by arbitrary multiplication with a factor X:

$$[\partial n/\partial(Xc)]^2 \; K(Xc)/\Delta R = (XM)^{-1}$$

As is seen above, X factors out and M is precisely given in terms of whatever particular concentration units for c have been used.

SUMMARY

We have, in the present seminar, discussed some of the problems which arise in the light scattering and equilibrium sedimentation analysis of multicomponent charge carrying (polyelectrolyte) macromolecular systems. We have, in particular, shown how these experiments relate to the determination of osmotic pressure, and have emphasized the insight gained by establishing this connection. We have shown that molecular weights are not directly derived but rather by way of the particular concentration units used, and that in multicomponent systems the use of refractive index and density increments at constant chemical potentials of diffusible solutes formally reduces the analysis of the systems to that of the much simpler two component systems. References given mostly refer to review articles and therefore allow the newcomer to this field to establish connections with the pertinent literature. Two examples discussed in this seminar referred to light scattering studies on the influence of base stacking on the conformation of polynucleotide systems (Eisenberg and Felsenfeld, 1967) and on the reversible association of the enzyme glutamate dehydrogenase to form long rods (Reisler and Eisenberg, 1971).

Extensions to larger number of components, heterogeneous macromolecular systems, particles whose dimensions are commensurate with the wavelength of the light [compare for instance the complementarity (Finch and Holmes, 1967) in studies of the angular dependence of scattering of light ($\lambda \sim 5000$ Å) and X-rays ($\lambda \sim 1.5$ Å) at small angles] do not present major difficulties.

REFERENCES

R.W. ARMSTRONG and U.P. STRAUSS, Polyelectrolytes, in "Encyclopedia of Polymer Science and Technology", H.F. Mark, N.G. Gaylord and N.M. Bikales, editors, Interscience Publishers, New York, 1969; Volume *10*, page 781.

E.F. CASASSA and H. EISENBERG, Thermodynamic Analysis of Multicomponent Systems, *Adv.Protein Chem.*, *19*, 287 (1964).

G. COHEN and H. EISENBERG, Deoxyribonucleate Solutions: Sedimentation in a Density Gradient, Partial Specific Volumes, Density and Refractive Index Increments and Preferential Interactions, *Biopolymers*, *6*, 1077 (1968).

J.M. CREETH and R.H. PAIN, The Determination of Molecular Weights of Biological Macromolecules by Ultracentrifuge Methods, *Progr. Biophys.Mol.Biol.*, *17*, 217 (1967).

H. EISENBERG, Multicomponent Polyelectrolyte Solutions. Part I. Thermodynamic Equations for Light Scattering and Sedimentation. *J.Chem.Phys.*, *36*, 1837 (1962).

H. EISENBERG, Light Scattering and Some Aspects of Small Angle X-Ray Scattering in "Procedures in Nucleic Acid Research", G.L. Cantoni and D.R. Davies, editors, Harper and Row Publishers, New York, Volume *2*, 1971, page 137.

H. EISENBERG, "Multicomponent Solutions of Biological Macromolecules", Clarendon Press, Oxford, in preparation, 1974.

H. EISENBERG and G. FELSENFELD, Studies of the Temperature Dependent Conformation and Phase Separation of Polyriboadenylic Acid Solutions at Neutral pH, *J.Mol.Biol.*, *30*, 17 (1967).

H. EISENBERG and E. REISLER, Angular Dependence of Scattered Light, Rotary Frictional Coefficients, and Distribution of Sizes of Associated Oligomers in Solutions of Bovine Liver Glutamate Dehydrogenase, *Biopolymers*, *10*, 2363 (1971).

H. EISENBERG and D. WOODSIDE, Multicomponent Polyelectrolyte Solutions, Part II: Excluded Volume Study of Polyvinylsulfonate Alkali Halide Systems, *J.Chem.Phys.*, *36*, 1844 (1962).

J.T. FINCH and K.C. HOLMES, in "Methods in Virology", K. Marmarosch and H. Koprowski, editors, Academic Press, Volume *3*, 1967, page 351.

P.J. FLORY, Principles of Polymer Chemistry, Cornell University Press, Ithaca, New York, 1953.

L.M. GROSS and U.P. STRAUSS, Interactions of Polyelectrolytes with Simple Electrolytes. I. Theory of Electrostatic Potential and Donnan Equilibrium for a Cylindrical Rod Model: The Effect of Site Binding, in "Chemical Physics of Ionic Solutions", B.E. Conway and R.G. Barradas, editors, John Wiley and Sons, Publishers, New York, 1966, page 361.

E.A. GUGGENHEIM, Thermodynamics,North Holland Publishing Company, Amsterdam, 5th edition.

H.S. HARNED and B.B. OWEN, The Physical Chemistry of Electrolyte Solutions, Reinhold Publishing Corporation, New York, 1958.

T.L. HILL, An Introduction to Statistical Thermodynamics, Addison-Wesley, Publishing Company, Reading, Massachusetts, 1960.

A. KATCHALSKY, Polyelectrolytes, *Pure and Appl. Chem. 26,* 327 (1971).

A. KATCHALSKY, Z. ALEXANDROWICZ and O. KEDEM, Polyelectrolyte Solutions, in "Chemical Physics and Ionic Solutions", B.E. Conway and R.G. Barradas, editors, John Wiley and Sons, Publishers, New York, 1966, page 295.

G.S. MANNING, Polyelectrolytes, *Ann.Rev.Phys.Chem. 23,* 117 (1972).

A. RAZIEL and H. EISENBERG, Excluded Volume Study of Potassium Polystyrenesulfonate Solutions, *Israel J.Chem., 11,* 183 (1973).

E. REISLER and H. EISENBERG, Interpretation of Equilibrium Sedimentation Measurements of Proteins in Guanidine HCl Solutions: Partial Volumes, Density Increments and Molecular Weight of the Subunits of Rabbit Muscle Aldolase, *Biochemistry, 8,* 4572 (1969).

W.H. STOCKMAYER, Light Scattering in Multicomponent Systems, *J.Chem.Phys., 18,* 58 (1950).

C. TANFORD, Physical Chemistry of Macromolecules, John Wiley and Sons, New York, 1961.

S.N. TIMASHEFF and R. TOWNEND, in "Physical Principles and Techniques of Protein Chemistry", Academic Press, New York, Part B, Chapter 3, 1970, page 147.

INFORMAL SEMINARS

A session of informal seminars was held at the Institute in which participants presented brief discussions of their current research. The titles and authors of all contributions are listed below. Abstracts are included if provided by the author.

1. INCOHERENT AND CHAOTIC SOURCES

Christine Benard
Universite de Paris-Sud
Orsay, France

2. MAXIMUM LIKELIHOOD ESTIMATES OF THE AVERAGE COUNT RATE AND THE LINEWIDTH OF GAUSSIAN-LORENTZIAN LIGHT BASED ON TIMES OF ARRIVAL OF PHOTON COUNTS

Bahaa Saleh
University of Santa Catarina
Santa Catarina, Brazil

3. THE DOPPLER-DIFFERENCE METHOD IN LASER ANEMOMETRY AND ITS APPLICATION TO SUPERSONIC GAS FLOWS

J. B. Abiss
Royal Aircraft Establishment
Farnborough, England

In the Doppler-difference method, the laser beam is split into two equal parts which are made to intersect at the point of interest in the flow. The frequency difference between the two Doppler-shifted light signals scattered by a moving particle is independent of the scattering angle, so that high collecting efficiency can be achieved by using a lens of large aperture.

Speeds up to 570 ms^{-1} have been measured at Farnborough on unseeded gas glows in back scatter using a Malvern photon correlator; with light seeding, autocorrelation functions can be acquired in under 1 second, yielding information on both mean velocity and turbulence intensity.

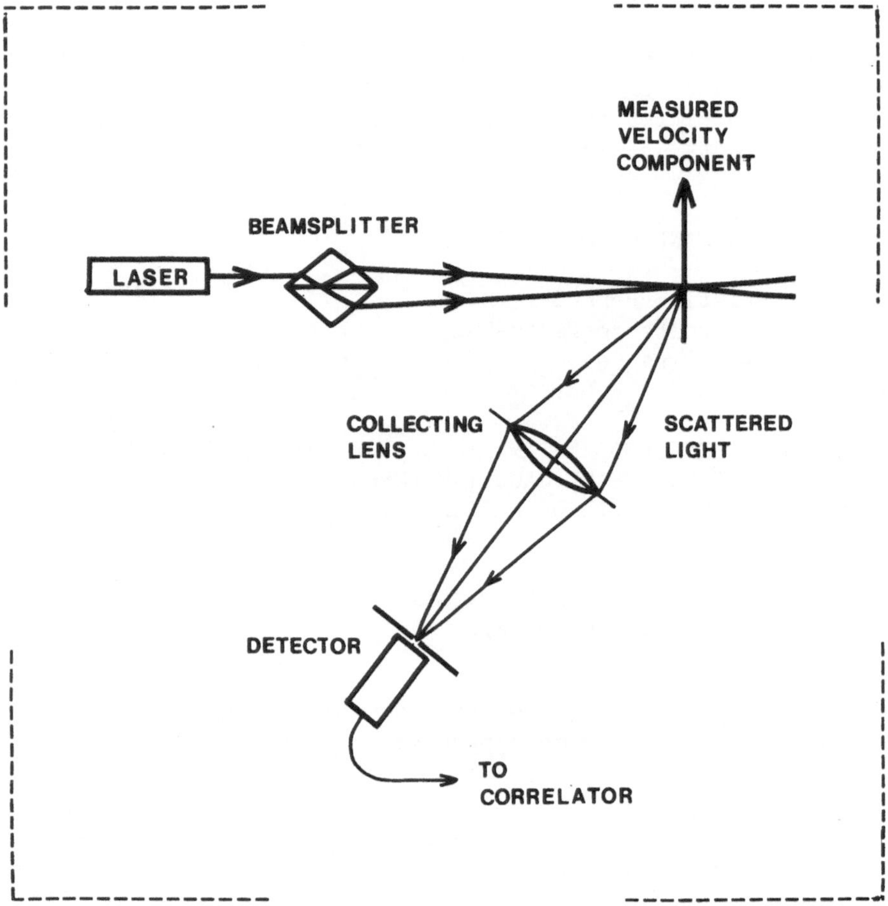

Figure 1. Schematic of the Doppler-difference laser anemometer.

4. A LASER DOPPLER VELOCIMETER FOR THE MEASUREMENT OF VELOCITY IN A TURBULENT ROTATING FLOW

William W. Fowlis
Geophysical Fluid Dynamics Institute
and Department of Meteorology
Florida State University
Tallahassee, Florida

A dual-scatter laser Doppler velocimeter (LDV) for the study of rotating flows has been successfully mounted on a rotating turntable. The capability of this LDV has been demonstrated by measurements of the relative azimuthal flow produced in a cylinder of water, which was also mounted on the turntable, by a small change in the rotation rate of the turntable. The LDV had an angle between the split beams (θ) of 5° and a zero-crossing technique was used to determine the Doppler frequency. A flow speed error of ±1% for speeds of 0.1 cm/sec was achieved. The spatial resolution was about 0.5 cm and the temporal resolution about 2 sec. With the objective of measuring velocity in a turbulent rotating flow, the Kolmogorov microscales for the turbulent flow between differentially rotating cylinders for a Reynolds number of 5,000 were calculated. The values for velocity, length and time are 0.3 cm/sec, 0.02 cm and 0.06 sec, respectively. By substantially increasing θ and reducing the sample time of the zero-crossing counter, it is possible to achieve LDV resolutions close to these microscales.

5. THE APPLICATION OF TIME TO HEIGHT CONVERTERS TO THE STUDY OF FLUCTUATING SUPERSONIC VELOCITIES

D. A. Jackson
Physics Laboratory, University of Kent
Canterbury, Kent, England

It is planned to measure rapid velocity fluctuations in a supersonic flow using a time of flight technique. A time to height converter (THC) is used to measure the transit time of flow borne particles between adjacent fringes formed by focussed crossed laser beams. At a mean flow rate of 500 meters/sec. the transit times will be $\sim 10^{-7}$ sec. The range of times which can be measured with the (THC) are $\sim 10^{-11} - 10^{-5}$ secs. with a dead time of $\sim 10^{-6}$ secs. The light signals detected using fast photomultipliers are standardized using low jitter constant fraction peak detecting discriminators.

Preliminary experiments in a high speed water flow show that the system does not suffer from 'drop-out' problems often encountered using analogue frequency trackers. Its sensitivity is comparable to that of a digital correlator with the advantage that the output voltage is related directly to the velocity rather than its correlation function.

6. CORRELATION AND SPECTRUM ANALYSIS - EXISTING CLASSES AND FUTURE TRENDS

Ira Langenthal
SAICOR/Honeywell
Hauppauge, New York

Correlation and spectrum analysis techniques have begun to play a central role in the field of light beating spectroscopy. Since the correlation function and the spectrum are Fourier transform pairs, equivalent information can be obtained in either domain.

Digital correlators which accept either analog signals or so-called photon counting (digital) signals, parallel digital correlators, real time or time compression spectrum analyzers, all digital fast Fourier transform (FFT) analyzers and digital filters are but a few of the existing classes of devices which can be utilized effectively.

The direction of the technology in this area of signal processing was reviewed with particular emphasis on the extension of the range, accuracy and speed of analysis. The conclusion is that both now and in the future, the researcher can select from several devices those that best suit his specific experiment.

7. THE MULTIPASSED INTERFEROMETER-CONSTRUCTION AND USE

J. R. Sandercock
R.C.A. - Zurich

8. FLUORESCENCE FLUCTUATION SPECTROSCOPY OF LATERAL DIFFUSION IN LIPID BILAYER MEMBRANES

Thomas J. Herbert
Cornell University
Ithaca, New York

9. DYNAMIC CHANGES IN THE SIZE AND SHAPE OF ADRENAL MEDULLA CHROMAFFIN GRANULES

Stephen J. Morris
Max-Planck-Institut fur Biophysikalische Chemie
D-3400 Göttingen-Nikolausberg
Göttingen, Germany

The chromaffin granules of the adrenal medulla cells store and release catecholamines (epinepherine and nor-epinepherine), ATP and protein. These subcellular organelles can be isolated and purified from the whole tissue by density gradient centrifugation techniques and appear in electron micrographs as polydisperse, electron dense, membrane bound spheres of 2000-5000Å diameter.

If the granule suspension is rapidly diluted at neutral pH and examined in a classical light scattering photometer as a function of time, the changes in intensity and dissymmetry suggest that the granules are breaking up into a series of smaller spheres. In the presence of divalent cations (Ca^{+2}, Mg^{+2}) or low pH the granules appear to aggregate as well. These models are supported by the electron microscopic observations.

10. MOTION IN BRAIN AND MOTILE CELLS

R. W. Piddington
Dept. of Zoology and Comparative Physiology
Queen Mary College
Mile End Road
London E1, England

632.8 nM light was scattered at right angles from locust brains. Modulation (at 1/5 the average scattered intensity) was reversibly doubled in isotonic K^+ (n=20) with no change in intensity nor in Γ (10 -20 Hz). Cyanide (1 mM) did not reduce K^+ modulation nor did 20% alcohol, but 100% abolished it. The living polarized cell appears gell-like (Shaw). Active/inactive <u>Nitella</u> (normal to incident and scattered directions) and motile bacteria gave similar Γ and comparable modulation changes (for <u>Nitella</u>, at constant Γ, n=2). Biological motion operates within an order of magnitude of the Brownian motion. These cells, sperm, and muscle half-sarcomeres all peak at 10-100 M/sec.

Spectra of active and inactive <u>Nitella</u> are shown in the figure.

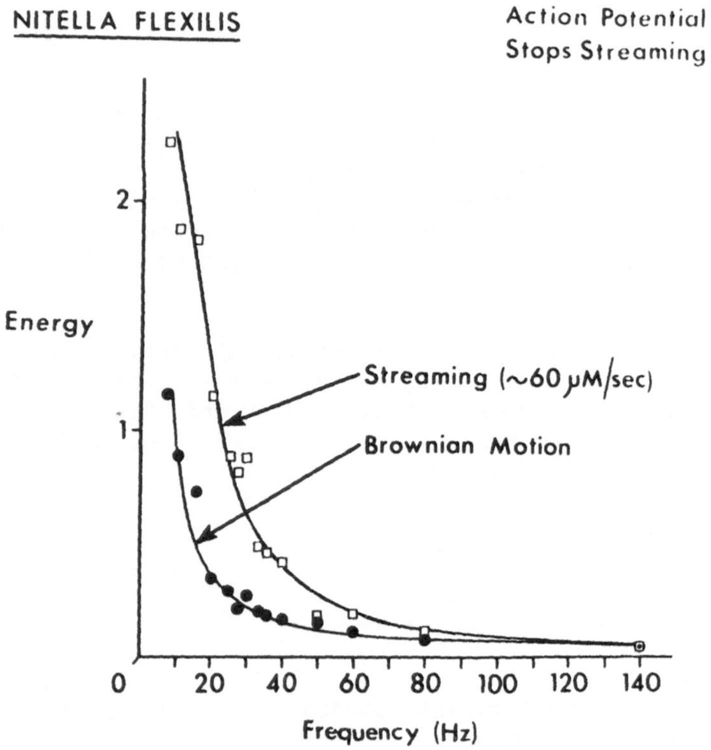

Figure 1. Light beating spectra of Nitella in active state (squares) and inactive state (circles).

11. PRELIMINARY MEASUREMENTS ON EXTREMELY COMPLEX DETERGENT SYSTEMS

A. J. Hyde
Chemistry Department
University of Strathclyde
Glascow C1, Scotland

PARTICIPANTS

Abbiss, J.	Royal Aircraft Establishment, Bramshot Golf House, Fleet, Aldershot, Hants., U.K.
Adam, M.	Commissariat à l'Énergie Atomique, Service de Physique du Solide et de Résonance Magnétique, Orme des Merisiers, B. P. No. 2, Gif-sur-Yvette 91190, France.
Asher, I. M.	107 University Park, Rochester, N.Y. 14620, U.S.A.
Baas, F.	Kamerlingh Onnes Laboratorium der Rijksuniversiteit Te Leiden Nieuwsteeg 18, Leiden, The Netherlands.
Baharudin, B. Y.	Physics Department, University of Kent, Canterbury, Kent, U.K.; National University of Malaysia, Singapore.
Barnett, W. A.	Chemistry Department, University of Birmingham, P.O. Box 363, Birmingham B15 2TT, U.K.
Bénard, C.	Laboratoire d'Étude des Phénomènes Aléatoires, Université de Paris-Sud, Batiment 210, 91405 Orsay, France.
Bertolotti, M.	Istituto di Fisica, Facoltà di Ingegneria, Università di Roma, Città Universitaria P. Le Scienze 5, Rome, Italy.
Boon, J -P.	Faculté des Sciences, Universite Libre de Bruxelles, 50, Avenue F. D. Roosevelt, 1050 Bruxelles, Belgium.
Brown, J. C.	Physics Department, Wellesley College, Wellesley, Mass., U.S.A.
Calmettes, P.	Commissariat à l'Énergie Atomique, Service de Physique du Solide et de Résonance Magnétique, Orme des Merisiers, B.P. No. 2, Gif-sur-Yvette 91190, France.
Caroline, D.	School of Physical and Molecular Sciences, University College of North Wales, Bangor, Caernarvonshire LL57 2UW, U.K.
Chang, R. F.	Department of Physics and Astrophysics, University of Maryland, College Park, Md., U.S.A.

Chen, S. H.	Nuclear Engineering Department, Massachusetts Institute of Technology, Cambridge, Mass. 02139, U.S.A.
Cialdea, R.	Istituto di Fisica Superiore dell'Universita, Rome, Italy.
Cowen, J.	Physics Department, Michigan State University, East Lansing, Michigan 48823, U.S.A.
Crosignani, B.	Fondazione U. Bordoni, Viale Travestere, 00100 Rome, Italy.
Cummins, H. Z.	Physics Department, New York University, 4 Washington Place, New York, N.Y. 10003, U.S.A.
Dietel, K.	Institut für Theoretische Physik, Universität Stuttgart, 7 Stuttgart 1, Germany.
Di Porto, P.	Fondazione U. Bordoni, Viale Travestere, 00100 Rome, Italy.
Dultz, W.	Fachbereich Physik, Universität Regensburg, 8400 Regensburg, Universitätsstrasse 31-Postfach, W. Germany.
Earnshaw, J. C.	Department of Pure and Applied Physics, Queen's University, Belfast BT7 1NN, N. Ireland, U.K.
Eisenberg, H.	Polymer Department, The Weizmann Institute of Science, Rehovot, Israel.
Fowlis, W. W.	Geophysical Fluid Dynamics Institute, Florida State University, Tallahassee, Fla. 32306, U.S.A.
Fraser, A.	Department of Biophysics-Jenkins, The Johns Hopkins University, Homewood Campus, Baltimore, Md. 21218, U.S.A.
French, M.J.	School of Chemistry, University of Bradford, Bradford BD7 1DP, Yorks., U.K.
Friedman, C.	Office National d'Études et de Recherches Aérospatiales, 29, Avenue de la Division Leclerc, 92 Chatillon, France.
Galerne, Y.	Laboratoire de Physique des Solides, Université Paris-Sud, Bâtiment 510, 91405 Orsay, France.
Gethner, J. S.	Chemistry Department, Columbia University, New York, N.Y. 10027, U.S.A.
Haken, H.	Institut für Theoretische Physik, Universität Stuttgart, Azenbergstr. 12, 7 Stuttgart 1, W. Germany.
Hanson, S.	Electronics Department, Research Establishment Risø, Danish Atomic Energy Commission, DK-4000 Roskilde, Denmark.

Harley, R.	Clarendon Laboratory, Parks Road, Oxford University, Oxford OX1 3PU, U.K.
Heisel, F.	Laboratoire de Physique des Rayonnements et d'Electronique Nucléaire, Centre de Recherches Nucléaires, Rue du Loess, 67 Strasbourg-Cronenbourg, France.
Hendrix, J.	Max-Planck-Institut für Biophysikalische Chemie, D-3400 Göttingen-Nikolausberg, Postfach 968, Germany.
Herbert, T.J.	School of Applied and Engineering Physics, Cornell University, Ithaca, N.Y. 14850, U.S.A.
Hillenkamp, F.	Abteilung für Kohärente Optik, Gesellschaft für Strahlen-und Umweltforschung mbH., Ingolstädter Str. 1, 8042 Neuherberg, Munchen, W. Germany.
Hyde, A.J.	Department of Pure and Applied Chemistry, University of Strathclyde, 295 Cathedral Street, Glasgow G1 1XL, Scotland, U.K.
Jackson, D.A.	Physics Department, University of Kent, Canterbury, Kent, U.K.
Jackson, H.E.	Physics Department, McMaster University, Hamilton, Ontario, Canada.
Jakeman, E.	Royal Radar Establishment, St. Andrews Road, Great Malvern, Worcs WR14 3PS, U.K.
Jolly, D.J.	Polymer Department, Weizmann Institute of Science, Rehovot, Israel.
Jones, D.L.	Mathematics Department, University of Manchester Institute of Science and Technology, P.O. Box No. 88 Sackville Street, Manchester, M60 1QD, U.K.
Lallemand, P.	Laboratoire de Physique, Université de Paris, École Normale Supérieure, 24, Rue Lhomond, Paris 5e, France
Langenthal, I.M.	Signal Analysis Operation, 595 Old Willets Path, Hauppauge, N.Y. 11787, U.S.A.
Layec, Y.	Laboratoire d'Hydrodynamique Moleculaire-PRg, Université de Bretagne Occidentale, 6, Avenue le Gorgeu, 29283 Brest-cedex, France.
Levy, C.	Laboratoire de Physique Thermique, Ecole Supérieure de Physique et Chimie Industrielles de la Ville de Paris, 10 Rue Vauquelin, 75231 Paris, France
Litovits, T.A.	Physics Department, Catholic University of America, Washington, D.C. 20017, U.S.A.
Litster, D.	Physics Department, Massachusetts Institute of Technology, Cambridge, Mass., 02139, U.S.A.
McAdam, J.D.G.	Schuster Laboratory, University of Manchester, Brunswick Street, Manchester M13 9PL, U.K.

Martellucci, S.	Facoltà di Ingegneria, Plasma Physics Laboratory, Universita di Roma, Rome, Italy.
Mathiez, P.	Départment de Physique, Luminy, 70, Route Léon Lachamp, 13 Marseille 9e, France.
Meystre, P.	Laboratoire d'Optique Physique, Ecole Polytechnique Fédérale, 2, Avenue Ruchonnet, 1003 Lausanne, Switzerland.
Migliardo, P.	Istituto di Fisica dell'Università di Messina, Via Dei Verdi, 98100 Messina, Italy.
Morris, S.J.	Max Planck Institute for Biophysical Chemistry, D-3400 Göttingen-Nikolausberg, W. Germany.
Munch, J.-P.	Laboratoire d'Acoustique Moléculaire, Institut de Physique, Université Louis Pasteur, 67070 Strasbourg Cedex, 4, Rue Blaise Pascal, France.
Oliver, C.J.	Royal Radar Establishment, St. Andrews Road, Great Malvern, Worcs WR14 3PS, U.K.
Ostrowsky, N.	Laboratoire de Physique, Université de Paris, École Normale Supérieure, 24, Rue Lhomond, Paris 5e, France.
Palin, C.	Physics Department, University of Birmingham, PO Box 363, Birmingham B15 2TT, U.K.
Parsons, R.R.	Physics Department, University of British Columbia, Vancouver 8, B.C., Canada.
Perrot, F.	Laboratoire d'Étude des Phénomènes Aléatoires, Université de Paris-Sud, Bâtiment 210, 91405 Orsay, France.
Piddington, R.W.	Department of Zoology and Comparative Physiology, Queen Mary College, Mile End Road, London E.1, U.K.
Pike, E.R.	Royal Radar Establishment, St. Andrews Road, Great Malvern, Worcs. WR14 3PS, U.K.
Postol, T.A.	NWB-241 Nuclear Reactor, Massachusetts Institute of Technology, Cambridge, Mass. 02139, U.S.A.
Prost, J.	Centre de Recherche Paul Pascal, Domaine Universitaire, 33-405 Talence, France.
Pusey, P.N.	Royal Radar Establishment, St Andrews Road, Great Malverm Worcs. WR14 3PS, U.K.
Raman, K.	Physics Department, Wesleyan University, Middletown, Conn. 06457, U.S.A.
Razzetti, C.	Istituto di Fisica, Università di Parma, Via M. d'Azeglio 75, 43100 Parma, Italy.

Reich, S.	Polymer Department, Weizmann Institute of Science, Rehovot, Israel.
Rouch, J.	Laboratoire d'Optique Moléculaire, Faculté des Sciences de Bordeaux, 351, Cours de la Libération, 33 Talence, France.
Saleh, B.E.A.	University of Santa Catarina, Cx.P. 429, Florianopolis, SC, Brazil.
Sandercock, J.	R.C.A., Ltd., Badenerstrasse 569, 8048 Zurich, Switzerland.
Scudieri, F.	Istituto di Fisica, Facoltà di Ingegneria, Universita degli Studi, Rome, Italy.
Searby, G.	Commissariat à l'Énergie Atomique, Service de Physique du Solide et de Résonance Magnétique, Orme des Merisiers, B.P. No. 2, Gif-sur-Yvette 91190, France.
Shigenari, T.	Institute J. Stefan, University of Ljubljana, P.O.B. 199/IV, 61001 Ljubljana, Yugoslavia; University of Electro-Communications, Tokyo.
Sixou, P.	Laboratoire de Physique Électronique, Universite de Paris, Bâtiment 220, 91405 Orsay, France.
Solimeno, S.	Istituto Electtrotecnica, Università di Napoli, Naples, Italy.
Stock, G.B.	Department of Biophysics-Jenkins, The Johns Hopkins University, Homewood Campus, Baltimore, Md. 21218, U.S.A.
Swinney, H.L.	Physics Department, New York University, 4 Washington Place, New York, N.Y. 10003, U.S.A.
Tartaglia, P.	Istituto di Fisica, Facoltà di Ingegneria, Università di Roma, Città Universitaria P. Le Scienze 5, Rome, Italy.
Van Dael, W.	Laboratorium voor Molekuulfysika, Universiteit te Leuven, Celestijnenlaan, 200D, 3030 Heverlee, Belgium.
Vaughan, J.M.	Royal Radar Establishment, St. Andrews Road, Great Malvern, Worcs. WR14 3PS, U.K.
Winchester, Jr., L.W.	Physics Department, University of Illinois at Chicago Circle,Chicago, Circle, Chicago, Ill. 60680, USA.
Wohrstein, H.-G.	Institüt fur Theoretische Physik, Universität Stuttgart, Azenbergstr. 12, 7 Stuttgart 1, Germany.

Zoppi, M.	Laboratorio di Elettronica Quantistica, Consiglio Nazionale dello Ricerche, Via Panciatichi 56/30, 50127 Firenze, Italy.
Zulauf, M.R.	Biozentrum, Universität Basel, Klingbergstr. 70, CH-4056 Basel, Switzerland.

SUBJECT INDEX